# KALMAN-BUCY FILTERS

***The Artech House Communication and Electronic Defense Library***

*Principles of Secure Communication Systems* by Don J. Torrieri
*Introduction to Electronic Warfare* by D. Curtis Schleher
*Electronic Intelligence: The Analysis of Radar Signals* by Richard G. Wiley
*Electronic Intelligence: The Interception of Radar Signals* by Richard G. Wiley
*Pulse Train Analysis Using Personal Computers* by Richard G. Wiley and Michael B. Szymanski
*RGCALC: Radar Range Detection Software and User's Manual* by John E. Fielding and Gary D. Reynolds
*Signal Theory and Random Processes* by Harry Urkowitz
*Signal Theory and Processing* by Frederic de Coulon
*Digital Image Signal Processing* by Murat Kunt
*Analysis and Synthesis of Logic Systems* by Olis Rubin
*Advanced Mathematics for Practicing Engineers* by Kurt Arbenz and Alfred Wohlhauser
*Mathematical Methods of Information Transmission* by Kurt Arbenz and Jean-Claude Martin
*Codes for Error-Control and Synchronization* by Djimitri Wiggert
*Machine Cryptography and Modern Cryptanalysis* by Cipher A. Deavours and Louis Kruh
*Cryptology Yesterday, Today, and Tomorrow,* Cipher A. Deavours et al., eds.
*Microcomputer Tools for Communications Engineering* by S.T. Li, J.C. Logan, J.W. Rockway, and D.W.S. Tam
*Detectability of Spread Spectrum Signals* by Robin A. Dillard and George M. Dillard
*Principles of Signals and Systems: Deterministic Signals* by Bernard Picinbono
*Introduction to Sensor Systems* Shahan A. Hovanessian
*Analog Automatic Control Loops in Radar and EW* by Richard Smith Hughes
*Numerical Analysis of Linear Networks and Systems* by Hermann Kremer
*Analysis and Synthesis of Logic Systems* by Daniel Marge
*Feedback Maximization* by Boris J. Lurie

# KALMAN-BUCY FILTERS

Karl Brammer
Electronik-System-Gesellschaft, Munich

Gerhard Siffling
University of Karlsruhe

Artech House

**Library of Congress Cataloging-in-Publication Data**

Brammer, Karl
[Kalman-Bucy-Filter. English]
Kalman-JBucy filters/Karl Brammer and Gerhard Siffling.

p. cm.
Translation of: Kalman-Bucy-Filter and Stochastische Grundlagen des Kalman-Bucy-Filters.
Bibliography: p.
Includes index.
ISBN 0-89006-294-3
1. Control theory. 2. Kalman filtering. I. Siffling, Gerhard. II. Brammer, Karl. Stochastische Grundlagen des Kalman-Bucy-Filters. English. 1989. III. Title.
QA402.B63913 1989
629.8'312—dc19 88-36727
CIP


**ARTECH HOUSE, INC.**
**685 Canton Street**
**Norwood, MA 02062**



**International Standard Book Number: 0-89006-294-3**
**Library of Congress Catalog Card Number: 88-36803**

**10 9 8 7 6 5 4 3 2 1**

**Originally published in German as two volumes,** *Kalman-Bucy-Filter* and *Stochastiche Grundlagen des Kalman-Bucy-Filters*, by R. Oldenbourg Verlag München, 1985 and 1986.

# *Contents*

# *Preface*

The time behavior of a dynamical system can be calculated if its mathematical model is given and if, apart from the input variables, the initial state is known. Knowledge of the state vector is therefore of fundamental importance for the analysis of the output of a dynamical system with given input. The state vector is equally important for the synthesis of any control input which is to influence the system output in a desired manner.

However, the state vector is often not accessible to direct measurement. It is then calculated from those variables of the system that are externally accessible. Due to measurement errors, the result is generally not exact, but only an approximate value: the so-called *estimate of the state vector*.

If this problem is solved by deterministic methods (i.e., if the random measurement errors are not taken into account explicitly), we speak of the *method of least squares* or *deterministic observation*. However, if we have some statistical information on the measurement errors (e.g., if we know their mean values and standard deviations), better or even optimal estimates can be obtained by using probabilistic methods. The measurement errors of the sensors as well as the unknown input variables (disturbances) are then modeled as random processes (stochastic processes) and we speak of *stochastic filtering*. The linear filter problem, originally treated by Wiener and Kolmogorov for special cases, has been comprehensively solved by Kalman and Bucy. Numerous applications prove the success of their theory.

For a thorough understanding of the Kalman-Bucy filter, it is essential to master certain basics of probability theory and the theory of random processes. However, newcomers in this field often have conceptual difficulties in going from the familiar deterministic point of view to the stochastic way of modeling.

Conversely, it is tedious to select the tools necessary for Kalman-Bucy filtering from the extensive literature on probability theory, especially insofar as only parts of it are needed. Therefore, the initial three chapters, written by G. Siffling, were included in the book.

In the first chapter, the description of dynamical systems by state variables is treated in a concise manner. The second chapter introduces the reader to probability theory and the theory of stochastic processes. No prior knowledge of this material is required. Using the example of a pair of dice, the concept of chance is explained and defined and the laws governing random behavior are discussed in detail. The concept of probability is introduced by means of the axioms of Kolmogorov, and the relationship with experimental results is shown. On this basis, probability theory is gradually established leading to the random process and such complicated concepts as the autocorrelation function and the covariance matrix. The fundamental concepts and results are presented in an elementary but rigorous manner. Only in the proofs was clarity of exposition given preference over mathematical rigor and completeness. The selection of the material was driven by the goal of deriving the filter equations. However, so many topics were involved that the material turned out to be complete in itself and is suitable to handle other problems as well.

In the third chapter we show how the properties of a random process are altered when it passes through a linear dynamic system and how these modified properties can be calculated.

The final three chapters, written by K. Brammer, deal with the deterministic observation and stochastic filtering problem proper. Again, the material is gradually developed and treated by elementary methods to the extent possible.

In the fourth chapter the observation problem for linear systems in discrete and continuous time is considered and the concepts of observability, duality and controllability are introduced.

The fifth chapter constitutes the main part of the book. It starts from the recursive implementation of the Gaussian method of least squares. By taking into account statistical *a priori* knowledge in terms of means and variances of the random errors and disturbances, the Gauss-Markov solution is obtained, providing a weighted least-squares estimate. Further, by adding statistical *a priori* knowledge on the initial state of the observed system, we obtain the minimum-variance estimate. The recursive version of this procedure is shown to be equivalent to the Kalman filter which applies to linear systems in discrete time. For continuous time the Kalman-Bucy filter is derived from this by a limiting transition using the matrix Wiener-Hopf equation. Despite the limited scope, we have attempted to give a complete and exact presentation of the basic concepts, derivations and results to provide a complete picture of the theory.

The sixth chapter examines selected problems related to the subject: observer in a closed loop system and algebraic separation, filter in a closed loop system and stochastic separation, stationary case and Wiener filter, colored noise, shaping filter and reduction of the filter order. Finally, we consider the Itô (calculus) and nonlinear filters.

The book is supplemented by an appendix on matrix calculus written by K. Brammer, which serves as a refresher and an easy reference for the required relations and formulas.

Here we wish to thank all the people who contributed to the book, especially those doing the typing, the drawings and the index.

We also express our gratitude for the care and patience of the publishers of the two German volumes and this English edition, and to David Oliver who did the English translation. Finally, we are still obliged to Professor Otto Föllinger who conceived the original idea for this book and provided the initial encouragement and support.

KARL BRAMMER
MUNICH

AUGUST 1988

GERHARD SIFFLING
KARLSRUHE

# Chapter 1
# Description of Dynamical Systems Using State Variables

In the third and subsequent chapters of this book, we will describe dynamical systems in terms of state variables. The concepts involved are introduced here. The associated formulas and calculation rules will not be derived point by point but will be summarized with short explanations to serve as a reminder to the reader. However, an exception is made in the case of the canonical forms which will be treated in greater detail. Reference is made to the textbooks by Zadeh and Desoer (1963), Ogata (1967), Landgraf and Schneider (1970), Thoma (1973) and Unbehauen (1983) for the derivations of the formulas and rules.

## 1.1 INTRODUCTION OF STATE VARIABLES

We consider a transfer system with $p$ input variables (e.g., control variables) $u_1(t), \ldots, u_p(t)$ and $m$ output variables (measurement variables) $y_1(t), \ldots, y_m(t)$ (Figure 1.1). If we want a mathematical description of the properties or the behavior of this "dynamical system" we need the relationships between the input and output variables as well as the relationships between the output variables and some special signals $x_1(t), \ldots, x_n(t)$ occurring inside the system. The physical laws governing the system usually yield a number of differential equations and ordinary equations, which together are called the *mathematical model* of the dynamical system. In what follows, it will always be assumed that only ordinary differential equations are present and that no partial differential equations occur.

It is often possible to transform the differential equations of the mathematical model into first-order differential equations by a skillful choice of internal signals or by introducing a sufficiently large number of internal signals. In doing this we define the mathematical model of the dynamical system in the following form:

$$\dot{x}_i(t) = f_i\{x_1(t), \ldots, x_n(t); u_1(t), \ldots, u_p(t); t\} \quad i = 1, \ldots, n \tag{1.1}$$

$$y_j(t) = g_j\{x_1(t), \ldots, x_n(t); u_1(t), \ldots, u_p(t); t\} \quad j = 1, \ldots, m \tag{1.2}$$

It should be noted that the differential equations (1.1) contain only the derivatives $\dot{x}_1(t), \ldots, \dot{x}_n(t)$ and not the derivatives of the input variables $u_1(t), \ldots, u_p(t)$. This is a significant advantage of the above representation because the input variables, which can be either actuating signals for the required control of the output variables or undesirable disturbance signals, are often nondifferentiable functions of time.

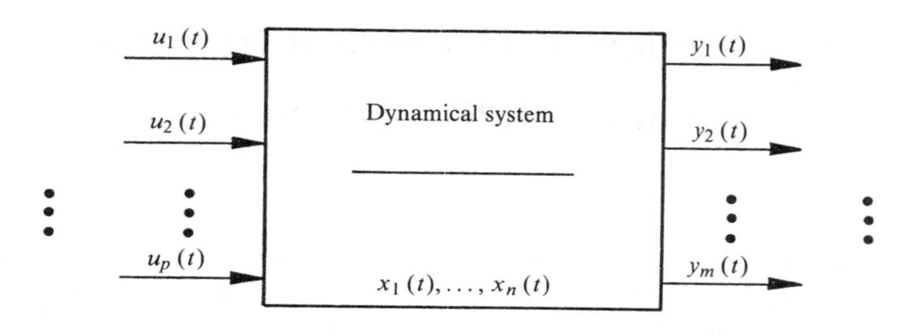


**Figure 1.1** Block diagram of a dynamical system with $p$ input variables, $m$ output variables and $n$ state variables.

The *state* of a dynamical system at a certain point in time $t = t_0$ is defined as the set of numerical values which, together with a knowledge of the mathematical model and the input variables $u_1(t), \ldots, u_p(t)$ for $t \geq t_0$, makes it possible to determine the time behavior of the dynamical system completely for $t \geq t_0$. Therefore if we know the state of a dynamical system at time $t = t_0$, the behavior of the system at $t \geq t_0$ depends only on the magnitude of these numerical values and not on the manner in which these numerical values were obtained by actuation on the dynamical system over a time interval $t < t_0$. In other words, if the present state is known, the past of the dynamical system can be ignored without prejudicing the investigation of its future behavior.

We now consider equations (1.1) and (1.2) from this point of view, assuming that solutions of the differential equation system (1.1) exist and are unique (these assumptions will always be satisfied in what follows). The numerical values $x_1(t_0), \ldots, x_n(t_0)$, taken as initial values, together with the input variables $u_1(t), \ldots, u_p(t)$ for $t \geq t_0$ define a unique solution of the differential equation (1.1). When these solutions $x_1(t), \ldots, x_n(t)$ are used in equation (1.2), the output functions $y_1(t), \ldots, y_n(t)$ are also uniquely defined for $t \geq t_0$ if unique functions $g_j$, $j = 1, \ldots, m$ are also assumed. Hence it is clear that the internal signals $x_1(t), \ldots, x_n(t)$, taken at time $t = t_0$, represent the state of the dynamical system at time $t_0$ in the sense of the definition given above. The functions $x_1(t), \ldots, x_n(t)$ are therefore called *state variables*.

Every set of $n$ linear combinations of the values $x_1(t_0), \ldots, x_n(t_0)$

$$z_1(t_0) = k_{11}x_1(t_0) + \cdots + k_{1n}x_n(t_0)$$
$$\vdots$$
$$z_n(t_0) = k_{n1}x_1(t_0) + \cdots + k_{nn}x_n(t_0)$$

also describes the state of the dynamical system at time $t = t_0$, provided that the determinant (system determinant) of this equation system is nonzero. For this reason the state of the dynamical system is not uniquely defined (i.e., there is not just one set of state variables for a dynamical system). The state variables will be chosen differently depending on the formulation of the problem but they do not have to be measurable physical quantities. They can be purely abstract variables without any physical meaning.

The smallest number of numerical values which is required to determine the state of a dynamical system, and hence the smallest number of state variables required, specifies the order of the dynamical system. This order is independent of the choice of state variables.

We now introduce the following column vectors:

$\boldsymbol{x}(t) = [x_1(t), \ldots, x_n(t)]'$ state vector

$\boldsymbol{u}(t) = [u_1(t), \ldots, u_p(t)]'$ input vector (control vector)

$\boldsymbol{y}(t) = [y_1(t), \ldots, y_m(t)]'$ output vector (measurement vector)

$\boldsymbol{f} = [f_1, \ldots, f_n]'$

$\boldsymbol{g} = [g_1, \ldots, g_m]'$

The notation of equations (1.1) and (1.2) can be considerably simplified by using the relationship

$$\dot{\boldsymbol{x}}(t) = [\dot{x}_1(t), \ldots, \dot{x}_n(t)]'$$

Thus the $n$ differential equations (1.1) are converted to a single vector differential equation

$$\dot{\boldsymbol{x}}(t) = \boldsymbol{f}\{\boldsymbol{x}(t); \boldsymbol{u}(t); t\} \tag{1.3}$$

which is called the *state differential equation*. Likewise the $m$ equations (1.2) are converted to the vector equation

$$\boldsymbol{y}(t) = \boldsymbol{g}\{\boldsymbol{x}(t); \boldsymbol{u}(t); t\} \tag{1.4}$$

which is called the *output equation*. Equations (1.3) and (1.4) are called the *state equations*. Together with the initial vector $\boldsymbol{x}(t_0)$ of the state differential equation they can be used to specify a block diagram for the mathematical model of the above dynamical system (Figure 1.2).

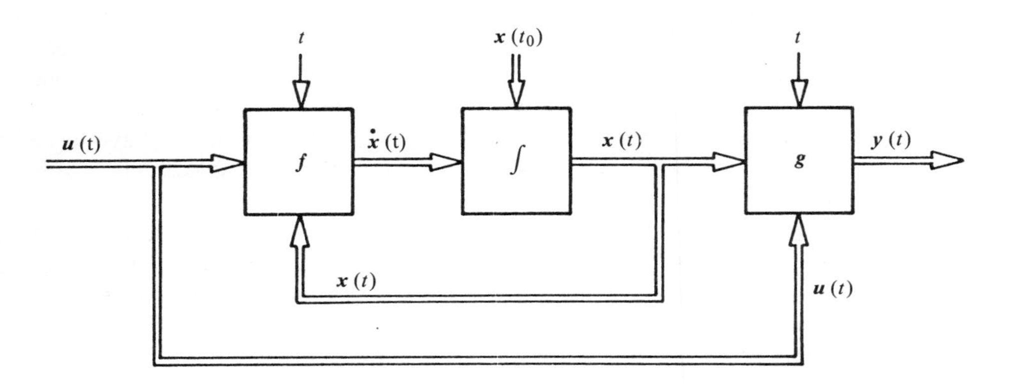


**Figure 1.2** Block diagram of the state equations of a dynamical system.

At every time $t$, the state vector $\boldsymbol{x}(t)$ can be interpreted as a point in the $n$-dimensional Euclidean space $\mathcal{R}_n$. The $n$-dimensional curve obtained by plotting these points by running through the values of the parameter $t$ is called the *trajectory* of the dynamical system in *state space*. If time is added as an extra dimension to the state space $\mathcal{R}_n$, we have an $(n + 1)$-dimensional *motion space* $\mathcal{R}_{n+1}$.

## 1.2 LINEAR SYSTEMS IN CONTINUOUS TIME

So far we have not considered the nature of the functions $f_1, \ldots, f_n$ and $g_1, \ldots, g_m$. An important special case occurs when all of these functions are linear in the components of both $\boldsymbol{x}(t)$ and $\boldsymbol{u}(t)$. We then have a linear dynamical system whose associated state equations are

$$\dot{\boldsymbol{x}}(t) = \boldsymbol{A}(t)\boldsymbol{x}(t) + \boldsymbol{B}(t)\boldsymbol{u}(t) \tag{1.5}$$

$$\boldsymbol{y}(t) = \boldsymbol{C}(t)\boldsymbol{x}(t) + \boldsymbol{D}(t)\boldsymbol{u}(t) \tag{1.6}$$

Only such linear dynamical systems will be considered in the remainder of this chapter.

In equations (1.5) and (1.6) $\boldsymbol{A}(t)$ is an $n \times n$ matrix (system matrix), $\boldsymbol{B}(t)$ is an $n \times p$ matrix (control or input matrix), $\boldsymbol{C}(t)$ is an $m \times n$ matrix (measurement or output matrix) and $\boldsymbol{D}(t)$ is an $m \times p$ matrix (feed-forward matrix). Often, $\boldsymbol{u}(t)$ does not act on $\boldsymbol{y}(t)$ directly but indirectly via the state vector. Then, $\boldsymbol{D}(t) \equiv \boldsymbol{0}$. If the elements of all four matrices are independent of time, we have a *time-invariant* dynamical system; otherwise, we have a *time-varying* dynamical system. If we assume that the elements of $\boldsymbol{A}(t)$ and $\boldsymbol{B}(t)$ are continuous functions of time and that the elements of $\boldsymbol{u}(t)$ are piecewise continuous, i.e., they are functions which have at the most a finite number of discontinuities in any finite time interval and are continuous everywhere else, there exists exactly one solution $\boldsymbol{x}(t)$ of the state differential equation (1.5) for every given initial vector $\boldsymbol{x}(t_0)$. The variation of $\boldsymbol{x}(t)$ can then also be calculated for $t < t_0$ (Landgraf and Schneider, 1970). In this way the existence of a uniquely defined solution of the state differential equation can be assumed for almost all linear dynamical systems.

The state equations and the initial vector $\boldsymbol{x}(t_0)$ of the state differential equation for a linear dynamical system can also be represented by a block diagram (Figure 1.3).

### 1.2.1 The Solution of the State Differential Equation

The time behavior of a linear dynamical system, i.e., the dependence of the output variables $y_1(t), \ldots, y_m(t)$ on the input variables $u_1(t), \ldots, u_p(t)$, can be explicitly calculated if it is possible to solve the state differential equation (1.5). When the integral $\boldsymbol{x}(t)$ is obtained, $\boldsymbol{y}(t)$ can be calculated from the output equation (1.6). The general solution of the linear state differential equation (1.5) is

$$\boldsymbol{x}(t) = \boldsymbol{\phi}(t, t_0)\boldsymbol{x}(t_0) + \int_{t_0}^{t} \boldsymbol{\phi}(t, \tau)\boldsymbol{B}(\tau)\boldsymbol{u}(\tau)\, \mathrm{d}\tau \tag{1.7}$$

where $\boldsymbol{x}(t_0)$ is the given initial value of the state variable $\boldsymbol{x}(t)$. It follows from this and equation (1.6) that the output vector $\boldsymbol{y}(t)$ is given by

$$\boldsymbol{y}(t) = \boldsymbol{C}(t)\boldsymbol{\phi}(t, t_0)\boldsymbol{x}(t_0) + \int_{t_0}^{t} \boldsymbol{C}(t)\boldsymbol{\phi}(t, \tau)\boldsymbol{B}(\tau)\boldsymbol{u}(\tau)\, \mathrm{d}\tau + \boldsymbol{D}(t)\boldsymbol{u}(t) \tag{1.8}$$

According to equation (1.7), the solution $\boldsymbol{x}(t)$ of the linear state differential equation (1.5) is composed of the sum of the general solution

$$\boldsymbol{x}_h(t) = \boldsymbol{\phi}(t, t_0)\boldsymbol{x}(t_0) \tag{1.9}$$

of the homogeneous state differential equation (i.e., $\boldsymbol{u}(t) \equiv \boldsymbol{0}$)

$$\dot{\boldsymbol{x}}(t) = \boldsymbol{A}(t)\boldsymbol{x}(t) \tag{1.10}$$

and the particular solution

$$\boldsymbol{x}_p(t) = \int_{t_0}^{t} \boldsymbol{\phi}(t, \tau)\boldsymbol{B}(\tau)\boldsymbol{u}(\tau)\, d\tau \tag{1.11}$$

of the inhomogeneous state differential equation (i.e., $\boldsymbol{x}(t_0) = \boldsymbol{0}$). The particular solution $\boldsymbol{x}_p(t)$ can be obtained from the homogeneous solution $\boldsymbol{x}_h(t)$ by the "variation of constants" method, as for scalar linear differential equations.

Equation (1.7) contains an unknown $n \times n$ matrix $\boldsymbol{\phi}(t, t)$ or $\boldsymbol{\phi}(t, \tau)$ in addition to the known initial value $\boldsymbol{x}(t_0)$, the given control vector $\boldsymbol{u}(t)$ and the control matrix $\boldsymbol{B}(t)$ of the dynamical system. This matrix is called the *transition matrix* of the linear dynamical system. This is because it describes the transition of the state $\boldsymbol{x}(t_0)$ to the state $\boldsymbol{x}(t)$ in the solution (1.9) of the homogeneous state differential equation (1.10). $\boldsymbol{\phi}$ is also known as the *fundamental matrix*.

There is no closed formula for calculating the transition matrix in the general case. Closed formulas can only be obtained for special cases, in particular for the time-invariant linear dynamical systems considered in Section 1.2.3. In the general case the transition matrix must be determined numerically from the solution of the homogeneous linear state differential equation (1.10) or by using a series approximation (Thoma, 1973). Another method of calculating the transition matrix as the solution of a differential equation system for linear time-varying dynamical systems is indicated in the next section where the properties of the transition matrix are considered.

### 1.2.2 Properties of the Transition Matrix

The main properties of the transition matrix, which are often useful for calculations with state variables, are given in this section.

(1) For $t = t_0$ the transition matrix $\boldsymbol{\phi}(t, t_0)$ is equal to the unit matrix $\boldsymbol{I}$:

$$\boldsymbol{\phi}(t_0, t_0) = \boldsymbol{I} \tag{1.12}$$

(2) For any points in time $t_0$, $t_1$ and $t_2$, we have

$$\boldsymbol{\phi}(t_2, t_1)\boldsymbol{\phi}(t_1, t_0) = \boldsymbol{\phi}(t_2, t_0) \tag{1.13}$$

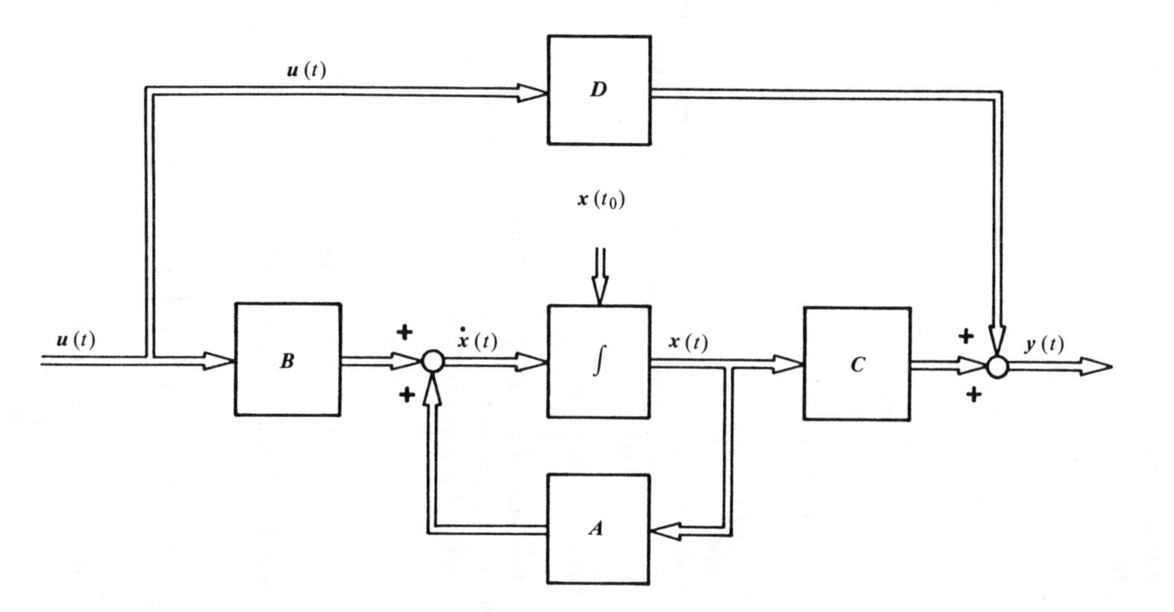


**Figure 1.3** Block diagram of the state equations of a linear dynamical system.

(3) The transition matrix $\boldsymbol{\phi}(t, t_0)$ is regular for all times $t$ and $t_0$ (i.e., it always has an inverse). This satisfies the equation

$$\boldsymbol{\phi}^{-1}(t, t_0) = \boldsymbol{\phi}(t_0, t) \tag{1.14}$$

(4) The transition matrix $\boldsymbol{\phi}(t, t_0)$ satisfies the homogeneous state differential equation (1.10):

$$\frac{\partial}{\partial t}\boldsymbol{\phi}(t, t_0) = \boldsymbol{A}(t)\boldsymbol{\phi}(t, t_0) \tag{1.15}$$

This differential equation, together with the initial value

$$\boldsymbol{\phi}(t_0, t_0) = \boldsymbol{I}$$

(equation (1.12)), provides another method of calculating the transition matrix. This is the general procedure adopted for linear time-varying dynamical systems.

### 1.2.3 Linear Time-Invariant Systems

If the elements of the four matrices $\boldsymbol{A}$, $\boldsymbol{B}$, $\boldsymbol{C}$ and $\boldsymbol{D}$ in the state equations for a linear dynamical system do not depend on time, this dynamical system is called *time invariant* because its dynamical properties do not vary with time:

$$\dot{\boldsymbol{x}}(t) = \boldsymbol{A}\boldsymbol{x}(t) + \boldsymbol{B}\boldsymbol{u}(t) \tag{1.16}$$

$$\boldsymbol{y}(t) = \boldsymbol{C}\boldsymbol{x}(t) + \boldsymbol{D}\boldsymbol{u}(t) \tag{1.17}$$

The solution of the state differential equation (1.16) is as follows:

$$\boldsymbol{x}(t) = \boldsymbol{\phi}(t - t_0)\boldsymbol{x}(t_0) + \int_{t_0}^{t} \boldsymbol{\phi}(t - \tau)\boldsymbol{B}\boldsymbol{u}(\tau)\, \mathrm{d}\tau \tag{1.18}$$

In contrast with the time-varying case, the transition matrix no longer depends on the two variables $t$ and $t_0$ (or $t$ and $\tau$) but on their difference $t - t_0$ (or $t - \tau$); i.e., the matrix is dependent on one variable only.

A closed formula can be derived for the calculation of the transition matrix of a time-invariant linear dynamical system:

$$\boldsymbol{\phi}(t) = \exp(\boldsymbol{A}t) \tag{1.19}$$

The matrix function $\exp(\boldsymbol{A}t)$ is defined by a power series:

$$\exp(\boldsymbol{A}t) = \sum_{\nu=0}^{\infty} \frac{(\boldsymbol{A}t)^{\nu}}{\nu!} = \boldsymbol{I} + \boldsymbol{A}\frac{t}{1!} + \boldsymbol{A}^2\frac{t^2}{2!} + \cdots \tag{1.20}$$

The convergence of this series is guaranteed (Föllinger, 1985). The explicit calculation of the transition matrix using equation (1.20) is generally very cumbersome and should only be used for small values of $t$. However, for the special case when the system matrix $\boldsymbol{A}$ is diagonal, $\exp(\boldsymbol{A}t)$ is also diagonal and can be written down immediately. We then have

$$\boldsymbol{A} = \begin{bmatrix} a_1 & & 0 \\ & \ddots & \\ 0 & & a_n \end{bmatrix} \rightarrow \exp(\boldsymbol{A}t) = \begin{bmatrix} \exp(a_1 t) & & 0 \\ & \ddots & \\ 0 & & \exp(a_n t) \end{bmatrix}$$

We shall now discuss another method of expressing and calculating the matrix $\boldsymbol{\phi}(t)$ in closed form. We consider the first term in equation (1.18) for the case when $t_0 = 0$:

$$\boldsymbol{x}(t) = \boldsymbol{\phi}(t)\boldsymbol{x}(0) \tag{1.21}$$

As already noted in Section 1.2.1 this is the general solution of the homogeneous state differential equation:

$$\dot{\boldsymbol{x}}(t) = \boldsymbol{A}\boldsymbol{x}(t) \tag{1.22}$$

Since $\boldsymbol{A}$ consists of constant elements only, this equation can be solved using the Laplace transformation. If we use the definition equation of the Laplace transform of a vector (see Appendix, (A.58a)) together with the differentiation rule for the Laplace transformation, equation (1.22) becomes

$$s\boldsymbol{X}(s) = \boldsymbol{x}(+0) = \boldsymbol{A}\boldsymbol{X}(s)$$

where $s$ is a complex variable and $\boldsymbol{X}(s)$ is a vector (not a matrix) representing the Laplace transform of the vector $\boldsymbol{x}(t)$. Solving for $\boldsymbol{X}(s)$, we obtain

$$\boldsymbol{X}(s) = (s\boldsymbol{I} - \boldsymbol{A})^{-1}\boldsymbol{x}(0) \tag{1.23}$$

Note that the continuity of the state variables is still assumed, i.e.,

$$\boldsymbol{x}(+0) = \boldsymbol{x}(-0) = \boldsymbol{x}(0)$$

If equation (1.23) is transformed back to the time domain, we have

$$\boldsymbol{x}(t) = \mathcal{L}^{-1}\{(s\boldsymbol{I} - \boldsymbol{A})^{-1}\}\boldsymbol{x}(0) \tag{1.24}$$

However, this is only an alternative way of writing equation (1.21). Comparison of equations (1.21) and (1.24) gives the closed expression for the transition matrix $\boldsymbol{\phi}(t)$:

$$\boldsymbol{\phi}(t) = \mathcal{L}^{-1}\{(s\boldsymbol{I} - \boldsymbol{A})^{-1}\} \tag{1.25}$$

The matrix $(s\boldsymbol{I} - \boldsymbol{A})$ can only be singular for those values of $s$ which coincide with the eigenvalues of the system matrix $\boldsymbol{A}$. However, there is a domain in the $s$-plane to the left of which all eigenvalues of $A$ lie. In this domain, the matrix $(s\boldsymbol{I} - \boldsymbol{A})$ is regular and its inverse $(s\boldsymbol{I} - \boldsymbol{A})^{-1}$ therefore exists.

It is worth noting the analogy between the relationship

$$\exp(\boldsymbol{A}t) = \mathcal{L}^{-1}\{(s\boldsymbol{I} - \boldsymbol{A})^{-1}\} \tag{1.26}$$

and the mapping

$$\exp(at) = \mathcal{L}^{-1}\{(s - a)^{-1}\}$$

from the theory of the Laplace transformation.

We can apply the Laplace transformation not only to the homogeneous state differential equation (1.22) but also to the complete linear state equations (1.16) and (1.17). For

$$\boldsymbol{x}(t_0) = \boldsymbol{x}(0) = \boldsymbol{0}$$

this gives a relationship between the transfer matrix $\boldsymbol{G}(s)$ of the linear time-invariant dynamical system and the four coefficient matrices $\boldsymbol{A}$, $\boldsymbol{B}$, $\boldsymbol{C}$ and $\boldsymbol{D}$ of the associated state equations:

$$\boldsymbol{G}(s) = \boldsymbol{C}(s\boldsymbol{I} - \boldsymbol{A})^{-1}\boldsymbol{B} + \boldsymbol{D} \tag{1.27}$$

with

$$\boldsymbol{Y}(s) = \boldsymbol{G}(s)\boldsymbol{U}(s) \tag{1.28}$$

The vector $\boldsymbol{U}(s)$ is the Laplace transform of the input vector $\boldsymbol{u}(t)$, and $\boldsymbol{Y}(s)$ is the Laplace transform of the output vector $\boldsymbol{y}(t)$. For single-input single-output systems, the elements in equation (1.27) become scalars and we obtain the classical transfer function $G(s)$. The Laplace transformation of the state equations is only feasible for linear time-invariant dynamical systems (i.e., all four coefficient matrices are independent of time). However, equations (1.19) and (1.25) for calculating the transition matrix $\boldsymbol{\phi}(t)$ remain valid provided that the elements of the system matrix $\boldsymbol{A}$ are time independent, even if the other coefficient matrices depend on time.

The properties of the transition matrix $\boldsymbol{\phi}(t, t_0)$ given in the last section also apply in an analogous manner to the transition matrix $\boldsymbol{\phi}(t)$. These properties are as follows:

$$\boldsymbol{\phi}(0) = \boldsymbol{I} \tag{1.29}$$

$$\boldsymbol{\phi}(t_1)\boldsymbol{\phi}(t_2) = \boldsymbol{\phi}(t_1 + t_2) \tag{1.30}$$

It immediately follows from this that

$$\boldsymbol{\phi}(t_1)\boldsymbol{\phi}(t_2) = \boldsymbol{\phi}(t_2)\boldsymbol{\phi}(t_1) \tag{1.31}$$

and

$$\boldsymbol{\phi}^k(t) = \boldsymbol{\phi}(kt) \tag{1.32}$$

In addition the matrix $\boldsymbol{\phi}(t)$ is regular for all points in time $t$ and we have

$$\boldsymbol{\phi}^{-1}(t) = \boldsymbol{\phi}(-t) \tag{1.33}$$

and

$$\frac{\mathrm{d}}{\mathrm{d}t}\boldsymbol{\phi}(t) = \boldsymbol{A}\boldsymbol{\phi}(t) \tag{1.34}$$

## 1.3 LINEAR SYSTEMS IN DISCRETE TIME

So far it has been tacitly assumed that the input variables $u_1(t), \ldots, u_p(t)$ of the linear dynamical system are continuous functions of time with functional values that can vary at any point in time. It has also been assumed that the output variables $y_1(t), \ldots, y_m(t)$ are of interest at every point in time.

In this section the state equations (1.5) and (1.6) of a linear dynamical system are rewritten and solved for the case when the functional values of the input variables can only vary at discrete times $t_0, t_1, t_2, \ldots$ and take constant values elsewhere. The times $t_0, t_1, t_2, \ldots$ do not need to be equally spaced. It is only necessary that the condition

$$t_{k+1} > t_k \quad k = 0, 1, \ldots$$

holds. Such a piecewise constant input vector $\bar{\boldsymbol{u}}(t)$ can be described by the equation

$$\bar{\boldsymbol{u}}(t) = \boldsymbol{u}(t_k) \quad \text{for} \quad t_k \leq t < t_{k+1} \quad k = 0, 1, \ldots \tag{1.35}$$

where $\boldsymbol{u}(t)$ is an input vector as defined previously.

The step components of $\bar{\boldsymbol{u}}(t)$ can be envisaged as the result of a sample-and-hold process applied to each of the corresponding continuous components of $\boldsymbol{u}(t)$ (Figure 1.4). The output variables of the dynamical system under consideration will only be calculated at times $t_0, t_1, t_2, \ldots$. First, the state vector $\boldsymbol{x}(t)$ is determined at the discrete times $t_0, t_1, t_2, \ldots$. To do this we replace $t_0$ and $t$ in equation (1.7), which applies to any times $t$ and $t_0$ for $t \geq t_0$, by $t_k$ and $t_{k+1}$ respectively. In this way we obtain

$$\boldsymbol{x}(t_{k+1}) = \boldsymbol{\phi}(t_{k+1}, t_k)\boldsymbol{x}(t_k) + \int_{t_k}^{t_{k+1}} \boldsymbol{\phi}(t_{k+1}, \tau)\boldsymbol{B}(\tau)\bar{\boldsymbol{u}}(\tau)\,\mathrm{d}\tau$$

According to equation (1.35), the input vector $\bar{\boldsymbol{u}}(t)$ is constant and equal to $\boldsymbol{u}(t_k)$ during the integration interval. Therefore the above equation becomes

$$\boldsymbol{x}(t_{k+1}) = \boldsymbol{\phi}(t_{k+1}, t_k)\boldsymbol{x}(t_k) + \int_{t_k}^{t_{k+1}} \boldsymbol{\phi}(t_{k+1}, \tau)\boldsymbol{B}(\tau)\,\mathrm{d}\tau\,\boldsymbol{u}(t_k) \tag{1.36}$$

Using the abbreviations

$$\boldsymbol{A}(k) := \boldsymbol{\phi}(t_{k+1}, t_k) \tag{1.37}$$

and

$$\boldsymbol{B}(k) := \int_{t_k}^{t_{k+1}} \boldsymbol{\phi}(t_{k+1}, \tau)\boldsymbol{B}(\tau)\,\mathrm{d}\tau \tag{1.38}$$

which must hold for $k = 0, 1, \ldots$, we obtain the vector difference equation

$$\boldsymbol{x}(t_{k+1}) = \boldsymbol{A}(k)\boldsymbol{x}(t_k) + \boldsymbol{B}(k)\boldsymbol{u}(t_k) \tag{1.39}$$

Since the output vector is only of interest at the discrete times $t_0, t_1, t_2, \ldots$, the output equation (1.6) for $t = t_k$ is

$$\boldsymbol{y}(t_k) = \boldsymbol{C}(k)\boldsymbol{x}(t_k) + \boldsymbol{D}(k)\boldsymbol{u}(t_k) \tag{1.40}$$

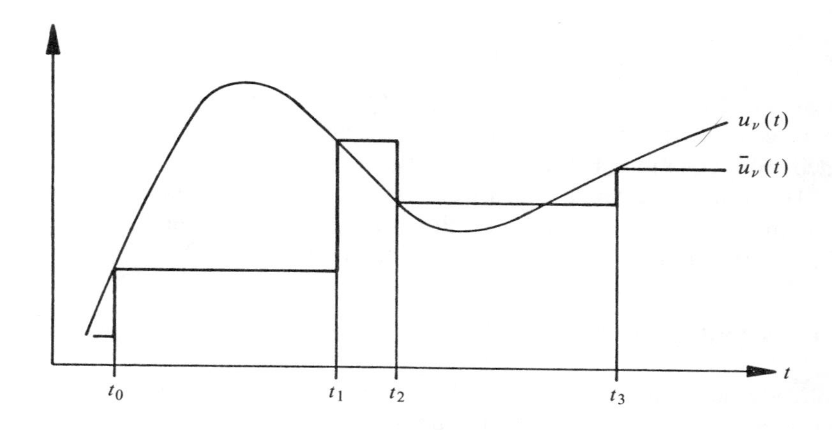


**Figure 1.4** Explanation of the effect of a sample-and-hold unit.

where the abbreviations

$$\boldsymbol{C}(k) := \boldsymbol{C}(t_k) \tag{1.41}$$

and

$$\boldsymbol{D}(k) := \boldsymbol{D}(t_k) \tag{1.42}$$

which must also hold for $k = 0, 1, \ldots,$ have been introduced. Equations (1.39) and (1.40) are the state equations of a linear dynamical system in discrete time. Their appearance is similar to that of the state equations of a linear dynamical system in continuous time.

Strictly speaking, an important rule has been broken in the definition equations (1.37) and (1.38). Different quantities have been denoted by the same symbols. The matrix $\boldsymbol{A}(k)$ in equation (1.37) is not identical with the system matrix $\boldsymbol{A}(t)$ at $t = k$ and the matrix $\boldsymbol{B}(k)$ in equation (1.38) is not identical with the control matrix $\boldsymbol{B}(t)$ at $t = k$ of the linear dynamical system in continuous time. However, there is risk of confusion only if dynamical systems in continuous time and their counterparts in discrete time occur in parallel, particularly in computer programs. In that case different symbols *must* be used! Here the risk is intentional to demonstrate the formal analogy between linear dynamical systems in continuous time and those in discrete time. This analogy is not restricted to the state equations of linear dynamical systems.

The solution of the vector difference equation (1.39) can be obtained by writing it for $k = 0$, $k = 1$ *et cetera*. The equations obtained are then successively substituted in each other, taking relationships (1.37) and (1.13) into consideration. From this it follows that

$$\boldsymbol{x}(t_k) = \boldsymbol{\phi}(t_k, t_0)\boldsymbol{x}(t_0) + \sum_{\nu=0}^{k-1} \boldsymbol{\phi}(t_k, t_{\nu+1})\boldsymbol{B}(\nu)\boldsymbol{u}(t_\nu) \tag{1.43}$$

It is assumed that the initial state $\boldsymbol{x}(t_0)$ is known. Once this result is obtained, the output vector $\boldsymbol{y}(t_k)$ for the desired discrete points in time can immediately be calculated from equation (1.40).

The results obtained in this section are simplified if the linear dynamical system is time invariant and if, in addition, times $t_0, t_1, t_2, \ldots$ are equally spaced (DeRusso *et al.*, 1966; Timothy and Bona, 1968; Thoma, 1973).

## 1.4 CANONICAL FORMS

As already mentioned in Section 1.1, the state variables and state equations which describe a dynamical system are not uniquely defined. Depending on the nature of the problem, one particular form of representation will be preferred. In the description of the physics of a dynamical system, some state variables and state equations will arise quite naturally. However, these can be converted to the desired state variables or the desired form of the state equations by a mathematical transformation.

We will now derive three different forms of the state equations for an $n$th-order linear time-invariant dynamical system with a single input variable $u(t)$ and a single output variable $y(t)$. This system is described by the ordinary linear differential equation with constant coefficients

$$\begin{aligned} a_0 y(t) + a_1 \dot{y}(t) + \cdots + a_{n-1} y^{(n-1)}(t) + y^{(n)}(t) \\ = b_0 u(t) + b_1 \dot{u}(t) + \cdots + b_n u^{(n)}(t) \end{aligned} \tag{1.44}$$

or by the transfer function

$$G(s) = \frac{Y(s)}{U(s)} = \frac{b_0 + b_1 s + \cdots + b_{n-1} s^{(n-1)} + b_n s^{(n)}}{a_0 + a_1 s + \cdots + a_{n-1} s^{(n-1)} + s^{(n)}} \tag{1.45}$$

Because of their frequent use, the three forms of state equations to be derived in this section are standard representations and are commonly called *canonical forms*.

When we refer here to different forms of the state equations, we only mean that the state variables are defined differently in each form and that the coefficient matrices $\boldsymbol{A}$, $\boldsymbol{B}$, $\boldsymbol{C}$ and $\boldsymbol{D}$ have different elements and different internal structures. It does *not* imply a change in the appearance of the vector state equations. In all cases, their appearance is the same and corresponds to equations (1.16) and (1.17). Because all of the different forms describe the same dynamical system completely, we say that they are equivalent to each other.

In (1.44) and (1.45), the coefficient $a_n$ for $y^n(t)$ and for $s^n$ in the denominator of $G(s)$ is assumed to be unity. This has the advantage of simplifying some of the results. The validity of the derivations is not restricted by this assumption because it is always possible to obtain a coefficient $a_n = 1$ by simple division. In addition, it must be assumed that the numerator and denominator in $G(s)$ do not have any common zeros.

As well as examining the different forms of the state equations for a linear time-invariant single-input single-output system, we also consider the general case of a linear time-invariant multiple-input multiple-output system (e.g., Figure 1.1). This is because the relationship that exists between each of the $m$ output variables and each of the $p$ input variables can be described by $mp$ differential equations according to (1.44) or by $mp$ transfer functions according to (1.45). First, a state representation is determined for each of these differential equations as shown in the following sections. Then it is only necessary to combine these state representations to obtain the state representation of the overall system. An example of this procedure is given by Föllinger (1985).

Canonical forms corresponding to those derived here for dynamical systems in continuous time can also be defined for linear dynamical systems in *discrete time* (Ackermann, 1983). Canonical forms for linear *time-varying* single-input single-output systems which are described by a differential equation such as equation (1.44), but now with time-dependent coefficients, are given by Zadeh and Desoer (1963) and Ogata (1967).

### 1.4.1 The Control Canonical Form

Using the abbreviation

$$N(s) = a_0 + a_1 s + \cdots + a_{n-1} s^{n-1} + s^n \tag{1.46}$$

for the denominator of the transfer function $G(s)$, we can write equation (1.45) as follows (Föllinger, 1985):

$$Y(s) = b_0 \frac{1}{N(s)} U(s) + b_1 \frac{s}{N(s)} U(s) + \ldots + b_{n-1} \frac{s^{n-1}}{N(s)} U(s) + b_n \frac{s^n}{N(s)} U(s) \tag{1.47}$$

Now the state variables $X_1(s), \ldots, X_n(s)$ in the complex frequency domain are given by

$$X_1(s) = \frac{1}{N(s)} U(s), \quad X_2(s) = \frac{s}{N(s)} U(s), \quad X_3(s) = \frac{s^2}{N(s)} U(s),$$

$$\ldots, X_{n-1}(s) = \frac{s^{n-2}}{N(s)} U(s), \quad X_n(s) = \frac{s^{n-1}}{N(s)} U(s) \tag{1.48}$$

and hence the following relationships hold:

$$sX_1(s) = X_2(s), \quad sX_2(s) = X_3(s), \quad \ldots, \quad sX_{n-1}(s) = X_n(s) \tag{1.49}$$

In the time domain (1.49) corresponds to the differential equations

$$\dot{x}_1(t) = x_2(t), \quad \dot{x}_2(t) = x_3(t), \quad \ldots, \quad \dot{x}_{n-1}(t) = x_n(t) \tag{1.50}$$

Thus we already have $n - 1$ of the required $n$ first-order differential equations. The $n$th differential equation is obtained from the equation:

$$X_1(s) = \frac{1}{N(s)} U(s) = \frac{1}{a_0 + a_1 s + \cdots + a_{n-1}s^{n-1} + s^n} U(s)$$

By multiplying both sides by the denominator $N(s)$ we obtain

$$a_0 X_1(s) + a_1 s X_1(s) + \cdots + a_{n-1}s^{n-1}X_1(s) + s^n X_1(s) = U(s)$$

Then, taking equation (1.49) into account, this equation can be converted to

$$a_0 X_1(s) + a_1 X_2(s) + \cdots + a_{n-1}X_n(s) + sX_n(s) = U(s)$$

or, in the time domain, to the differential equation:

$$\dot{x}_n(t) = -a_0 x_1(t) - a_1 x_2(t) - \cdots - a_{n-1}x_n(t) + u(t) \tag{1.51}$$

Equations (1.50) and (1.51) are the required state differential equations:

$$\begin{aligned}
\dot{x}_1(t) &= x_2(t) \\
\dot{x}_2(t) &= \qquad x_3(t) \\
&\vdots \\
\dot{x}_{n-1}(t) &= \qquad\qquad x_n(t) \\
\dot{x}_n(t) &= -a_0 x_1(t) - a_1 x_2(t) - a_2 x_3(t) - \cdots - a_{n-1}x_n(t) + u(t)
\end{aligned} \tag{1.52}$$

Substituting (1.48) into (1.47) we obtain the output variable $Y(s)$ in the complex frequency domain as

$$Y(s) = b_0 X_1(s) + b_1 X_2(s) + \cdots + b_{n-1} X_n(s) + b_n s X_n(s)$$

or, by transforming back into the time domain,

$$y(t) = b_0 x_1(t) + b_1 x_2(t) + \cdots + b_{n-1} x_n(t) + b_n \dot{x}_n(t) \tag{1.53}$$

Using this equation and (1.51) we can now obtain the output equation in the required form:

$$\begin{aligned} y(t) = {} & (b_0 - b_n a_0) x_1(t) + (b_1 - b_n a_1) x_2(t) + \cdots \\ & + (b_{n-1} - b_n a_{n-1}) x_n(t) + b_n u(t) \end{aligned} \tag{1.54}$$

Equations (1.52) and (1.54) are the state equations describing the dynamical system given by the $n$th order differential equation (1.44). If the matrix notation

$$\dot{\boldsymbol{x}}(t) = \boldsymbol{A}\boldsymbol{x}(t) + \boldsymbol{b}u(t) \tag{1.55}$$

$$y(t) = \boldsymbol{c}'\boldsymbol{x}(t) + du(t) \tag{1.56}$$

is used for this, the associated coefficient "matrices" are

$$\boldsymbol{A}_C = \begin{bmatrix} 0 & 1 & 0 & \ldots & 0 \\ 0 & 0 & 1 & & 0 \\ \vdots & & & & \vdots \\ 0 & 0 & 0 & & 1 \\ -a_0 & -a_1 & -a_2 & \ldots & -a_{n-1} \end{bmatrix} \tag{1.57}$$

$$\boldsymbol{b}_C = [0, 0, \ldots, 0, 1]' \tag{1.58}$$

$$\boldsymbol{c}_C' = [(b_0 - b_n a_0), \ldots, (b_{n-1} - b_n a_{n-1})] \tag{1.59}$$

$$d_C = b_n \tag{1.60}$$

Because a single-input single-output system has been assumed, the matrix $\boldsymbol{B}$ is converted to the column vector $\boldsymbol{b}$, the matrix $\boldsymbol{C}$ is converted to the row vector $\boldsymbol{c}'$ and the matrix $\boldsymbol{D}$ is converted to the scalar $d$. The subscript C refers to the control canonical form. As can be seen, this result is considerably simplified when $b_n = 0$.

The state equations (1.55) and (1.56) and their coefficient matrices (1.57)–(1.60) constitute the *control canonical form* for describing the dynamical system (1.44) by state variables. The name is derived from the fact that, in this notation for a controlled process, a controller can be designed in a simple manner if certain assumptions are made.

We again draw attention to the fact that there are no derivatives of the input function $u(t)$ in the state equations, whereas in the $n$th-order differential equation (1.44), which describes the same dynamical system, all derivatives of $u(t)$ up to $n$th order occur. All these derivatives must be assumed to exist if the differential equation (1.44) is to be solved directly. Equations (1.55)–(1.60) produce the block diagram for the control canonical form shown in Figure 1.5.

The characteristic equation of the matrix $\boldsymbol{A}_\mathrm{C}$ (see Appendix, Section A.5.1) coincides with the polynomial $N(s)$ in the denominator of the transfer function $G(s)$ in equation (1.45) as well as with the characteristic equation of the homogeneous part of the differential equation (1.44). In addition, the coefficients of the characteristic equation of $\boldsymbol{A}_\mathrm{C}$ occur as elements with negative signs in the matrix $\boldsymbol{A}_\mathrm{C}$. Such a matrix is called a *companion matrix* or a *Frobenius matrix* of the polynomial $N(s)$ (Zurmuehl and Falk, 1984). The control canonical form is also known as the *Frobenius canonical form*. Given state equations can be converted to the Frobenius canonical form by a transformation described by Schwarz (1971).

### 1.4.2 The Observer Canonical Form

Another state representation of the linear time-invariant dynamical system described by the differential equation (1.44) is obtained by transferring all terms of this differential equation to the right-hand side and combining derivatives of the same order of $y(t)$ and $u(t)$ in pairs:

$$\begin{aligned} 0 = & -a_0 y(t) + b_0 u(t) \\ & - a_1 \dot{y}(t) + b_1 \dot{u}(t) \\ & \vdots \\ & -a_{n-1} y^{(n-1)}(t) + b_{n-1} u^{(n-1)}(t) \\ & - y^{(n)}(t) + b_n u^{(n)}(t) \end{aligned} \tag{1.61}$$

The state variables are now introduced according to the following procedure:

$$\begin{array}{lllll} & -\ a_0 y(t) & +\ b_0 u(t) & =:\ \dot{x}_1(t) & \\ \dot{x}_1(t) & -\ a_1 \dot{y}(t) & +\ b_1 \dot{u}(t) & =:\ \ddot{x}_2(t) & \\ & & & \vdots & \\ x_{n-1}^{(n-1)}(t) & -\ a_{n-1} y^{(n-1)}(t) & +\ b_{n-1} u^{(n-1)}(t) & =:\ x_n^{(n)}(t) & \\ x_n^{(n)}(t) & -\ y^{(n)}(t) & +\ b_n u^{(n)}(t) & =\ 0 & \end{array} \tag{1.62}$$

The final equation in (1.62) is solved for $y^{(n)}(t)$:

$$y^{(n)}(t) = x_n^{(n)}(t) + b_n u^{(n)}(t)$$

The $n$-fold integration of this expression with respect to time, with all integration constants set equal to zero, gives the output equation of the required state equations:

$$y(t) = x_n(t) + b_n u(t) \tag{1.63}$$

We obtain the state differential equations from the first $n$ equations of (1.62) by taking the first equation, integrating the second equation once, integrating the third equation twice *et cetera*. $y(t)$ is expressed by using (1.63) and again all integration constants are set equal to zero. Thus we obtain

$$\begin{array}{llllll} \dot{x}_1(t) = & & -\ a_0 x_n(t) & +\ (b_0 & -\ b_n a_0) u(t) & \\ \dot{x}_2(t) = & x_1(t) & -\ a_1 x_n(t) & +\ (b_1 & -\ b_n a_1) u(t) & \\ \vdots & & & & & \\ \dot{x}_n(t) = & x_{n-1}(t) & -\ a_{n-1} x_n(t) & +\ (b_{n-1} & -\ b_n a_{n-1}) u(t) & \end{array} \tag{1.64}$$

Transforming to the matrix notation (1.55) and (1.56), we obtain the coefficient matrices:

$$\boldsymbol{A}_O = \begin{bmatrix} 0 & 0 & \dots & 0 & -a_0 \\ 1 & 0 & & 0 & -a_1 \\ 0 & 1 & & 0 & -a_2 \\ \vdots & & & & \vdots \\ 0 & 0 & \dots & 1 & -a_{n-1} \end{bmatrix} \tag{1.65}$$

$$\boldsymbol{b}_O = [(b_0 - b_n a_0), \dots, (b_{n-1} - b_n a_{n-1})]' \tag{1.66}$$

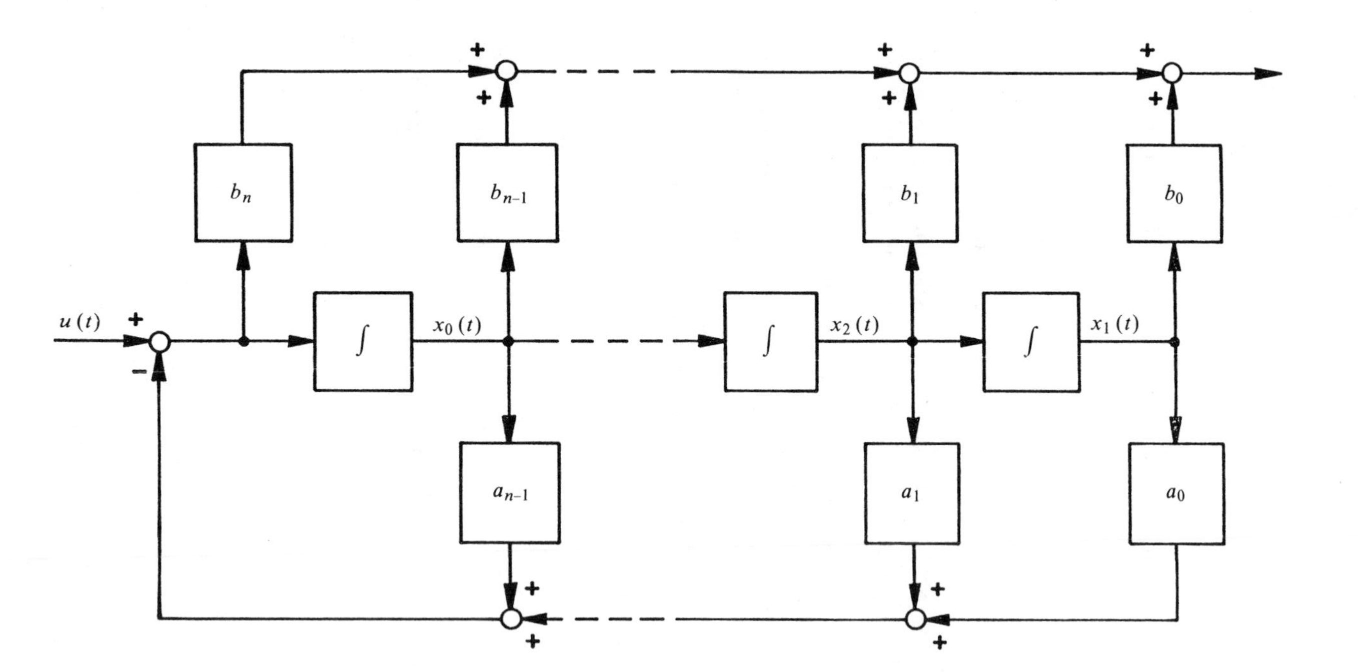


**Figure 1.5** Control canonical form of a linear time-invariant single-input single-output dynamical system.

$$c_O' = [0, 0, \ldots, 0, 1] \tag{1.67}$$

$$d_O = b_n \tag{1.68}$$

The coefficient matrices are identified by the subscript O (observer canonical form) in order to distinguish them from those for the control canonical form. Equations (1.55) and (1.56) together with the coefficient matrices (1.65)–(1.68) represent the *observer canonical form* for the dynamical system (1.44). The name was chosen because, when a control process is described by this type of state equation, an observer of the state variables of the process can be designed in a simple manner. A block diagram of the observer canonical form is shown in Figure 1.6.

There is a very close relationship between the coefficient matrices of the control canonical form and those of the observer canonical form:

$$\boldsymbol{A}_O = \boldsymbol{A}_C' \tag{1.69}$$

$$\boldsymbol{b}_O = (\boldsymbol{c}_C')' \tag{1.70}$$

$$\boldsymbol{c}_O' = \boldsymbol{b}_C' \tag{1.71}$$

$$\boldsymbol{d}_O = \boldsymbol{d}_C \tag{1.72}$$

However, from a physical point of view the state variables of the two forms have completely different meanings. This can easily be seen by comparing Figures 1.5 and 1.6.

### 1.4.3 The Partial Fraction Canonical Form

The third method of describing a linear time-invariant single-input single-output dynamical system by state variables is based on the decomposition of the transfer function $G(s)$ (equation (1.45)) into partial fractions. If, for simplicity, it is assumed that all poles $s_1, \ldots, s_n$ of $G(s)$ are single (distinct) and real, the partial fraction decomposition of $G(s)$ is

$$G(s) = r_0 + \sum_{\nu=1}^{n} \frac{r_\nu}{s - s_\nu} \tag{1.73}$$

The value $r_0$ and the residues $r_1, \ldots, r_n$ are functions of the coefficients $a_0, \ldots, a_n$ and $b_0, \ldots, b_n$. For $b_n = 0$, we always have $r_0 = 0$.

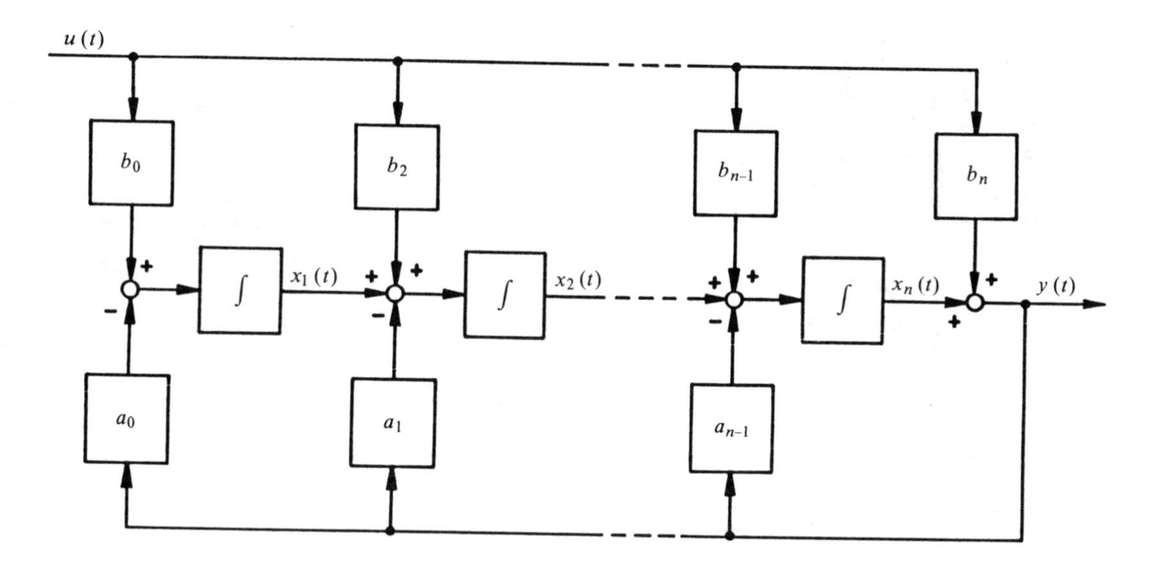


**Figure 1.6** Observer canonical form of a linear time-invariant single-input single-output dynamical system.

Using equation (1.73) we can write equation (1.45) in the form

$$Y(s) = \left\{ r_0 + \sum_{\nu=1}^{n} \frac{r_\nu}{s - s_\nu} \right\} U(s) \tag{1.74}$$

We now define the real state variables, which in the complex frequency domain are

$$X_\nu(s) = \frac{1}{s - s_\nu} U(s) \qquad \nu = 1, \ldots, n \tag{1.75}$$

In the time domain this corresponds to the set of differential equations

$$\dot{x}_\nu(t) = s_\nu x_\nu(t) + u(t) \qquad \nu = 1, \ldots, n \tag{1.76}$$

Thus we have the required state differential equations. It follows from equation (1.74) and the definition equation (1.75) that

$$Y(s) = \sum_{\nu=1}^{n} r_\nu X_\nu(s) + r_0 U(s)$$

or in the time domain

$$y(t) = \sum_{\nu=1}^{n} r_\nu x_\nu(t) + r_0 u(t) \tag{1.77}$$

This output equation, together with the state differential equations (1.76), gives the *partial fraction canonical form*. If we transform to matrix notation, equations (1.55) and (1.56) also apply, and the coefficient matrices have the form

$$\boldsymbol{A}_P = \begin{bmatrix} s_1 & 0 & \ldots & 0 \\ 0 & s_2 & & 0 \\ \vdots & & & \vdots \\ 0 & 0 & \ldots & s_n \end{bmatrix} \tag{1.78}$$

$$\boldsymbol{b}_P = [1, 1, \ldots, 1]' \tag{1.79}$$

$$\boldsymbol{c}_P' = [r_1, r_2, \ldots, r_n]' \tag{1.80}$$

$$d_P = r_0 \tag{1.81}$$

where the subscript $P$ indicates the partial fraction canonical form.

A particular advantage of this notation is that the system matrix $\boldsymbol{A}_P$ is diagonal (for distinct poles). The notation considerably simplifies calculations with state equations of this type. The diagonal nature of the system matrix $\boldsymbol{A}_P$ is essentially due to the special form of the state differential equations (1.76) in which every derivative $\dot{x}_\nu(t)$ is dependent only on $x_\nu(t)$ and on no other state variable. Therefore every state differential equation can be solved individually without considering the other state differential equations. In this case it is said that the state differential equations are decoupled.

A drawback of the partial fraction canonical form is that it is difficult to find the elements of the coefficient matrices. Unlike the control canonical form or the observer canonical form they cannot be obtained simply from the coefficients of the differential equation (1.44) or the coefficients of the transfer function $G(s)$ (equation (1.45)). In particular, the poles and residues of $G(s)$ must first be calculated. The state equations (1.55) and (1.56) with the coefficient matrices (1.78)–(1.81) can also be represented by a block diagram (Figure 1.7).

The above discussion is also valid when the poles of $G(s)$ are complex but are still single (distinct). However, in this case, complex elements occur in the coefficient matrices $\boldsymbol{A}_P$ and $\boldsymbol{c}'_P$ and in the associated block diagram. Furthermore, the state variables take complex values. Often we only want to work with real coefficients and real state variables, for example, in computer calculations. This can be achieved by transforming the state equations (Föllinger, 1985). Of course, this means that the decoupling of the state differential equations is lost and that the system matrix is no longer diagonal.

A partial fraction canonical form can also be obtained for multiple poles of $G(s)$. Then, however, the state differential equations are not decoupled and the system matrix $\boldsymbol{A}_P$ is no longer diagonal. The selection of the state variables in this case is discussed by Föllinger (1985) and DeRusso *et al.* (1966).

The system matrix occurring in the partial fraction canonical form is called a *Jordan matrix* (Zurmuehl and Falk, 1984). Hence the partial fraction canonical form is also known as the *Jordan canonical form*. Sometimes the term canonical form is used exclusively for the partial fraction canonical form. The Jordan canonical form for any given state equations can be found by a transformation (Föllinger, 1985).

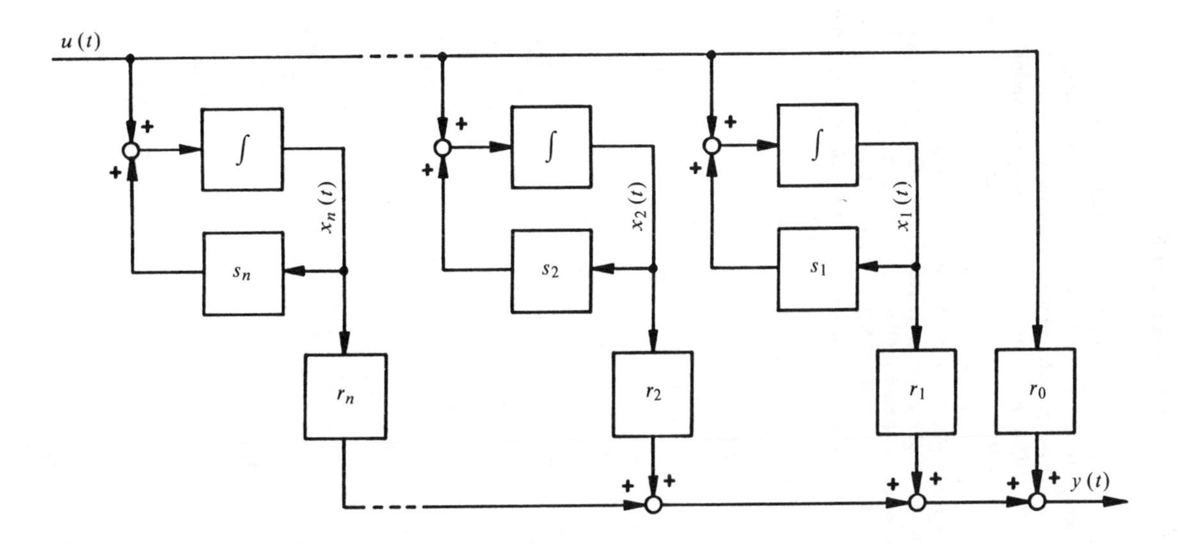


**Figure 1.7** Partial fraction canonical form of a linear time-invariant single-input single-output dynamical system.

# REFERENCES

Ackermann, J. (1983). *Abtastregelung,* Vol. 1, *Analyse und Synthese* (2nd ed.), Springer, Berlin.
DeRusso, P. M., Roy, R. J. and Close, C. M. (1966). *State Variables for Engineers,* Wiley, New York.
Föllinger, O. (1985). *Regelungstechnik* (5th ed.), Hüthig, Heidelberg.
Landgraf, C., and Schneider, G. (1970). *Elemente der Regelungstechnik,* Springer, Berlin.
Ogata, K. (1967). *State Space Analysis of Control Systems,* Prentice-Hall, Englewood Cliffs, NJ.
Schwarz, H. (1971). *Mehrfachregelungen,* Vol. 2, Springer, Berlin.
Thoma, M. (1973). *Theorie linearer Regelsysteme,* Vieweg, Braunschweig.
Timothy, L. K. and Bona, B. E. (1968). *State Space Analysis,* McGraw-Hill, New York.
Unbehauen, R. (1983). *Systemtheorie* (4th ed.), Oldenbourg, Munich.
Zadeh, L. A. and Desoer, C. A. (1963). *Linear System Theory,* McGraw-Hill, New York.
Zurmuehl, R. and Falk, S. (1984). *Matrizen und ihre Anwendungen,* Part 1, *Grundlagen* (5th ed.), Springer, Berlin.

# Chapter 2
# Elements of Probability Theory

In this chapter we introduce the concepts that are used to describe and to identify random signals. For filter problems, these random signals are functions of time with values that are determined by chance. Complicated concepts such as the covariance matrix or the autocorrelation function are explained not only in terms of their defining equations but are also derived step by step from the basic principles of probability theory.

Probability theory is the branch of mathematics that is concerned with the laws obeyed by random events. These problems were first examined in the 17th century (Fermat and Pascal) when attempts were made to calculate the probabilities that the bank or the gamblers would win in games of chance.

## 2.1 RANDOM EVENTS

The concept of the random (stochastic) event has developed from the concept of the elementary event. An elementary event can be defined in terms of the following simple example. When a die having its faces marked with one, two, three, four, five and six dots is thrown, it is impossible to predict what number of dots will be on the top face after the throw. The result of a throw depends on random phenomena. Therefore the entire process of throwing a die is called a *random experiment* and the result of the random experiment is an *elementary event*.

If, however, the arrangement shown in Figure 2.1 is considered, it can immediately be seen that, of the six lamps $L_1, \ldots, L_6$, only lamp $L_2$ lights up when a voltage is applied across terminals 1 and 2. The result of applying a voltage (i.e., the illumination of lamp $L_2$) is not random, and therefore this is not a random experiment but a *deterministic experiment*.

The concepts of random experiment and deterministic experiment introduced in these two examples are generally valid. An experiment is random when its result cannot be determined in advance and is deterministic when its result can be determined in advance.

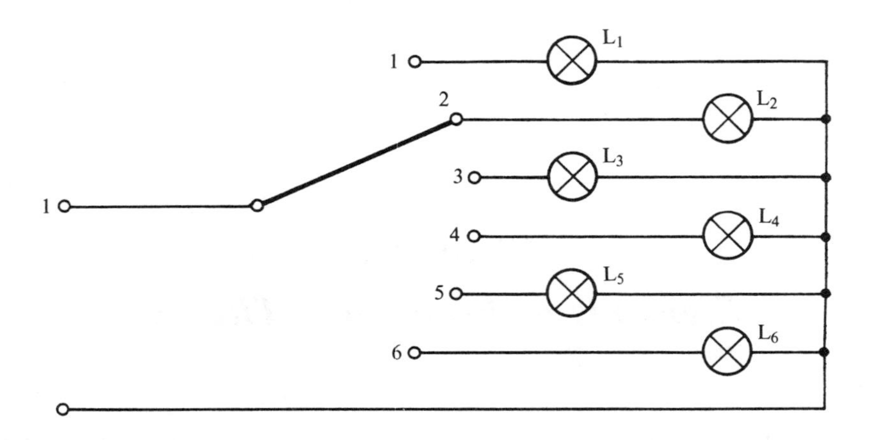


**Figure 2.1** Example of a deterministic experiment.

The difference between the two experiments becomes even clearer when we modify the above formulation: we can intervene in a deterministic experiment so that a desired result is certain to be obtained, provided that this result is allowed by the experiment. If, for example, in the circuit of Figure 2.1, we want lamp $L_5$ to illuminate after the voltage is applied, the switch arm S must be moved to position 5. In contrast, there is no way of influencing the throw of a die so that it is certain that the number five will appear on the top face, provided, of course, that the die is perfect and it has been thrown correctly. All random experiments have this property for the reasons explained below, using the above two examples for illustration.

Whereas in the deterministic experiment the relationship between cause (application of the voltage) and effect (illumination of a lamp) can immediately be seen from the position of the switch, in the experiment with the die this relationship is too complicated to allow the result of a throw to be calculated using a mathematical model. This result still depends on the position of the die at the beginning of the throw, the speed at which the die is thrown, the collision angle and the elasticity of the die, *et cetera*. These variables, which must be included as parameters or initial values in a mathematical model of dice throwing, cannot be adjusted so that the number predicted is certain to come up when the die is thrown.

Therefore the essential characteristic of chance is not that the law of cause and effect is violated or even that unfamiliar physical laws come into effect. The nature of chance lies purely in the fact that the initial values, parameters and relationships responsible for the result cannot all be described mathematically, and also that not all of the factors influencing the result are known.

Because the result of a random experiment cannot be predicted in advance, a single event (i.e., an elementary event) cannot be used to describe the properties of such an experiment. The next time the same experiment is carried out, a different elementary event may occur. For this reason, all elementary events which can occur are used to identify a random experiment. Subsequently, a nonnegative real number is assigned to each of these elementary events, and this number is a measure of the probability that this elementary event will occur. However, first we will examine the possible outcomes of a random experiment in more detail.

In the dice-throwing experiment the six elementary events "the one is on top" to "the six is on top" can occur. We can also say that a one or a two *et cetera* has been thrown. However, we can also ask whether an "even number" or a "number less than four" has occurred. Such complicated "results" of a random experiment are known as *random events*. The "even number" random event occurs when the two is thrown, but it also occurs when the four or the six is thrown. Thus the "even number" can occur in different ways. If the two is thrown, both the random event "even number" and the random event "number smaller than four" have occurred. Thus in this case two random events have occurred at the same time in the performance of the random experiment.

The elementary events "number one is on top" to "number six is on top" cannot occur in different ways but only in one way. It is also impossible for a number of elementary events to occur together during one throw with one die. It is said that the elementary events are *disjoint* or that they are *mutually exclusive*. This is why these six elementary events have a special significance for the dice experiment. They are, as will be shown below, the elements from which the random events can be formed; this is why they are called elementary events.

The elementary events or their properties, which so far have been discussed using the example of dice throwing, can be defined as follows in the general case.

> The possible results of a random experiment are called elementary events. Every elementary event can only occur in one way and cannot occur in combination with any other elementary event.

All the elementary events, denoted $\omega_1, \omega_2, \ldots$, taken as a whole, form the *set $\Omega$ of elementary events*; $\omega_i$ is an *element* of $\Omega$ (in symbols $\omega_i \in \Omega$). In the dice experiment, the set $\Omega$ consists of the six elementary events $\omega_1, \ldots, \omega_6$:

$$\Omega = \{\omega_1, \ldots, \omega_6\}$$

The random events "even number" and "number less than four" can also be described by sets. These sets consist of those elementary events for which their occurrence is responsible for the occurrence of the random event itself. If we denote the throwing of a one by $\omega_1$, the throwing of a two by $\omega_2$, *et cetera,* the random event "even number" consists of the set with elements $\omega_2$, $\omega_4$ and $\omega_6$:

$$\text{"even number"} = \{\omega_2, \omega_4, \omega_6\}$$

Similarly, we have

$$\text{“number less than four”} = \{\omega_1, \omega_2, \omega_3\}$$

If these considerations are generalized, the concept of the random event can be formulated as follows.

Every random event consists of a set of elementary events. A random event occurs when the elementary event that takes place as a result of the random experiment is an element of the set that exactly describes this random event.

The concept of the random event is therefore derived from the concept of the elementary event. However, in the sense of the above notation, every elementary event also represents a random event because the elementary event can also be regarded as a set which in this case, however, only contains a single element.

Every random event that is possible in a random experiment can be described as a set of elementary events and the set $\Omega$ contains all elementary events as elements, and thus every random event must be a subset of the set $\Omega$. By definition, a set A is called a *subset* of the set $\Omega$ (in symbols $A \subset \Omega$) when every element of A is also contained in $\Omega$. It follows from this definition that the set $\Omega$ must be a subset of itself. If, however, $\Omega$ is a subset of $\Omega$ and if a subset of $\Omega$ is called a random event, then $\Omega$ itself must be a random event.

A random event occurs when an elementary event from the set which defines this random event occurs. However, because $\Omega$ consists of all elementary events which are possible in a random experiment, the event $\Omega$ must occur with certainty every time the random experiment is carried out. Therefore $\Omega$ is called the *certain event*.

The event $\Omega$ is therefore not random in the sense in which the concept of "chance" was introduced at the beginning of this section. However, because of the set definition of a random event in probability theory, $\Omega$ is related to random events. To some extent, $\Omega$ represents the limiting case of a random event. Another limiting case is the event which certainly cannot occur in a random experiment. For example, in a dice-throwing experiment (with one die), a number greater than six can never be thrown. Such an occurrence is called an *impossible event* and is denoted by the symbol $\emptyset$. It consists of the "subset" of $\Omega$ which does not contain any elements (the empty set or the null set). In probability theory the impossible event is also related to random events.

All the random events of a random experiment taken as a whole (i.e., all subsets of the set $\Omega$ of elementary events), including the impossible event and the certain event, form the *set Z of random events* of a random experiment. Z also contains the elementary events as random events.

If the set $\Omega$ consists of $m$ elementary events, according to the rules of combinatorial algebra, the set Z of random events contains exactly $2^m$ elements. An elementary event can only assume two states: it can either occur or not occur. Therefore in the dice-throwing experiment with six elementary events, a total of $2^6 = 64$ random events is possible.

## 2.2 CALCULATION RULES FOR RANDOM EVENTS

Now that we have defined random events as sets (i.e., as sets of elementary events), we can use the operations of set theory to combine random events. Random events that are also elements of the set $Z$ are denoted by A, B, C, ... or by $A_1$, $A_2$, $A_3$, ... .

### 2.2.1 The Product (Intersection) of Events

Suppose that there are two random events A and B. Both A and B are subsets of $\Omega$ ($A \subset \Omega$, $B \subset \Omega$) and are also elements of the set Z of all random events of a random experiment (in symbols $A \in Z$, $B \in Z$).

The *product* (intersection) of the random events A and B is the random event which consists of all elementary events which belong to both the random event A and the random event B. This definition can be extended in a completely analogous manner to the combination of more than two random events. The new random event formed in this way is denoted by $A \cap B$ (in words A and B) or by $A \cap B \cap C \cap \ldots$ (or $A \cdot B$ or $A \cdot B \cdot C \ldots$). $A \cap B$ occurs when both the random event A and the random event B occur.

It was assumed in the definition of the product event that the random event A and the random event B are subsets of the set $\Omega$ ($A \subset \Omega$, $B \subset \Omega$). Therefore the elementary events which belong to both the random event A and the random event B (i.e., constitute the random event $A \cap B$ exactly), must also form a subset of $\Omega$; thus $A \cap B$ must be an element of the set Z ($(A \cap B) \in Z$).

The product of events is clearly illustrated in Figure 2.2, where the rectangles represent the set of all elementary events and the circles forming subsets of the set $\Omega$ symbolize the random events A, B and C. All the points in the plane of the diagram inside the rectangle can be interpreted as elementary events. The points in the plane of the diagram which belong to both the random event A and the random event B or to A, B and C (Figure 2.2(d)) form the random events $A \cap B$ and $A \cap B \cap C$ respectively. The product events are indicated by shading.

If the random events A and B have no common elementary event (Figure 2.2(b)), we say that events A and B are *disjoint* or that they are *mutually exclusive*. Then the product event $A \cap B$ can only be the impossible event $\emptyset$($A \cap B = \emptyset$). Because the elementary events are mutually exclusive, by definition, they must obey the equation

$$\omega_\nu \cap \omega_\mu = \emptyset \quad \text{for} \quad \nu \neq \mu \tag{2.1}$$

This property forms the basis for the simplifications involved in combining elementary events.

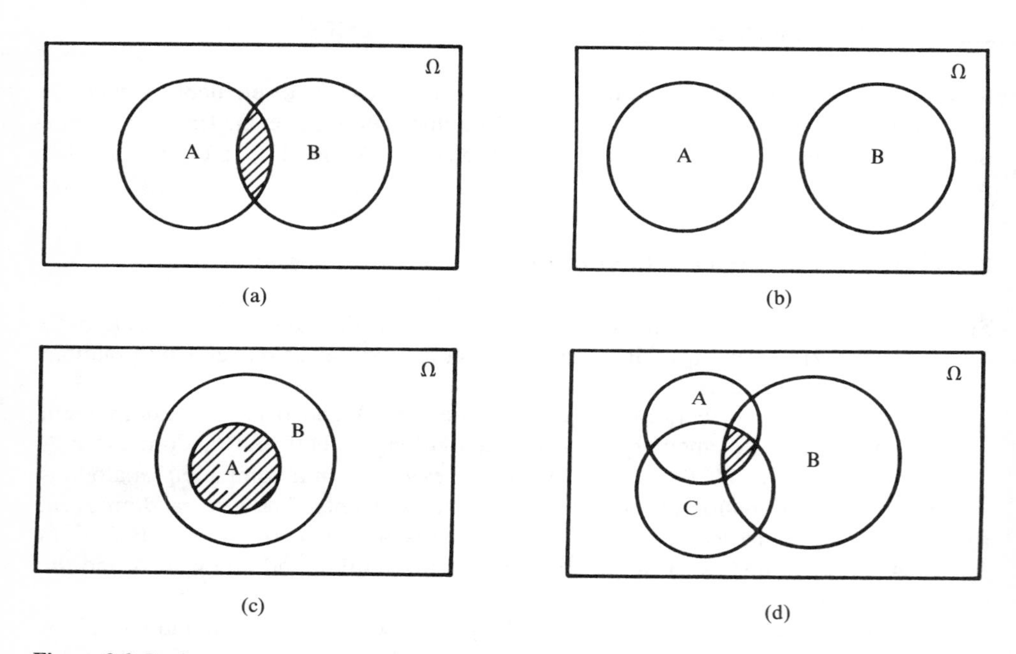


**Figure 2.2** Product events A ∩ B or A ∩ B ∩ C.

The following calculation rules follow directly from the definition of the product event or from Figure 2.2:

$$A \cap B = B \cap A \qquad \text{commutative law} \tag{2.2}$$

$$(A \cap B) \cap C = A \cap (B \cap C) = A \cap B \cap C \qquad \text{associative law} \tag{2.3}$$

$$A \cap A = A \tag{2.4}$$

$$A \cap \Omega = A \tag{2.5}$$

$$A \cap \emptyset = \emptyset \tag{2.6}$$

$$A \cap B = A \quad \text{if} \quad A \subset B \tag{2.7}$$

The condition that A is a subset of B ($A \subset B$) as a prerequisite for the validity of (2.7) means that all elementary events which form the random event A also belong to the random event B. We say that A is contained in B or A implies B, because the occurrence of the random event A means that the random event B must necessarily occur. The product of events corresponds to the *intersection set* in set theory.

### 2.2.2 The Sum (Union) of Events

Let A and B be two random events with $A \subset \Omega$, $B \subset \Omega$ and $A \in Z$, $B \in Z$. The elementary events that belong to at least one of the two random events A or B form a new random event which is called the *sum* (union) of the random events A and B. The sum event occurs when the random event A occurs or when the random event B occurs or when both random events A and B occur together. By using this definition, more than two random events can also be combined in a corresponding manner. The random event obtained in this way is denoted by $A \cup B$ (in words A or B) or by $A \cup B \cup C \cup \ldots$ (or $A + B$ or $A + B + C + \ldots$). $A \cup B$ is also an element of the set Z of all random events of a random experiment. In Figure 2.3 the shaded areas symbolize the sum events corresponding to different cases.

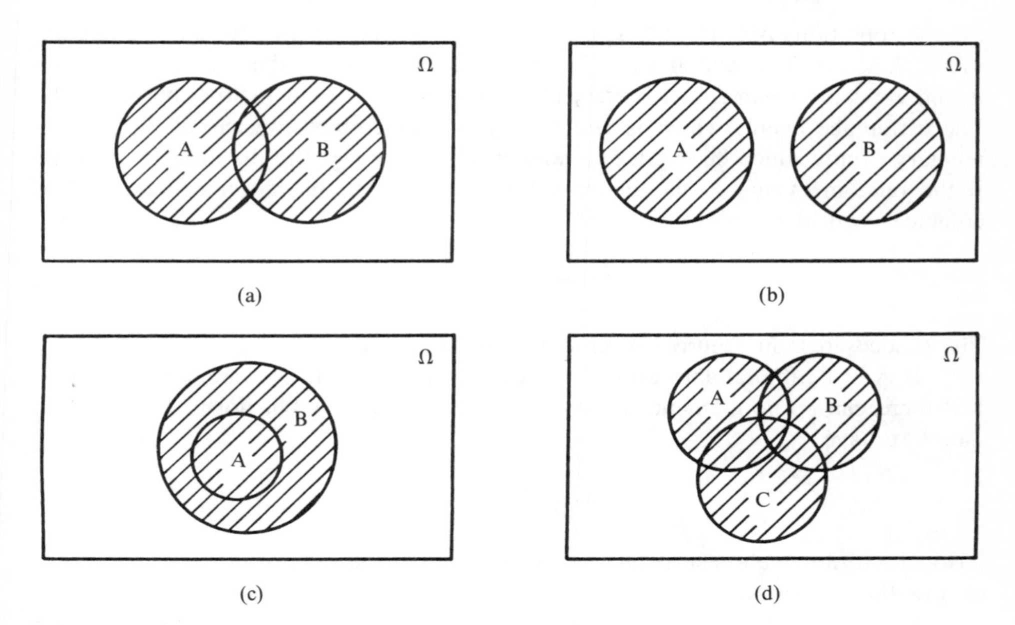


**Figure 2.3** Sum events $A \cup B$ and $A \cup B \cup C$.

Using either the definition of the sum of random events or Figure 2.3, we can obtain the following relationships:

$$A \cup B = B \cup A \qquad \text{commutative law} \tag{2.8}$$

$$(A \cup B) \cup C = A \cup (B \cup C) = A \cup B \cup C \qquad \text{associative law} \tag{2.9}$$

$$A \cup A = A \tag{2.10}$$

$$A \cup \Omega = \Omega \tag{2.11}$$

$$A \cup \emptyset = A \tag{2.12}$$

$$A \cup B = B \quad \text{if} \quad A \subset B \tag{2.13}$$

The sum event corresponds to the *union set* in set theory.

### 2.2.3 The Difference of Events

Let the conditions $A \subset \Omega$, $B \subset \Omega$ and $A \in Z$, $B \in Z$ be satisfied for the two random events A and B. The random event known as the *difference* of the two random events A and B occurs when an elementary event occurs which belongs to A but not to B. The difference event, which is denoted by $A \setminus B$, therefore consists of all the elementary events which belong to the random event A but not to the random event B. If the random events A and B are mutually exclusive, the difference event $A \setminus B$ coincides with the random event A:

$$A \setminus B = A \quad \text{if} \quad A \cap B = \emptyset \tag{2.14}$$

The shaded areas in Figure 2.4 represent difference events.

If $A \subset B$, the random event B necessarily occurs with the random event A so that there can be no elementary event which belongs to A but not to B. We must therefore have

$$A \setminus B = \emptyset \quad \text{if} \quad A \subset B \tag{2.15}$$

Also, the difference event as an element belongs to the set Z of all random events of a random experiment.

We now have to consider a fourth way of obtaining a new random event by a combination operation.

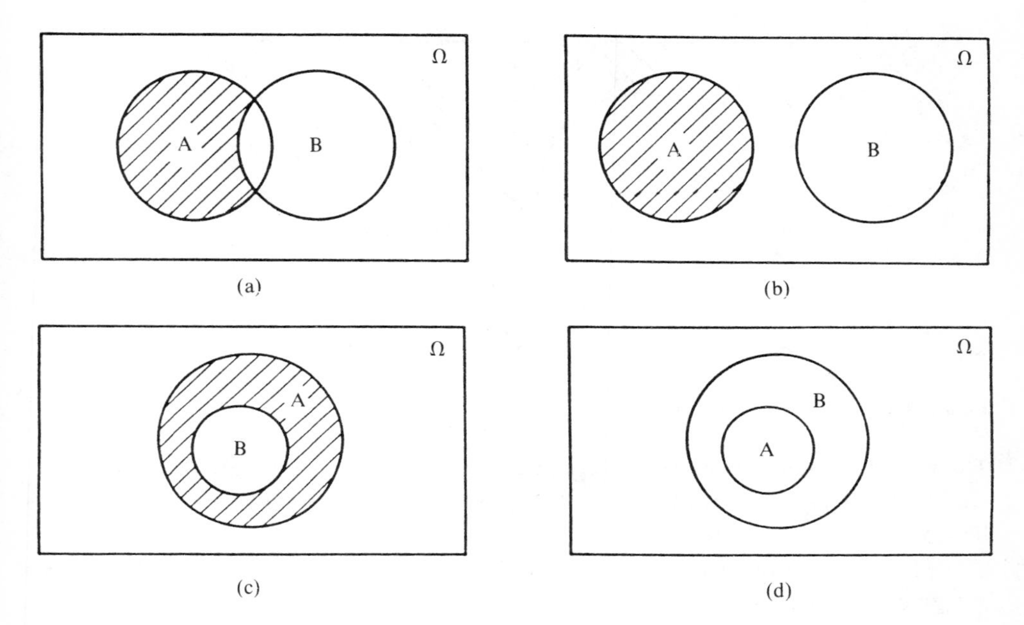


**Figure 2.4** Difference events A \ B.

### 2.2.4 The Complementary Event

The complementary event is a special case of the difference between two events. All the elementary events which belong to the certain event $\Omega$ but not to the random event A form the *complementary event* $\bar{A}$ with respect to the random event $A$:

$$\bar{A} = \Omega \setminus A \tag{2.16}$$

It follows from the assumption $A \subset \Omega$ and $A \in Z$ that $\bar{A} \in Z$ is also true in this case.

The following relationships can be derived from Figure 2.5:

$$\bar{\Omega} = \emptyset \tag{2.17}$$

$$\bar{\emptyset} = \Omega \tag{2.18}$$

$$\bar{\bar{A}} = A \tag{2.19}$$

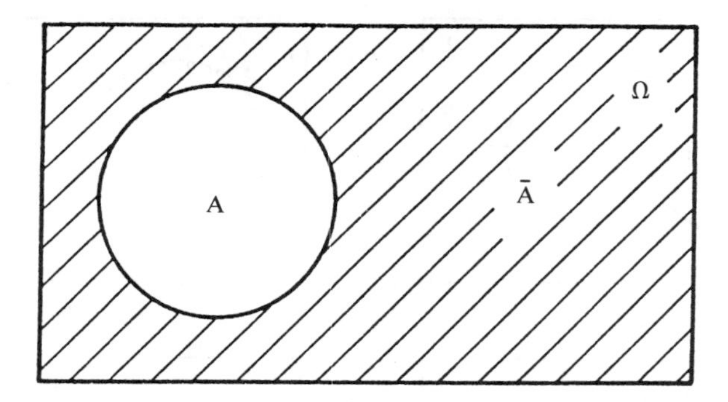


**Figure 2.5** Complementary event Ā.

The complementary event to the certain event is therefore the impossible event and *vice versa*. The complementary event corresponds to the *complementary set* in set theory.

To conclude this section, some additional calculation rules for random events are given without any derivation:

$$A \cap \bar{A} = \emptyset \tag{2.20}$$

$$A \cup \bar{A} = \Omega \tag{2.21}$$

$$(A \cap B) \cup C = (A \cup C) \cap (B \cup C) \quad \text{distributive law} \tag{2.22}$$

$$(A \cup B) \cap C = (A \cap C) \cup (B \cap C) \quad \text{distributive law} \tag{2.23}$$

$$\overline{A \cap B} = \bar{A} \cup \bar{B} \tag{2.24}$$

$$\overline{A \cup B} = \bar{A} \cap \bar{B} \tag{2.25}$$

Equations (2.24) and (2.25) are known as DeMorgan's formulas.

## 2.3 AXIOMATIC INTRODUCTION OF THE PROBABILITY CONCEPT

At the beginning of this chapter, it was stated that probability theory is the branch of mathematics concerned with the laws obeyed by random events. However, remembering the introduction of the concept "random event," it is clear that the laws mentioned above cannot refer to the relationship between the experimental conditions and the experimental result in a random experiment because it is just the lack of knowledge of this relationship which characterizes a random experiment and thus a random event. Therefore the laws involved in probability theory must have different characteristics.

These laws become clear if we perform a random experiment a large number of times under the *same experimental conditions*. Probability theory then provides us with a mathematical model for describing the empirical results. For these experimental results to be obtained, a further two conditions must be satisfied:

1. it must be possible to perform each random experiment any number of times under the same conditions;
2. it must be possible to decide unambiguously whether or not a specified random event has occurred as the result of a random experiment.

In what follows it is assumed that these conditions are always satisfied.

We will consider the example of throwing a coin. Throwing a symmetric coin is certainly a random experiment. We cannot predict in advance whether heads or tails will be obtained in a single throw. The random event "heads" is denoted by H and the random event "tails" is denoted by T. Insofar as the coin is assumed to be totally symmetric, we would intuitively expect that, for a large number of throws, heads and tails would occur almost equally. The assumed symmetry of the coin excludes the preference of one side of the coin over the other. This purely intuitive supposition can be confirmed experimentally. Before we give some results for this, two further concepts will be introduced.

If a random experiment is carried out $n$ times under the same conditions and the random event A occurs exactly $n_A$ times, the number $n_A$ is called the *frequency* $F_A$ and the ratio of $n_A$ to $n$ is called the *relative frequency* $f_A$ of the random event A. Both $F_A$ and $f_A$ depend on the number of experiments and are therefore functions of $n$:

$$F_A(n) := n_A \tag{2.26}$$

$$f_A(n) := \frac{n_A}{n} = \frac{F_A(n)}{n} \tag{2.27}$$

In particular, in the example of throwing a coin where the random events H and T are mutually exclusive, such that $H \cap T = \emptyset$, we must have*

$$n = n_H + n_T$$

*This equation is only correct if it is assumed that the rare case for which the coin remains standing on its edge after it has been thrown is not included in the experiment.

so that

$$f_{\mathrm{H}}(n) = \frac{n_{\mathrm{H}}}{n}$$

$$f_{\mathrm{T}}(n) = \frac{n_{\mathrm{T}}}{n} = \frac{n - n_{\mathrm{H}}}{n} = 1 - f_{\mathrm{H}}(n)$$

Table 2.1 gives some experimental results for the coin experiment for large values of $n$ (Kreyszig, 1979). It can be seen that the relative frequency $f_{\mathrm{H}}$ approaches 0.5 more closely as $n$ increases. This means that as $n$ increases, the frequencies $F_{\mathrm{H}}$ and $F_{\mathrm{T}}$ approach each other so that the intuitive assumption made above is confirmed.

**Table 2.1** Results for a Large Number $n$ of Coin-Throwing Experiments

| | $n$ | $F_H(n)$ | $f_H(n)$ |
|---|---|---|---|
| Buffon | 4 040 | 2 048 | 0.5069 |
| Pearson | 12 000 | 6 019 | 0.5016 |
| | 24 000 | 12 012 | 0.5005 |

If we plot the relative frequencies $f_{\mathrm{A}}(n)$ of the random event A obtained from an arbitrary random experiment for various values of $n$, we obtain the fundamental curve shown in Figure 2.6. The relative frequency $f_{\mathrm{A}}(n)$ fluctuates about a fixed end value $P(\mathrm{A})$, and as the number $n$ of experiments increases the deviations from this end value become smaller. This is known as the *stability of the relative frequency*. The magnitude of this numerical value $P(\mathrm{A})$, which is approximated by $f_{\mathrm{A}}(n)$ for large $n$, is a specific property of the random event A under investigation.

We shall now investigate some more properties of the relative frequency $f_{\mathrm{A}}(n)$.

(1) The relative frequency $f_{\mathrm{A}}(n)$ is defined as the number $n_{\mathrm{A}}$ of events A which have occurred, divided by the number $n$ of experiments (i.e., as the ratio of two positive numbers, and therefore must also be a positive number). If, for example, we assume a finite number $n$ of experiments, in one extreme case event A occurs in none of the $n$ experiments but in the other extreme case it occurs in all $n$ experiments.

Thus we have

$$0 \leqslant n_{\mathrm{A}} \leqslant n$$

or, for the frequency of the random event A,

$$0 \leqslant F_A(n) \leqslant n \tag{2.28}$$

Then, from equation (2.27), we have for the associated relative frequency:

$$0 \leqslant f_A(n) \leqslant 1 \tag{2.29}$$

(2) Because the certain event $\Omega$ occurs every time the random experiment is done, we must have

$$n_\Omega = n$$

so that

$$F_\Omega(n) = n_\Omega = n \tag{2.30}$$

or, in effect,

$$f_\Omega(n) = 1 \tag{2.31}$$

Hence the relative frequency $f_\Omega(n)$ of the certain event $\Omega$ is always equal to unity, regardless of the value of $n$.

(3) Consider a random experiment in which, among other events, two mutually exclusive random events A and B can occur:

$$A \cap B = \emptyset$$

For a finite number $n$ of experiments we assume that the random event A has occurred $n_A$ times, the random event B has occurred $n_B$ times and the random event $A \cup B$ has occurred $n_{A \cup B}$ times. Since the random events A and B are mutually exclusive by assumption, i.e., A and B cannot both occur at the same time, there is no experimental result possible which can be assigned to both the quantity $n_A$ and the quantity $n_B$. In other words there is no experimental result which can be counted twice. This means that

$$n_{A \cup B} = n_A + n_B \quad \text{if} \quad A \cap B = \emptyset$$

must be valid. Using equation (2.26) it follows from this that

$$F_{A \cup B}(n) = F_A(n) + F_B(n) \quad \text{if} \quad A \cap B = \emptyset \tag{2.32}$$

or, in terms of the relative frequency,

$$f_{A \cup B}(n) = f_A(n) + f_B(n) \quad \text{if} \quad A \cap B = \emptyset \tag{2.33}$$

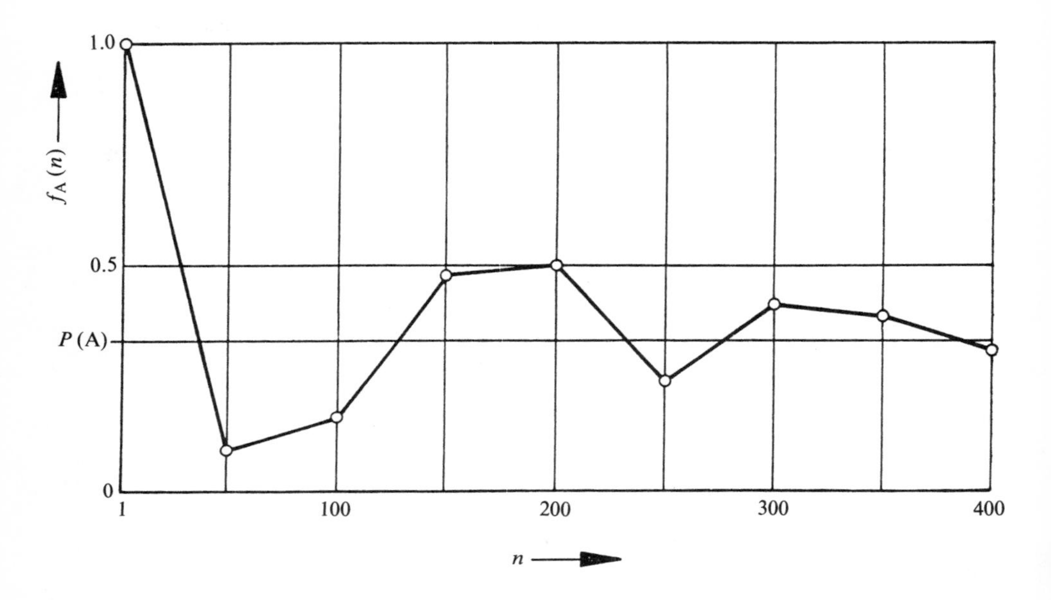


**Figure 2.6** Relative frequency $f_A(n)$ of a random event A.

### 2.3.1 The Kolmogorov Axioms

It has been found empirically that, for large $n$, the relative frequency $f_A(n)$ of a random event A fluctuates around a fixed value $P(A)$ which is a specific property of A (Figure 2.6). This remarkable property of $f_A(n)$ forms the point of departure for the definition of a number which gives a measure of the probability that this random event A will occur. Such a number can be defined implicitly by the following axioms which assign to this number analogous properties to those of the relative frequency.

*Axiom 1*. Every random event A is assigned a real nonnegative number $P(A)$, which is called the *probability* of the random event A:

$$P(A) \geq 0 \tag{2.34}$$

*Axiom 2*. The probability that the certain event $\Omega$ will occur is equal to unity:

$$P(\Omega) = 1 \tag{2.35}$$

*Axiom 3. (Addition Axiom).* The probability of the sum of two mutually exclusive random events A and B is equal to the sum of the probabilities of these two random events:

$$P(\mathrm{A} \cup \mathrm{B}) = P(\mathrm{A}) + P(\mathrm{B}) \quad \text{if} \quad \mathrm{A} \cap \mathrm{B} = \emptyset \tag{2.36}$$

If, in a random experiment, an infinite (countable) number of random events is possible, the addition axiom must be replaced by the extended addition axiom.

*Axiom 4. (Extended Addition Axiom).* If the random events $\mathrm{A}_1$, $\mathrm{A}_2$, $\mathrm{A}_3, \ldots$ are mutually exclusive in pairs,

$$\mathrm{A}_i \cap \mathrm{A}_j = \emptyset \quad i \neq j$$

then we have

$$P(\mathrm{A}_1 \cup \mathrm{A}_2 \cup \cdots) = P(\mathrm{A}_1) + P(\mathrm{A}_2) + \cdots \tag{2.37}$$

It is to the credit of A.N. Kolmogorov that he recognized that the entire edifice of probability theory can be erected completely logically and in agreement with the experimental results on the basis of these axioms together with the concepts of the elementary event and the random event introduced in Section 2.1. With the axiomatic introduction of the concept of the *mathematical probability* $P(\mathrm{A})$ of a random event A, we have found a mathematical model that is free from contradictions and can be used to describe the results of random experiments.

In Axiom 1 the probability $P(\mathrm{A})$ is introduced as a measure of the probability that the random event A will occur. It should be noted that in this statement the word "probability" has been used with two different meanings. In the latter case, the concept of "probability" is used in the sense of everyday speech (e.g., it is more probable to be 30 years old than 90 years old). In the former case, the concept of "probability" is used as the name for the number $P(\mathrm{A})$ that represents a mathematical specification or quantification of the concept of probability used in everyday speech.

So far every subset of $\Omega$ has been designated as a random event. However, if the above axioms are used to assign a probability to every random event defined in this way, difficulties arise if the set $\Omega$ consists of an infinite (uncountable) number of elementary events. The axioms can then no longer be applied to all subsets of $\Omega$, i.e., they can no longer be applied to all elements of Z but only to certain subsets of $\Omega$ which together define the set Z*. In the case of an uncountable infinite number of elementary events only the elements of Z* are designated random events (Papoulis, 1984). However, care should be taken that new random events obtained by combination operations from the random events $A_1 \in Z^*$, $A_2 \in Z^*, \ldots$ are themselves elements of $Z^*$.[†] Without this requirement, the system of axioms given above would lead to contradictions. For example, a probability $P(B)$ can always be assigned to a random event B produced by combination operations from the random events $A_1 \in Z^*$, $A_2 \in Z^*$, $A_3 \in Z^*, \ldots$ using the rules which will be derived in the next section. However, according to Axiom 1, no probability $P(B)$ is possible if $B \notin Z^*$.

It is not possible to pursue the discussion of these problems any further in this introductory text. They are of importance in the theory but are not generally involved in applications.

So far we have not considered how the probability $P(A)$, which according to Axiom 1 is assigned to random event A, can be determined numerically. Apart from a special case, which will be discussed in the next section, the only approach possible is to determine the value of $P(A)$ by using the property of the relative frequency $f_A(n)$ for a large number $n$ of experiments. However, in this way, $P(A)$ can only be determined as an approximation, although the quality of this approximation improves as the number of experiments increases. This method of measurement is not as unsatisfactory as it appears at first glance because fixed numerical values are used in physics in a similar manner to the calculation of $P(A)$ in probability theory. For example, the dimensions of a body are assigned numerical values, although in practice they can never be measured exactly.

In the next section, we shall consider some rules for performing calculations with probabilities. These rules follow from the axioms given above.

### 2.3.2 Deductions from the Axioms

(1) If $A \subset B$, the random event B must necessarily occur when the random event A occurs. Therefore B can be represented as the union of events as follows (Figure 2.7):

$$B = A \cup (\bar{A} \cap B) \tag{2.38}$$

---

[†] A set $Z^*$ the elements of which are subsets of the set $\Omega$ of all elementary events is called a *ring* (sigma-algebra, sigma-field, or Borel field) if $Z^*$ has the following properties:

(1) $Z^*$ contains $\Omega$ as an element.

(2) $Z^*$ contains the empty set $\emptyset$ as an element.

(3) If a finite or a countable infinite number of subsets $A_1, A_2, \ldots$ of $\Omega$ belong to the set $Z^*$, the product set $A_1 \cap A_2 \cap \ldots$ and the sum set $A_1 \cup A_2 \cup \ldots$ also belong to the set $Z^*$.

(4) $A_1$ and $A_2$ are elements of $Z^*$, and thus the difference set $A_1 \backslash A_2$ is also an element of set $Z^*$.

Further details are given by Papoulis (1984) and Fisz (1988).

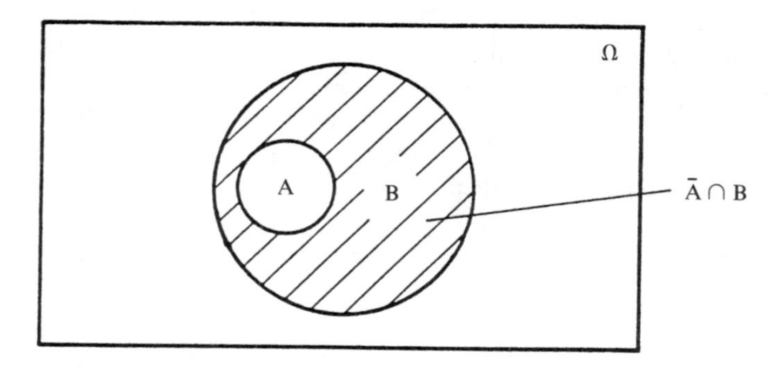


**Figure 2.7** Diagram for the derivation of equation (2.40).

The intersection $\bar{A} \cap B$ is represented by the shaded area in Figure 2.7 and the circular areas represent the random events A and B. We can also see from this figure that the random events A and $\bar{A} \cap B$ are mutually exclusive:

$$A \cap (\bar{A} \cap B) = \emptyset \tag{2.39}$$

This means that the addition theorem (2.36) can be applied to (2.38). We thus obtain

$$P(B) = P(A) + P(\bar{A} \cap B)$$

Because, according to Axiom 1, these probabilities are all greater than zero, we must have

$$P(A) \leq P(B) \quad \text{if} \quad A \subset B \tag{2.40}$$

Of course, the validity of relationships (2.38) and (2.39), which is illustrated in Figure 2.7, can also be proved mathematically by using the rules given in Section 2.2.

(2) Insofar as the relationship $A \subset \Omega$ applies to every random event A, according to (2.40) the probability of A must be given by

$$P(A) \leq P(\Omega) = 1$$

if we also take Axiom 2 (equation (2.35)) into consideration. Combining this result with (2.34) we obtain

$$0 \leq P(A) \leq 1 \tag{2.41}$$

This is another condition which, together with conditions (2.35) and (2.36) satisfied by the probability $P(\mathrm{A})$, exactly reflects the properties of the relative frequency $f_{\mathrm{A}}(n)$. To demonstrate the equivalence between the properties of the relative frequency $f_{\mathrm{A}}(n)$ (i.e., the empirical observations) and the properties of the probability $P(\mathrm{A})$ (i.e., the mathematical model for describing the experimental results), these six equations are summarized and compared in Table 2.2.

**Table 2.2** Comparison of the Properties of the Relative Frequency $f_{\mathrm{A}}(n)$ and the Probability $P(\mathrm{A})$ of a Random Event A

| *Properties of* $f_{\mathrm{A}}(n)$ | | *Properties of* $P(\mathrm{A})$ | |
|---|---|---|---|
| $0 \leqslant f_{\mathrm{A}}(n) \leqslant 1$ | (2.29) | $0 \leqslant P(\mathrm{A}) \leqslant 1$ | (2.41) |
| $f_{\Omega}(n) = 1$ | (2.31) | $P(\Omega) = 1$ | (2.35) |
| $f_{\mathrm{A} \cup \mathrm{B}}(n) = f_{\mathrm{A}}(n) + f_{\mathrm{B}}(n)$ if $\mathrm{A} \cap \mathrm{B} = \emptyset$ | (2.33) | $P(\mathrm{A} \cup \mathrm{B}) = P(\mathrm{A}) + P(\mathrm{B})$ if $\mathrm{A} \cap \mathrm{B} = \emptyset$ | (2.36) |

(3) According to equation (2.21) the certain event $\Omega$ can always be expressed as the union of the random events A and $\bar{\mathrm{A}}$:

$$\Omega = \mathrm{A} \cup \bar{\mathrm{A}} \tag{2.42}$$

However, the random events A and $\bar{\mathrm{A}}$ are mutually exclusive ($\mathrm{A} \cap \bar{\mathrm{A}} = \emptyset$). Therefore the addition axiom (2.36) can be applied to the preceding equation and we obtain

$$P(\Omega) = P(\mathrm{A}) + P(\bar{\mathrm{A}})$$

and hence, from equation (2.35),

$$P(\mathrm{A}) = 1 - P(\bar{\mathrm{A}}) \tag{2.43}$$

This equation is often useful if we want to find $P(\mathrm{A})$ but it is easier to determine $P(\bar{\mathrm{A}})$.

(4) If $\mathrm{A} = \emptyset$ is substituted in equation (2.43) and if we also take into account the fact that $\emptyset = \Omega$ according to (2.18) and Axiom 2 (equation (2.35)), $P(\Omega) = 1$, it follows that

$$P(\emptyset) = 1 - P(\Omega) = 1 - 1$$

$$P(\emptyset) = 0 \tag{2.44}$$

Therefore the probability of the impossible event has been found to be zero as a special case of (2.43). However, the inverse of equation (2.44) is not correct. If, in a random experiment, we have found that the probability $P(\mathrm{A})$ for the random event A is zero, we cannot conclude that A must be the impossible event $\emptyset$. This can be explained by using the random experiment of throwing a coin, which will be examined somewhat more closely than was done at the beginning of Section 2.3. There we only considered the elementary events "heads" and "tails." Here we shall also consider the extremely rare but still possible elementary event that the coin lands on its edge after being thrown and stays there. Let the three possible elementary events be H, T and E. In terms of the three elementary events the certain event $\Omega$ can be expressed by

$$\Omega = \mathrm{H} \cup \mathrm{T} \cup \mathrm{E}$$

and the impossible event $\emptyset$ by

$$\emptyset = \bar{\Omega} = \bar{\mathrm{H}} \cap \bar{\mathrm{T}} \cap \bar{\mathrm{E}}$$

The equation for $\emptyset$ is a generalization of DeMorgan's rule (2.25). Because the impossible event $\emptyset$ (neither heads nor tails nor edge) cannot occur in principle, the frequency $F_{\emptyset}(n)$ is always zero independent of the number $n$ of experiments. This is also true for the relative frequency $f_{\emptyset}(n)$. The relative frequency $f_{\mathrm{E}}(n)$ will have a very small value (almost zero) for large $n$ because the coin only rarely remains on its edge. Therefore in theory the probability $P(\mathrm{E})$ takes a very small value but for practical purposes it has a value of zero. According to this, not only the impossible event but also a very rare random event has a probability of zero. Thus it is shown that relationship (2.44) cannot be inverted. In this connection, the difference between the impossible event $\emptyset$ which cannot occur in principle and the very rare but still possible random event is important. Correspondingly, we obtain for $P(\mathrm{H} \cup \mathrm{T})$ a numerical value which is very close to unity and almost equal to unity. Again, we cannot deduce from $P(\mathrm{H} \cup \mathrm{T}) = 1$ that the random event "heads or tails" must occur for every throw (i.e., is the certain event). However, this means that the inverse of Axiom 2 (equation (2.35)) is also forbidden.

(5) To determine $P(A \cup B)$ if the random events A and B are not mutually exclusive (i.e., if it is impossible to apply the addition axiom (2.36)), an attempt must be made to split the sum event $A \cup B$ into disjoint random events. For this, the following relationships can be derived from Figure 2.8 (their validity can be proved mathematically using the rules of Section 2.2):

$$A \cup B = A \cup (\bar{A} \cap B) \tag{2.45}$$

with

$$A \cap (\bar{A} \cap B) = \emptyset \tag{2.46}$$

and

$$B = (A \cap B) \cup (\bar{A} \cap B) \tag{2.47}$$

with

$$(A \cap B) \cap (\bar{A} \cap B) = \emptyset \tag{2.48}$$

As a result of (2.46) and (2.48), the addition theorem (2.36) can be applied to both (2.45) and (2.47). We then obtain

$$P(A \cup B) = P(A) + P(\bar{A} \cap B)$$
$$P(B) = P(A \cap B) + P(\bar{A} \cap B)$$

If we eliminate $P(\bar{A} \cap B)$ between the first and second equations, it follows that

$$P(A \cup B) = P(A) + P(B) - P(A \cap B) \tag{2.49}$$

This result is a generalization of the addition axiom (2.36) and contains it as a special case because we have

$$P(A \cap B) = P(\emptyset) = 0$$

for the disjoint random events A and B.

(6) We now express the random event A as the union of the random events $(A \backslash B)$ and $(A \cap B)$ as shown in Figure 2.9:

$$A = (A \backslash B) \cup (A \cap B) \tag{2.50}$$

Then the addition axiom (2.36) can be applied to this relationship because the random events $(A \backslash B)$ and $(A \cap B)$ are mutually exclusive:

$$(A \backslash B) \cap (A \cap B) = \emptyset \tag{2.51}$$

We then obtain the following expression for the probability of the difference event $(A \backslash B)$:

$$P(A \backslash B) = P(A) - P(A \cap B) \tag{2.52}$$

According to equation (2.7), we have for the special case $B \subset A$

$$A \cap B = B$$

and from (2.52) it follows that

$$P(A \backslash B) = P(A) - P(B) \quad \text{if} \quad B \subset A. \tag{2.53}$$

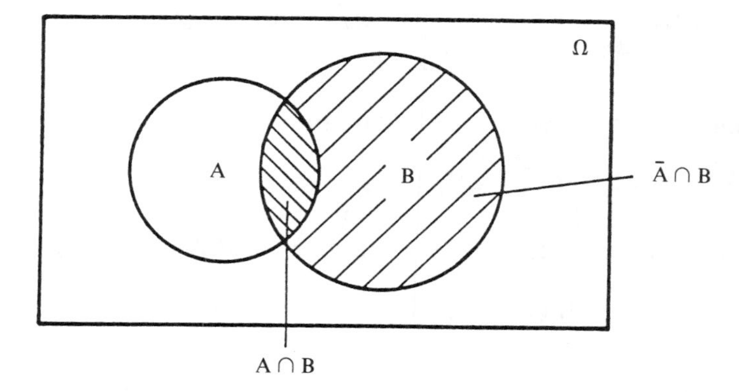


**Figure 2.8** Diagram for the derivation of equation (2.49).

(7) To familiarize ourselves with the rules for calculating probabilities derived so far, we will determine the probability of a sum event consisting of three random events. We make no assumptions as to whether these three random events are disjoint. If we consider the associative law (2.9), the generalized addition axiom (2.49) can be applied to

$$P(\mathrm{A} \cup \mathrm{B} \cup \mathrm{C}) = P\{\mathrm{A} \cup (\mathrm{B} \cup \mathrm{C})\}$$

We then obtain

$$P(\mathrm{A} \cup \mathrm{B} \cup \mathrm{C}) = P(\mathrm{A}) + P(\mathrm{B} \cup \mathrm{C}) - P\{\mathrm{A} \cap (\mathrm{B} \cup \mathrm{C})\}$$

The generalized addition axiom (2.49) can also be applied to the second term on the right-hand side of this equation:

$$\begin{aligned} P(\mathrm{A} \cup \mathrm{B} \cup \mathrm{C}) = {} & P(\mathrm{A}) + P(\mathrm{B}) + P(\mathrm{C}) - P(\mathrm{B} \cap \mathrm{C}) \\ & - P\{\mathrm{A} \cap (\mathrm{B} \cup \mathrm{C})\} \end{aligned}$$

If we rewrite the random event in the last term of this equation using the rule given by (2.23),

$$P\{\mathrm{A} \cap (\mathrm{B} \cup \mathrm{C})\} = P\{(\mathrm{A} \cap \mathrm{B}) \cup (\mathrm{A} \cap \mathrm{C})\}$$

the generalized addition axiom (2.49) can also be applied to this expression (Figure 2.10) and we obtain

$$\begin{aligned} P(\mathrm{A} \cup \mathrm{B} \cup \mathrm{C}) = {} & P(\mathrm{A}) + P(\mathrm{B}) + P(\mathrm{C}) - P(\mathrm{B} \cap \mathrm{C}) \\ & - \{P(\mathrm{A} \cap \mathrm{B}) + P(\mathrm{A} \cap \mathrm{C}) - P(\mathrm{A} \cap \mathrm{B} \cap \mathrm{A} \cap \mathrm{C})\} \end{aligned}$$

From equations (2.2) and (2.4), we must also have

$$\mathrm{A} \cap \mathrm{B} \cap \mathrm{A} \cap \mathrm{C} = \mathrm{A} \cap \mathrm{B} \cap \mathrm{C}$$

so that we now have the following expression for $P(\mathrm{A} \cup \mathrm{B} \cup \mathrm{C})$:

$$\begin{aligned} P(\mathrm{A} \cup \mathrm{B} \cup \mathrm{C}) = {} & P(\mathrm{A}) + P(\mathrm{B}) + P(\mathrm{C}) - P(\mathrm{A} \cap \mathrm{B}) \\ & - P(\mathrm{A} \cap \mathrm{C}) - P(\mathrm{B} \cap \mathrm{C}) + P(\mathrm{A} \cap \mathrm{B} \cap \mathrm{C}) \end{aligned} \tag{2.54}$$

For the special case of pairwise disjoint events, for

$$\mathrm{A} \cap \mathrm{B} = \emptyset \quad \mathrm{A} \cap \mathrm{C} = \emptyset \quad \mathrm{B} \cap \mathrm{C} = \emptyset$$

it follows from (2.54) and relationship (2.6) that

$$P(\mathrm{A} \cup \mathrm{B} \cup \mathrm{C}) = P(\mathrm{A}) + P(\mathrm{B}) + P(\mathrm{C}) \tag{2.55}$$

if A, B and C are pairwise disjoint.

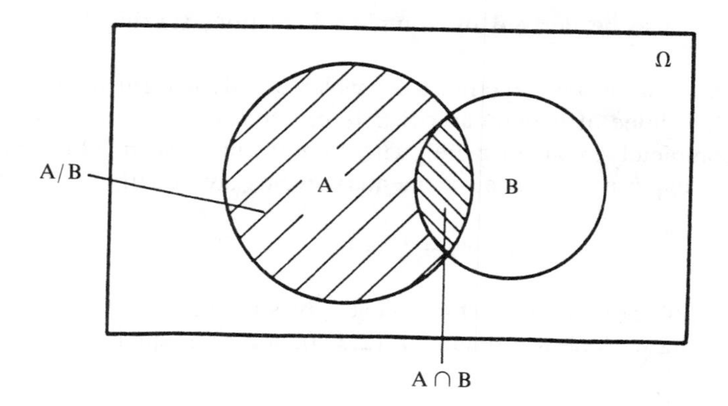


**Figure 2.9** Diagram for the derivation of equation (2.52).

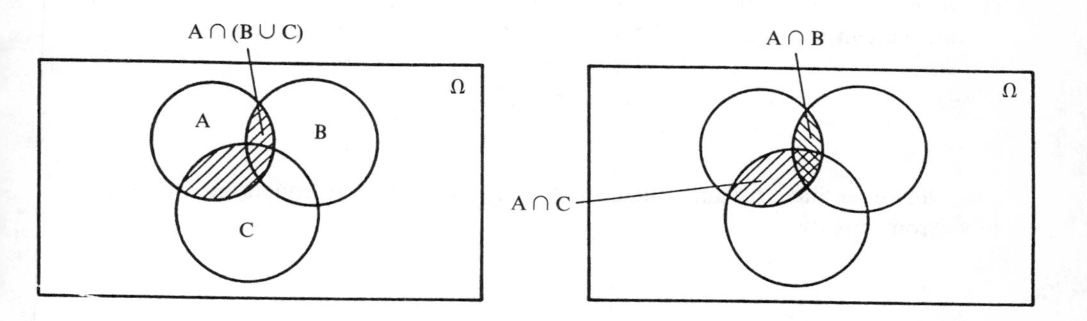


**Figure 2.10** Diagram for the derivation of equation (2.54).

Thus addition axiom (2.36) has been generalized for three random events. The probability of a sum event for more than three random events can be derived in the same way. We then obtain the following generalization of (2.55):

$$P(\mathrm{A}_1 \cup \mathrm{A}_2 \cup \cdots \cup \mathrm{A}_n) = P(\mathrm{A}_1) + P(\mathrm{A}_2) + \cdots + P(\mathrm{A}_n)$$
$$\text{if} \quad \mathrm{A}_i \cap \mathrm{A}_j = \emptyset \quad \text{for} \quad i \neq j$$

or, in abbreviated form:

$$P\left(\bigcup_{i=1}^{n} \mathrm{A}_i\right) = \sum_{i=1}^{n} P(\mathrm{A}_i) \qquad (2.56)$$
$$\text{if} \quad \mathrm{A}_j \cap \mathrm{A}_k = \emptyset \quad \text{for} \quad j \neq k$$

This result can also be derived by complete induction, starting from the addition axiom (2.36).

(8) Consider a random experiment for which exactly $m$ elementary events $\omega_1, \ldots, \omega_m$ are possible, none of which can occur in preference to the others (for example, throwing a completely symmetric die). Therefore it can be assumed from the physics of the random experiment that all elementary events are equally likely; thus,

$$P(\omega_1) = P(\omega_2) = \cdots = P(\omega_m)$$

The sum event of these $m$ elementary events gives the certain event $\Omega$ whose probability, according to Axiom 2 (equation (2.35)), must be equal to unity:

$$P\left(\bigcup_{i=1}^{m} \omega_i\right) = P(\Omega) = 1$$

However, insofar as the elementary events are mutually exclusive in pairs by definition, we can use equation (2.56) to transform the above equation to

$$\sum_{i=1}^{m} P(\omega_i) = 1$$

Since we have assumed that all elementary events are equally possible, it follows from this that

$$mP(\omega_i) = 1$$

or

$$P(\omega_i) = \frac{1}{m} \quad i = 1, \ldots, m \tag{2.57}$$

Thus, for the first time, we have found a formula which allows the probability $P(\omega_i)$ of an elementary event $\omega_i$ to be calculated numerically. However, it is important to note that the essential condition under which this result holds is the equal probability of all $m$ possible elementary events $\omega_i$, $i = 1, \ldots, m$.

Equation (2.57) can also be used to calculate the probability of a random event G consisting of $g$ elementary events in the above random experiment. By renaming, it can always be ensured that these $g$ elementary events are the first $g$ elementary events of the total of $m$ elementary events. The random event G occurs when at least one of the $g$ elementary events $\omega_1, \ldots, \omega_g$ occurs so that we have

$$G = \omega_1 \cup \omega_2 \cup \cdots \cup \omega_g$$

Because all the $\omega_i$ are disjoint, equation (2.56) can be applied:

$$P(G) = P(\omega_1) + P(\omega_2) + \cdots + P(\omega_g)$$

It then follows from equation (2.57) that

$$P(G) = \frac{g}{m} \tag{2.58}$$

The probability for the occurrence of the random event G is thus given by the ratio of the number $g$ of favorable elementary events for G to the total number $m$ of possible elementary events for this random experiment. An elementary event is favorable for G when its occurrence implies the occurrence of the random event G.

Equations (2.57) and (2.58) can be derived in other ways than on the basis of the $m$ elementary events. The same results can be obtained if the certain event $\Omega$ is divided into $m$ random events which are pairwise mutually exclusive and are all equally likely. The partition of $\Omega$ into elementary events represents the finest possible subdivision of the certain event into random events.

Equation (2.58) is the definition of the *classical probability* of a random event. The equation was obtained here as a special case of the axiomatic probability concept. Classical probability therefore satisfies the three axioms of probability theory and the rules derived in points 1–7 are also valid. The essential condition for the correctness of (2.57) and (2.58), and this will be emphasized again here, is the equal probability of the pairwise mutually exclusive random events into which the certain event $\Omega$ is partitioned.

For example, the probability of throwing a four with a symmetric die can be given immediately as 1/6 by using (2.58) because a total of $m = 6$ elementary events is possible for throwing with just one die, of which only one (i.e., a four) is a favorable elementary event ($g = 1$). The probability that an even number is thrown with a single throw and just one die is $3/6 = 1/2$ because three elementary events (i.e., throwing a two, a four or a six) are favorable events for the random event "even number" ($g = 3$).

However, if an asymmetric die is thrown or, in a coin experiment, the random event "coin remains on its edge" is also considered, the basis for equal probability has been violated and the classical definition of probability can no longer be used. We find ourselves in the same situation in the investigation of random experiments in technical systems. These random experiments are usually so complex that in general it is impossible to split the certain event into a finite number of random events which are pairwise mutually exclusive and in addition are also equally likely. Furthermore, much effort is made to ensure that, for example, an automatic lathe rarely produces faulty screws. This means that the probability for the perfect product "screws" is considerably larger than that for faulty screws. Here, we have to resort to a more general concept of probability, that being axiomatic probability.

In conclusion, it should be emphasized that the definition equation (2.58) of classical probability introduces an exact concept of probability that can be used to solve actual problems if the relevant conditions are met.

### 2.3.3 Conditional Probability

In a random experiment carried out $n$ times under the same conditions, let the random event A occur exactly $n_A$ times, the random event B exactly $n_B$ times and the product event A ∩ B exactly $n_{A \cap B}$ times. These three values $n_A$, $n_B$ and $n_{A \cap B}$ give the frequencies

$$F_A(n) = n_A, \; F_B(n) = n_B$$

and

$$F_{A \cap B}(n) = n_{A \cap B}$$

of the random events A, B and A ∩ B. The associated relative frequencies can then be derived as follows:

$$f_A(n) = \frac{F_A(n)}{n} = \frac{n_A}{n} \tag{2.59}$$

$$f_B(n) = \frac{F_B(n)}{n} = \frac{n_B}{n} \tag{2.60}$$

$$f_{A \cap B}(n) = \frac{F_{A \cap B}(n)}{n} = \frac{n_{A \cap B}}{n} \tag{2.61}$$

If the random event A does not occur as a result when the random experiment is done, this result need no longer be investigated to see whether the product event A ∩ B has occurred because A ∩ B can only occur when both the random event A and the random event B occur. Thus, on examining the $n$ results for determining the value $n_{A \cap B}$ a preselection can be made by considering only results for which the random event A has occurred. Therefore, in addition to the relative frequency $f_{A \cap B}$ of (2.61), another ratio can be introduced; namely,

$$\frac{n_{A \cap B}}{n_A} = \frac{F_{A \cap B}(n)}{F_A(n)}$$

This ratio also describes a relative frequency for the occurrence of the product event $A \cap B$ but now under the condition that only results for which the random event A has occurred are considered. However, under this condition, the occurrence of the random event B is identical with the occurrence of the product event $A \cap B$. This ratio is therefore called the *conditional relative frequency* of the random event B under condition A, or more concisely the conditional relative frequency, and is denoted by $f_{B|A}(n)$:

$$f_{B|A}(n) := \frac{F_{A \cap B}(n)}{F_A(n)} \quad \text{if} \quad F_A(n) > 0 \tag{2.62}$$

This expression is only meaningful for $F_A(n) > 0$.

If the numerator and the denominator of (2.62) are divided by the number $n$ of experiments and equations (2.59) and (2.61) are used, the conditional relative frequency $f_{B|A}(n)$ can be expressed in terms of the ordinary relative frequencies $f_{A \cap B}(n)$ and $f_A(n)$:

$$\begin{aligned} f_{B|A}(n) &= \frac{F_{A \cap B}(n)/n}{F_A(n)/n} \\ f_{B|A}(n) &= \frac{f_{A \cap B}(n)}{f_A(n)} \quad \text{if} \quad f_A(n) > 0 \end{aligned} \tag{2.63}$$

In a similar manner, we obtain the following expression for the conditional relative frequency $f_{A|B}(n)$ of the random event A under condition B:

$$f_{A|B}(n) = \frac{f_{A \cap B}(n)}{f_B(n)} \quad \text{if} \quad f_B(n) > 0 \tag{2.64}$$

Now, experience shows that the relative frequency fluctuates around a fixed value for a large number $n$ of experiments, and naturally this must apply to the ratios

$$\frac{f_{A \cap B}(n)}{f_A(n)} \quad \text{and} \quad \frac{f_{A \cap B}(n)}{f_B(n)}$$

Hence, this is applicable to the conditional relative frequencies $f_{B|A}(n)$ and $f_{A|B}(n)$. We denote the values around which $f_{B|A}$ and $f_{A|B}$ stabilize for large $n$ by $P(B|A)$ and $P(A|B)$. These quantities must obey the equations

$$P(B|A) := \frac{P(A \cap B)}{P(A)} \quad \text{if} \quad P(A) > 0 \tag{2.65}$$

and

$$P(A|B) := \frac{P(A \cap B)}{P(B)} \quad \text{if} \quad P(B) > 0 \tag{2.66}$$

because for a large number $n$ of experiments the probability $P(A \cap B)$ is approximated by the relative frequency $f_{A \cap B}(n)$ whereas the probabilities $P(A)$ and $P(B)$ are approximated by the relative frequencies $f_A(n)$ and $f_B(n)$.

The following formula for calculating the probability of the product event A $\cap$ B can be derived from the last two equations:

$$P(A \cap B) = P(A)P(B|A) = P(B)P(A|B) \tag{2.67}$$

In equations (2.65) and (2.66) we have introduced a new concept of probability. $P(B|A)$ is the probability of the random event B under the condition that the random event A has occurred or more concisely the *conditional probability* of the random event B. In a corresponding manner, we call $P(A|B)$ the conditional probability for the random event A. As in the case of (2.63) and (2.64), which are only meaningful under the conditions $f_A(n) > 0$ or $f_B(n) > 0$ respectively, the denominators of (2.65) and (2.66) must be greater than zero.

The conditional probability defined by equations (2.65) and (2.66), which also represent methods for calculating the conditional probability, must still be examined to determine whether it fits into the structure of probability theory. Hence, we must verify that the conditional probability satisfies the three axioms of probability theory.

Before we investigate this, it will be emphasized once more that $P(A|B)$ is the conditional probability for the occurrence of the random event A and that $P(B|A)$ is the conditional probability for the occurrence of the random event B. Therefore confusion should not arise because of the appearance of two random events A and B in the representations $P(A|B)$ and $P(B|A)$.

(1) The conditional probability $P(A|B)$ is given by

$$P(A|B) = \frac{P(A \cap B)}{P(B)} \quad \text{if} \quad P(B) > 0$$

Because, according to Axiom 1 of probability theory (relation (2.34)), $P(A \cap B) \geqslant 0$ and $P(B) \geqslant 0$ and as the case $P(B) = 0$ is excluded by the condition $P(B) > 0$ in equation (2.66), the ratio $P(A \cap B)/P(B)$ must always have a meaningful value which is greater than or equal to zero. Therefore

$$P(\mathrm{A}|\mathrm{B}) \geq 0 \tag{2.68}$$

(2) In view of the commutative law (2.2) for products of random events, the equation $\Omega \cap \mathrm{B} = \mathrm{B}$ follows from (2.5). Therefore

$$P(\Omega|\mathrm{B}) = \frac{P(\Omega \cap \mathrm{B})}{P(\mathrm{B})} = \frac{P(\mathrm{B})}{P(\mathrm{B})}$$

or

$$P(\Omega|\mathrm{B}) = 1 \tag{2.69}$$

(3) Let $\mathrm{A}_1$ and $\mathrm{A}_2$ be two disjoint random events (i.e., $\mathrm{A}_1 \cap \mathrm{A}_2 = \emptyset$). If B is a random event belonging to the same random experiment, both product events $\mathrm{A}_1 \cap \mathrm{B}$ and $\mathrm{A}_2 \cap \mathrm{B}$ are also disjoint as can easily be seen from Figure 2.11:

$$(\mathrm{A}_1 \cap \mathrm{B}) \cap (\mathrm{A}_2 \cap \mathrm{B}) = \emptyset \tag{2.70}$$

We shall now consider the conditional probability $P\{(\mathrm{A}_1 \cup \mathrm{A}_2)|\mathrm{B}\}$:

$$P\{(\mathrm{A}_1 \cup \mathrm{A}_2)|\mathrm{B}\} = \frac{P\{(\mathrm{A}_1 \cup \mathrm{A}_2) \cap \mathrm{B}\}}{P(\mathrm{B})} = \frac{P\{(\mathrm{A}_1 \cap \mathrm{B}) \cup (\mathrm{A}_2 \cap \mathrm{B})\}}{P(\mathrm{B})}$$

If the addition axiom (2.36) is applied to the numerator that has been previously rearranged by using the distributive law (2.23), which is allowed because of (2.70), it follows that

$$P\{(\mathrm{A}_1 \cup \mathrm{A}_2)|\mathrm{B}\} = \frac{P(\mathrm{A}_1 \cap \mathrm{B}) + P(\mathrm{A}_2 \cap \mathrm{B})}{P(\mathrm{B})} = \frac{P(\mathrm{A}_1 \cap \mathrm{B})}{P(\mathrm{B})} + \frac{P(\mathrm{A}_2 \cap \mathrm{B})}{P(\mathrm{B})}$$

or

$$P\{(\mathrm{A}_1 \cup \mathrm{A}_2)|\mathrm{B}\} = P(\mathrm{A}_1|\mathrm{B}) + P(\mathrm{A}_2|\mathrm{B}) \quad \text{if} \quad \mathrm{A}_1 \cap \mathrm{A}_2 = \emptyset \tag{2.71}$$

The three equations (2.68), (2.69) and (2.71) show that conditional probability also satisfies the three axioms of probability theory. Thus all the rules derived in Section 2.3.2 for ordinary probability apply in an analogous manner to conditional probability.

The concept of conditional probability is a generalization of the ordinary or unconditional probability concept and contains it as a special case. This relationship

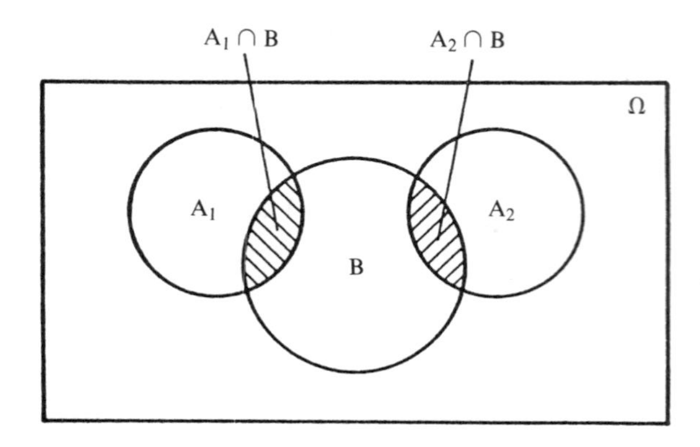


**Figure 2.11** Diagram for the derivation of equation (2.70).

immediately becomes clear if the random event B is replaced by the certain event $\Omega$ in (2.66). We then obtain

$$P(\mathrm{A}|\Omega) = \frac{P(\mathrm{A} \cap \Omega)}{P(\Omega)}$$

Because $P(\Omega) = 1$, according to Axiom 2 (equation (2.35)), and $A \cap \Omega = A$, according to (2.5), it follows that

$$P(\mathrm{A}|\Omega) = P(\mathrm{A}) \tag{2.72}$$

so that the above statement is confirmed.

### 2.3.4 Independent Random Events

Now that we have introduced conditional probability, we can assign two numerical values $P(\mathrm{A})$ and $P(\mathrm{A}|\mathrm{B})$ to the random event A. These are both called probabilities and provide a measure of the probability that the random event A occurs. We again note the two meanings of the word "probability" used here. In general, the two values $P(\mathrm{A})$ and $P(\mathrm{A}|\mathrm{B})$ differ in magnitude. For example, consider a random experiment of selecting a letter blindfolded from any page of text of a book written in English. Let the random event Q occur if the letter "q" is guessed. Let the random event U occur if the letter immediately after the guessed letter is a "u." Whereas the probability $P(\mathrm{U})$ will certainly be less than unity ($P(\mathrm{U}) < 1$), the conditional probability $P(\mathrm{U}|\mathrm{Q})$ must be equal to unity ($\mathrm{P}(\mathrm{U}|\mathrm{Q}) = 1$) because in English the letter "q" is always followed by the letter "u."

An important special case arises when both probabilities are the same. We then have

$$P(\mathrm{A}|\mathrm{B}) = P(\mathrm{A}) \tag{2.73}$$

This equality means that the probability for the occurrence of the random event A is completely independent of the occurrence of random event B. If $P(\mathrm{A}) = P(\mathrm{A}|\mathrm{B})$ is true, we therefore say that the random event A is *independent* (stochastically independent) of the random event B.

If, using equation (2.66), we eliminate the probability $P(\mathrm{A} \cap \mathrm{B})$ in (2.65), we obtain

$$P(\mathrm{B}|\mathrm{A}) = \frac{P(\mathrm{A}|\mathrm{B})P(\mathrm{B})}{P(\mathrm{A})}$$

if $P(\mathrm{A}) > 0$ and $P(\mathrm{B}) > 0$. If, from equation (2.73), we substitute the conditional probability $P(\mathrm{A}|\mathrm{B})$ by $P(\mathrm{A})$, we obtain

$$P(\mathrm{B}|\mathrm{A}) = P(\mathrm{B}) \tag{2.74}$$

Therefore if the random event A is independent of the random event B (equation (2.73)), it follows from this that B is also independent of A (equation (2.74)). Hence it is sufficient to speak of the independence of two random events A and B.

The two relationships (2.73) and (2.74) (it should be noted that only one equation is needed to define two independent random events) can be used to transform the double equation (2.67) into a single relationship:

$$P(\mathrm{A} \cap \mathrm{B}) = P(\mathrm{A})P(\mathrm{B}) \tag{2.75}$$

if A is independent of B. According to this, (2.75) is a necessary condition for the independence of two random events A and B. However, if we substitute equation (2.75) in (2.65) and (2.66), we again obtain equations (2.73) and (2.74) (i.e., the definition equations for two independent random events). Therefore (2.75) is a necessary and sufficient condition for A and B to be independent random events.

Equation (2.67) and consequently equation (2.75) have been derived from equations (2.65) and (2.66). However, whereas the latter equations are only meaningful for $P(\mathrm{A}) > 0$ or $P(\mathrm{B}) > 0$, the former also apply for $P(\mathrm{A}) = 0$ and $P(\mathrm{B}) = 0$ respectively.

The important concept of independence must now be extended to the case of $n$ possible random events $\mathrm{A}_1, \mathrm{A}_2, \ldots, \mathrm{A}_n$ in a random experiment. The $n$ random events $A_1, A_2, \ldots, A_n$ are called *independent events* if the probability of *every* product event which can be formed from $2, 3, \ldots, n$ of the total $n$ random events $A_1, A_2, \ldots, A_n$ is equal to the product of the probabilities of the random events occurring in the corresponding combination. In the sense of this definition, the three random events $A_1$, $A_2$ and $A_3$ are only independent if the equations

$$P(A_1 \cap A_2) = P(A_1)P(A_2)$$

$$P(A_1 \cap A_3) = P(A_1)P(A_3)$$

$$P(A_2 \cap A_3) = P(A_2)P(A_3)$$

are satisfied, and if

$$P(A_1 \cap A_2 \cap A_3) = P(A_1)P(A_2)P(A_3)$$

It is therefore not sufficient for the independence of $A_1, A_2, \ldots, A_n$ if these random events are only pairwise independent of each other. This statement will be verified by using a simple example (Gnedenko, 1978).

Consider a regular tetrahedron with four colored faces. The first face is red, the second is green, the third is blue and the fourth is red, green and blue. Four elementary events are possible when the tetrahedron is thrown. The elementary events $\omega_1$, $\omega_2$ and $\omega_3$ occur when the tetrahedron falls on the completely red face, the completely green face and the completely blue face, respectively, and event $\omega_4$ occurs when it falls on the face of three colors.

Since the tetrahedron is assumed to be regular, the condition for equal probability for the four elementary events $\omega_1, \ldots, \omega_4$ is satisfied for a correct throw. The probability of these elementary events can immediately be written down from (2.57):

$$P(\omega_1) = P(\omega_2) = P(\omega_3) = P(\omega_4) = \frac{1}{4}$$

In this random experiment we next investigate whether the three random events $A_1$, $A_2$ and $A_3$ are independent. Let $A_1$ occur if the face pointing downward after the throw is completely or partly red. Correspondingly, let $A_2$ or $A_3$ occur when the face at the bottom is completely or partly green or blue, respectively. It should be noted that none of the three random events $A_1$, $A_2$ and $A_3$ is identical with one of the four elementary events $\omega_1, \ldots, \omega_4$. Rather, they are related as follows:

$$A_1 = \omega_1 \cup \omega_4 \qquad A_2 = \omega_2 \cup \omega_4 \qquad A_3 = \omega_3 \cup \omega_4$$

Then, remembering that elementary events are mutually exclusive by definition, it follows from this equation and the addition axiom (2.36) that

$$P(A_1) = P(\omega_1) + P(\omega_4) = \frac{1}{4} + \frac{1}{4} = \frac{1}{2}$$

Correspondingly, we also have

$$P(A_2) = P(A_3) = \frac{1}{2}$$

Furthermore, we have

$$A_1 \cap A_2 = \omega_4 \quad A_1 \cap A_3 = \omega_4 \quad A_2 \cap A_3 = \omega_4$$

and so

$$P(A_1 \cap A_2) = P(A_1 \cap A_3) = P(A_2 \cap A_3) = P(\omega_4) = \frac{1}{4}$$

These results prove the validity of the following equations:

$$P(A_1 \cap A_2) = P(A_1)P(A_2) = \frac{1}{4}$$

$$P(A_1 \cap A_3) = P(A_1)P(A_3) = \frac{1}{4}$$

$$P(A_2 \cap A_3) = P(A_2)P(A_3) = \frac{1}{4}$$

thus, the three random events $A_1$, $A_2$ and $A_3$ are pairwise independent. By using the above three equations and equation (2.65), we obtain a second set of equations:

$$P(A_1|A_2) = P(A_1|A_3) = P(A_1) = \frac{1}{2}$$

$$P(A_2|A_1) = P(A_2|A_3) = P(A_2) = \frac{1}{2}$$

$$P(A_3|A_1) = P(A_3|A_2) = P(A_3) = \frac{1}{2}$$

which also confirm the pairwise independence of the three random events $A_1$, $A_2$ and $A_3$.

However, to prove the independence of the three random events $A_1$, $A_2$ and $A_3$, not just their pairwise independence, the probability $P(A_1 \cap A_2 \cap A_3)$ must also be investigated. It follows from

$$A_1 \cap A_2 \cap A_3 = \omega_4$$

that the required probability is

$$P(A_1 \cap A_2 \cap A_3) = P(\omega_4) = \frac{1}{4}$$

Because

$$P(A_1)P(A_2)P(A_3) = \frac{1}{2} \cdot \frac{1}{2} \cdot \frac{1}{2} = \frac{1}{8}$$

we now have

$$P(A_1 \cap A_2 \cap A_3) \neq P(A_1)\, P(A_2)P(A_3)$$

thus, according to the above definition, the three random events $A_1$, $A_2$ and $A_3$ are not independent.

In practical problems we often check the dependence or independence of random events not by using the corresponding definition equations but from our knowledge of the physics of the random experiment or from our experience. However, this question is sometimes very difficult to answer.

## 2.4 RANDOM VARIABLES, DISTRIBUTION FUNCTIONS AND DENSITY FUNCTIONS

### 2.4.1 Random Variables

Every possible random event resulting from a random experiment can be represented as a subset of the set $\Omega$, where $\Omega$ contains all possible elementary events in this random experiment as elements. So far we have described these elementary events verbally for each random experiment. For example, in the dice experiment (Section 2.1), we defined $\omega_1$ = "number one on top," $\omega_2$ = "number two on top" *et cetera*. In the coin experiment (Section 2.3) the two elementary events were expressed as $\omega_1$ = "heads" and $\omega_2$ = "tails." However, every elementary event can be assigned a real number to identify it. Frequently the way in which this number is assigned follows automatically from the variables of interest in the elementary event. For example, if screws are manufactured in very large quantities by an automatic lathe and any screw is withdrawn to check its dimensional accuracy, the sampling can be regarded as an elementary event. This elementary event can be identified by the number $\lambda$ if a length of $\lambda$ mm has been measured on the sampled screw. In a similar manner it makes sense to designate the elementary event "the die face with number one on top" by the number 1 *et cetera*.

Sometimes it is essential to designate an elementary event by a number. For example, consider a card index for the inhabitants of a town. To accommodate as much data as possible, such as name, sex, year of birth, profession, *et cetera*, on the punched card that exists for every citizen of the town, these data are coded. Assume that, in the card column provided for sex, a 1 is punched if the card belongs to a man and a 0 is punched if the card belongs to a woman. Then the selection of any card from the file is regarded as an elementary event which, if we are only interested in the sex of the person concerned, is identified by a 0 or a 1 depending on what is punched in the appropriate column. The two numbers cannot occur together.

We now symbolize the assignment procedure by $\psi$. In a random experiment $\psi$ assigns a fixed real number $\xi^{(i)}$ to every elementary event $\omega_i$. The relationships described above can then be expressed by the equation

$$\xi^{(i)} = \psi(\omega_i) \quad \omega_i \in \Omega \tag{2.76}$$

At this point, we must clarify the notation of a function or mapping. The notation $b = f(a)$, where $a$ is the independent variable and $b$ is the dependent variable, is generally used with two different meanings:

(1) $b = f(a)$ describes a procedure which assigns a value $b \in B$ where $B$ is the range of values of the function to *every* $a \in A$ where $A$ is the definition domain of the function $f$;

(2) $b = f(a)$ is the element of the set $B$ which is assigned to a *quite specific* element $a$ from the set $A$ according to the procedure $f$.

Since in subsequent discussions which are based on this section a strict distinction must be made between a variable and the numerical value which this variable can take, they will be represented by different symbols. The variable which can take the numerical values $\xi^{(i)}$, $i = 1, 2, \ldots$, is denoted $x$ and $\omega$ as a variable can take all $\omega_i$, $\omega_i \in \Omega$.

These new symbols can be used to generalize equation (2.76), which means that the numerical value $\xi^{(i)}$ belongs to the elementary event $\omega_i$, as a mapping procedure:

$$x = \psi(\omega) \tag{2.77}$$

This mapping assigns a point on the number axis or in one-dimensional Euclidean space $\mathcal{R}_1$ to every element of the set $\Omega$. Since only a single point on the number axis belongs to every elementary event, the range of values cannot contain more points than the number of elementary events possible in the random experiment.

So far an elementary event $\omega_1$ always has a single real numerical value $\xi^{(i)}$ assigned to it. If, for example, we were to take a punched card from the inhabitant card index of a town to find out not only the sex of the arbitrarily selected person but also the year of birth and profession, the elementary event that has occurred by withdrawing the card is identified by *three* numerical values. In general, every elementary event $\omega_i$ can be assigned $n$ real numbers $\xi_1^{(i)}, \xi_2^{(i)}, \ldots, \xi_n^{(i)}$ by a mapping procedure. These $n$ numbers can be combined to form an $n$-dimensional vector:

$$\boldsymbol{\xi}^{(i)} := [\xi_1^{(i)}, \ldots, \xi_n^{(i)}]'$$

In a similar manner to equations (2.76) and (2.77), this multidimensional assignment can be expressed by

$$\xi_k^{(i)} = \psi_k(\omega_i) \quad k = 1, \ldots, n, \quad \omega_i \in \Omega$$

or by

$$\boldsymbol{\xi}^{(i)} = \boldsymbol{\psi}(\omega_i) := [\psi_1^{(i)}(\omega_i), \ldots, \psi_n^{(i)}(\omega_i)]', \quad \omega_i \in \Omega \tag{2.78}$$

or

$$\boldsymbol{x} := [x_1, \ldots, x_n]' = \boldsymbol{\psi}(\omega) \tag{2.79}$$

In this way, every elementary event $\omega_i$ from the set $\Omega$ is assigned a point in $n$-dimensional Euclidean space $\mathcal{R}_n$. This fixed assignment is illustrated in Figure 2.12 (Wunsch, 1972; Wunsch and Schreiber, 1984).

The fixed assignment of numbers or vectors to elementary events described so far in this section only represents a renaming of the elementary event. Therefore, when the elementary event $\omega_i$ occurs, the variable $x$ must take the value $\xi^{(i)}$ or the vector $\boldsymbol{x}$ must take the value $\boldsymbol{\xi}^{(i)}$. In view of the condition that different $\xi^{(i)}$ or $\boldsymbol{\xi}^{(i)}$ belong to different $\omega_i$, this means that the probability for the occurrence of the elementary event $\omega_i$ must be the same as the probability that the variable $x$ takes the numerical value $\xi^{(i)}$ or the vector $\boldsymbol{x}$ takes the value $\boldsymbol{\xi}^{(i)}$. Therefore we have

$$P(\omega_i) = P(x = \xi^{(i)}) \quad \omega_i \in \Omega \tag{2.80a}$$

or

$$P(\omega_i) = P(\boldsymbol{x} = \boldsymbol{\xi}^{(i)}) \quad \omega_i \in \Omega \tag{2.80b}$$

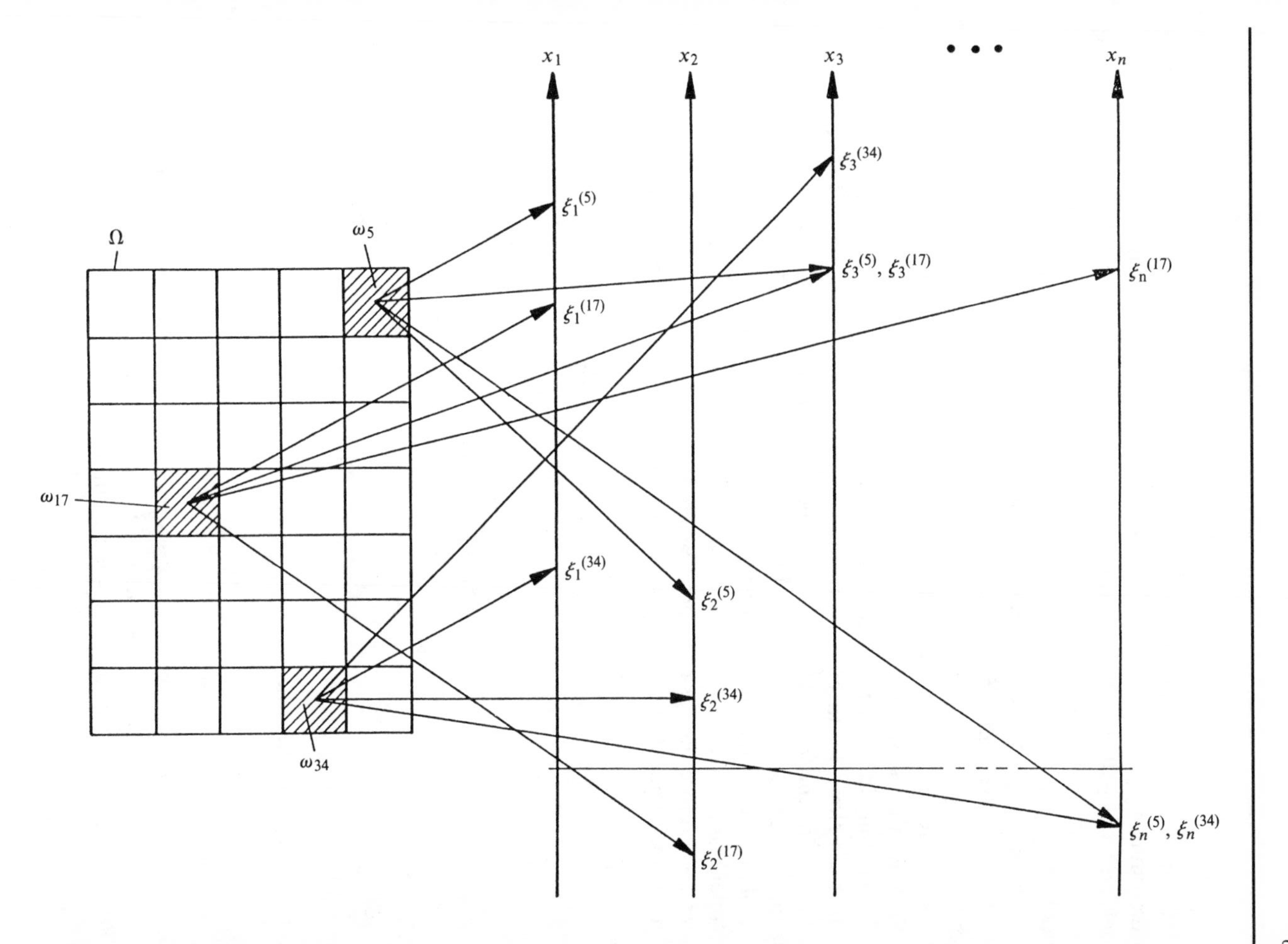


**Figure 2.12** Assignment of an $n$-dimensional vector $\boldsymbol{\xi}^{(i)}$ to an elementary event $\omega_i$.

However, since the occurrence or nonoccurrence of a particular elementary event $\omega_i$ is determined by chance, this must also apply to the values which $x$ or $\boldsymbol{x}$ can take. Therefore $x$ is called a *scalar or one-dimensional random variable* (stochastic variable) and correspondingly $\boldsymbol{x}$ is called a *vector or multidimensional random variable* (stochastic variable).

In this way every elementary event $\omega_i$ is assigned two types of numerical values which should be kept strictly separate. The value $\xi^{(i)}$ of the scalar random variable $x$ or the value $\boldsymbol{\xi}^{(i)}$ of the vector random variable $\boldsymbol{x}$ only represents another designation of the elementary event, and the other numerical value $P(\omega_i)$ is a measure of the probability that $\omega_i$ will occur.

The function $\psi$ or the vector function $\boldsymbol{\psi}$ is often not injective (i.e., equal values of $x$ or $\boldsymbol{x}$ belong to different elementary events). The number of values $\xi^{(i)}$ or $\boldsymbol{\xi}^{(i)}$ that are different from each other will therefore be smaller than the number of elementary events $\omega_i$. For example, there are only *two* numerical values of the random variable that describes the sex of the inhabitants belonging to *all* punched cards of the inhabitant card index of a town and therefore belonging to *all* possible elementary events. Therefore the case can arise where the random variable $x$ takes the value $\xi^{(i)}$ if either the elementary event $\omega_i$ or the elementary event $\omega_j$ occurs. Then, because

$$\xi^{(i)} = \psi(\omega_i) \quad \xi^{(j)} = \psi(\omega_j)$$

for $\xi^{(j)} = \xi^{(i)}$, we have

$$P(x = \xi^{(i)}) = P(\omega_i \cup \omega_j) = P(\omega_i) + P(\omega_j)$$

or, from (2.34),

$$P(\omega_i) < P(x = \xi^{(i)})$$

Correspondingly, the case

$$P(\omega_i) < P(\boldsymbol{x} = \boldsymbol{\xi}^{(i)})$$

can arise if the values of the vector $\boldsymbol{x}$ assigned to several elementary events coincide.

Because the functions $\psi$ and $\boldsymbol{\psi}$ need not be one-to-one (injective), equation (2.80) can be expressed in generalized form as

$$P(\omega_i) \leqslant P(x = \xi^{(i)}) \quad \omega_i \in \Omega$$

and

$$P(\omega_i) \leqslant P(\boldsymbol{x} = \boldsymbol{\xi}^{(i)}) \quad \omega_i \in \Omega$$

The equality holds when the number of elementary events coincides with the number of $\xi^{(i)}$ or $\boldsymbol{\xi}^{(i)}$ that are different from each other.

### 2.4.2 The Distribution Function for Scalar Random Variables

In introducing the concept of random variables in Section 2.4.1, one of the examples used was the manufacture of screws on an automatic lathe. In this example, which will be considered again here from a different viewpoint, an elementary event consists in taking any screw from the current batch. Every elementary event is assigned the length $\lambda$ of the screw selected as the numerical value of the random variable $l$ (length).

We now define a quality criterion for assessing the precision of the lathe. For example, we can use the probability that the screw length $l$ lies within the production tolerance. If the numerical value of this probability is very close to unity, we are dealing with a high precision lathe. The precision of the lathe decreases as the value of the probability falls. If, for example, a screw length of $50 \pm 0.2$ mm is required, we must have $49.8 \leqslant l \leqslant 50.2$ for a usable screw. However, it makes more sense for subsequent considerations to specify this range in the form $49.8 \leqslant l < 50.21$ which, however, is only correct when a measurement accuracy of 0.01 mm is assumed. Now, the numerical values of $l$ are determined by chance and, in addition, every elementary event is assigned a fixed numerical value $\lambda$. Thus it is clear that the specification $49.8 \leqslant l < 50.21$ uniquely defines a random event which consists of all the elementary events which have numerical values of the random variable $l$ falling within the range $49.8 \leqslant l < 50.21$. In this way, it has become possible to describe a random event using random variables, i.e., in a different manner from that used previously.

In view of the above considerations, the quality criterion defined for the screws can be written in the form $P(49.8 \leqslant l < 50.21)$. However, the numerical value of this quality criterion depends not only on the quality of the lathe but also on the choice of the two limits $\lambda_1$ and $\lambda_2$ within which the values of the random variable $l$ should remain. If these limits are brought closer together, the numerical value of the quality criterion for the same lathe is reduced. In contrast, if the limits are pushed wider apart the probability must approach unity because increasingly more screws satisfy the requirement $\lambda_1 \leqslant l < \lambda_2$. Therefore the number of elementary events forming the random event $\lambda_1 \leqslant l < \lambda_2$ varies with the limits $\lambda_1$ and $\lambda_2$, as of course does the random event itself. Thus the probability $P(\lambda_1 \leqslant l < \lambda_2)$ must be a function of $\lambda_1$ and $\lambda_2$.

It will be shown later in this section that the probability $P(\lambda_1 \leqslant l < \lambda_2)$ can be expressed in terms of the probabilities $P(l < \lambda_1)$ and $P(l < \lambda_2)$, where $l < \lambda_1$ and $l < \lambda_2$ are two random events in the above sense. $P(l < \lambda_1)$ is a function of $\lambda_1$ and $P(l < \lambda_2)$ is a function of $\lambda_2$.

At this point we leave the above example and denote the random variable in general by $x$ and a fixed numerical value of this variable by $\xi$. We then have

$$P(x < \xi) =: F(\xi) \tag{2.81}$$

i.e., the probability $P(x < \xi)$ is a function of the limit $\xi$. This equation is the defining equation of the *distribution function* $F(\xi)$ of the scalar random variable $x$ or, more concisely, the distribution function $F(\xi)$, which is an important concept in probability theory.[‡] Figure 2.13 gives a clear illustration of the dependence of the random event $x < \xi$ on $\xi$ and also of the dependence of $P(x < \xi)$ on $\xi$. The shaded area shows the random event described by $x < \xi$ (Wunsch, 1972; Wunsch and Schreiber, 1984).

Using a simple example, we shall now derive the typical curve of the distribution function $F(\xi)$ of a scalar random variable $x$. We consider a random experiment in which the set $\Omega$ of all elementary events consists of just two elementary events $\omega_1$ and $\omega_2$:

$$\Omega = \{\omega_1, \omega_2\}$$

If we assign the value of zero to the elementary event $\omega_1$ and the value of unity to the elementary event $\omega_2$, the random variable $x$ can only take the two values zero and unity. However, as we shall now show, this does not mean that the distribution function $F(\xi)$ is defined only at the two points $\xi = 0$ and $\xi = 1$. Because the two different values of the random variable $x$ belong to the two elementary events, the function $\psi$ of (2.76) is a one-to-one function. We therefore have

$$P(\omega_1) = P(x = 0)$$

and

$$P(\omega_2) = P(x = 1)$$

It is also assumed that

$$P(\omega_1) = P(x = 0) = p$$

---

[‡]Some authors define the distribution function $F(\xi)$ by

$$P(x \leqslant \xi) =: F(\xi)$$

Special care should be taken regarding the definition of the distribution function, particularly in the calculation of the functional values of $F(\xi)$ at discontinuity points.

and

$$P(\omega_2) = P(x = 1) = q$$

For the certain event $\Omega$ we can write

$$\Omega = \omega_1 \cup \omega_2$$

Then, according to the addition axiom (2.36), we have

$$P(\Omega) = P(\omega_1) + P(\omega_2)$$

Using (2.35) and the above definitions it follows from this that

$$1 = p + q$$

In this way, all the assumptions required for calculating the distribution function are known.

(1) $\xi < 0$. According to (2.81),

$$F(\xi) = P(x < \xi)$$

and, hence, for the range $\xi < 0$,

$$F(\xi) = P(x < \xi < 0)$$

Because the random event $x < \xi < 0$ does not contain any elementary events, it must be the impossible event $\emptyset$ and, from (2.44), it follows that

$$F(\xi) = P(\emptyset) = 0 \quad \text{for} \quad \xi < 0$$

(2) $\xi = 0$. From

$$F(\xi) = P(x < \xi = 0) = P(x < 0)$$

we obtain the value of the distribution function for both $\xi = 0$ and $\xi < 0$ as the probability of the impossible event $\emptyset$; i.e.,

$$F(\xi = 0) = 0$$

Therefore in this example the distribution function $F(\xi)$ takes a value of zero for $\xi \leqslant 0$.

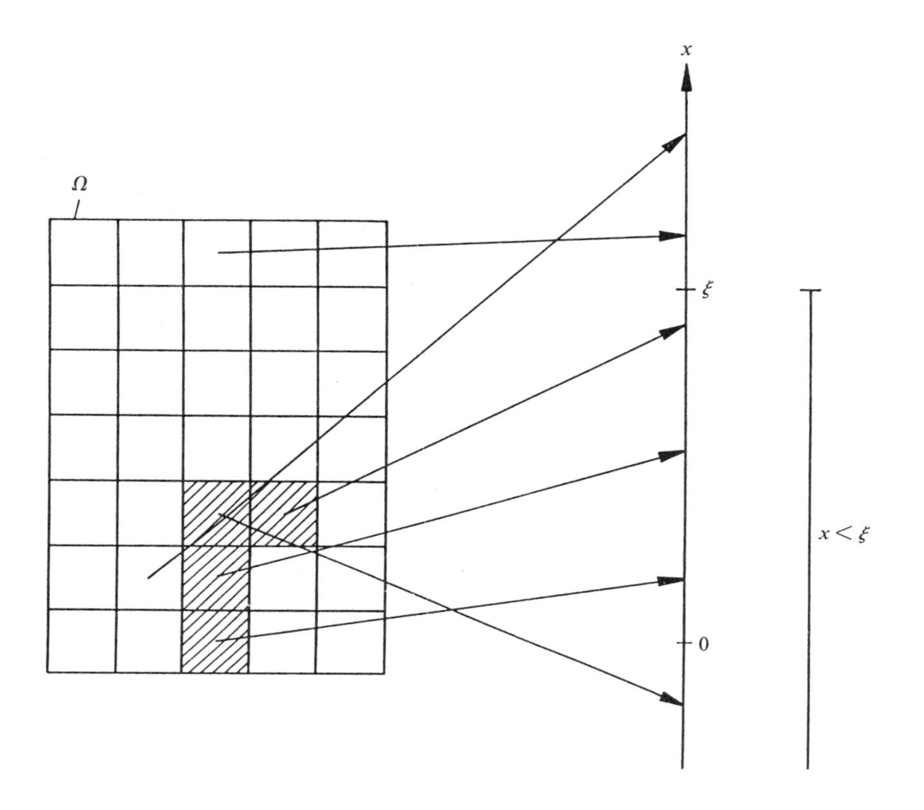


**Figure 2.13** The random event $x < \xi$.

(3) $0 < \xi \leqslant 1$. In this range of $\xi$ the random event $x < \xi$ consists of only the elementary event $\omega_1$; $x < \xi|_{0<\xi\leqslant 1}$ therefore occurs if and only if $\omega_i$ occurs. We must then have

$$F(\xi) = P(x < \xi)|_{0<\xi\leqslant 1} = P(\omega_1)$$

so that

$$F(\xi) = p \quad \text{for} \quad 0 < \xi \leqslant 1$$

(4) $\xi > 1$. Because

$$F(\xi) = P(x < \xi > 1)$$

the random event $x < \xi > 1$ must be investigated more closely here. It consists of the two elementary events $\omega_1$ and $\omega_2$, and therefore it occurs if either $\omega_1$ or $\omega_2$ occurs. We must then have

$$F(\xi) = P(\omega_1 \cup \omega_2) = P(\omega_1) + P(\omega_2)$$

or

$$F(\xi) = p + q$$

However, because $\omega_1 \cup \omega_2$ is the certain event $\Omega$, we must also have

$$F(\xi) = 1 \quad \text{for} \quad \xi > 1$$

We can now draw the distribution function $F(\xi)$ as shown in Figure 2.14. The circles in Figure 2.14 indicate that the distribution function $F(\xi)$ takes the limiting value on the left-hand side at both discontinuity points.

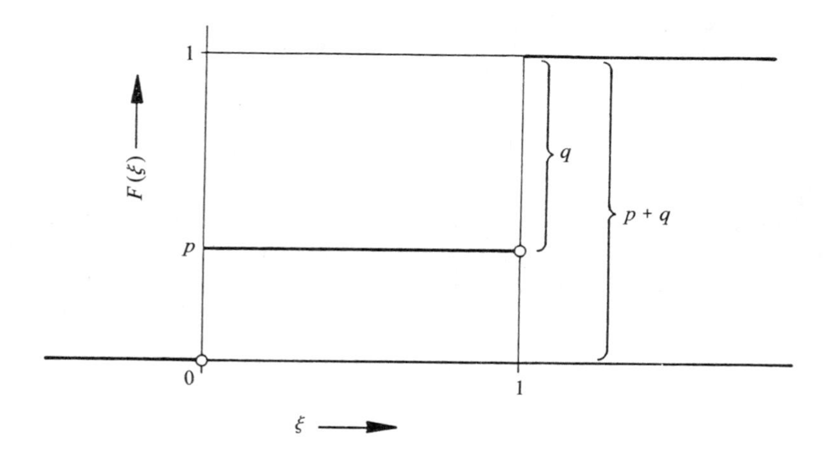


**Figure 2.14** Distribution function $F(\xi)$ of a random variable $x$ which can only take the values zero and unity.

The distribution function $F(\xi)$ found in this way is a nondecreasing function which only takes values between zero and unity and which for $\xi = 0$ or $\xi = 1$ jumps by $P(x = 0) = p$ or $P(x = 1) = q$ respectively, i.e., by the probability of an elementary event in each case.

It will now be shown that the properties found in this example apply quite generally to distribution functions.

### 2.4.2.1 *Properties of the Distribution Function* $F(\xi)$

(1) If we let the $\xi$ in

$$P(x < \xi) = F(\xi)$$

increase, then the number of elementary events which form the random event $x < \xi$ must also increase until in the limiting case for $\xi = +\infty$ all possible elementary events belong to $x < \xi$. However, all possible elementary events for a random experiment form the certain event $\Omega$; thus according to (2.35) we must have

$$\lim_{\xi \to +\infty} F(\xi) = F(+\infty) = P(\Omega) = 1 \tag{2.82a}$$

Similarly, the number of elementary events forming the random event $x < \xi$ decreases as $\xi$ decreases until in the limiting case for $\xi = -\infty$ there are no elementary events belonging to $x < \xi$; $x < \xi$ is then the impossible event $\emptyset$ and, according to (2.44), we have

$$\lim_{\xi \to -\infty} F(\xi) = F(-\infty) = P(\emptyset) = 0 \tag{2.82b}$$

(2) Two random events $x < \xi^{(a)}$ and $x < \xi^{(b)}$ belong to the two numerical values $\xi^{(a)}$ and $\xi^{(b)}$ of the random variable $x$. If we assume that $\xi^{(a)} < \xi^{(b)}$, all elementary events which form the random event $x < \xi^{(a)}$ also belong to the random event $x < \xi^{(b)}$. Therefore if $x < \xi^{(a)}$ occurs, $x < \xi^{(b)}$ must necessarily also occur. Because

$$(x < \xi^{(a)}) \subset (x < \xi^{(b)})$$

we then have, using (2.40):

$$P(x < \xi^{(a)}) \leqslant P(x < \xi^{(b)})$$

or, using the defining equation (2.81) of the distribution function:

$$F(\xi^{(a)}) \leqslant F(\xi^{(b)}) \quad \text{if} \quad \xi^{(a)} < \xi^{(b)} \tag{2.83}$$

The functional values of the distribution function $F(\xi)$ can therefore remain the same or increase as $\xi$ increases; thus $F(\xi)$ is a nondecreasing function which can only take values between zero and unity.

(3) The random event $\xi^{(a)} \leqslant x < \xi^{(b)}$ can be written as the difference event (see Section 2.2)

$$(\xi^{(a)} \leqslant x < \xi^{(b)}) = (x < \xi^{(b)}) \backslash (x < \xi^{(a)})$$

However, insofar as the notation $\xi^{(a)} \leqslant x < \xi^{(b)}$ tacitly contains the assumption $\xi^{(a)} < \xi^{(b)}$, we must also have

$$(x < \xi^{(a)}) \subset (x < \xi^{(b)})$$

According to (2.53) the probability of the random event $\xi^{(a)} \leqslant x < \xi^{(b)}$ is given by

$$P(\xi^{(a)} \leqslant x < \xi^{(b)}) = P\{(x < \xi^{(b)}) \backslash (x < \xi^{(a)})\} = P(x < \xi^{(b)}) - P(x < \xi^{(a)})$$

or, using (2.81):

$$P(\xi^{(a)} \leqslant x < \xi^{(b)}) = F(\xi^{(b)}) - F(\xi^{(a)}) \tag{2.84}$$

We can use this relationship to calculate the probability of every possible random event $\xi^{(a)} \leqslant x < \xi^{(b)}$ in a random experiment and of course the probability of a random event formed from such events if we know the distribution function $F(\xi)$ of the random experiment. If we substitute $\xi^{(a)} = -\infty$ and $\xi^{(b)} = \xi$ into (2.84), it is transformed into the defining equation of the distribution function $F(\xi)$, given that equation (2.82b) is taken into consideration.

(4) Let the set $\Omega$ consist of $m$ elementary events $\omega_1, \ldots, \omega_m$. Each of these elementary events is assigned a numerical value $\xi^{(i)}$ by the equation $\xi^{(i)} = \psi(\omega_i)$. If $\psi$ is a one-to-one function, all the $\omega^{(1)}, \ldots, \omega^{(m)}$ are different and the random variable $x$ can take $m$ different values. If, however, $\psi$ is not a one-to-one function, i.e. some of the values $\xi^{(1)}, \ldots, \xi^{(m)}$ have the same magnitude, which is allowed in the following, the random variable $x$ can only take $m^* < m$ different values. For both cases together, we have

$$m^* \leq m$$

In what follows we shall assume that the elementary events are numbered, so that the first $m^*$ of the total of $m$ numerical values $\xi^{(1)}, \ldots, \xi^{(m)}$ are all different and we have

$$\xi^{(i+1)} > \xi^{(i)} \quad i = 1, \ldots, m^* - 1$$

A probability $P(\omega_1), \ldots, P(\omega_m)$ is associated with each of the $m$ elementary events $\omega_1, \ldots, \omega_m$ and a probability $P(x = \xi^{(1)}), \ldots, P(x = \xi^{(m^*)})$ is associated with each of the different $m^*$ numerical values $\xi^{(1)}, \ldots, \xi^{(m^*)}$ which the random variable $x$ can take. As was shown in Section 2.4.1, the inequalities

$$P(\omega_i) \leq P(x = \xi^{(i)}) \quad i = 1, \ldots, m$$

apply. If we express the certain event $\Omega$ as the union of all $m$ elementary events,

$$\Omega = \bigcup_{i=1}^{m} \omega_i$$

then, because $P(\Omega) = 1$, we must have

$$P\left(\bigcup_{i=1}^{m} \omega_i\right) = 1$$

Because the elementary events are mutually exclusive pairwise by definition, this equation can be transformed into

$$\sum_{i=1}^{m} P(\omega_i) = 1 \tag{2.85a}$$

if relationship (2.56) is taken into consideration. However, this random experiment can be completely described not only by the $m$ elementary events but also by the $m^*$ random events $x = \xi^{(1)}, \ldots, x = \xi^{(m^*)}$. The latter are also mutually exclusive pairwise. This can easily be seen from Figure 2.12. Then, for the certain event $\Omega$, we must have

$$\Omega = \bigcup_{i=1}^{m^*} (x = \xi^{(i)})$$

as more than $m^*$ random events $x = \xi^{(i)}$ are not possible. Because $P(\Omega) = 1$ it follows from this, using equation (2.56), that

$$\sum_{i=1}^{m^*} P(x = \xi^{(i)}) = 1 \tag{2.85b}$$

This result contains (2.85a) as a special case because, for a one-to-one function $\psi$, we have $m^* = m$ and $P(\omega_i) = P(x = \xi^{(i)})$, $i = 1, \ldots, m$.

After this preliminary work, the distribution function $F(\xi)$ is now examined more closely in the neighborhood of the point $\xi = \xi^{(i)}$. For $\xi = \xi^{(i)}$, the random event $x < \xi^{(i)}$ consists of the events $x = \xi^{(1)}, \ldots, x = \xi^{(i-1)}$ because we have assumed that all numerical values $\xi^{(1)}, \ldots, \xi^{(i-1)}$ are smaller than $\xi^{(i)}$. However, we have

$$F(\xi^{(i)}) = P(x < \xi^{(i)}) = P\left\{\bigcup_{k=1}^{i-1} (x = \xi^{(k)})\right\}$$

or, according to (2.56),

$$F(\xi^{(i)}) = \sum_{k=1}^{i-1} P(x = \xi^{(k)})$$

For $\xi = \xi^{(i)} + \varepsilon$, where $\varepsilon$ must satisfy the conditions

$$0 < \varepsilon < \xi^{(i+1)} - \xi^{(i)}$$

but otherwise may be arbitrarily small, the random event $x < \xi^{(i)} + \varepsilon$ contains exactly the random event $x = \xi^{(i)}$ more than the random event $x < \xi^{(i)}$. The associated probability is therefore obtained in an analogous manner to that described above as

$$P(x < \xi^{(i)} + \varepsilon) = P\left\{\bigcup_{k=1}^{i} (x = \xi^{(k)})\right\} = \sum_{k=1}^{i} P(x = \xi^{(k)}) = F(\xi^{(i)}) + P(x = \xi^{(i)})$$

Hence, the distribution function is given by

$$F(\xi^{(i)} + \varepsilon) = F(\xi^{(i)}) + P(x = \xi^{(i)})$$

$F(\xi)$ therefore jumps by the value $P(x = \xi^{(i)})$ at the point $\xi = \xi^{(i)}$. In the range $\xi^{(i)} < \xi \leqslant \xi^{(i+1)}$ the number of random events $x = \xi^{(1)}, \ldots, x = \xi^{(i)}$ which form the random event $x < \xi$ does not change; for $\xi^{(i)} < \xi \leqslant \xi^{(i+1)}$, the value of $F(\xi)$ remains constant and equal to $F(\xi^{(i)}) + P(x = \xi^{(i)})$ and then jumps at the point $\xi = \xi^{(i+1)}$ by $P(x = \xi^{(i+1)})$, although at the discontinuity point itself we still have

$$F(\xi^{(i+1)}) = F(\xi^{(i)}) + P(x = \xi^{(i)})$$

In summary, we can say that in a random experiment with $m$ possible elementary events where the random variable $x$ can take $m^*$ different values $\xi^{(1)}, \ldots, \xi^{(m^*)}$ the distribution function $F(\xi)$ varies in a stepwise manner. It jumps at the points $\xi = \xi^{(i)}$, $i = 1, \ldots, m^*$, by $P(x = \xi^{(i)}) \geqslant P(\omega_i)$. The sum of all $m^*$ step heights is unity. At the discontinuity points, $F(\xi)$ takes the limiting value on the left-hand side. Thus $F(\xi)$ must have the appearance shown in Figure 2.15.

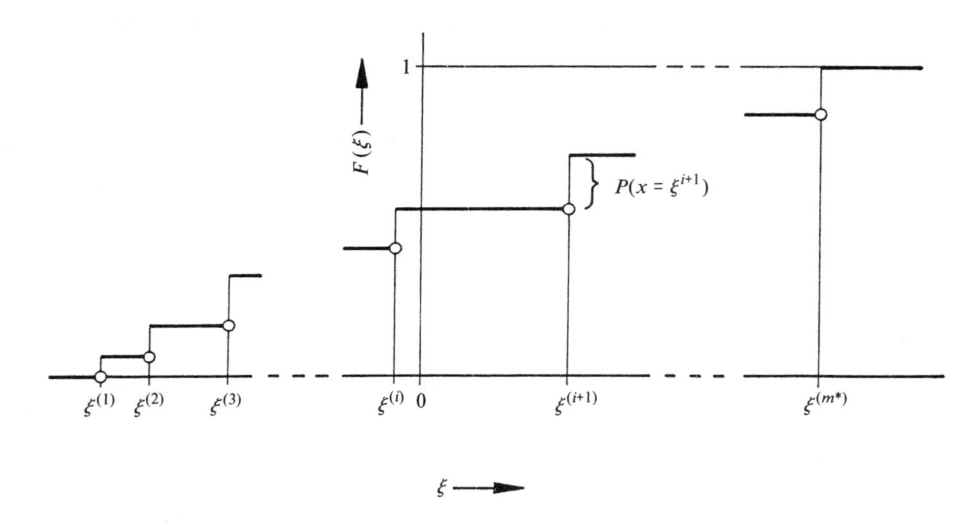


**Figure 2.15** Distribution function $F(\xi)$ for a random experiment with $m$ possible elementary events and with $m^* \leq m$ different values of the random variable $x$.

A random variable $x$ which can only take $m^* \leq m$ values $\xi^{(1)}, \ldots, \xi^{(m^*)}$ belongs to a random experiment with $m$ possible elementary events $\omega_1, \ldots, \omega_m$. Quite generally, a random variable which can only take discrete values is called a *discrete random variable*.

Now, according to equation (2.85b),

$$\sum_{i=1}^{m^*} P(x = \xi^{(i)}) = 1$$

the step heights $P(x = \xi^{(i)})$, $i = 1, \ldots, m^*$, in $F(\xi)$ must become smaller as the number $m^*$ of possible values of the random variable $x$ increases. In the limiting case, when the random variable $x$ can take all values on the $\xi$ axis, we have a *continuous random variable* and the step distribution function becomes a continuous distribution function (Figure 2.16).

A continuous distribution function also has the properties derived in this section. It is a nondecreasing function with $F(-\infty) = 0$ and $F(+\infty) = 1$. Because, in the derivation of equation (2.84),

$$P(\xi^{(a)} \leq x < \xi^{(b)}) = F(\xi^{(b)}) - F(\xi^{(a)})$$

no assumptions were made regarding the nature of the random variable $x$, this equation must also hold for continuous random variables.

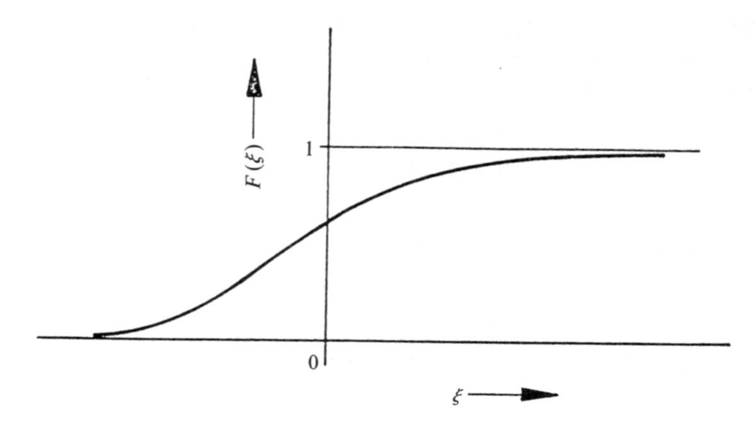


**Figure 2.16** Distribution function of a continuous random variable.

Distribution functions which have sections corresponding to Figure 2.16 but are interrupted by discontinuity points will be excluded from the discussion below.

### 2.4.3 The Density Function for Scalar Random Variables

We now compare equation (2.84)

$$P(\xi^{(a)} \leq x < \xi^{(b)}) = F(\xi^{(b)}) - F(\xi^{(a)})$$

with the following relationship from integral calculus:

$$\int_{\xi^{(a)}}^{\xi^{(b)}} f(\xi)\,\mathrm{d}\xi = F(\xi^{(b)}) - F(\xi^{(a)}) \quad \text{if} \quad f(\xi) = \frac{\mathrm{d}}{\mathrm{d}\xi} F(\xi)$$

Because the right-hand sides of these two equations are the same, the following conclusion can be drawn if $F(\xi)$ in the second equation is regarded as a distribution function: the derivative $f(\xi)$ of the distribution function $F(\xi)$ can also be used to determine the probability of every possible random event $\xi^{(a)} \leqslant x < \xi^{(b)}$ in the associated random experiment. Then the probability of any random event formed from such events can also be calculated. This means that the derivative

$$f(\xi) := \frac{\mathrm{d}}{\mathrm{d}\xi} F(\xi) \tag{2.86}$$

of the distribution function $F(\xi)$ is as important as the distribution function itself. $f(\xi)$ is known as the *density function* of the one-dimensional random variable $x$.

If we substitute in equation (2.84) the generally valid expression

$$\int_{\xi^{(a)}}^{\xi^{(b)}} f(\xi)\,\mathrm{d}\xi = F(\xi^{(b)}) - F(\xi^{(a)})$$

where $F(\xi)$ is a distribution function and $f(\xi)$ is the associated density function, it follows that

$$P(\xi^{(a)} \le x < \xi^{(b)}) = F(\xi^{(b)}) - F(\xi^{(a)}) = \int_{\xi^{(a)}}^{\xi^{(b)}} f(\xi)\,\mathrm{d}\xi \tag{2.87}$$

For $\xi^{(a)} = -\infty$ and $\xi^{(b)} = \xi$, this equation becomes

$$P(x < \xi) = F(\xi) = \int_{-\infty}^{\xi} f(v)\,\mathrm{d}v \tag{2.88}$$

This equation can be used to calculate the distribution function $F(\xi)$ when the density function is known, and equation (2.86) can be used to calculate the density function $f(\xi)$ when the distribution function is known. In (2.88) the integration variable is denoted $v$ in order to avoid confusion with the upper integration limit. The existence or, equivalently, the convergence of the integral with an infinite lower integration limit is always assumed.

The division or distribution of the unit probability of the certain event $\Omega$ over the entire $\xi$ axis is described by the variation of the distribution function $F(\xi)$ as well as by the variation of the density function $f(\xi)$. As can be seen from equation (2.84), $P(\xi^{(a)} \leqslant x < \xi^{(b)})$ takes a large value in regions where the distribution function $F(\xi)$ increases rapidly and a small value in regions where $F(\xi)$ increases slowly. According to equation (2.86) $f(\xi)$ is large for rapidly increasing $F(\xi)$ and hence, according to equation (2.87), $P(\xi^{(a)} \leqslant x < \xi^{(b)})$ is also large. This is often referred to as the *probability distribution* or, more concisely, the *distribution* of a random variable.

#### 2.4.3.1 Properties of the Density Function $f(\xi)$

(1) As the distribution function $F(\xi)$ is a nondecreasing function, its derivative (i.e., the density function $f(\xi)$), must be a nonnegative function:

$$f(\xi) \ge 0 \tag{2.89}$$

(2) If we substitute $\xi^{(a)} = -\infty$ and $\xi^{(b)} = +\infty$ into equation (2.87), it follows that, for $F(-\infty) = 0$ and $F(+\infty) = 1$,

$$\int_{-\infty}^{+\infty} f(\xi)\mathrm{d}\xi = 1 \tag{2.90}$$

If the integral $\int_{\xi(a)}^{\xi(b)} f(\xi)\mathrm{d}\xi$ is interpreted as the area under the density function $f(\xi)$, according to equation (2.87), this area between $\xi = \xi^{(a)}$ and $\xi = \xi^{(b)}$ corresponds to the probability that the random event $\xi^{(a)} \leqslant x < \xi^{(b)}$ will occur. According to equation (2.88) the area under $f(\xi)$ to the left of $\xi = \xi^{(0)}$ corresponds to the probability that the random event $x < \xi^{(0)}$ will occur (Figure 2.17). Because the random event $-\infty \leqslant x < +\infty$ contains the values of the random variable of all possible elementary events, it must be the certain event $\Omega$; in other words, the associated area (i.e., the total area under $f(\xi)$) corresponds to the probability of the certain event and must therefore be equal to unity as has already been shown by equation (2.90).

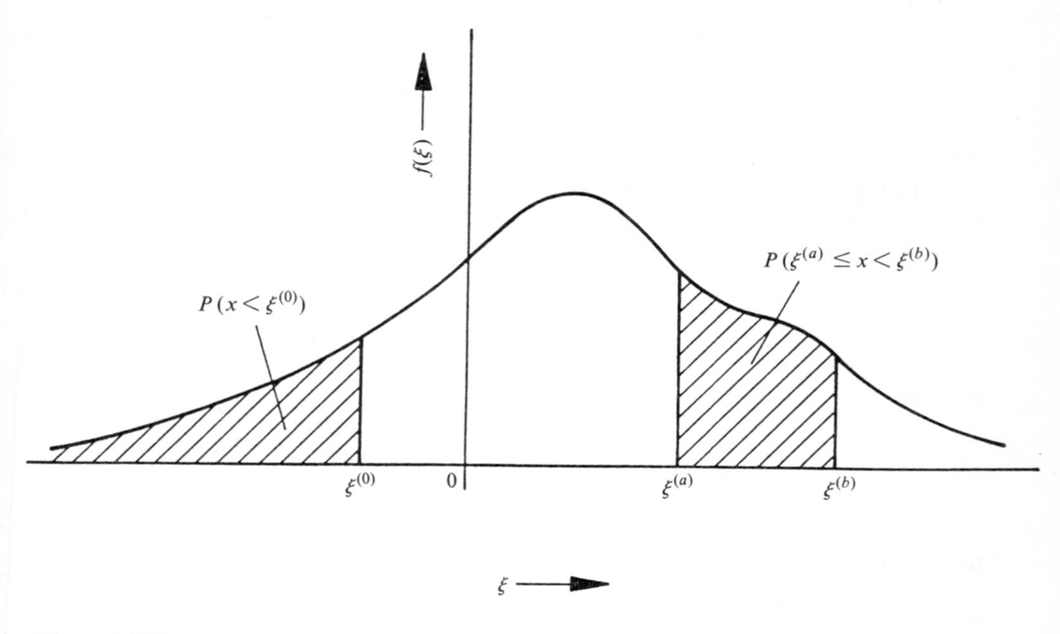


**Figure 2.17** Density function and probability.

(3) For a continuous random variable $x$ with a continuous distribution function $F(\xi)$, there is no difficulty in calculating the density function $f(\xi)$ from equation (2.86). However, equation (2.86) does not give a useful result for a discrete random

variable with a step distribution function if the differentiation is defined in the classical sense. According to this the derivative of a step distribution function like that in Figure 2.15 is zero everywhere except at the discontinuity points where it is not defined. In order to use equation (2.86) to calculate the density function $f(\xi)$ from $F(\xi)$ and equation (2.88) to calculate the distribution function $F(\xi)$ from $f(\xi)$ for discrete random variables, a generalized concept of differentiation must be used. However, here we shall only concern ourselves with one-dimensional random variables for the questions associated with this. In subsequent sections of this chapter, only continuous random variables will be considered.

We now introduce as a generalized derivative of the step function:

$$\sigma(\xi) = \begin{cases} 0 & \text{for } \xi \le 0 \\ 1 & \text{for } \xi > 0 \end{cases}$$

the $\delta$ function (variously known as the delta function, Dirac function, impulse function) $\delta(\xi)$ with the property

$$\delta(\xi) = \begin{cases} +\infty & \text{for } \xi = 0 \\ 0 & \text{for } \xi \neq 0 \end{cases}$$

$\delta(\xi)$ coincides with the classical derivative of the step function $\sigma(\xi)$ for all $\xi \neq 0$, but when $\xi = 0$, where the classical derivative is not defined, it takes the value infinity. If we now use the masking property of the $\delta$ function

$$\int_{-\infty}^{+\infty} f(\xi)\delta(\xi - \xi^{(0)})\,\mathrm{d}\xi = f(\xi^{(0)})$$

or in the special case

$$\int_{\xi^{(a)}}^{\xi^{(b)}} f(\xi)\delta(\xi - \xi^{(0)})\,\mathrm{d}\xi = \begin{cases} f(\xi^{(0)}) & \text{for} \quad \xi^{(a)} < \xi^{(0)} < \xi^{(b)} \\ 0 & \text{otherwise} \end{cases}$$

the above problem can be solved.

The step distribution function $F(\xi)$ of a discrete random variable $x$ (see Figure 2.15) can be represented in terms of the step function by

$$F(\xi) = \sum_{i=1}^{m^*} P(x = \xi^{(i)})\sigma(\xi - \xi^{(i)}) \tag{2.91}$$

Then, from equation (2.86), if we perform a generalized differentiation of equation (2.91), we obtain the density function $f(\xi)$ of a discrete random variable $x$ as

$$f(\xi) = \sum_{i=1}^{m^*} P(x = \xi^{(i)})\delta(\xi - \xi^{(i)}) \tag{2.92}$$

where it is assumed that $P(x = \xi^{(i)})$ has a fixed numerical value. The graphical representation of this function $f(\xi)$ does not contain any information regarding the corresponding random experiment because it has the same appearance for different sets of probabilities $P(x = \xi^{(1)}), \ldots, P(x = \xi^{(m^*)})$. $f(\xi)$ is a comb function which has a value of $+\infty$ at the points $\xi = \xi^{(i)}$, $i = 1, \ldots, m^*$, and is zero everywhere else. Therefore in the case of a discrete random variable $x$ the *probability function*

$$\tilde{f}(\xi) := \begin{cases} P(x = \xi^{(i)}) & \text{for } \xi = \xi^{(i)},\ i = 1, \ldots, m^* \\ 0 & \text{otherwise} \end{cases} \tag{2.93}$$

is plotted instead of the density function $f(\xi)$. The probability function $\tilde{f}(\xi)$ gives a particularly clear representation of the probability distribution concept introduced above or, more concisely, of the distribution of a random variable which describes the division of unit probability over the different events which are possible in a random experiment, in this case over the random events $x = \xi^{(i)}$, $i = 1, \ldots, m^*$. It should be noted that $\tilde{f}(\xi)$ is not identical with the density function $f(\xi)$.

We can now determine the density function $f(\xi)$ from the distribution function $F(\xi)$ for a discrete random variable, but we still require to show how $F(\xi)$ can be calculated from $f(\xi)$ using equation (2.88):

$$\begin{aligned} F(\xi) = \int_{-\infty}^{\xi} f(v)\,\mathrm{d}v &= \int_{-\infty}^{\xi} \sum_{i=1}^{m^*} P(x = \xi^{(i)})\delta(v - \xi^{(i)})\,\mathrm{d}v \\ &= \sum_{i=1}^{m^*} P(x = \xi^{(i)}) \int_{-\infty}^{\xi} \delta(v - \xi^{(i)})\,\mathrm{d}v \end{aligned}$$

Because of the masking property of the $\delta$ function, the value of the integral

$$\int_{-\infty}^{\xi} \delta(v - \xi^{(i)})\mathrm{d}v = \int_{-\infty}^{\xi} 1 \times \delta(v - \xi^{(i)})\mathrm{d}v$$

must be unity for $-\infty < \xi^{(i)} < \xi$ and zero otherwise. Thus the expression for $F(\xi)$ becomes

$$F(\xi) = \sum_{\xi^{(i)}<\xi} P(x = \xi^{(i)}) = \sum_{\xi^{(i)}<\xi} \tilde{f}(\xi^{(i)}) \tag{2.94}$$

where the subscripts under the summation sign indicate that, in order to calculate the distribution function $F$ at the point $\xi$ of a discrete random variable $x$, all probabilities $P(x = \xi^{(i)}) = \tilde{f}(\xi^{(i)})$ of the random events $x = \xi^{(i)}$ for which $\xi^{(i)} < \xi$ holds must be summed. Equation (2.94), which is constructed completely analogously to equation (2.88), could have been obtained from Figure 2.15 by interpreting the integration as the limiting case of the summation and by assuming, as will be shown below, that $f(\xi)\mathrm{d}\xi$ is a probability. However, it should be shown that, if a generalized concept of differentiation is used, equations (2.86) and (2.88) are also valid for discrete random variables.

(4) Equation (2.84) gives the probability of the random event $\xi \leqslant x < \xi + \Delta\xi$, $|\Delta\xi| \ll |\xi|$, as

$$P(\xi \leq x < \xi + \Delta\xi) = F(\xi + \Delta\xi) - F(\xi) = \frac{F(\xi + \Delta\xi) - F(\xi)}{\Delta\xi}\Delta\xi$$

Because

$$\frac{F(\xi + \Delta\xi) - F(\xi)}{\Delta\xi} \approx \lim_{\Delta\xi\to 0} \frac{F(\xi + \Delta\xi) - F(\xi)}{\Delta\xi} = \frac{\mathrm{d}}{\mathrm{d}\xi}F(\xi) = f(\xi)$$

we obtain the expression:

$$P(\xi \leq x < \xi + \Delta\xi) \approx f(\xi)\Delta\xi \tag{2.95}$$

from which the density function $f(\xi)$ can be approximated point by point if the probability $P(\xi \leqslant x < \xi + \Delta\xi)$ is known (for example, experimentally from the relative frequency of the random event $\xi \leqslant x < \xi + \Delta\xi$).

### 2.4.4 The Distribution Function and the Density Function for Vector Random Variables

Just as for a one-dimensional random variable $x$ the probability that the random event $x < \xi$ will occur depends on the limit $\xi$ for a two-dimensional random variable $\boldsymbol{x} = [x_1, x_2]'$, the probability that the random event $(x_1 < \xi_1) \cap (x_2 < \xi_2)$ will occur must depend on the two limits $\xi_1$ and $\xi_2$. Therefore

$$P\{(x_1 < \xi_1) \cap (x_2 < \xi_2)\} =: F(\xi_1, \xi_2) \tag{2.96}$$

By analogy with the distribution function $F(\xi)$, the function $F(\xi_1, \xi_2)$ is known as the distribution function (or the joint distribution function) of the two-dimensional random variable $\boldsymbol{x} = [x_1, x_2]'$.

The random event $(x_1 < \xi_1) \cap (x_2 < \xi_2)$ consists of all the elementary events $\omega_i \in \Omega$ of the random experiment in question, and a vector $\boldsymbol{\xi}^{(i)} = [\xi_1^{(i)}, \xi_2^{(i)}]'$ is assigned to this as the value of the two-dimensional random variable $\boldsymbol{x} = [x_1, x_2]'$. The conditions $\xi_1^{(i)} < \xi_1$ and $\xi_2^{(i)} < \xi_2$ apply concurrently to this vector $\boldsymbol{\xi}^{(i)}$ (Figure 2.18).

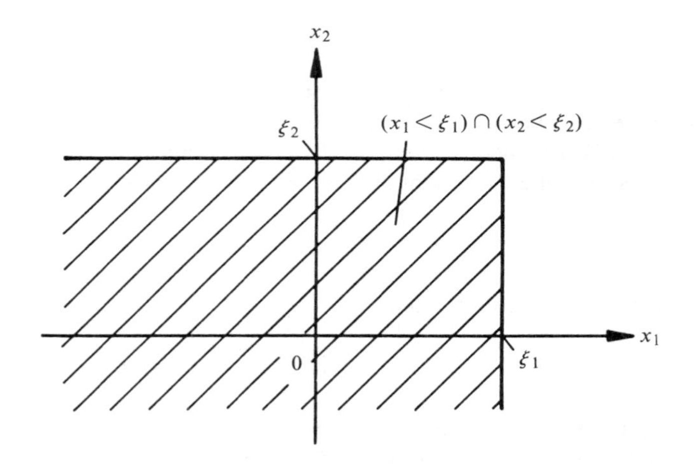


**Figure 2.18** For the definition of the distribution function $F(\xi_1, \xi_2) := P\{(x_1 < \xi_1) \cap (x_2 < \xi_2)\}$ of a two-dimensional random variable $\boldsymbol{x} = [x_1, x_2]'$.

In the literature, the defining equation for $F(\xi_1, \xi_2)$ is almost always written in the form:

$$P(x_1 < \xi_1, x_2 < \xi_2) =: F(\xi_1, \xi_2)$$

or

$$P(x_1 < \xi_1; x_2 < \xi_2) =: F(\xi_1, \xi_2)$$

However, because this does not clearly identify the AND operation on the two random events $x_1 < \xi_1$ and $x_2 < \xi_2$, these notations will not be used here.

The distribution function of an $n$-dimensional random variable $\boldsymbol{x} = [x_1, x_2, \ldots, x_n]'$ can be defined, similarly to equation (2.96), as

$$P\{(x_1 < \xi_1) \cap (x_2 < \xi_2) \cap \cdots \cap (x_n < \xi_n)\} =: F(\xi_1, \xi_2, \ldots, \xi_n) \quad (2.97)$$

We shall restrict our investigation of the properties of the distribution function of an $n$-dimensional random variable to the case $n = 2$. The results found for this can easily be generalized to dimensions $n > 2$.

(1) For the limiting case $\xi_1 = +\infty$ and $\xi_2 = +\infty$, the random event $(x_1 < \xi_1 = +\infty) \cap (x_2 < \xi_2 = +\infty)$ must contain all possible elementary events for this random experiment so that it must be the certain event $\Omega$ (i.e., for $P(\Omega) = 1$) we have

$$F(+\infty, +\infty) = P(\Omega) = 1 \quad (2.98)$$

However, for $\xi_2 = -\infty$ the random event $(x_2 < \xi_2 = -\infty)$ is the impossible event $\emptyset$ because there can be no elementary event $\omega_i$ to which a vector $\boldsymbol{\xi}^{(i)} = [\xi_1^{(i)}, \xi_2^{(i)}]'$ with $\xi_2^{(i)} < -\infty$ is assigned. Since $A \cap \emptyset = \emptyset$ the random event $(x_1 < \xi_1) \cap (x_2 < \xi_2 < -\infty)$ must also be the impossible event $\emptyset$. In exactly the same way we have

$$(x_1 < \xi_1 = -\infty) \cap (x_2 < \xi_2) = \emptyset \cap (x_2 < \xi_2) = \emptyset$$

Insofar as $P(\emptyset) = 0$, it follows that

$$F(\xi_1, -\infty) = F(-\infty, \xi_2) = 0 \quad (2.99)$$

However, equations (2.98) and (2.99) are not symmetric in the same way as (2.82a) and (2.82b). The distribution function $F(\xi_1, \xi_2)$ of a two-dimensional random variable is zero if at least one of the two variables tends to $-\infty$, with the other variable kept constant, and is unity if both $\xi_1$ and $\xi_2$ tend to $+\infty$.

(2) For $\xi_1^{(a)} < \xi_1^{(b)}$, we have

$$\{(x_1 < \xi_1^{(a)}) \cap (x_2 < \xi_2)\} \subset \{(x_1 < \xi_1^{(b)}) \cap (x_2 < \xi_2)\}$$

as the number of elementary events contained in the random event $(x_1 < \xi_1^{(b)}) \cap (x_2 < \xi_2)$ is equal to or greater than the number of elementary events contained in the random event $(x_1 < \xi_1^{(a)}) \cap (x_2 < \xi_2)$. According to equation (2.40), it follows from this that

$$P\{(x_1 < \xi_1^{(a)}) \cap (x_2 < \xi_2)\} \leq P\{(x_1 < \xi_1^{(b)}) \cap (x_2 < \xi_2)\}$$

or, using equation (2.96),

$$F(\xi_1^{(a)}, \xi_2) \leq F(\xi_1^{(b)}, \xi_2) \quad \text{if} \quad \xi_1^{(a)} < \xi_1^{(b)} \quad (2.100a)$$

For the same reason the following relationships must hold:

$$F(\xi_1, \xi_2^{(a)}) \le F(\xi_1, \xi_2^{(b)}) \quad \text{if} \quad \xi_2^{(a)} < \xi_2^{(b)} \tag{2.100b}$$

and

$$F(\xi_1^{(a)}, \xi_2^{(a)}) \le F(\xi_1^{(b)}, \xi_2^{(b)}) \quad \text{if} \quad \xi_1^{(a)} < \xi_1^{(b)} \quad \text{and} \quad \xi_2^{(a)} < \xi_2^{(b)} \tag{2.100c}$$

The distribution function $F(\xi_1, \xi_2)$ of a two-dimensional random variable is therefore a nondecreasing function of each of the two variables $\xi_1$ and $\xi_2$.

(3) We now split the random event $(\xi_1^{(a)} \leqslant x_1 < \xi_1^{(b)}) \cap (\xi_2^{(a)} \leqslant x_2 < \xi_2^{(b)})$ into the difference between the two events $(x_1 < \xi_1^{(b)}) \cap (x_2 < \xi_2^{(b)})$ and $(A_1 \cup A_2)$ (the random events $A_1$ and $A_2$ are shown by the shaded areas in Figure 2.19). Then, because

$$(\xi_1^{(a)} \le x_1 < \xi_1^{(b)}) \cap (\xi_2^{(a)} \le x_2 < \xi_2^{(b)}) = \{(x_1 < \xi_1^{(b)}) \cap (x_2 < \xi_2^{(b)})\} \backslash (A_1 \cup A_2)$$

as a result of

$$(A_1 \cup A_2) \subset \{(x_1 < \xi_1^{(b)}) \cap (x_2 < \xi_2^{(b)})\}$$

we obtain, using equation (2.53),

$$\begin{aligned} &P\{(\xi_1^{(a)} \le x_1 < \xi_1^{(b)}) \cap (\xi_2^{(a)} \le x_2 < \xi_2^{(b)})\} \\ &\quad = P\{(x_1 < \xi_1^{(b)}) \cap (x_2 < \xi_2^{(b)})\} - P(A_1 \cup A_2) \\ &\quad = F(\xi_1^{(b)}, \xi_2^{(b)}) - P(A_1 \cup A_2) \end{aligned} \tag{2.101}$$

Furthermore, according to equation (2.49),

$$P(A_1 \cup A_2) = P(A_1) + P(A_2) - P(A_1 \cap A_2)$$

Substituting this expression into equation (2.101) and using the equations

$$P(A_1) = P\{(x_1 < \xi_1^{(a)}) \cap (x_2 < \xi_2^{(b)})\} = F(\xi_1^{(a)}, \xi_2^{(b)})$$

$$P(A_2) = P\{(x_1 < \xi_1^{(b)}) \cap (x_2 < \xi_2^{(a)})\} = F(\xi_1^{(b)}, \xi_2^{(a)})$$

$$P(A_1 \cap A_2) = P\{(x_1 < \xi_1^{(a)}) \cap (x_2 < \xi_2^{(a)})\} = F(\xi_1^{(a)}, \xi_2^{(a)})$$

(see Figure 2.19), we obtain the relationship

$$\begin{aligned} &P\{(\xi_1^{(a)} \le x_1 < \xi_1^{(b)}) \cap (\xi_2^{(a)} \le x_2 < \xi_2^{(b)})\} \\ &\quad = F(\xi_1^{(b)}, \xi_2^{(b)}) - F(\xi_1^{(a)}, \xi_2^{(b)}) - F(\xi_1^{(b)}, \xi_2^{(a)}) + F(\xi_1^{(a)}, \xi_2^{(a)}) \end{aligned} \tag{2.102}$$

Thus it is clear that the distribution function $F(\xi_1, \xi_2)$ of a two-dimensional random variable $\boldsymbol{x} = [x_1, x_2]'$ plays the same role in the calculation of probabilities as is played by the distribution function $F(\xi)$ for a one-dimensional random variable $x$.

(4) If $m$ elementary events are possible in a random experiment, the distribution function $F(\xi_1, \xi_2)$ of the associated two-dimensional random variable $\boldsymbol{x}$ has a step variation in both the $\xi_1$ and the $\xi_2$ directions. Since the variation in $F(\xi_1, \xi_2)$ over the $\xi_1$, $\xi_2$ plane can be described in three dimensions, the distribution function $F(\xi_1, \xi_2)$ can be considered as a terrace function.

Such a terrace function can be constructed using a simple example. Let there be four possible elementary events $\omega_1, \ldots, \omega_4$ in a random experiment to which the vectors $\boldsymbol{\xi}^{(1)} = [0, 0]'$, $\boldsymbol{\xi}^{(2)} = [0, 1]'$, $\boldsymbol{\xi}^{(3)} = [1, 0]'$ and $\boldsymbol{\xi}^{(4)} = [1, 1]'$ are assigned in a fixed manner as values of a two-dimensional random variable $\boldsymbol{x}$. Since all four vectors $\boldsymbol{\xi}^{(1)}, \ldots, \boldsymbol{\xi}^{(4)}$ are different, $m^* = m = 4$ ($m^*$ is defined in Section 2.4.2.1). The following probabilities are assumed:

$$P(\omega_1) = P\{(x_1 = 0) \cap (x_2 = 0)\} = P(\boldsymbol{x} = \boldsymbol{\xi}^{(1)}) = 9/16$$

$$P(\omega_2) = P\{(x_1 = 0) \cap (x_2 = 1)\} = P(\boldsymbol{x} = \boldsymbol{\xi}^{(2)}) = 3/16$$

$$P(\omega_3) = P\{(x_1 = 1) \cap (x_2 = 0)\} = P(\boldsymbol{x} = \boldsymbol{\xi}^{(3)}) = 3/16$$

$$P(\omega_4) = P\{(x_1 = 1) \cap (x_2 = 1)\} = P(\boldsymbol{x} = \boldsymbol{\xi}^{(4)}) = 1/16$$

Of course, equation (2.85a) must also be satisfied here; thus,

$$\sum_{i=1}^{4} P(\omega_i) = \frac{9}{16} + \frac{3}{16} + \frac{3}{16} + \frac{1}{16} = \frac{16}{16} = 1$$

This information can now be used to plot the distribution function $F(\xi_1, \xi_2)$ (Figure 2.20).

In contrast with the distribution function $F(\xi)$, the distribution function $F(\xi_1, \xi_2)$ for discrete random variables does not jump at every point $\boldsymbol{\xi} = \boldsymbol{\xi}^{(i)}$ by the value $P(\boldsymbol{x} = \boldsymbol{\xi}^{(i)})$. Because of the "less than" signs in the defining equation of $F(\xi_1, \xi_2)$, the function $F(\xi_1, \xi_2)$ always takes the limiting value on the left-hand side at the discontinuity points if one of the two variables is kept constant and the other is allowed to vary.

The distribution function $F(\xi_1, \xi_2)$ which has just been examined in detail for a discrete two-dimensional random variable also becomes a continuous function for the case of a continuous two-dimensional random variable if $\boldsymbol{x}$ can take all points in the $\xi_1$, $\xi_2$ plane.

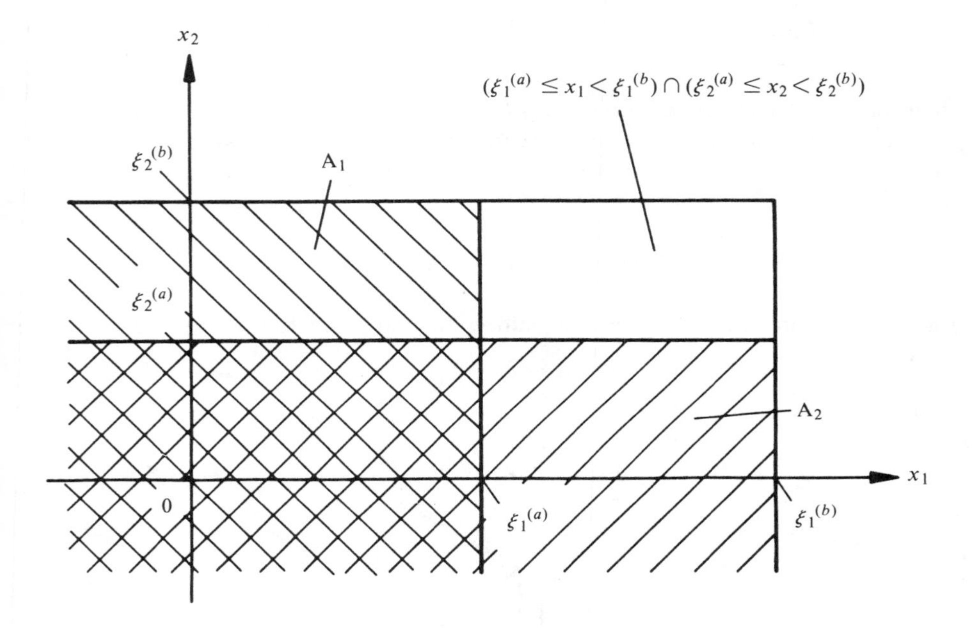


**Figure 2.19** Diagram for the derivation of equation (2.102).

We must now consider the density function for vector random variables and derive its properties. In the generalization of equation (2.86) the mixed type of second-order partial derivative of the distribution function $F(\xi_1, \xi_2)$ is introduced as the density function (joint density function) $f(\xi_1, \xi_2)$ of a two-dimensional random variable:

$$f(\xi_1, \xi_2) := \frac{\partial^2}{\partial \xi_1 \, \partial \xi_2} F(\xi_1, \xi_2) \tag{2.103}$$

As in equation (2.87) the probability that the random event $(\xi_1^{(a)} \leqslant x_1 < \xi_1^{(b)}) \cap (\xi_2^{(a)} \leqslant x_2 < \xi_2^{(b)})$ will occur can be calculated using the density function $f(\xi_1, \xi_2)$:

$$P\{(\xi_1^{(a)} \leq x_1 < \xi_1^{(b)}) \cap (\xi_2^{(a)} \leq x_2 < \xi_2^{(b)})\} = \int_{\xi_1^{(a)}}^{\xi_1^{(b)}} \int_{\xi_2^{(a)}}^{\xi_2^{(b)}} f(\xi_1, \xi_2) \, d\xi_1 \, d\xi_2 \tag{2.104}$$

If we now substitute $\xi_1^{(a)} = -\infty$, $\xi_2^{(a)} = -\infty$ and $\xi_1^{(b)} = \xi_1$, $\xi_2^{(b)} = \xi_2$ here, we obtain the relationship

$$P\{(x_1 < \xi_1) \cap (x_2 < \xi_2)\} = F(\xi_1, \xi_2) = \int_{-\infty}^{\xi_1} \int_{-\infty}^{\xi_2} f(v_1, v_2) \mathrm{d}v_1 \mathrm{d}v_2 \quad (2.105)$$

Therefore the density function $f(\xi_1, \xi_2)$ for a two-dimensional random variable $\boldsymbol{x} = [x_1, x_2]'$ can be calculated from the distribution function $F(\xi_1, \xi_2)$ (equation (2.103)) and $F(\xi_1, \xi_2)$ can be calculated from $f(\xi_1, \xi_2)$ (equation (2.105)). In the determination of

$$P\{(\xi_1^{(a)} \leqslant x_1 < \xi_1^{(b)}) \cap (\xi_2^{(a)} \leqslant x_2 < \xi_2^{(b)})\}$$

from (2.102) and (2.104), the distribution function and the density function are equivalent.

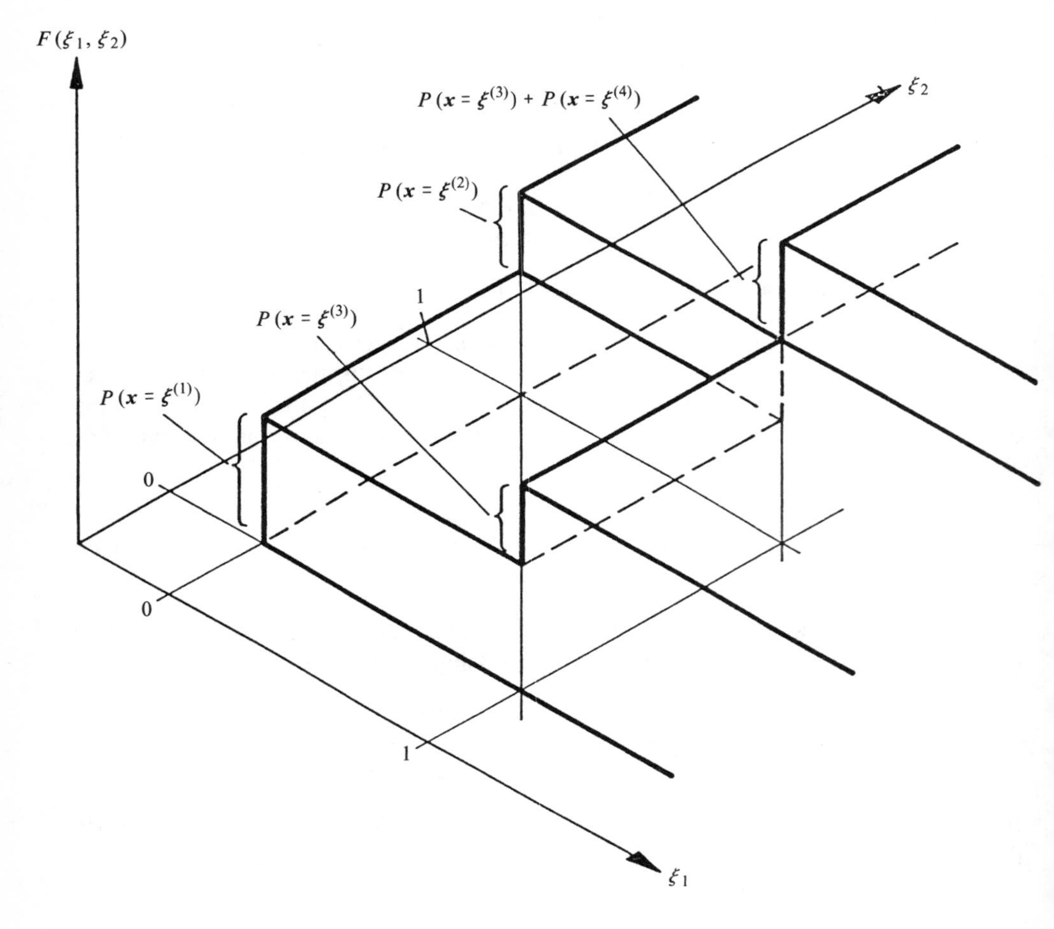


**Figure 2.20** Distribution function $F(\xi_1, \xi_2)$ of a two-dimensional discrete random variable $\boldsymbol{x}$.

Equations (2.103), (2.104) and (2.105) can easily be extended to three-dimensional and multidimensional random variables. However, the properties of the density function, like the properties of the distribution function, will only be considered for two-dimensional random variables.

(5) Because $F(\xi_1, \xi_2)$ is nondecreasing in each of the two variables, we must have

$$f(\xi_1, \xi_2) \geq 0 \tag{2.106}$$

(6) If the integration in equation (2.105) extends over the entire $\xi_1$, $\xi_2$ plane, then the probability of the certain event $\Omega$ is calculated and, because $P(\Omega) = 1$, we have

$$F(+\infty, +\infty) = \int_{-\infty}^{+\infty} \int_{-\infty}^{+\infty} f(\xi_1, \xi_2) \mathrm{d}\xi_1 \mathrm{d}\xi_2 = 1 \tag{2.107}$$

If the probabilities in equations (2.104) and (2.105) are interpreted as volumes under the density function $f(\xi_1, \xi_2)$, then according to equation (2.107) the total volume under $f(\xi_1, \xi_2)$ must be equal to unity.

(7) If $\xi_1^{(a)} = \xi_1$, $\xi_1^{(b)} = \xi_1 + \Delta\xi_1$, $\xi_2^{(a)} = \xi_2$ and $\xi_2^{(b)} = \xi_2 + \Delta\xi_2$, equation (2.102) becomes

$$\begin{aligned} &P\{(\xi_1 \leq x_1 < \xi_1 + \Delta\xi_1) \cap (\xi_2 \leq x_2 < \xi_2 + \Delta\xi_2)\} \\ &\quad = F(\xi_1 + \Delta\xi_1, \xi_2 + \Delta\xi_2) - F(\xi_1, \xi_2 + \Delta\xi_2) - F(\xi_1 + \Delta\xi_1, \xi_2) + F(\xi_1, \xi_2) \end{aligned}$$

Using the abbreviation

$$S := (\xi_1 \leq x_1 < \xi_1 + \Delta\xi_1) \cap (\xi_2 \leq x_2 < \xi_2 + \Delta\xi_2)$$

we can rewrite this expression as

$$\begin{aligned} P(S) = {} & \frac{F(\xi_1 + \Delta\xi_1, \xi_2 + \Delta\xi_2) - F(\xi_1 + \Delta\xi_1, \xi_2)}{\Delta\xi_2} \Delta\xi_2 \\ & - \frac{F(\xi_1, \xi_2 + \Delta\xi_2) - F(\xi_1, \xi_2)}{\Delta\xi_2} \Delta\xi_2 \end{aligned}$$

or

$$\begin{aligned} P(S) &\approx \left\{ \frac{\partial}{\partial\xi_2} P(\xi_1 + \Delta\xi_1, \xi_2) - \frac{\partial}{\partial\xi_2} F(\xi_1, \xi_2) \right\} \Delta\xi_2 \\ &= \frac{(\partial/\partial\xi_2)F(\xi_1 + \Delta\xi_1, \xi_2) - (\partial/\partial\xi_2)F(\xi_1, \xi_2)}{\Delta\xi_1} \Delta\xi_1\Delta\xi_2 \end{aligned}$$

If the difference ratio in this expression is approximated by the differential ratio, we obtain the equation

$$P\{(\xi_1 \leq x_1 < \xi_1 + \Delta\xi_1) \cap (\xi_2 \leq x_2 < \xi_2 + \Delta\xi_2)\}$$
$$\approx \frac{\partial}{\partial \xi_1}\left\{\frac{\partial}{\partial \xi_2} F(\xi_1, \xi_2)\right\}\Delta\xi_1\Delta\xi_2 = f(\xi_1, \xi_2)\Delta\xi_1\Delta\xi_2 \qquad (2.108)$$

for the experimental determination of $f(\xi_1, \xi_2)$. This equation is completely equivalent in appearance to (2.95). The abbreviation $S$ introduced above is eliminated in equation (2.108).

### 2.4.5 Marginal Distributions

In Section 2.4.4 we found the following properties for the distribution function $F(\xi_1, \xi_2)$ of a two-dimensional random variable $\boldsymbol{x} = [x_1, x_2]'$:

$$F(+\infty, +\infty) = 1$$

and

$$F(\xi_1, -\infty) = F(-\infty, \xi_2) = 0$$

Here we shall examine the case $F(\xi_1, +\infty)$ more closely.

From the defining equation (2.96),

$$F(\xi_1, \xi_2) := P\{(x_1 < \xi_1) \cap (x_2 < \xi_2)\}$$

it follows that, for $\xi_2 = +\infty$,

$$F(\xi_1, +\infty) = P\{(x_1 < \xi_1) \cap (x_2 < +\infty)\}$$

However, the random event $x_2 < +\infty$ is a certain event $\Omega$ because it occurs with every possible elementary event. Therefore relationship (2.5), $A \cap \Omega = A$, can be applied to the previous equation so that we have

$$F(\xi_1, +\infty) = P(x_1 < \xi_1)$$

We now have the distribution function for the one-dimensional random variable $x_1$ on the right-hand side:

$$F(\xi_1, +\infty) = P(x_1 < \xi_1) =: F_1(\xi_1) \qquad (2.109)$$

Therefore, in the limiting case $\xi_2 = +\infty$, the distribution function $F(\xi_1, \xi_2)$ of a two-dimensional random variable $\boldsymbol{x} = [x_1, x_2]'$ is converted to the distribution function $F_1(\xi_1)$ of the one-dimensional random variable $x_1$. However, there must also be a connection between the associated density functions. To derive this connection we express the distribution function $F(\xi_1, +\infty)$ by the density function $f(\xi_1, \xi_2)$, according to equation (2.105):

$$F_1(\xi_1) = F(\xi_1, +\infty) = \int_{-\infty}^{\xi_1} \int_{-\infty}^{+\infty} f(v_1, v_2)\,\mathrm{d}v_1\,\mathrm{d}v_2 \tag{2.110}$$

or

$$F_1(\xi_1) = \int_{-\infty}^{\xi_1} \left\{ \int_{-\infty}^{+\infty} f(v_1, v_2)\,\mathrm{d}v_2 \right\} \mathrm{d}v_1$$

By differentiating this equation with respect to the upper limit of the outer integral, we obtain the density function

$$f_1(\xi_1) := \frac{\mathrm{d}}{\mathrm{d}\xi_1} F_1(\xi_1) \tag{2.111}$$

for the distribution function $F_1(\xi_1)$ as

$$f_1(\xi_1) = \int_{-\infty}^{+\infty} f(\xi_1, v_2)\,\mathrm{d}v_2$$

or, if the integration variable $v_2$ is replaced by $\xi_2$,

$$f_1(\xi_1) = \int_{-\infty}^{+\infty} f(\xi_1, \xi_2)\mathrm{d}\xi_2 \tag{2.112}$$

This result allows us to express $F_1(\xi_1)$ in terms of $f_1(\xi_1)$ as an inversion of (2.109):

$$F_1(\xi_1) = \int_{-\infty}^{\xi_1} f_1(v_1)\,\mathrm{d}v_1 \tag{2.113}$$

If $F(+\infty, \xi_2)$ is considered instead of $F(\xi_1, \infty)$, we arrive at completely analogous results:

$$F(+\infty, \xi_2) = P(x_2 < \xi_2) = \int_{-\infty}^{+\infty} \int_{-\infty}^{\xi_2} f(v_1, v_2)\,\mathrm{d}v_1\,\mathrm{d}v_2 =: F_2(\xi_2) \tag{2.114}$$

$$f_2(\xi_2) := \frac{d}{d\xi_2} F_2(\xi_2) = \int_{-\infty}^{+\infty} f(\xi_1, \xi_2) d\xi_1 \tag{2.115}$$

$$F_2(\xi_2) = \int_{-\infty}^{\xi_2} f_2(v_2) dv_2 \tag{2.116}$$

Again, it is assumed that the integrals with one or two integration limits of infinity converge.

The two probability distributions described by the functions $F_1(\xi_1)$, $f_1(\xi_1)$ or $F_2(\xi_2)$, $f_2(\xi_2)$ are called the *marginal distributions* of the one-dimensional random variables $x_1$ or $x_2$ with respect to the two-dimensional random variable $\boldsymbol{x} = [x_1, x_2]'$. In particular, $F(\xi_1, +\infty) = F_1(\xi_1)$ and $F(+\infty, \xi_2) = F_2(\xi_2)$ are called the *distribution functions* of the marginal distributions and $f_1(\xi_1)$ and $f_2(\xi_2)$ are called the *density functions* of the marginal distributions.

As with a two-dimensional random variable, marginal distributions can also be obtained for higher-dimensional random variables. Because in the distribution function $F(\xi_1, \xi_2, \ldots, \xi_n)$ of an $n$-dimensional random variable for $n > 2$, not only a single component of $\boldsymbol{\xi} = [\xi_1, \xi_2, \ldots, \xi_n]'$ can take a value of $+\infty$ but also 2, 3, ..., $n - 1$ components in the limiting case, the number of possible marginal distributions rapidly increases with increasing $n$.

### 2.4.6 Conditional Distribution Function and Conditional Density Function

So far the distribution functions $F(\xi) := P(x < \xi)$, $F(\xi_1, \xi_2) := P\{(x_1 < \xi_1) \cap (x_2 < \xi_2)\}$ *et cetera* have always been defined in terms of the ordinary probability of the random events $x < \xi$, $(x_1 < \xi_1) \cap (x_2 < \xi_2)$ *et cetera*. In this section we shall show that distribution functions and associated density functions based on conditional probability can be introduced. These are known as the *conditional distribution* and *conditional density functions*. Our considerations will be restricted to two-dimensional random variables. However, we will first recall the concept of conditional probability.

We can now use equation (2.104) to express the ordinary probabilities in the numerator and denominator of the conditional probability $P(x_1 < \xi_1 | \xi_2 \le x_2 < \xi_2 + \Delta\xi_2)$ in terms of the density function $f(\xi_1, \xi_2)$ of the random variable $\boldsymbol{x} = [x_1, x_2]'$. Thus we obtain

$$P(x_1 < \xi_1 | \xi_2 \le x_2 < \xi_2 + \Delta\xi_2) = \frac{\displaystyle\int_{-\infty}^{\xi_1} \int_{\xi_2}^{\xi_2+\Delta\xi_2} f(v_1, v_2) dv_1 dv_2}{\displaystyle\int_{-\infty}^{+\infty} \int_{\xi_2}^{\xi_2+\Delta\xi_2} f(\xi_1, v_2) d\xi_1 dv_2}$$

$$= \frac{\displaystyle\int_{-\infty}^{\xi_1} \left\{ \int_{\xi_2}^{\xi_2+\Delta\xi_2} f(v_1, v_2) \mathrm{d}v_2 \right\} \mathrm{d}v_1}{\displaystyle\int_{-\infty}^{+\infty} \left\{ \int_{\xi_2}^{\xi_2+\Delta\xi_2} f(\xi_1, v_2) \mathrm{d}v_2 \right\} \mathrm{d}\xi_1} \tag{2.117}$$

This expression is now converted to the conditional probability $P(x_1 < \xi_1 | x_2 = \xi_2)$ by applying the limiting process $\Delta\xi_2 \to 0$.

It is not possible to determine $P(x_1 < \xi_1 | x_2 = \xi_2)$ from equation (2.66), because for a continuous random variable $x_2$ we must always have $P(x_2 = \xi_2) = 0$ (see, for example, (2.87)). For small $\Delta\xi_2$ and continuous $f(\xi_1, \xi_2)$, the two inner integrals in (2.117) can be approximated by rectangular functions:

$$\int_{\xi_2}^{\xi_2+\Delta\xi_2} f(v_1, v_2) \mathrm{d}v_2 \approx f(v_1, \xi_2)\Delta\xi_2$$

$$\int_{\xi_2}^{\xi_2+\Delta\xi_2} f(\xi_1, v_2) \mathrm{d}v_2 \approx f(\xi_1, \xi_2)\Delta\xi_2$$

We therefore have

$$P(x_1 < \xi_1 | \xi_2 \leq x_2 < \xi_2 + \Delta\xi_2) \approx \frac{\Delta\xi_2 \displaystyle\int_{-\infty}^{\xi_1} f(v_1, \xi_2) \mathrm{d}v_1}{\Delta\xi_2 \displaystyle\int_{-\infty}^{+\infty} f(\xi_1, \xi_2) \mathrm{d}\xi_1}$$

Because the $\Delta\xi_2$ can be cancelled here and the approximation improves as $\Delta\xi_2$ becomes smaller, it follows that as $\Delta\xi_2 \to 0$:

$$\lim_{\Delta\xi_2 \to 0} P(x_1 < \xi_1 | \xi_2 \leq x_2 < \xi_2 + \Delta\xi_2) = \frac{\displaystyle\int_{-\infty}^{\xi_1} f(v_1, \xi_2) \mathrm{d}v_1}{\displaystyle\int_{-\infty}^{+\infty} f(\xi_1, \xi_2) \mathrm{d}\xi_1} =: F(\xi_1 | \xi_2) \tag{2.118}$$

This limiting value, the existence of which is assumed, defines the *conditional distribution function* $F(\xi_1|\xi_2)$ of the random variable $x_1$ for the condition $x_2 = \xi_2$.

According to equation (2.115), the denominator in (2.118) is the density function $f_2(\xi_2)$ of the marginal distribution of the one-dimensional random variable $x_2$ with respect to the two-dimensional random variable $\boldsymbol{x} = [x_1, x_2]'$. However, $f_2(\xi_2)$

does not depend on the integration variable $v_1$ so that it can be included in the integrand in the numerator of equation (2.118). Thus we have found an expression corresponding to equation (2.88):

$P(\mathrm{A}|\mathrm{B})$ is the probability that the random event A will occur if the random event B has already occurred. Equation (2.66)

$$P(\mathrm{A}|\mathrm{B}) := \frac{P(\mathrm{A} \cap \mathrm{B})}{P(\mathrm{B})} \quad \text{if} \quad P(\mathrm{B}) > 0$$

expresses the conditional probability in terms of the ordinary probability.

We will now consider a continuous two-dimensional random variable $\boldsymbol{x} = [x_1, x_2]'$ which can occupy all points in two-dimensional Euclidean space $\mathcal{R}_2$. It has the distribution function $F(\xi_1, \xi_2)$ and the density function $f(\xi_1, \xi_2)$. According to (2.66), and remembering that $\Omega \cap \mathrm{A} = \mathrm{A}$, the conditional probability $P(x_1 < \xi_1|\xi_2 \leqslant x_2 < \xi_2 + \Delta\xi_2)$ is given by

$$\begin{aligned} P(x_1 < \xi_1|\xi_2 \leq x_2 < \xi_2 + \Delta\xi_2) &= \frac{P\{(x_1 < \xi_1) \cap (\xi_2 \leq x_2 < \xi_2 + \Delta\xi_2)\}}{P(\xi_2 \leq x_2 < \xi_2 + \Delta\xi_2)} \\ &= \frac{P\{(x_1 < \xi_1) \cap (\xi_2 \leq x_2 < \xi_2 + \Delta\xi_2)\}}{P\{(x_1 < +\infty) \cap (\xi_2 \leq x_2 < \xi_2 + \Delta\xi_2)\}} \end{aligned}$$

As in equation (2.66), the above ratio is only meaningful if its denominator is non-zero.

$$F(\xi_1|\xi_2) = \int_{-\infty}^{\xi_1} \frac{f(v_1, \xi_2)}{f_2(\xi_2)} \mathrm{d}v_1 \tag{2.119}$$

The integrand in this equation at the upper integration limit gives the *conditional density function* $f(\xi_1|\xi_2)$ associated with $F(\xi_1|\xi_2)$:

$$f(\xi_1|\xi_2) := \frac{\partial}{\partial \xi_1} F(\xi_1|\xi_2) = \frac{f(\xi_1, \xi_2)}{f_2(\xi_2)} \tag{2.120}$$

In the same way we obtain the following relationships for the conditional distribution function $F(\xi_2|\xi_1)$ of the random variable $x_2$ under the condition $x_1 = \xi_1$:

$$\lim_{\Delta\xi_1 \to 0} P(x_2 < \xi_2|\xi_1 \leq x_1 < \xi_1 + \Delta\xi_1) = \frac{\displaystyle\int_{-\infty}^{\xi_2} f(\xi_1, v_2)\mathrm{d}v_2}{\displaystyle\int_{-\infty}^{+\infty} f(\xi_1, \xi_2)\mathrm{d}\xi_2} =: F(\xi_2|\xi_1) \tag{2.121}$$

$$F(\xi_2|\xi_1) = \int_{-\infty}^{\xi_2} \frac{f(\xi_1, v_2)}{f_1(\xi_1)} \, dv_2 \tag{2.122}$$

The conditional density function $f(\xi_2|\xi_1)$ associated with the conditional distribution function $F(\xi_2|\xi_1)$ is given by

$$f(\xi_2|\xi_1) := \frac{\partial}{\partial \xi_2} F(\xi_2|\xi_1) = \frac{f(\xi_1, \xi_2)}{f_1(\xi_1)} \tag{2.123}$$

### 2.4.7 Independent Random Variables

The concept of the independence of two random events introduced in Section 2.3.4 will now be transferred to random variables. The necessary and sufficient condition for the independence of two random events A and B (i.e., the probability that one of these events will occur is independent of whether the other event has occurred) is given by (equation (2.75))

$$P(\mathrm{A} \cap \mathrm{B}) = P(\mathrm{A}) \, P(\mathrm{B})$$

We will now consider a continuous two-dimensional random variable $\boldsymbol{x} = [x_1, x_2]'$ with distribution function $F(\xi_1, \xi_2)$, density function $f(\xi_1, \xi_2)$, marginal distribution functions $F_1(\xi_1)$, $F_2(\xi_2)$ and marginal density functions $f_1(\xi_1)$, $f_2(\xi_2)$. The two random variables $x_1$ and $x_2$ are independent of each other if the random events $(x_1 < \xi_1)$ and $(x_2 < \xi_2)$ are said to be *independent* of each other for all real $\xi_1$ and $\xi_2$. Then the distribution function $F(\xi_1, \xi_2)$ (equation (2.96)) for the independent random variables $x_1$ and $x_2$ is given by

$$F(\xi_1, \xi_2) = P\{(x_1 < \xi_1) \cap (x_2 < \xi_2)\} = P(x_1 < \xi_1) \, P(x_2 < \xi_2)$$

or, if the probabilities $P(x_1 < \xi_1)$ and $P(x_2 < \xi_2)$ are written as marginal distribution functions according to (2.109) and (2.114),

$$F(\xi_1, \xi_2) = F_1(\xi_1) F_2(\xi_2) \tag{2.124}$$

It follows from this and equations (2.115) and (2.111) that the density function

$$f(\xi_1, \xi_2) := \frac{\partial^2}{\partial \xi_1 \partial \xi_2} F(\xi_1, \xi_2)$$

of two independent random variables $x_1$ and $x_2$ is given by

$$f(\xi_1, \xi_2) = \frac{\partial}{\partial \xi_1}\left\{\frac{\partial}{\partial \xi_2} F_1(\xi_1)F_2(\xi_2)\right\} = \frac{\partial}{\partial \xi_1}\{F_1(\xi_1)f_2(\xi_2)\}$$

or

$$f(\xi_1, \xi_2) = f_1(\xi_1)f_2(\xi_2) \tag{2.125}$$

This initially appears to be only a necessary condition for the independence of $x_1$ and $x_2$. However, if we substitute equation (2.125) into (2.105) and use equations (2.116) and (2.113), we obtain

$$\begin{aligned} F(\xi_1, \xi_2) &= \int_{-\infty}^{\xi_1}\int_{-\infty}^{\xi_2} f_1(v_1)f_2(v_2)\mathrm{d}v_1\mathrm{d}v_2 \\ &= \int_{-\infty}^{\xi_1} f_1(v_1)\left\{\int_{-\infty}^{\xi_2} f_2(v_2)\mathrm{d}v_2\right\}\mathrm{d}v_1 \\ &= F_2(\xi_2)\int_{-\infty}^{\xi_1} f_1(v_1)\mathrm{d}v_1 = F_1(\xi_1)F_2(\xi_2) \end{aligned}$$

Therefore, since

$$F(\xi_1, \xi_2) = F_1(\xi_1)F_2(\xi_2)$$

is sufficient for $x_1$ and $x_2$ to be independent, and equation (2.125), which must hold for all real $\xi_1$ and $\xi_2$, is a necessary and sufficient condition for the independence of the random variables $x_1$ and $x_2$.

For independent random variables $x_1$ and $x_2$, equation (2.119) becomes

$$F(\xi_1|\xi_2) = \int_{-\infty}^{\xi_1} \frac{f_1(v_1)f_2(\xi_2)}{f_2(\xi_2)}\mathrm{d}v_1 = \int_{-\infty}^{\xi_1} f_1(v_1)\mathrm{d}v_1$$

The marginal distribution function $F_1(\xi_1)$ of (2.113) is on the right-hand side here, so that we have

$$F(\xi_1|\xi_2) = F_1(\xi_1) \tag{2.126}$$

if $x_1$ and $x_2$ are independent random variables. Furthermore, it follows from equation (2.120) that, for independent $x_1$ and $x_2$,

$$f(\xi_1|\xi_2) = \frac{f(\xi_1, \xi_2)}{f_2(\xi_2)} = \frac{f_1(\xi_1)f_2(\xi_2)}{f_2(\xi_2)} = f_1(\xi_1) \tag{2.127}$$

Therefore the conditional distribution function $F(\xi_1|\xi_2)$ and the conditional density function $f(\xi_1|\xi_2)$ for the independent random variables $x_1$ and $x_2$ do not depend on $\xi_2$.

Correspondingly, we also have

$$F(\xi_2|\xi_1) = F_2(\xi_2) \tag{2.128}$$

and

$$f(\xi_2|\xi_1) = f_2(\xi_2) \tag{2.129}$$

for the independent random variables $x_1$ and $x_2$. The random events $(\xi_1^{(a)} \leqslant x_1 < \xi_1^{(b)})$ and $(\xi_2^{(a)} \leqslant x_2 < \xi_2^{(b)})$ are also independent. This can easily be proved using (2.102) if we remember that (2.124) must hold for all real $\xi_1$ and $\xi_2$:

$$\begin{aligned} &P\{(\xi_1^{(a)} \leq x_1 < \xi_2^{(b)}) \cap (\xi_2^{(a)} \leq x_2 < \xi_2^{(b)})\} \\ &\quad = F(\xi_1^{(b)}, \xi_2^{(b)}) - F(\xi_1^{(a)}, \xi_2^{(b)}) - F(\xi_1^{(b)}, \xi_2^{(a)}) + F(\xi_1^{(a)}, \xi_2^{(a)}) \\ &\quad = F_1(\xi_1^{(b)})F_2(\xi_2^{(b)}) - F_1(\xi_1^{(a)})F_2(\xi_2^{(b)}) - F_1(\xi_1^{(b)})F_2(\xi_2^{(a)}) + F_1(\xi_1^{(a)})F_2(\xi_2^{(a)}) \\ &\quad = \{F_1(\xi_1^{(b)}) - F_1(\xi_1^{(a)})\}\{F_2(\xi_2^{(b)}) - F_2(\xi_2^{(a)})\} \end{aligned}$$

Application of equation (2.84) to the marginal distributions gives

$$\begin{aligned} &P\{(\xi_1^{(a)} \leq x_1 < \xi_1^{(b)}) \cap (\xi_2^{(a)} \leq x_2 < \xi_2^{(b)})\} \\ &\quad = P(\xi_1^{(a)} \leq x_1 < \xi_1^{(b)})\, P(\xi_2^{(a)} \leq x_2 < \xi_2^{(b)}) \end{aligned} \tag{2.130}$$

and hence the independence of the random events is proved.

The random variable components $x_1, x_2, \ldots, x_n$ of an $n$-dimensional random variable $\boldsymbol{x} = [x_1, x_2, \ldots, x_n]'$ with $n \geqslant 2$ are said to be independent of each other if, for any real values $\xi_1, \xi_2, \ldots, \xi_n$,

$$\begin{aligned} F(\xi_1, \xi_2, \ldots, \xi_n) &= F(\xi_1, +\infty, \ldots, +\infty)F(+\infty, \xi_2, +\infty, \ldots, +\infty) \ldots \\ &\quad \ldots F(+\infty, \ldots, +\infty, \xi_n) \\ &= F_1(\xi_1)F_2(\xi_2) \ldots F_n(\xi_n) \end{aligned} \tag{2.131a}$$

It follows from this that the associated density function of the $n$-dimensional random variable $\boldsymbol{x}$ is given by

$$f(\xi_1, \xi_2, \ldots, \xi_n) = f_1(\xi_1)f_2(\xi_2) \ldots f_n(\xi_n) \tag{2.131b}$$

If all $n$ components $x_1, \ldots, x_n$ of the vector random variable $\boldsymbol{x} = [x_1, \ldots, x_n]'$ are independent, then $s$ of the total of $n$ components ($s \leqslant n$) are independent (Fisz, 1988).

Two vector random variables $\boldsymbol{x} = [x_1, \ldots, x_n]'$ and $\boldsymbol{z} = [z_1, \ldots, z_m]'$ are said to be independent if, for all real values $\xi_1, \ldots, \xi_n, \zeta_1, \ldots, \zeta_m$,

$$F(\xi_1, \ldots, \xi_n, \zeta_1, \ldots, \zeta_m) = F_x(\xi_1, \ldots, \xi_n)F_z(\zeta_1, \ldots, \zeta_m) \qquad (2.132a)$$

$F$ is the distribution function of the vector random variable $[\boldsymbol{x}', \boldsymbol{z}']'$, and $F_x$ and $F_z$ are respectively the distribution functions of the marginal distributions of $\boldsymbol{x}$ with respect to $[\boldsymbol{x}', \boldsymbol{z}']'$ and $\boldsymbol{z}$ with respect to $[\boldsymbol{x}', \boldsymbol{z}']'$.

By taking the $(n + m)$th derivative with respect to $\xi_1, \ldots, \xi_n, \zeta_1, \ldots, \zeta_m$ for the corresponding density functions, we obtain the following relationship from equation (2.132a):

$$f(\xi_1, \ldots, \xi_n, \zeta_1, \ldots, \zeta_m) = f_x(\xi_1, \ldots, \xi_n)f_z(\zeta_1, \ldots, \zeta_m) \qquad (2.132b)$$

If the two vector random variables $\boldsymbol{x}$ and $\boldsymbol{z}$ are independent of each other, then every vector or scalar random variable which consists of $s$ of the total of $n$ components of $\boldsymbol{x}$ ($1 \leqslant s < n$) is also independent of every vector or scalar random variable which consists of $r$ of the total of $m$ components of $\boldsymbol{z}$ ($1 \leqslant r < m$). This statement can easily be proved by a corresponding integration of (2.132b), i.e., an integration in the same sense in which, for example, the density function $f_1(\xi_1)$ was derived from the density function $f(\xi_1, \xi_2)$ in (2.112).

Because $\boldsymbol{x}$ and $\boldsymbol{z}$ are independent, every component of $\boldsymbol{x}$ must be independent of every component of $\boldsymbol{z}$. However, the independence of $\boldsymbol{x}$ and $\boldsymbol{z}$ does not necessarily imply that all the components of $\boldsymbol{x}$ are independent of each other or that all the components of $\boldsymbol{z}$ are independent of each other. In other words, the validity of equation (2.132a) implies, for example, the independence of $x_1$ and $z_2$ but not the independence of $x_1$ and $x_2$.

### 2.4.8 The Normal Distribution (Gaussian Distribution)

A continuous scalar random variable $x$ has a normal or Gaussian distribution when its density function is given by

$$f(\xi) = \frac{1}{(2\pi)^{1/2}\sigma} \exp\left\{ -\frac{1}{2}\left(\frac{\xi - m}{\sigma}\right)^2 \right\} \qquad (2.133)$$

where $m$ is any real number and $\sigma$ is a positive real number ($\sigma \neq 0$). The density function $f(\xi)$ is symmetric with respect to the straight line $\xi = m$ and has a bell-shaped curve (Figure 2.21) with a maximum of $1/(2\pi)^{1/2}\sigma$ at $\xi = m$. The normal

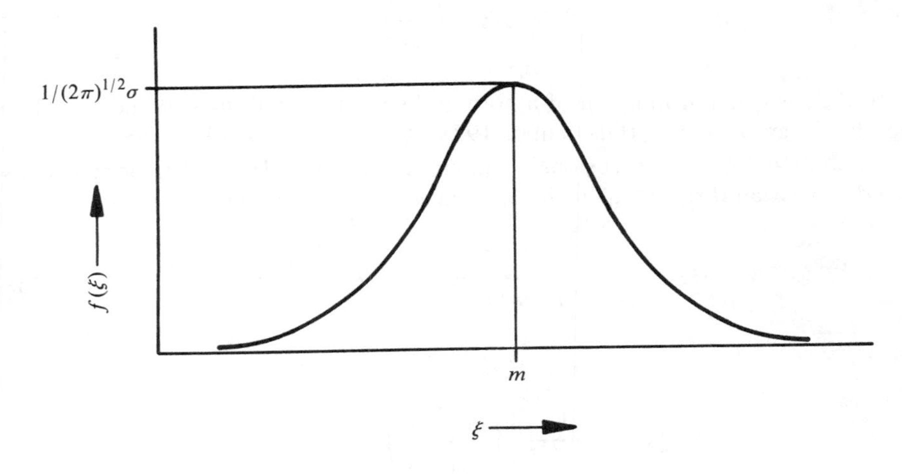


**Figure 2.21** Density function $f(\xi)$ of the normal distribution.

distribution is of great importance in probability theory because in many practical problems the random variables are normally distributed or can be approximated by a normal distribution. If a random variable is known to be normally distributed, its associated density function $f(\xi)$ does not have to be determined point by point. It is quite sufficient to determine the two parameters $m$ and $\sigma$ which uniquely define $f(\xi)$ through equation (2.133).

The integral involved in calculating the distribution function

$$F(\xi) = \frac{1}{(2\pi)^{1/2}\sigma} \int_{-\infty}^{\xi} \exp\left\{ -\frac{1}{2}\left(\frac{v - m}{\sigma}\right)^2 \right\} dv \tag{2.134}$$

from the density function $f(\xi)$ using equation (2.88) can only be solved in the closed form for $\xi = +\infty$. For $\xi < +\infty$, the substitution

$$\frac{v - m}{\sigma} = u \tag{2.135}$$

is used to reduce $F(\xi)$ to

$$\phi(z) = \frac{1}{(2\pi)^{1/2}} \int_{-\infty}^{z} \exp\left( -\frac{1}{2} u^2 \right) du \tag{2.136}$$

where

$$z = \frac{\xi - m}{\sigma} \tag{2.137}$$

i.e., to the distribution function of a normal distribution with $m = 0$ and $\sigma = 1$ for which tables are available (Gnedenko, 1978; Kreyszig, 1979; Fisz, 1988).

A continuous two-dimensional random variable $\boldsymbol{x} = [x_1, x_2]'$ is normally distributed (Gaussian distributed) if the associated density function

$$f(\xi_1, \xi_2) = \frac{1}{2\pi\sigma_1\sigma_2(1 - \rho^2)^{1/2}} \exp\left\{ -\frac{1}{2} Q(\xi_1, \xi_2) \right\} \tag{2.138}$$

where

$$Q(\xi_1, \xi_2) = \frac{1}{1 - \rho^2} \left\{ \left( \frac{\xi_1 - m_1}{\sigma_1} \right)^2 - 2\rho \frac{(\xi_1 - m_1)(\xi_2 - m_2)}{\sigma_1\sigma_2} + \left( \frac{\xi_2 - m_2}{\sigma_2} \right)^2 \right\} \tag{2.139}$$

holds; $f(\xi_1, \xi_2)$ is uniquely defined by the five parameters $m_1$, $m_2$, $\sigma_1$, $\sigma_2$ and $\rho$ where the following relationship exists between $\sigma_1$, $\sigma_2$ and $\rho$:

$$\rho = \frac{\mu_{11}}{\sigma_1\sigma_2} \tag{2.140}$$

The meanings of all these parameters as well as the meanings of the parameters $m$ and $\sigma$ in the density function $f(\xi)$ of the one-dimensional normal distribution will be explained in Section 2.5.

To ensure that $f(\xi_1, \xi_2)$ is a density function with the property $f(\xi_1, \xi_2) \geqslant 0$, we must have $\sigma_1\sigma_2 > 0$. The factor in front of the exponential function in equation (2.138) is chosen so that the second requirement for $f(\xi_1, \xi_2)$, i.e.,

$$\int_{-\infty}^{+\infty} \int_{-\infty}^{+\infty} f(\xi_1, \xi_2) d\xi_1 d\xi_2 = 1$$

is satisfied. The same conditions must be applied to the density function $f(\xi)$ for the one-dimensional normal distribution. Furthermore, we have $-1 < \rho < 1$.

For the special case $\rho = 0$, the density function $f(\xi_1, \xi_2)$ of the two-dimensional normal distribution can be split into a product in which one factor depends only on $\xi_1$ and the other depends only on $\xi_2$:

$$f(\xi_1, \xi_2) = \frac{1}{(2\pi)^{1/2}\sigma_1}\exp\left\{-\frac{1}{2}\left(\frac{\xi_1 - m_1}{\sigma_1}\right)^2\right\}\frac{1}{(2\pi)^{1/2}\sigma_2}\exp\left\{-\frac{1}{2}\left(\frac{\xi_2 - m_2}{\sigma_2}\right)^2\right\}$$
$$= f_1(\xi_1)f_2(\xi_2) \tag{2.141}$$

Therefore $\rho = 0$ implies that the two random variables $x_1$ and $x_2$ are independent of each other and are also normally distributed.

We will now rearrange equations (2.138) and (2.139) so as to make it easy to generalize them in the $n$-dimensional normal distribution. To do so, the parameters $\sigma_1$, $\sigma_2$ and $\mu_{11}$ of $f(\xi_1, \xi_2)$ are combined to form a matrix:

$$\boldsymbol{N} = \begin{bmatrix} \sigma_1^2 & \mu_{11} \\ \mu_{11} & \sigma_2^2 \end{bmatrix} \tag{2.142}$$

Then, using (2.140), equations (2.138) and (2.139) become

$$f(\xi_1, \xi_2) = \frac{1}{2\pi|\boldsymbol{N}|^{1/2}}\exp\left\{-\frac{1}{2}Q(\xi_1, \xi_2)\right\}$$

and

$$Q(\xi_1, \xi_2) = \frac{1}{|\boldsymbol{N}|}\{\sigma_2^2(\xi_1 - m_1)^2 - 2\mu_{11}(\xi_1 - m_1)(\xi_2 - m_2) + \sigma_1^2(\xi_2 - m_2)^2\}$$

where $|\boldsymbol{N}|$ is the determinant of the matrix $\boldsymbol{N}$. We can express $Q(\xi_1, \xi_2)$ in a more compact form by using the inverse matrix:

$$\boldsymbol{N}^{-1} = \frac{1}{|\boldsymbol{N}|}\begin{bmatrix} \sigma_2^2 & -\mu_{11} \\ -\mu_{11} & \sigma_1^2 \end{bmatrix}$$

and combining the variables $\xi_1$ and $\xi_2$ to form the vector $\boldsymbol{\xi} = [\xi_1, \xi_2]'$ and the parameters $m_1$ and $m_2$ to form the vector $\boldsymbol{m} = [m_1, m_2]'$. We thus obtain

$$Q(\xi_1, \xi_2) = (\boldsymbol{\xi} - \boldsymbol{m})'\boldsymbol{N}^{-1}(\boldsymbol{\xi} - \boldsymbol{m})$$

Then the density function of the two-dimensional normal distribution can be written as

$$f(\xi_1, \xi_2) = \frac{1}{2\pi|\boldsymbol{N}|^{1/2}}\exp\left\{-\frac{1}{2}(\boldsymbol{\xi} - \boldsymbol{m})'\boldsymbol{N}^{-1}(\boldsymbol{\xi} - \boldsymbol{m})\right\} \tag{2.143}$$

Therefore, like $f(\xi), f(\xi_1, \xi_2)$ is uniquely defined by "two parameters" i.e., the vector $\boldsymbol{m}$ and the matrix $\boldsymbol{N}$.

Equation (2.143) can easily be extended to the general $n$-dimensional case. The $n$-dimensional random variable $\boldsymbol{x} = [x_1, x_2, \ldots, x_n]'$ is normally distributed if its density function is given by

$$f(\xi_1, \ldots, \xi_n) = \frac{1}{(2\pi)^{n/2}|\boldsymbol{N}|^{1/2}} \exp\left\{ -\frac{1}{2}(\boldsymbol{\xi} - \boldsymbol{m})'\boldsymbol{N}^{-1}(\boldsymbol{\xi} - \boldsymbol{m}) \right\} \tag{2.144}$$

where $\boldsymbol{\xi} = [\xi_1, \xi_2, \ldots, \xi_n]'$ is an $n$-dimensional variable vector, $\boldsymbol{m} = [m_1, \ldots, m_n]'$ is an $n$-dimensional parameter vector and $\boldsymbol{N}$ is an $n \times n$ parameter matrix which must be regular and its elements will be described in Section 2.5. It will also be shown in Section 2.5 that, from the definition of $\boldsymbol{N}$, equation (2.144) is valid not only for $n = 2, 3, \ldots$ but also for $n = 1$; i.e., it goes over to (2.133) for $n = 1$.

## 2.5 EXPECTATIONS

If, for a random experiment containing a one-dimensional random variable $x$, we know the distribution function $F(\xi)$ or its derivative, the density function $f(\xi)$, then, using equation (2.87), we can derive the distribution of the unit probability of the certain event over the axis of the random variable $x$. To do so, we can, for example, divide the $x$ axis into equally spaced intervals and calculate the probabilities of the random events defined by these intervals. This gives us the probability distribution of the random variable $x$, i.e., the random experiment is completely described in terms of probability theory.

In exactly the same way, a random experiment with a two-dimensional random variable is completely described if the probability distribution of the random variable $\boldsymbol{x} = [x_1, x_2]'$ is known over the entire $x_1, x_2$ plane. This can be done by using equation (2.102) or (2.104) if we know the distribution function $F(\xi_1, \xi_2)$ or the density function $f(\xi_1, \xi_2)$ respectively. The same treatment can be used for random experiments with three- or higher dimensional random variables.

In practical problems involving random events or random phenomena in general it is often unnecessary to know the complete probability distribution as it is possible to solve these problems by using certain characteristic values that define the distribution. In this section, such characteristic values are introduced as simple or complicated average values of the random variables. These average values are known as the *expectations* (or *moments*) of the random variables. Thus we have a similar situation to mechanics where, even if the spatial distribution of the mass of a body is unknown, many dynamic problems can be solved if the center of gravity and one or more moments of inertia are known.

## 2.5.1 Expectations for Scalar Random Variables

### *2.5.1.1 Expectation for a Discrete Random Variable*

The concept of expectation will be introduced for a random experiment with $m$ possible elementary events $\omega_1, \ldots, \omega_m$ to which the numerical values $\xi^{(1)}, \ldots \xi^{(m)}$ of the scalar random variable $x$ are permanently assigned by the mapping $\xi^{(i)} = \psi(\omega_i)$. It is assumed that $\psi$ is not a one-to-one function. In this case the $m$ values $\xi^{(1)}, \ldots, \xi^{(m)}$ need not all have different magnitudes. However, the $m$ elementary events should be numbered so that the first $m^* \leqslant m$ of the total of $m$ numerical values $\xi^{(1)}, \ldots, \xi^{(m)}$ are different.

Let the probabilities that the $m$ elementary events occur be $P(\omega_i)$, $i = 1, \ldots, m$, and the probabilities that the $m^*$ random events $x = \xi^{(i)}$, $i = 1, \ldots, m^*$, occur be $P(x = \xi^{(i)})$, $i = 1, \ldots, m^*$. Then, from Section 2.4.2, we have

$$P(\omega_i) \leqslant P(x = \xi^{(i)}) \quad i = 1, \ldots, m$$

Furthermore, according to equations (2.85a) and (2.85b) the following relationships hold:

$$\sum_{i=1}^{m} P(\omega_i) = 1$$

and

$$\sum_{i=1}^{m^*} P(x = \xi^{(i)}) = 1 \tag{2.145}$$

The random experiment with these properties is now performed $N$ times ($N \geqslant m$) with, for example, the following result:

$$\begin{aligned} x &= \xi^{(1)} && \text{exactly } N_1 \text{ times} \\ x &= \xi^{(2)} && \text{exactly } N_2 \text{ times} \\ & && \vdots \\ x &= \xi^{(m^*)} && \text{exactly } N_{m^*} \text{ times} \end{aligned}$$

The condition

$$N_1 + N_2 + \cdots + N_{m^*} = N$$

must also be obeyed because the random events $x = \xi^{(i)}$, $i = 1, \ldots, m^*$, are mutually exclusive pairwise.

The arithmetic mean for all $N$ experimental results is now obtained:

$$\bar{\xi} := \frac{1}{N}(\xi^{(1)}N_1 + \xi^{(2)}N_2 + \cdots + \xi^{(m^*)}N_{m^*}) = \frac{1}{N}\sum_{i=1}^{m^*} \xi^{(i)}N_i$$

or

$$\bar{\xi} := \sum_{i=1}^{m^*} \xi^{(i)} \frac{N_i}{N} \tag{2.146}$$

According to equation (2.27) each term in this sum contains the relative frequency

$$f_i(N) = \frac{N_i}{N}$$

of the random event $x = \xi^{(i)}$. However, the relative frequency $f_i(N)$ for a large number $N$ of experiments approximately coincides with the probability that the random event $x = \xi^{(i)}$ will occur:

$$f_i(N) \approx P(x = \xi^{(i)}) \quad i = 1, \ldots, m^*$$

Since this approximation improves as the number $N$ of experiments increases, we must have

$$\bar{\xi} = \sum_{i=1}^{m^*} \xi^{(i)} \frac{N_i}{N} \approx \sum_{i=1}^{m^*} \xi^{(i)} P(x = \xi^{(i)}) \tag{2.147}$$

if $N$ is not too small. The right-hand side of this relationship, into which the probability function $\tilde{f}(\xi)$ of equation (2.93) can still be introduced, is called the *mean value* or the *expectation* $E\{x\}$ of the discrete random variable $x$:

$$E\{x\} := \sum_{i=1}^{m^*} \xi^{(i)} \tilde{f}(\xi^{(i)}) \tag{2.148}$$

Relationship (2.147) which can now be written

$$\bar{\xi} \approx E\{x\}$$

is a "measurement prescription" for determining the expectation $E\{x\}$ for a discrete scalar random variable. However, an exact conceptual distinction must be made between the mean value $\bar{\xi}$ of the $N$ experimentally determined numerical values and the mean value $E\{x\}$ of the random variable $x$. Whereas $\bar{\xi}$ can take a different numerical value depending on how often the random experiment is performed, $E\{x\}$ is a fixed numerical value specific to the random experiment. Furthermore, it should be noted that $\bar{\xi}$ is not identical with the arithmetic mean

$$\frac{1}{m^*}\sum_{i=1}^{m^*} \xi^{(i)}$$

of the $m^*$ possible numerical values $\xi^{(1)}, \ldots, \xi^{(m^*)}$ of the random variable $x$. However, this does not mean that these two mean values cannot coincide in special cases.

Another mean value, in addition to the mean value $\bar{\xi}$, can be calculated from the above $N$ experimental results, and that is the mean square deviation of the $N$ numerical values of the random variable $x$ from the mean value $\bar{\xi}$:

$$\overline{(\xi - \bar{\xi})^2} := \frac{1}{N}\{(\xi^{(1)} - \bar{\xi})^2 N_1 + \cdots + (\xi^{(m^*)} - \bar{\xi})^2 N_{m^*}\} = \frac{1}{N}\sum_{i=1}^{m^*} (\xi^{(i)} - \bar{\xi})^2 N_i$$

or

$$\overline{(\xi - \bar{\xi})^2} := \sum_{i=1}^{m^*} (\xi^{(i)} - \bar{\xi})^2 \frac{N_i}{N} \tag{2.149}$$

This mean value can be approximated in the same way as $\bar{\xi}$:

$$\overline{(\xi - \bar{\xi})^2} = \sum_{i=1}^{m^*} (\xi^{(i)} - \bar{\xi})^2 \frac{N_i}{N} \approx \sum_{i=1}^{m^*} (\xi^{(i)} - E\{x\})^2 \tilde{f}(\xi^{(i)}) \tag{2.150}$$

The right-hand side of relationship (2.150) is called the *variance*, the *dispersion* or the *scattering* of the discrete random variable $x$. The positive square root of the variance is called the *standard deviation*. Some authors designate the standard deviation as the scattering, and therefore care should be taken when using the concept of scattering. Because the variance is equal to the mean value of $(x - E\{x\})^2$, it can also be written as the expectation $E\{(x - E\{x\})^2\}$:*

$$E\{(x - E\{x\})^2\} := \sum_{i=1}^{m^*} (\xi^{(i)} - E\{x\})^2 \tilde{f}(\xi^{(i)}) \tag{2.151}$$

*The variance is often denoted by the symbol $\sigma^2$ or $D^2(x)$.

If the random variable $x$ can take numerical values $\xi^{(i)}$ which are scattered far from the mean value $E\{x\}$, a large value is obtained for the variance. If, however, the numerical values $\xi^{(i)}$ are all near $E\{x\}$, the variance has a small value. The relationship

$$\overline{(\xi - \bar{\xi})^2} \approx E\{(x - E\{x\})^2\}$$

can be defined as a measurement prescription for determining the variance.

For a random experiment, the probability distribution determines which random events occur frequently and which occur infrequently if the experiment is performed $N$ times ($N \gg 1$). Then the two expectations $E\{x\}$ (equation (2.148)) and $E\{(x - E\{x\})^2\}$ (equation (2.151)) which, for a random experiment with one discrete random variable (which can only take $m^*$ different values), characterize the distribution of the $N$ experimental results over the $m^*$ numerical values $\xi^{(1)}, \ldots, \xi^{(m^*)}$ of the random variable $x$ through the relations

$$E\{x\} \approx \bar{\xi} \quad \text{and} \quad E\{(x - E\{x\})^2\} \approx \overline{(\xi - \bar{\xi})^2}$$

must also be characteristic values of the probability distribution for a discrete random variable.

We shall look a little more closely at these relationships (which are not only valid for discrete random variables) by using the above example which is representative of all other cases. To do so, we use the following analogy from mechanics. A rod of zero mass has point masses $m(\xi^{(1)}), m(\xi^{(2)}), \ldots, m(\xi^{(m^*)})$ spaced at intervals $\xi^{(1)}, \xi^{(2)}, \ldots, \xi^{(m^*)}$ from an arbitrarily chosen zero point (Figure 2.22). The center of mass or the center of gravity of these $m^*$ masses follows from the equation (Gerthsen *et al.*, 1986):

$$\xi^{(s)} = \frac{\sum_{i=1}^{m^*} \xi^{(i)}\, m(\xi^{(i)})}{\sum_{i=1}^{m^*} m(\xi^{(i)})}$$

This expression is now compared with (2.148) which, from equation (2.145), can also be written in the form:

$$E\{x\} = \frac{\sum_{i=1}^{m^*} \xi^{(i)} \tilde{f}(\xi^{(i)})}{\sum_{i=1}^{m^*} \tilde{f}(\xi^{(i)})}$$

if we also set

$$P(x = \xi^{(i)}) = \tilde{f}(\xi^{(i)}) \quad i = 1, \ldots, m^*$$

according to equation (2.93). By analogy with the equations for $\xi^{(s)}$ and $E\{x\}$, we can draw the following conclusion: if the probabilities $\tilde{f}(\xi^{(i)})$ are interpreted as the masses over the $x$ axis, the expectation $E\{x\}$ corresponds to the center of gravity of this arrangement. In exactly the same way as the center of gravity $\xi^{(s)}$ is a characteristic value defining the distribution of the masses $m(\xi^{(i)})$ over the $x$ axis, the expectation $E\{x\}$ of the discrete random variable $x$ characterizes the distribution of the probabilities $\tilde{f}(\xi^{(i)})$ over the $x$ axis. $E\{x\}$ is therefore not just the mean value of the random variable $x$, but is also a numerical value for the overall characterization of its probability distribution.

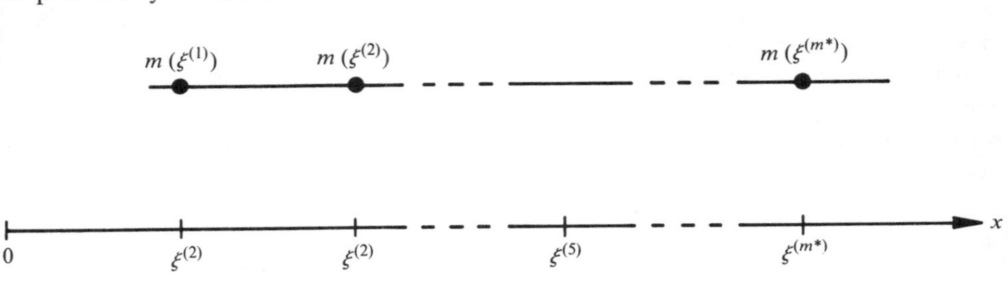


**Figure 2.22** Mechanical analogy for the probability function $\tilde{f}$ (ξ).

We will now determine the moment of inertia $\theta$ of the above system of $m^*$ masses $m(\xi^{(i)})$, $i = 1, \ldots, m^*$, about the center of gravity $\xi^{(s)}$. In Figure 2.22 the axis of rotation is perpendicular to the plane of the paper and passes through the rod with zero mass. Thus we have (Gerthsen *et al.*, 1986):

$$\theta = \sum_{i=1}^{m^*} (\xi^{(i)} - \xi^{(s)})^2 m(\xi^{(i)})$$

If this expression is compared with equation (2.151) it immediately becomes clear that the variance of the discrete random variable $x$ (i.e., the expectation $E\{(x - E\{x\})^2\}$ is also a characteristic value for the distribution of the probabilities $\tilde{f}(\xi^{(i)})$ over the $x$ axis. In other words it must be a parameter for identifying the probability distribution in exactly the same way as the moment of inertia $\theta$ in mechanics characterizes the distribution of masses.

Now that we have used the example of a discrete random variable which can only take a finite number of values to show that the expectations $E\{x\}$ and $E\{(x - E\{x\})^2\}$, which were introduced as mean values of the random variable, can also be used to characterize the probability distribution, we shall leave this simple example and introduce the expectation for continuous scalar random variables.

### *2.5.1.2 Expectation for a Continuous Random Variable*

Once more we examine the defining equation (2.148) for the expectation $E(x)$ of a discrete scalar random variable $x$:

$$E\{x\} := \sum_{i=1}^{m^*} \xi^{(i)} \tilde{f}(\xi^{(i)})$$

If the probability function $\tilde{f}(\xi)$, which is only meaningful for discrete random variables, can be expressed in terms of the density function $f(\xi)$, which is valid for both discrete and continuous random variables, the range of validity for the above defining equation can be extended to the case of continuous random variables. To do this $\xi^{(i)}$, which is a single value of all the possible numerical values $\xi$ which the random variable $x$ can take, is expressed in terms of $\xi$ using the masking property of the $\delta$ function:

$$\xi^{(i)} = \int_{-\infty}^{+\infty} \xi \delta(\xi - \xi^{(i)}) d\xi$$

Thus it follows that

$$E\{x\} = \sum_{i=1}^{m^*} \int_{-\infty}^{+\infty} \xi \delta(\xi - \xi^{(i)}) d\xi \tilde{f}(\xi^{(i)})$$

$$= \int_{-\infty}^{+\infty} \xi \left\{ \sum_{i=1}^{m^*} \tilde{f}(\xi^{(i)}) \delta(\xi - \xi^{(i)}) \right\} d\xi$$

If we now use the relationship

$$\tilde{f}(\xi^{(i)}) = P(x = \xi^{(i)}) \quad i = 1, \ldots, m^*$$

as has already been done several times in this section, we can see that the factor in braces in the integrand of the above integral is equal to the density function $f(\xi)$ of the discrete random variable $x$ given by equation (2.92). Thus we have

$$E\{x\} := \int_{-\infty}^{+\infty} \xi f(\xi) d\xi \tag{2.152}$$

This equation which, as its derivation shows, is just an alternative way of writing equation (2.148), for a continuous scalar random variable $x$ also gives a numerical value $E\{x\}$ which, if

$$\int_{-\infty}^{+\infty} |\xi| f(\xi) \mathrm{d}\xi < \infty \tag{2.153}$$

holds (Fisz, 1988), characterizes the probability distribution just as well as in the discrete case. Thus equation (2.152) is the general defining equation for the mean value or the expectation $E\{x\}$ of a scalar random variable $x$. Equation (2.152) contains (2.148) as a special case.

The defining equation (2.151) of the variance $E\{(x - E\{x\})^2\}$ of a discrete scalar random variable $x$ can be generalized in the same way as above. In this way we obtain

$$E\{(x - E\{x\})^2\} := \int_{-\infty}^{+\infty} (\xi - E\{x\})^2 f(\xi) \mathrm{d}\xi \tag{2.154}$$

as the generally valid defining equation of the variance of a scalar random variable $x$ or the expectation $E\{(x - E\{x\})^2\}$ provided that the integral is convergent. Equation (2.154) contains (2.151) as a special case.

We shall now use equations (2.152) and (2.154) to calculate the *mean value and variance of a normally distributed scalar random variable* $x$ with the density function (equation (2.133))

$$f(\xi) = \frac{1}{(2\pi)^{1/2}\sigma} \exp\left\{ -\frac{1}{2}\left(\frac{\xi - m}{\sigma}\right)^2 \right\}$$

It should be noted that the parameter $m$ in $f(\xi)$ is not identical with the number of elementary events in the random experiment considered at the beginning of this section.

(a)

$$E\{x\} = \int_{-\infty}^{+\infty} \xi f(\xi) \mathrm{d}\xi = \frac{1}{(2\pi)^{1/2}\sigma} \int_{-\infty}^{+\infty} \xi \exp\left\{ -\frac{1}{2}\left(\frac{\xi - m}{\sigma}\right)^2 \right\} \mathrm{d}\xi$$

When the substitution

$$\frac{\xi - m}{\sigma} = u$$

is made, this integral is converted into the sum of two integrals:

$$E\{x\} = \frac{1}{(2\pi)^{1/2}} \left\{ \sigma \int_{-\infty}^{+\infty} u \exp\left(-\frac{1}{2}u^2\right) du + m \int_{-\infty}^{+\infty} \exp\left(-\frac{1}{2}u^2\right) du \right\}$$

The first integral vanishes because its integrand is an odd function. From tables the value of the second integral is

$$\int_{-\infty}^{+\infty} \exp\left(-\frac{1}{2}u^2\right) du = (2\pi)^{1/2}$$

Therefore we have

$$E\{x\} = m$$

(b)

$$E\{(x - E\{x\})^2\} = \int_{-\infty}^{+\infty} (\xi - E\{x\})^2 f(\xi) d\xi$$

$$= \frac{1}{(2\pi)^{1/2}\sigma} \int_{-\infty}^{+\infty} (\xi - m)^2 \exp\left\{-\frac{1}{2}\left(\frac{\xi - m}{\sigma}\right)^2\right\} d\xi$$

When the same substitution as above is made, this expression becomes

$$E\{(x - E\{x\})^2\} = \frac{\sigma^2}{(2\pi)^{1/2}} \int_{-\infty}^{+\infty} u^2 \exp\left(-\frac{1}{2}u^2\right) du$$

From tables we obtain

$$\int_{-\infty}^{+\infty} u^2 \exp\left(-\frac{1}{2}u^2\right) du = (2\pi)^{1/2}$$

and therefore we have

$$E\{(x - E\{x\})^2\} = \sigma^2$$

The two parameters $m$ and $\sigma$ which unambiguously define the density function $f(\xi)$ of a normally distributed scalar random variable $x$ are therefore identical with the mean value and the square root of the variance of $x$. The square root of the variance is the standard deviation.

#### 2.5.1.3 Expectation for a Random Variable g(x)

In introducing the variance $E\{(x - E\{x\})^2\}$ through the mean value of the squares of the distances $(\xi^{(i)} - E\{x\})^2$, $i = 1, \ldots, m^*$, the latter were tacitly interpreted as the numerical values of a random variable

$$z = (x - E\{x\})^2$$

Since every possible value $\xi^{(i)}$ of the random variable $x$ has a fixed numerical value

$$\zeta^{(i)} = (\xi^{(i)} - E\{x\})^2 \quad i = 1, \ldots, m^*$$

assigned to it and since the occurrence or nonoccurrence of $x = \xi^{(i)}$ is determined by chance, the occurrence or nonoccurrence of $z = \zeta^{(i)}$ must also be random. In other words, $z$ is actually a random variable and $\zeta^{(i)}$ is a numerical value of this random variable.

It should be noted that the mean value or the expectation $E\{z\}$ of the random variable $z$ which, according to equation (2.152), must be given by

$$E\{z\} = \int_{-\infty}^{+\infty} \zeta \bar{f}(\zeta) \mathrm{d}\zeta \tag{2.155}$$

where $\bar{f}(\zeta)$ is the density function associated with the random variable $z$, can be expressed in terms of the density function $f(\xi)$ of the random variable $x$. However, according to equation (2.154), we have

$$E\{z\} = E\{(x - E\{x\})^2\} = \int_{-\infty}^{+\infty} (\xi - E\{x\})^2 f(\xi) \mathrm{d}\xi$$

We will now determine whether this result can be generalized for the case when the random variable $z$ is a function $z = g(x)$ of the random variable $x$.

Since the density function $\bar{f}(\zeta)$ occurs in the expression for the expectation $E(z)$ (equation (2.155)), we must first try to express $\bar{f}(\zeta)$ in terms of the density function $f(\xi)$ of the random variable $x$ using the equation $z = g(x)$. It will be assumed that the assignment prescription $z = g(x)$ is a one-to-one correspondence; i.e., exactly one value $\zeta = g(\xi)$ of $z$ is associated with each value $\xi$ of $x$ and *vice versa*. Let the inverse function be $x = h(z)$ or, expressed in terms of the corresponding numerical values, $\xi = h(\zeta)$.

We now define a random event A which consists of all values $\xi$ for which $\xi^{(a)} \leq \xi < \xi^{(b)}$ is true. This random event A has a corresponding random event B which consists of all values $\zeta$ with $\zeta^{(a)} \leq \zeta < \zeta^{(b)}$ (Figure 2.23). If the relationships $\zeta^{(a)} = g(\xi^{(a)})$, $\zeta^{(b)} = g(\xi^{(b)})$ and $\xi^{(a)} = h(\zeta^{(a)})$, $\xi^{(b)} = h(\zeta^{(b)})$ are true, and if the random event

$\zeta^{(a)} \leqslant \zeta < \zeta^{(b)}$ consists only of points $\zeta$ to which a value $\xi$ is assigned by the relationship $\xi = h(\zeta)$, where $\xi^{(a)} \leqslant \xi < \xi^{(b)}$, and if in addition the random event $\xi^{(a)} \leqslant \xi < \xi^{(b)}$ consists only of points $\xi$ to which a value $\zeta$ is assigned using the relationship $\zeta = g(\xi)$, where $\zeta^{(a)} \leqslant \zeta < \zeta^{(b)}$, then if the random event A occurs, the random event B must also necessarily occur and *vice versa*. We therefore have

$$P(\mathrm{A}) = P(\mathrm{B})$$

Using equation (2.87) we can rewrite this as

$$P(\mathrm{A}) = \int_{\xi^{(a)}}^{\xi^{(b)}} f(\xi)\mathrm{d}\xi = \int_{\zeta^{(a)}}^{\zeta^{(b)}} \bar{f}(\zeta)\mathrm{d}\zeta = P(\mathrm{B})$$

Using the substitution

$$\xi = h(\zeta) \quad \mathrm{d}\xi = h'(\zeta)\mathrm{d}\zeta$$

we now introduce the variable $\zeta$ into the left-hand integral:

$$P(\mathrm{A}) = \int_{\zeta^{(a)}}^{\zeta^{(b)}} f\{h(\zeta)\}h'(\zeta)\mathrm{d}\zeta = \int_{\zeta^{(a)}}^{\zeta^{(b)}} \bar{f}(\zeta)\mathrm{d}\zeta = P(\mathrm{B})$$

This equation can be satisfied for any $f\{h(\zeta)\}$ provided that

$$\bar{f}(\zeta) = f\{h(\zeta)\}h'(\zeta) \tag{2.156}$$

It was assumed in the derivation of this result that the relationship $\zeta^{(a)} < \zeta^{(b)}$ follows from $\xi^{(a)} < \xi^{(b)}$ (see Figure 2.23). For a one-to-one function $\xi = h(\zeta)$, this means that $h'(\zeta) > 0$. This guarantees that the density function $\bar{f}(\zeta)$ satisfies condition (2.89), i.e., $\bar{f}(\zeta) \geqslant 0$.

However, if from $\xi^{(a)} < \xi^{(b)}$ that $\zeta^{(a)} > \zeta^{(b)}$, under the above assumptions $h'(\zeta) < 0$ is true and, because $P(\mathrm{A}) = P(\mathrm{B})$, we have

$$P(\mathrm{A}) = \int_{\xi^{(a)}}^{\xi^{(b)}} f(\xi)\mathrm{d}\xi = \int_{\zeta^{(b)}}^{\zeta^{(a)}} \bar{f}(\zeta)\mathrm{d}\zeta = P(\mathrm{B})$$

After substituting for $\xi$ as described above, this equation becomes

$$P(\mathrm{A}) = \int_{\zeta^{(a)}}^{\zeta^{(b)}} f\{h(\zeta)\}h'(\zeta)\mathrm{d}\zeta = \int_{\zeta^{(b)}}^{\zeta^{(a)}} \bar{f}(\zeta)\mathrm{d}\zeta = P(\mathrm{B})$$

Because $h'(\zeta) < 0$, if we also set

$$h'(\zeta) = -|h'(\zeta)|$$

it follows that

$$-\int_{\zeta^{(a)}}^{\zeta^{(b)}} f\{h(\zeta)\}|h'(\zeta)|\mathrm{d}\zeta = \int_{\zeta^{(b)}}^{\zeta^{(a)}} f\{h(\zeta)\}|h'(\zeta)|\mathrm{d}\zeta = \int_{\zeta^{(b)}}^{\zeta^{(a)}} \bar{f}(\zeta)\mathrm{d}\zeta$$

or

$$\bar{f}(\zeta) = f\{h(\zeta)\}|h'(\zeta)| \tag{2.157}$$

where $\bar{f}(\zeta) \geqslant 0$. Equation (2.157) contains (2.156) as a special case.

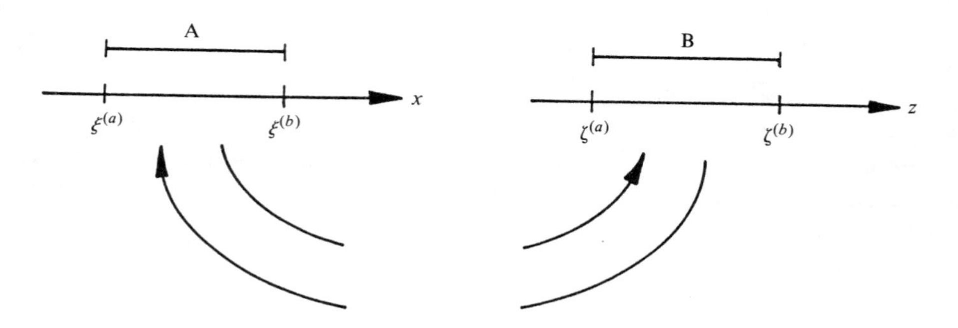


**Figure 2.23** Diagram for the derivation of equation (2.156).

Having obtained this result, we now return to equation (2.155) which can be rewritten as follows:

$$E\{z = g(x)\} = \int_{-\infty}^{+\infty} \zeta\bar{f}(\zeta)\mathrm{d}\zeta = \int_{-\infty}^{+\infty} \zeta f\{h(\zeta)\}|h'(\zeta)|\mathrm{d}\zeta \tag{2.158}$$

Then, using the relationships

$$\zeta = g(\xi) \quad \text{and} \quad \xi = h(\zeta) \quad \mathrm{d}\xi = h'(\zeta)\mathrm{d}\zeta$$

the integration variable $\zeta$ can be replaced by $\xi$, and again two cases have to be distinguished. If the values $\xi^{(1)} = -\infty$ and $\xi^{(2)} = +\infty$ correspond to $\zeta^{(1)} = -\infty$ and $\zeta^{(2)} = +\infty$ through $\xi = h(\zeta)$, we have $h'(\zeta) > 0$, i.e., $|h'(\zeta)| = h'(\zeta)$, and hence

$$E\{g(x)\} = \int_{-\infty}^{+\infty} g(\xi)f(\xi)\mathrm{d}\xi \tag{2.159}$$

However, if the values $\xi^{(1)} = +\infty$ and $\xi^{(2)} = -\infty$ correspond to $\zeta^{(1)} = -\infty$ and $\zeta^{(2)} = +\infty$ respectively, we must have $h'(\zeta) < 0$, i.e., $|h'(\zeta)| = -h'(\zeta)$, and (2.158) becomes

$$E\{g(x)\} = -\int_{+\infty}^{-\infty} g(\xi)f(\xi)h'(\zeta)\mathrm{d}\zeta = \int_{-\infty}^{+\infty} g(\xi)f(\xi)\mathrm{d}\xi$$

However, this is the same as equation (2.159) with which it was possible to show that it is feasible to calculate the mean value of the random variable $g(x)$ from the density function $f(\xi)$ of the random variable $x$.

If we assume that (Fisz, 1988):

$$\int_{-\infty}^{+\infty} |g(\xi)|f(\xi)\mathrm{d}\xi < \infty \tag{2.160}$$

the expectation $E\{g(x)\}$ represents the most general form of a characteristic value for identifying the probability distribution of a random experiment with one (continuous or discrete) scalar random variable $x$. This also holds when $z = g(x)$ is not a one-to-one function. $E\{g(x)\}$ contains the mean value and the variance of the random variable $x$ as special cases.

#### 2.5.1.4 Generalization of the Expectation of a Vector Random Variable

Consider a problem where, for example, $n$ random variables $g_1(x), \ldots, g_n(x)$ occur, which all depend on the same random variable $x$. If we combine these $n$ random variables to form a vector $\mathbf{g}(x)$

$$\mathbf{g}(x) := [g_1(x), \ldots, g_n(x)]' \tag{2.161}$$

and define

$$E\{\mathbf{g}(x)\} = E\left\{\begin{bmatrix} g_1(x) \\ \vdots \\ g_n(x) \end{bmatrix}\right\} := \begin{bmatrix} E\{g_1(x)\} \\ \vdots \\ E\{g_n(x)\} \end{bmatrix} \tag{2.162a}$$

the $n$ expectations $E\{g_1(x)\}, \ldots, E\{g_n(x)\}$ can be written in the abbreviated form $E\{\mathbf{g}(x)\}$. By simple renaming,

$$y_i = g_i(x) \quad i = 1, \ldots, n$$

equation (2.162a) becomes

$$E\{\mathbf{y}\} = E\left\{\begin{bmatrix} y_1 \\ \vdots \\ y_n \end{bmatrix}\right\} := \begin{bmatrix} E(y_1) \\ \vdots \\ E(y_n) \end{bmatrix} \tag{2.162b}$$

and hence the equation defining the expectation of a vector random variable $\mathbf{y} = [y_1, \ldots, y_n]'$ has been found.

#### 2.5.1.5 *kth-Order Moment, Central Moment, Ordinary Moment*

Equation (2.159) enables us to introduce in a simple manner some additional characteristics for defining the probability distribution of a random experiment with a scalar random variable $x$. For example, we define the mean value (expectation) of the random variable $g(x) = x^k$ as

$$E\{x^k\} = \int_{-\infty}^{+\infty} \xi^k f(\xi) \mathrm{d}\xi \tag{2.163}$$

where $k$, which is a nonnegative integer, is the *kth-order moment* of the random variable $x$. The zero-order moment, for which

$$E\{x^0\} = \int_{-\infty}^{+\infty} 1 \times f(\xi) \mathrm{d}\xi = 1$$

must hold (equation (2.90)), cannot be used to characterize the probability distribution because it always has the same value whatever the density function $f(\xi)$, i.e., whatever the probability distribution. According to equation (2.152) the first-order moment $E\{x\}$ of the random variable $x$ is identical with the mean value of the random variable $x$.

The mean value of the random variable $g(x) = (x - c)^k$ is designated as the $k$th-order moment of the random variable $x$ *with respect to the point* $c$:

$$E\{(x - c)^k\} = \int_{-\infty}^{+\infty} (\xi - c)^k f(\xi) \mathrm{d}\xi \tag{2.164}$$

where $c$ is an arbitrary constant. The moments with respect to the mean value of the random variable $x$, i.e., for $c = E\{x\}$, are particularly important. These mean values of the random variable $g(x) = (x - E\{x\})^k$ are called the *central moments of kth order* of the random variable $x$:

$$E\{(x - E\{x\})^k\} = \int_{-\infty}^{+\infty} (\xi - E\{x\})^k f(\xi)\mathrm{d}\xi \tag{2.165}$$

For the case $c = 0$ or $E\{x\} = 0$, both the moments with respect to the point $c$ and the central moments become the moments $E\{x^k\}$ of the random variable $x$. These moments $E\{x^k\}$ are also called *ordinary moments*.

We will now look a little more closely at the first three central moments of the random variable $x$. The zero- and first-order moments are given by

$$E\{(x - E\{x\})^0\} = \int_{-\infty}^{+\infty} 1 \times f(\xi)\mathrm{d}\xi = 1$$

$$E\{(x - E\{x\})^1\} = \int_{-\infty}^{+\infty} (\xi - E\{x\})f(\xi)\mathrm{d}\xi = \int_{-\infty}^{+\infty} \xi f(\xi)\mathrm{d}\xi - E\{x\} \int_{-\infty}^{+\infty} f(\xi)\mathrm{d}\xi$$

Because the mean value $E\{x\}$ of the random variable $x$ is a constant, $E\{x\}$ can be brought outside the integral. The first integral gives the mean value $E\{x\}$ of the random variable $x$ according to equation (2.163) or (2.152). From equation (2.90) the second integral is equal to unity. Thus the first-order central moment of the random variable $x$ can be expressed as

$$E\{(x - E\{x\})^1\} = E\{x\} - E\{x\} = 0$$

The second-order central moment is

$$\begin{aligned} E\{(x - E\{x\})^2\} &= \int_{-\infty}^{+\infty} (\xi - E\{x\})^2 f(\xi)\mathrm{d}\xi \\ &= \int_{-\infty}^{+\infty} (\xi^2 - 2E\{x\}\xi + E^2\{x\})f(\xi)\mathrm{d}\xi \\ &= \int_{-\infty}^{+\infty} \xi^2 f(\xi)\mathrm{d}\xi - 2E\{x\} \int_{-\infty}^{+\infty} \xi f(\xi)\mathrm{d}\xi + E^2\{x\} \int_{-\infty}^{+\infty} f(\xi)\mathrm{d}\xi \end{aligned}$$

Using (2.90) and (2.163) we obtain from this

$$E\{(x - E\{x\})^2\} = E\{x^2\} - 2E\{x\}E\{x\} + E^2\{x\} = E\{x^2\} - E^2\{x\}$$

The second-order central moment of the random variable $x$, which is identical with the variance of the random variable $x$, can therefore be expressed in terms of the ordinary moments of the random variable $x$. This result is also valid for higher-order central moments.

The central moments $E\{(x - E\{x\})^0\}$ and $E\{(x - E\{x\})^1\}$, like the moment $E\{x^0\}$, are unsuitable for characterizing the probability distribution of a random experiment with a scalar random variable $x$ because they always give the same numerical values regardless of the probability distribution.

In conclusion we recall that the absolute convergence of the relevant integrals is a necessary condition for the existence of all the moments described here (equation (2.160)).

#### *2.5.1.6 Absolute Moment, Remarks on Expectation*

In the literature the various moments possible for a one-dimensional random variable $x$ are often designated as follows:

$$E\{x^k\} = \alpha_k \quad \text{or} \quad m_k$$

$$E\{|x|^k\} = \int_{-\infty}^{+\infty} |\xi|^k f(\xi) \mathrm{d}\xi = \beta_k$$

$E\{|x|^k\}$ is the *absolute moment of kth-order* of the random variable $x$:

$$E\{(x - c)^k\} = \gamma_k$$

Finally

$$E\{(x - E\{x\})^k\} = \mu_k$$

We shall dispense with this multiplicity of symbols in favor of the notation $E\{\ \}$ because, according to equation (2.159), if the moments, the absolute moment, the central moment *et cetera* are designated as expectations, only a single symbol $E$ with corresponding "arguments" is necessary. The advantage of this method of designating moments is particularly obvious for vector random variables where, in addition to the type and order of the moment, the symbol must also indicate to which component or components of the vector random variable the moment in question belongs. This information can also be included in the argument of the expectation $E\{\ \}$.

Another reason for using the expectation as a means of identifying the various moments is provided by the calculation rules which will be derived in Section 2.5.4.

It is shown there that the expectations can be handled in a completely similar manner to the complex functions in the Laplace transformation. The analogy between the two becomes clear if equation (2.159) and the Laplace integral are juxtaposed:

$$E\{g(x)\} = \int_{-\infty}^{+\infty} g(\xi)f(\xi)\mathrm{d}\xi$$

$$\mathcal{L}\{g(t)\} = \int_{0}^{+\infty} g(t)\exp(-st)\mathrm{d}t = \int_{0}^{+\infty} g(\xi)\exp(-s\xi)\mathrm{d}\xi$$

In each case a new variable $E\{g(x)\}$ or $\mathcal{L}\{g(t)\}$ is assigned to a function $g(x)$ (the random variable $g(x)$ is a function of the random variable $x$ or $g(t)$ by means of an integral transformation. In the first case $E\{g(x)\}$ is a numerical value and in the second case, because of the parameter $s$ in the integrand, $\mathcal{L}\{g(t)\}$ is a function of $s$. The analogy between the last two equations becomes even clearer if the random variable $x$ characterizes a random process (a stochastic process). The density function $f(\xi)$ will in addition depend on a parameter, generally the time $t$, which is not an integration variable, and then the expectation $E\{g(x)\}$ will also be a function of $t$.

### 2.5.2 Expectations for Vector Random Variables

In this section we will first investigate the expectations of two-dimensional random variables. The results found in this way will then be transferred to $n$-dimensional random variables. Only continuous random variables will be considered.

#### *2.5.2.1 Expectation for a Two-Dimensional Random Variable*

The most general characteristic value for defining the probability distribution of a random experiment with a two-dimensional random variable $\boldsymbol{x} = [x_1, x_2]'$ is the mean value or the expectation of the random variable $z = g(x_1, x_2)$ for which there is a completely analogous relationship to equation (2.159):

$$E\{g(x_1, x_2)\} = \int_{-\infty}^{+\infty}\int_{-\infty}^{+\infty} g(\xi_1, \xi_2)f(\xi_1, \xi_2)\mathrm{d}\xi_1\mathrm{d}\xi_2 \tag{2.166}$$

if this double integral converges absolutely; i.e., if

$$\int_{-\infty}^{+\infty}\int_{-\infty}^{+\infty} |g(\xi_1, \xi_2)|f(\xi_1, \xi_2)\mathrm{d}\xi_1\mathrm{d}\xi_2 < \infty \tag{2.167}$$

Let the function $g(x_1, x_2)$ or $g(\xi_1, \xi_2)$ be a single-valued function; $f(\xi_1, \xi_2)$ is the density function of the random variable $\boldsymbol{x} = [x_1, x_2]'$.

In investigating the mean value of the random variable $z = g(x_1, x_2)$, we have also included the case where a *two*-dimensional random variable $\boldsymbol{z} = [z_1, z_2]'$ is assigned to the two-dimensional random variable $\boldsymbol{x}$ by

$$\boldsymbol{z} = \boldsymbol{g}(x_1, x_2) := \begin{bmatrix} g_1(x_1, x_2) \\ g_2(x_1, x_2) \end{bmatrix}$$

because for every component of $\boldsymbol{z}$ we again have $z_\nu = g_\nu(x_1, x_2)$.

As in the case of scalar random variables, we call the mean value

$$E\{x_1^{k_1} x_2^{k_2}\} = \int_{-\infty}^{+\infty} \int_{-\infty}^{+\infty} \xi_1^{k_1} \xi_2^{k_2} f(\xi_1, \xi_2) \mathrm{d}\xi_1 \mathrm{d}\xi_2 \qquad (2.168)$$

of the random variable $g(x_1, x_2) = x_1^{k_1} x_2^{k_2}$ the $(k_1 + k_2)$*th-order moment*[†] of the random variable $\boldsymbol{x} = [x_1, x_2]'$, where $k_1$ and $k_2$ are nonnegative integers. Here also, the zero-order moment has a value of unity because of (2.107):

$$E\{x_1^0 x_2^0\} = \int_{-\infty}^{+\infty} \int_{-\infty}^{+\infty} 1 \times 1 \times f(\xi_1, \xi_2) \mathrm{d}\xi_1 \mathrm{d}\xi_2 = 1$$

Hence it is independent of the density function $f(\xi_1, \xi_2)$ and so is also independent of the associated probability distribution, i.e., $E\{x_1^0, x_2^0\} = 1$ is useless as a characteristic value.

We will now examine the moments

$$E\{x_1^{k_1} x_2^0\} = \int_{-\infty}^{+\infty} \int_{-\infty}^{+\infty} \xi_1^{k_1} f(\xi_1, \xi_2) \mathrm{d}\xi_1 \mathrm{d}\xi_2$$
$$= \int_{-\infty}^{+\infty} \xi_1^{k_1} \left\{ \int_{-\infty}^{+\infty} f(\xi_1, \xi_2) \mathrm{d}\xi_2 \right\} \mathrm{d}\xi_1$$

more closely. Because, through equation (2.112), the integral in braces is equal to the density function $f_1(\xi_1)$ of the marginal distribution of the one-dimensional random

[†]The $(k_1 + k_2)$th-order moment is sometimes denoted $\alpha_{k_1 k_2}$ or $m_{k_1 k_2}$.

variable $x_1$ with respect to the two-dimensional random variable $\boldsymbol{x} = [x_1, x_2]'$, we have

$$E\{x_1^{k_1}x_2^0\} = \int_{-\infty}^{+\infty} \xi_1^{k_1} f_1(\xi_1)\mathrm{d}\xi_1$$

However, this is just the moment $E\{x_1^{k_1}\}$ of the random variable $x_1$ or, more explicitly, the moment $E\{x_1^{k_1}\}$ of the random variable $x_1$ which characterizes the marginal distribution of the one-dimensional random variable $x_1$ with respect to the two-dimensional random variable $\boldsymbol{x} = [x_1, x_2]'$. Therefore we have

$$E\{x_1^{k_1}x_2^0\} = \int_{-\infty}^{+\infty}\int_{-\infty}^{+\infty} \xi_1^{k_1} f(\xi_1, \xi_2)\mathrm{d}\xi_1\mathrm{d}\xi_2 = E\{x_1^{k_1}\} \tag{2.169}$$

It should be noted that the notation for the expectations anticipates this result. This is because the first factor in the integrand of the integral in equation (2.166), i.e., the function $g$, also occurs in the argument of the associated expectation. Thus the relationship

$$E\{x_1^0x_2^{k_2}\} = \int_{-\infty}^{+\infty}\int_{-\infty}^{+\infty} \xi_2^{k_2} f(\xi_1, \xi_2)\mathrm{d}\xi_1\mathrm{d}\xi_2 = E\{x_2^{k_2}\} \tag{2.170}$$

which corresponds to equation (2.169), must also apply. Then, from equations (2.169) and (2.170), we can derive the two first-order moments as follows:

$$E\{x_1^1x_2^0\} = \int_{-\infty}^{+\infty}\int_{-\infty}^{+\infty} \xi_1 f(\xi_1, \xi_2)\mathrm{d}\xi_1\mathrm{d}\xi_2 = E\{x_1\} \tag{2.171}$$

$$E\{x_1^0x_2^1\} = \int_{-\infty}^{+\infty}\int_{-\infty}^{+\infty} \xi_2 f(\xi_1, \xi_2)\mathrm{d}\xi_1\mathrm{d}\xi_2 = E\{x_2\} \tag{2.172}$$

Thus the two first-order moments of the two-dimensional random variable $\boldsymbol{x} = [x_1, x_2]'$ are identical with the mean values of the components $x_1$ and $x_2$ of the random variable $\boldsymbol{x} = [x_1, x_2]'$.

The mean value of the random variable

$$g(x_1, x_2) = (x_1 - E\{x_1^1x_2^0\})^{k_1}(x_2 - E\{x_1^0x_2^1\})^{k_2}$$

is known as the *central moment of order* $k_1 + k_2$ of the random variable $\boldsymbol{x} = [x_1, x_2]'$, as in the case of a one-dimensional random variable:

$$E\{(x_1 - E\{x_1^1x_2^0\})^{k_1}(x_2 - E\{x_1^0x_2^1\})^{k_2}\}$$
$$= \int_{-\infty}^{+\infty}\int_{-\infty}^{+\infty} (\xi_1 - E\{x_1^1x_2^0\})^{k_1}(\xi_2 - E\{x_1^0x_2^1\})^{k_2} f(\xi_1, \xi_2)\mathrm{d}\xi_1\mathrm{d}\xi_2 \quad (2.173)$$

This equation can be simplified using equations (2.171) and (2.172):

$$E\{(x_1 - E\{x_1\})^{k_1}(x_2 - E\{x_2\})^{k_2}\}$$
$$= \int_{-\infty}^{+\infty}\int_{-\infty}^{+\infty} (\xi_1 - E\{x_1\})^{k_1}(\xi_2 - E\{x_2\})^{k_2} f(\xi_1, \xi_2)\mathrm{d}\xi_1\mathrm{d}\xi_2 \quad (2.174)$$

The central moment of order $k_1 + k_2$ is equal to the ordinary moment of order $k_1 + k_2$ for $E\{x_1\} = E\{x_2\} = 0$.‡

Like the zero-order moment $E\{x_1^0x_2^0\}$, the zero-order central moment is also unsuitable as a characteristic value because, owing to (2.107),

$$E\{(x_1 - E\{x_1\})^0(x_2 - E\{x_2\})^0\} = \int_{-\infty}^{+\infty}\int_{-\infty}^{+\infty} 1 \times 1 \times f(\xi_1, \xi_2)\mathrm{d}\xi_1\mathrm{d}\xi_2 = 1$$

also applies here. Furthermore, equations which are completely analogous to those for the ordinary moments apply to the central moments:

$$E\{(x_1 - E\{x_1\})^{k_1}(x_2 - E\{x_2\})^0\} = \int_{-\infty}^{+\infty}\int_{-\infty}^{+\infty} (\xi_1 - E\{x_1\})^{k_1} f(\xi_1, \xi_2)\mathrm{d}\xi_1\mathrm{d}\xi_2$$
$$= E\{(x_1 - E\{x_1\})^{k_1}\} \quad (2.175)$$

$$E\{(x_1 - E\{x_1\})^0(x_2 - E\{x_2\})^{k_2}\} = \int_{-\infty}^{+\infty}\int_{-\infty}^{+\infty} (\xi_2 - E\{x_2\})^{k_2} f(\xi_1, \xi_2)\mathrm{d}\xi_1\mathrm{d}\xi_2$$
$$= E\{(x_2 - E\{x_2\})^{k_2}\} \quad (2.176)$$

These central moments describe the marginal distributions of the scalar random variables $x_1$ and $x_2$ with respect to the two-dimensional random variable $\boldsymbol{x} = [x_1, x_2]'$. Because

$$E\{(x - E\{x\})^1\} = 0$$

‡The central moment of order $k_1 + k_2$ is also denoted $\mu_{k_1k_2}$.

(see Section 2.5.1.5), the two first-order central moments of the two-dimensional random variable $\boldsymbol{x} = [x_1, x_2]'$ must also be zero:

$$E\{(x_1 - E\{x_1\})^1(x_2 - E\{x_2\})^0\} = E\{(x_1 - E\{x_1\})^1\} = 0$$
$$E\{(x_1 - E\{x_1\})^0(x_2 - E\{x_2\})^1\} = E\{(x_2 - E\{x_2\})^1\} = 0$$

The second-order central moments are particularly important for the characterization of the probability distribution. They can be obtained from (2.175) and (2.176) as

$$E\{(x_1 - E\{x_1\})^2(x_2 - E\{x_2\})^0\} = E\{(x_1 - E\{x_1\})^2\}$$

which is the *variance of the component* $x_1$ of the two-dimensional random variable $\boldsymbol{x} = [x_1, x_2]'$, and

$$E\{(x_1 - E\{x_1\})^0(x_2 - E\{x_2\})^2\} = E\{(x_2 - E\{x_2\})^2\}$$

which is the *variance of the component* $x_2$ of the two-dimensional random variable $\boldsymbol{x} = [x_1, x_2]'$. The square root of the variance $E\{(x_1 - E\{x_1\})^2\}$ or $E\{(x_2 - E\{x_2\})^2\}$ is called the *standard deviation** of the random variable $x_1$ or $x_2$ respectively.

Another second-order central moment, i.e., the expectation $E\{(x_1 - E\{x_1\})(x_2 - E\{x_2\})\}$, can be obtained for a two-dimensional random variable $\boldsymbol{x} = [x_1, x_2]'$ in addition to the two variances. This characteristic value is called the *covariance* of the random variables $x_1$ and $x_2$ and is abbreviated $\text{cov}(x_1, x_2)$:

$$\begin{aligned}\text{cov}(x_1, x_2) &:= E\{(x_1 - E\{x_1\})(x_2 - E\{x_2\})\} \\ &= \int_{-\infty}^{+\infty}\int_{-\infty}^{+\infty} (\xi_1 - E\{x_1\})(\xi_2 - E\{x_2\})f(\xi_1, \xi_2)\mathrm{d}\xi_1\mathrm{d}\xi_2 \end{aligned} \tag{2.177}$$

It follows immediately from this defining equation that

$$\text{cov}(x_1, x_2) = \text{cov}(x_2, x_1) \tag{2.178}$$

Moreover, equation (2.177) can be used to interpret the variances of the components $x_1$ and $x_2$ of a two-dimensional random variable $\boldsymbol{x} = [x_1, x_2]'$ as special cases of the

*The standard deviations of the random variables $x_1$ and $x_2$ are often denoted $\sigma_1$ and $\sigma_2$ and correspondingly the variances are denoted $\sigma_1^2$ and $\sigma_2^2$.

covariance:

$$\mathrm{cov}(x_1, x_1) = E\{(x_1 - E\{x_1\})^2\}$$
$$\mathrm{cov}(x_2, x_2) = E\{(x_2 - E\{x_2\})^2\}$$

It was shown in Section 2.5.1.5 that the variance (i.e., the second-order central moment of the scalar random variable $x$), can be expressed in terms of ordinary moments. In so doing, the equation

$$E\{(x - E\{x\})^2\} = E\{x^2\} - E^2\{x\}$$

was obtained. In exactly the same way, the variances of the components $x_1$ and $x_2$ of a two-dimensional random variable $\boldsymbol{x} = [x_1, x_2]'$ satisfy the relationships

$$E\{(x_1 - E\{x_1\})^2\} = E\{x_1^2\} - E^2\{x_1\} \tag{2.179}$$

$$E\{(x_2 - E\{x_2\})^2\} = E\{x_2^2\} - E^2\{x_2\} \tag{2.180}$$

The covariance $\mathrm{cov}(x_1, x_2)$ can also be written as the sum of ordinary moments. If the product of the first two factors in the integrand of equation (2.177) is multiplied out and the integral of the sum is split into the sum of integrals, it follows that $\mathrm{cov}(x_1, x_2)$ is given by

$$\begin{aligned} \mathrm{cov}(x_1, x_2) = & \int_{-\infty}^{+\infty}\int_{-\infty}^{+\infty} \xi_1\xi_2 f(\xi_1, \xi_2)\mathrm{d}\xi_1\mathrm{d}\xi_2 \\ & - E\{x_1\}\int_{-\infty}^{+\infty}\int_{-\infty}^{+\infty} \xi_2 f(\xi_1, \xi_2)\mathrm{d}\xi_1\mathrm{d}\xi_2 \\ & - E\{x_2\}\int_{-\infty}^{+\infty}\int_{-\infty}^{+\infty} \xi_1 f(\xi_1, \xi_2)\mathrm{d}\xi_1\mathrm{d}\xi_2 \\ & + E\{x_1\}E\{x_2\}\int_{-\infty}^{+\infty}\int_{-\infty}^{+\infty} f(\xi_1, \xi_2)\mathrm{d}\xi_1\mathrm{d}\xi_2 \end{aligned}$$

Using equations (2.107) and (2.168–2.170) we obtain from this

$$\begin{aligned} \mathrm{cov}(x_1, x_2) = & E\{x_1x_2\} - E\{x_1\}E\{x_2\} \\ & - E\{x_2\}E\{x_1\} + E\{x_1\}E\{x_2\} \times 1 \end{aligned}$$

or

$$\mathrm{cov}(x_1, x_2) = E\{(x_1 - E\{x_1\})(x_2 - E\{x_2\})\} = E\{x_1x_2\} - E\{x_1\}E\{x_2\} \quad (2.181)$$

Having introduced the characteristics which define the probability distribution for a two-dimensional random variable, we can now interpret the five parameters which define the density function of a normally distributed two-dimensional random variable (see equations (2.138) and (2.139)). The following relationships hold:

$$E\{x_1\} = \int_{-\infty}^{+\infty}\int_{-\infty}^{+\infty} \xi_1 f(\xi_1, \xi_2)\mathrm{d}\xi_1\mathrm{d}\xi_2 = m_1$$

$$E\{x_2\} = \int_{-\infty}^{+\infty}\int_{-\infty}^{+\infty} \xi_2 f(\xi_1, \xi_2)\mathrm{d}\xi_1\mathrm{d}\xi_2 = m_2$$

$$E\{(x_1 - E\{x_1\})^2\} = \int_{-\infty}^{+\infty}\int_{-\infty}^{+\infty} (\xi_1 - E\{x_1\})^2 f(\xi_1, \xi_2)\mathrm{d}\xi_1\mathrm{d}\xi_2 = \sigma_1^2$$

$$E\{(x_2 - E\{x_2\})^2\} = \int_{-\infty}^{+\infty}\int_{-\infty}^{+\infty} (\xi_2 - E\{x_2\})^2 f(\xi_1, \xi_2)\mathrm{d}\xi_1\mathrm{d}\xi_2 = \sigma_2^2$$

$$E\{(x_1 - E\{x_1\})(x_2 - E\{x_2\})\} = \int_{-\infty}^{+\infty}\int_{-\infty}^{+\infty} (\xi_1 - E\{x_1\})(\xi_2 - E\{x_2\}) f(\xi_1, \xi_2)\mathrm{d}\xi_1\mathrm{d}\xi_2 = \mu_{11}$$

A full derivation of these results is given by Heinhold and Gaede (1979). According to these relationships, the density function $f(\xi_1, \xi_2)$ of the two-dimensional normal distribution is uniquely defined by the two first-order moments and the three second-order central moments of the random variable $\boldsymbol{x} = [x_1, x_2]'$. In addition, according to equation (2.140), we also define the parameter

$$\rho = \frac{\mu_{11}}{\sigma_1\sigma_2}$$

which is known as the *correlation coefficient*. It characterizes the strength of the statistical dependence or independence between the two random variables $x_1$ and $x_2$. The matrix $\boldsymbol{N}$ introduced for the two-dimensional normal distribution (equation (2.142)) is known as the *covariance matrix*. The concept of the covariance matrix will be discussed further in Section 2.5.2.3.

### 2.5.2.2 Expectation for an n-Dimensional Random Variable

For an $n$-dimensional continuous random variable $\boldsymbol{x} = [x_1, \ldots, x_n]'$ with the density function $f(\xi_1, \ldots, \xi_n)$, the expectation

$$E\{g(x_1, \ldots, x_n)\} = \int_{-\infty}^{+\infty} \cdots \int_{-\infty}^{+\infty} g(\xi_1, \ldots, \xi_n) f(\xi_1, \ldots, \xi_n) \mathrm{d}\xi_1 \ldots \mathrm{d}\xi_n \quad (2.182\mathrm{a})$$

of the random variable $z = g(x_1, \ldots, x_n)$ is the most general characteristic value for defining the associated probability distribution if the $n$-fold integral is absolutely convergent. By making the substitutions

$$g(\boldsymbol{x}) := g(x_1, \ldots, x_n)$$
$$g(\boldsymbol{\xi}) := g(\xi_1, \ldots, \xi_n)$$
$$f(\boldsymbol{\xi}) := f(\xi_1, \ldots, \xi_n)$$

we can rewrite equation (2.182a) in a more compact form as

$$E\{g(\boldsymbol{x})\} = \int_{-\infty}^{+\infty} \cdots \int_{-\infty}^{+\infty} g(\boldsymbol{\xi}) f(\boldsymbol{\xi}) \mathrm{d}\xi_1 \ldots \mathrm{d}\xi_n \quad (2.182\mathrm{b})$$

Sometimes the notation

$$E\{g(\boldsymbol{x})\} = \int_{-\infty}^{+\infty} g(\boldsymbol{\xi}) f(\boldsymbol{\xi}) \mathrm{d}\boldsymbol{\xi} \quad (2.182\mathrm{c})$$

is used where the substitution

$$\mathrm{d}\boldsymbol{\xi} := \mathrm{d}\xi_1 \ldots \mathrm{d}\xi_n$$

has been made and the vector integration limits indicate the $n$-fold integration.

As in the case of a two-dimensional random variable, we obtain the moment of the $n$-dimensional random variable $\boldsymbol{x} = [x_1, \ldots, x_n]'$ from equation (2.182a) as the mean value or the expectation of the random variable $z = g(x_1, \ldots, x_n) = x_1^{k_1} \ldots x_n^{k_n}$:

$$E\{x_1^{k_1} \cdots x_n^{k_n}\} = \int_{-\infty}^{+\infty} \cdots \int_{-\infty}^{+\infty} \xi_1^{k_1} \cdots \xi_n^{k_n} f(\xi_1, \ldots, \xi_n) \mathrm{d}\xi_1 \cdots \mathrm{d}\xi_n \quad (2.183)$$

This is the (ordinary) *moment of order* $\Sigma_{\nu=1}^{n} k_\nu$. In this case also the corresponding moment can be taken as the characteristic value of a marginal distribution if at least one of the exponents $k_\nu$ is zero. For example, we have

$$E\{x_1^1 x_2^0 \cdots x_n^0\} = \int_{-\infty}^{+\infty} \cdots \int_{-\infty}^{+\infty} \xi_1 f(\xi_1, \ldots, \xi_n) \mathrm{d}\xi_1 \cdots \mathrm{d}\xi_n = E\{x_1^1\}$$

Then the expectation

$$E\{x_1^1, x_2^0, \ldots, x_n^0\} = E\{x_1^1\}$$

is the mean value of the component $x_1$ of the vector random variable $\boldsymbol{x} = [x_1, \ldots, x_n]'$, i.e., a characteristic value which describes the marginal distribution of the random variable $x_1$ with respect to the random variable $\boldsymbol{x}$. The same argument applies to the central moments of an $n$-dimensional random variable.

The mean value of the random variable

$$z = g(x_1, \ldots, x_n) = (x_1 - E\{x_1\})^{k_1} \ldots (x_n - E\{x_n\})^{k_n}$$

is known as the *central moment of order* $\Sigma_{\nu=1}^{n} k_\nu$ of the random variable $\boldsymbol{x} = [x_1, \ldots, x_n]'$:

$$\begin{aligned} &E\{(x_1 - E\{x_1\})^{k_1} \ldots (x_n - E\{x_n\})^{k_n}\} \\ &= \int_{-\infty}^{+\infty} \cdots \int_{-\infty}^{+\infty} (\xi_1 - E\{x_1\})^{k_1} \ldots \\ &(\xi_n - E\{x_n\})^{k_n} f(\xi_1, \ldots, \xi_n) \mathrm{d}\xi_1 \cdots \mathrm{d}\xi_n \end{aligned} \tag{2.184}$$

Two special cases of this, which will subsequently be used frequently, will be examined more closely.

(a) The first case,

$$\begin{aligned} &E\{(x_1 - E\{x_1\})^2 (x_2 - E\{x_2\})^0 \ldots (x_n - E\{x_n\})^0\} \\ &= \int_{-\infty}^{+\infty} \cdots \int_{-\infty}^{+\infty} (\xi_1 - E\{x_1\})^2 f(\xi_1, \ldots, \xi_n) \mathrm{d}\xi_1 \cdots \mathrm{d}\xi_n \\ &= E\{(x_1 - E\{x_1\})^2\} \end{aligned}$$

is the variance of the component $x_1$ of the random variable $\boldsymbol{x} = [x_1, \ldots, x_n]'$, i.e., the second-order central moment for characterizing the marginal distribution of the random variable $x_1$ with respect to the random variable $\boldsymbol{x} = [x_1, \ldots, x_n]'$. We obtain the variances of the other components of $\boldsymbol{x}$ in a similar manner.

(b) The second case,

$$E\{(x_1 - E\{x_1\})^1(x_2 - E\{x_2\})^1(x_3 - E\{x_3\})^0 \cdots (x_n - E\{x_n\})^0\}$$
$$= \int_{-\infty}^{+\infty} \cdots \int_{-\infty}^{+\infty} (\xi_1 - E\{x_1\})(\xi_2 - E\{x_2\}) f(\xi_1, \ldots, \xi_n) \mathrm{d}\xi_1 \cdots \mathrm{d}\xi_n$$
$$= E\{(x_1 - E\{x_1\})(x_2 - E\{x_2\})\}$$
$$= \operatorname{cov}(x_1, x_2)$$

is the covariance of the two components $x_1$ and $x_2$ of the random variable $\boldsymbol{x} = [x_1, \ldots, x_n]'$, i.e., the second-order central moment which characterizes the marginal distribution of the random variables $x_1$ and $x_2$ with respect to the random variable $\boldsymbol{x} = [x_1, \ldots, x_n]'$. In a corresponding manner, all covariances $\operatorname{cov}(x_i, x_j)$, $i \neq j$, $i, j = 1, \ldots, n$, possible here are defined in this way.

### *2.5.2.3 The Covariance Matrix*

We now introduce the covariance matrix, first for a two-dimensional random variable. To do so, as a generalization of the defining equation (2.162a), we first define the expectation of a matrix $\boldsymbol{G}(x_1, x_2)$ with elements $g_{ij}(x_1, x_2)$ as the matrix of the expectations $E\{g_{ij}(x_1, x_2)\}$:

$$E\{\boldsymbol{G}(x_1, x_2)\} = E\left\{\begin{bmatrix} g_{11}(x_1, x_2) & \ldots & g_{1m}(x_1, x_2) \\ \vdots & & \vdots \\ g_{n1}(x_1, x_2) & \ldots & g_{nm}(x_1, x_2) \end{bmatrix}\right\}$$
$$:= \begin{bmatrix} E\{g_{11}(x_1, x_2)\} & \ldots & E\{g_{1m}(x_1, x_2) \\ \vdots & & \vdots \\ E\{g_{n1}(x_1, x_2)\} & \ldots & E\{g_{nm}(x_1, x_2)\} \end{bmatrix} \quad (2.185)$$

Instead of using this explicit notation together with the generalization to the $n$-dimensional random variable $\boldsymbol{x} = [x_1, \ldots, x_n]'$, we can define the expectation of a matrix $\boldsymbol{G}(\boldsymbol{x}) = (g_{ij}(\boldsymbol{x}))$ in the abbreviated form:

$$E\{\boldsymbol{G}(\boldsymbol{x})\} = E\{(g_{ij}(\boldsymbol{x}))\}$$
$$:= (E\{g_{ij}(\boldsymbol{x})\}) \quad i = 1, \ldots, n, \quad j = 1, \ldots, m \quad (2.186a)$$

Here,

$$G(\boldsymbol{x}) := G(x_1, \ldots, x_n)$$

and

$$g_{ij}(\boldsymbol{x}) := g_{ij}(x_1, \ldots, x_n)$$

By simply changing symbols

$$y_{ij} = g_{ij}(\boldsymbol{x}) \quad i = 1, \ldots, n, \quad j = 1, \ldots, m$$

we obtain from equation (2.186a) the defining equation of the expectation of a matrix $Y = (y_{ij})$ the elements $y_{ij}$ of which are random variables:

$$\begin{aligned} E\{Y\} &= E\{(y_{ij})\} \\ &:= (E\{y_{ij}\}) \quad i = 1, \ldots, n, \quad j = 1, \ldots, m \end{aligned} \tag{2.186b}$$

This defining equation contains the defining equation (2.162b) as a special case.

Using equation (2.185) and the expression:

$$E\{\boldsymbol{x}\} = E\left\{\begin{bmatrix} x_1 \\ x_2 \end{bmatrix}\right\} = \left\{\begin{bmatrix} E\{x_1\} \\ E\{x_2\} \end{bmatrix}\right\}$$

according to equation (2.162b), we can combine all possible second-order central moments for a two-dimensional random variable $\boldsymbol{x} = [x_1, x_2]'$ to form a closed expression. For this, we will first consider the dyadic product:

$$\begin{aligned} (\boldsymbol{x} - E\{\boldsymbol{x}\})(\boldsymbol{x} - E\{\boldsymbol{x}\})' &= \begin{bmatrix} x_1 - E\{x_1\} \\ x_2 - E\{x_2\} \end{bmatrix} [x_1 - E\{x_1\}, x_2 - E\{x_2\}] \\ &= \begin{bmatrix} (x_1 - E\{x_1\})^2 & (x_1 - E\{x_1\})(x_2 - E\{x_2\}) \\ (x_2 - E\{x_2\})(x_1 - E\{x_1\}) & (x_2 - E\{x_2\})^2 \end{bmatrix} \end{aligned}$$

If the expectation is formed from this product, we obtain a matrix having elements that are all possible second-order central moments for a two-dimensional random variable $\boldsymbol{x} = [x_1, x_2]'$:

$$\begin{aligned} &E\{(\boldsymbol{x} - E\{\boldsymbol{x}\})(\boldsymbol{x} - E\{\boldsymbol{x}\})'\} \\ &= \begin{bmatrix} E\{(x_1 - E\{x_1\})^2\} & E\{(x_1 - E\{x_1\})(x_2 - E\{x_2\})\} \\ E\{(x_2 - E\{x_2\})(x_1 - E\{x_1\})\} & E\{(x_2 - E\{x_2\})^2\} \end{bmatrix} \end{aligned} \tag{2.187}$$

The main diagonal of this matrix contains the variances of the components of the vector $\boldsymbol{x} = [x_1, x_2]'$. The other matrix elements are the covariances $\text{cov}(x_1, x_2)$ and $\text{cov}(x_2, x_1)$ which, according to equation (2.178), must be equal. The above matrix is therefore symmetric. If we now recall that variances are special cases of the covariance, then the matrix (2.187) contains only covariances as elements. This matrix is therefore called the *covariance matrix* and is abbreviated as $\text{COV}(\boldsymbol{x}, \boldsymbol{x})$. Sometimes, $\text{COV}(\boldsymbol{x}, \boldsymbol{x})$ is also called the *moment matrix* or the *dispersion matrix*. The covariance matrix is defined as follows:

$$\text{COV}(\boldsymbol{x}, \boldsymbol{x}) := E\{(\boldsymbol{x} - E\{\boldsymbol{x}\})(\boldsymbol{x} - E\{\boldsymbol{x}\})'\} \tag{2.188}$$

So far the covariance matrix has only been considered for a two-dimensional random variable $\boldsymbol{x} = [x_1, x_2]'$. However, the defining equation (2.188) is also valid for an $n$-dimensional random variable $\boldsymbol{x} = [x_1, \ldots, x_n]'$:

$$\begin{aligned}
&\text{COV}(\boldsymbol{x}, \boldsymbol{x}): \\
&= E\{(\boldsymbol{x} - E\{\boldsymbol{x}\})(\boldsymbol{x} - E\{\boldsymbol{x}\})'\} \\
&= \begin{bmatrix} E\{(x_1 - E\{x_1\})^2\} & \ldots & E\{(x_1 - E\{x_1\})(x_n - E\{x_n\})\} \\ \vdots & & \vdots \\ E\{(x_n - E\{x_n\})(x_1 - E\{x_1\})\} & \ldots & E\{(x_n - E\{x_n\})^2\} \end{bmatrix}
\end{aligned} \tag{2.189}$$

Here, too, the matrix $\text{COV}(\boldsymbol{x}, \boldsymbol{x})$ contains all possible second-order central moments for the $n$-dimensional random variable $\boldsymbol{x} = [x_1, \ldots, x_n]'$. The variances of the $n$ components of $\boldsymbol{x}$ lie on the main diagonal, whereas all the possible covariances $\text{cov}(x_i, x_j)$, $i \neq j$, $i, j = 1, \ldots, n$ lie outside the main diagonal. As in the two-dimensional case, the covariance matrix $\text{COV}(\boldsymbol{x}, \boldsymbol{x})$ of an $n$-dimensional random variable $\boldsymbol{x}$ is a symmetric matrix.

The defining equation of $\text{COV}(\boldsymbol{x}, \boldsymbol{x})$ is considerably simplified if $E\{\boldsymbol{x}\} = \boldsymbol{0}$. In this case, equation (2.189) becomes

$$\text{COV}(\boldsymbol{x}, \boldsymbol{x}) = E\{\boldsymbol{x}\boldsymbol{x}'\} = \begin{bmatrix} E\{x_1^2\} & \cdots & E\{x_1 x_n\} \\ \vdots & & \vdots \\ E\{x_n x_1\} & \cdots & E\{x_n^2\} \end{bmatrix} \tag{2.190}$$

If the vector mean value $E\{\boldsymbol{x}\}$ for a vector random variable $\boldsymbol{x}$ is not equal to zero, we can use the linear transformation (substitution) $\boldsymbol{y} = \boldsymbol{x} - E\{\boldsymbol{x}\}$ to obtain from the random variable $\boldsymbol{x}$ a new random variable $\boldsymbol{y}$ of the same dimension for which

$$E\{\boldsymbol{y}\} = E\{\boldsymbol{x} - E\{\boldsymbol{x}\}\} = \boldsymbol{0}$$

because

$$E\{(x_i - E\{x_i\})^1\} = 0 \quad i = 1, \ldots, n$$

for every component $y_i = x_i - E\{x_i\}$, $i = 1, \ldots, n$, of $\boldsymbol{y}$. This can easily be seen from (2.184). Therefore the covariance matrix for the new random variable $\boldsymbol{y}$ can be expressed in the simpler form:

$$\mathrm{COV}(\boldsymbol{y}, \boldsymbol{y}) = E\{\boldsymbol{y}, \boldsymbol{y}'\}$$

In addition to the symmetry, another two properties of the covariance matrix $\mathrm{COV}(\boldsymbol{x}, \boldsymbol{x})$ will be considered.

(a) By analogy with the property

$$E\{(x - E\{x\})^2\} = E\{x^2\} - E^2\{x\}$$

of the variance of a one-dimensional random variable $x$ (see Section 2.5.1.5), the relationship

$$\mathrm{COV}(\boldsymbol{x}, \boldsymbol{x}) = E\{\boldsymbol{x}\boldsymbol{x}'\} - E\{\boldsymbol{x}\}E\{\boldsymbol{x}'\} \tag{2.191}$$

applies to the covariance matrix $\mathrm{COV}(\boldsymbol{x}, \boldsymbol{x})$ of an $n$-dimensional random variable $\boldsymbol{x}$. This equation will now be proved for the case of a two-dimensional random variable. The general proof will be given in Section 2.5.4, which deals with calculation rules for expectations, so that the computing requirements in the two cases can be compared.

For a two-dimensional random variable $\boldsymbol{x} = [x_1, x_2]'$ we have

$$\mathrm{COV}(\boldsymbol{x}, \boldsymbol{x}) = \begin{bmatrix} E\{(x_1 - E\{x_1\})^2\} & E\{(x_1 - E\{x_1\})(x_2 - E\{x_2\})\} \\ E\{(x_2 - E\{x_2\})(x_1 - E\{x_1\})\} & E\{(x_2 - E\{x_2\})^2\} \end{bmatrix}$$

From equations (2.179)–(2.181) this expression becomes

$$\begin{aligned} \mathrm{COV}(\boldsymbol{x}, \boldsymbol{x}) &= \begin{bmatrix} E\{x_1^2\} - E^2\{x_1\} & E\{x_1x_2\} - E\{x_1\}E\{x_2\} \\ E\{x_2x_1\} - E\{x_2\}E\{x_1\} & E\{x_2^2\} - E^2\{x_2\} \end{bmatrix} \\ &= \begin{bmatrix} E\{x_1x_1\} & E\{x_1x_2\} \\ E\{x_2x_1\} & E\{x_2x_2\} \end{bmatrix} - \begin{bmatrix} E\{x_1\}E\{x_1\} & E\{x_1\}E\{x_2\} \\ E\{x_2\}E\{x_1\} & E\{x_2\}E\{x_2\} \end{bmatrix} \end{aligned}$$

Then, according to the defining equation of the expectation of a matrix (equation (2.186a)), the first matrix in the above difference can be written as $E\{\boldsymbol{x}\boldsymbol{x}'\}$. The second matrix is identical with $E\{\boldsymbol{x}\}E\{\boldsymbol{x}'\}$. Thus we have proved equation (2.191) for a two-dimensional random variable $\boldsymbol{x} = [x_1, x_2]'$.

(b) The covariance matrix $\mathrm{COV}(\boldsymbol{x}, \boldsymbol{x})$ is positive semidefinite. To prove this statement we examine the quadratic form $\boldsymbol{u}'\mathrm{COV}(\boldsymbol{x}, \boldsymbol{x})\boldsymbol{u}$ of the symmetric matrix $\mathrm{COV}(\boldsymbol{x}, \boldsymbol{x})$ ($\boldsymbol{u} = [u_1, \ldots, u_n]'$ is an $n$-dimensional vector whose components do not depend on chance, i.e., they are not random variables):

$$\boldsymbol{u}'\mathrm{COV}(\boldsymbol{x}, \boldsymbol{x})\boldsymbol{u} = \boldsymbol{u}'E\{(\boldsymbol{x} - E\{\boldsymbol{x}\})(\boldsymbol{x} - E\{\boldsymbol{x}\})'\}\boldsymbol{u}$$

Because $\boldsymbol{u}$ is not a random variable, the right-hand side of this equation does not change when $\boldsymbol{u}'$ and $\boldsymbol{u}$ are included in the argument of the expectation (this statement will be proved in Section 2.5.4):

$$\boldsymbol{u}'\mathrm{COV}(\boldsymbol{x}, \boldsymbol{x})\boldsymbol{u} = E\{\boldsymbol{u}'(\boldsymbol{x} - E\{\boldsymbol{x}\})(\boldsymbol{x} - E\{\boldsymbol{x}\})'\boldsymbol{u}\}$$

Because, for the scalar product of two vectors,

$$\boldsymbol{u}'(\boldsymbol{x} - E\{\boldsymbol{x}\}) = (\boldsymbol{x} - E\{\boldsymbol{x}\})'\boldsymbol{u}$$

it follows that

$$\begin{aligned}\boldsymbol{u}'\mathrm{COV}(\boldsymbol{x}, \boldsymbol{x})\boldsymbol{u} &= E\{[\boldsymbol{u}'(\boldsymbol{x} - E\{\boldsymbol{x}\})]^2\} \\ &= E\{[u_1(x_1 - E\{x_1\}) + \cdots + u_n(x_n - E\{x_n\})]^2\}\end{aligned}$$

However, the expectation or the mean value of a square is always greater than or equal to zero. Therefore we have

$$\boldsymbol{u}'\mathrm{COV}(\boldsymbol{x}, \boldsymbol{x})\boldsymbol{u} \geqslant 0 \tag{2.192}$$

for any $\boldsymbol{u}$. The quadratic form $\boldsymbol{u}'\mathrm{COV}(\boldsymbol{x}, \boldsymbol{x})\boldsymbol{u}$ can become zero for $\boldsymbol{u} = \boldsymbol{0}$ or if

$$u_1(x_1 - E\{x_1\}) + \cdots + u_n(x_n - E\{x_n\}) = 0$$

i.e., if there is a stochastic dependence between the components of the random variable $\boldsymbol{x} = [x_1, \ldots, x_n]'$ corresponding to the preceding equation. This proves that the quadratic form $\boldsymbol{u}'\mathrm{COV}(\boldsymbol{x}, \boldsymbol{x})\boldsymbol{u}$ and thus the covariance matrix $\mathrm{COV}(\boldsymbol{x}, \boldsymbol{x})$ are positive semidefinite.

Now that we have considered these two properties of $COV(\boldsymbol{x}, \boldsymbol{x})$, the concept of the covariance matrix will be generalized a little more.

Just as the covariance matrix $COV(\boldsymbol{x}, \boldsymbol{x})$ of the vector random variable $\boldsymbol{x}$ can be formed, a covariance matrix $COV(\boldsymbol{x}, \boldsymbol{y})$ can be formed from two vector random variables $\boldsymbol{x} = [x_1, \ldots, x_n]'$ and $\boldsymbol{y} = [y_1, \ldots, y_m]'$:

$$COV(\boldsymbol{x}, \boldsymbol{y}) := E\{(\boldsymbol{x} - E\{\boldsymbol{x}\})(\boldsymbol{y} - E\{\boldsymbol{y}\})'\}$$
$$= \begin{bmatrix} E\{(x_1 - E\{x_1\})(y_1 - E\{y_1\})\} & \cdots & E\{(x_1 - E\{x_1\})(y_m - E\{y_m\})\} \\ \vdots & & \vdots \\ E\{(x_n - E\{x_n\})(y_1 - E\{y_1\})\} & \cdots & E\{(x_n - E\{x_n\})(y_m - E\{y_m\})\} \end{bmatrix} \quad (2.193)$$

Again, $COV(\boldsymbol{x}, \boldsymbol{y})$ can be substantially simplified if $E\{\boldsymbol{x}\} = \boldsymbol{0}$ and $E\{\boldsymbol{y}] = \boldsymbol{0}$. Then we have

$$COV(\boldsymbol{x}, \boldsymbol{y}) = E\{\boldsymbol{x}\,\boldsymbol{y}'\} = \begin{bmatrix} E\{x_1y_1\} & E\{x_1y_2\} & \cdots & E\{x_1y_m\} \\ E\{x_2y_1\} & E\{x_2y_2\} & \cdots & E\{x_2y_m\} \\ \vdots & \vdots & & \vdots \\ E\{x_ny_1\} & E\{x_ny_2\} & \cdots & E\{x_ny_m\} \end{bmatrix} \quad (2.194)$$

It should be noted that $COV(\boldsymbol{x}, \boldsymbol{y})$ contains no variances but all the covariances $cov(x_i, y_j)$, $i = 1, \ldots, n$, $j = 1, \ldots, m$, which can be formed from the elements of $\boldsymbol{x}$ and $\boldsymbol{y}$. Whereas $COV(\boldsymbol{x}, \boldsymbol{x})$ is always square and symmetric, $COV(\boldsymbol{x}, \boldsymbol{y})$ is generally not square and even for the special case where $m = n$ is not symmetric. This can easily be seen from the defining equation (2.194).

#### *2.5.2.4 Expectation with Independent and Uncorrelated Random Variables*

If the components $x_1$ and $x_2$ of a two-dimensional random variable $\boldsymbol{x} = [x_1, x_2]'$ are independent of each other, then the associated density function is given by (equation (2.125))

$$f(\xi_1, \xi_2) = f_1(\xi_1)f_2(\xi_2)$$

where $f_1(\xi_1)$ and $f_2(\xi_2)$ are the density functions of the marginal distributions of the one-dimensional random variables $x_1$ and $x_2$ with respect to the two-dimensional random variable $\boldsymbol{x} = [x_1, x_2]'$. It also follows from equation (2.125) that the moment of order $k_1 + k_2$ of the random variable $\boldsymbol{x} = [x_1, x_2]'$ is given by

$$E\{x_1^{k_1}x_2^{k_2}\} = \int_{-\infty}^{+\infty}\int_{-\infty}^{+\infty} \xi_1^{k_1}\xi_2^{k_2} f(\xi_1, \xi_2)\mathrm{d}\xi_1\mathrm{d}\xi_2$$

$$= \int_{-\infty}^{+\infty} \int_{-\infty}^{+\infty} \xi_1^{k_1} \xi_2^{k_2} f_1(\xi_1) f_2(\xi_2) \mathrm{d}\xi_1 \mathrm{d}\xi_2$$

$$= \int_{-\infty}^{+\infty} \xi_1^{k_1} f_1(\xi_1) \mathrm{d}\xi_1 \int_{-\infty}^{+\infty} \xi_2^{k_2} f_2(\xi_2) \mathrm{d}\xi_2$$

From equation (2.163), the two integrals in the final expression are the moments $E\{x_1^{k_1}\}$ and $E\{x_2^{k_2}\}$ of the random variables $x_1$ and $x_2$ respectively. Therefore

$$E\{x_1^{k_1} x_2^{k_2}\} = E\{x_1^{k_1}\} E\{x_2^{k_2}\} \tag{2.195}$$

if $x_1$ and $x_2$ are independent of each other. In the same way, the central moment of order $k_1 + k_2$ of the random variable $\boldsymbol{x} = [x_1, x_2]'$ is given by

$$E\{(x_1 - E\{x_1\})^{k_1} (x_2 - E\{x_2\})^{k_2}\} = E\{(x_1 - E\{x_1\})^{k_1}\} E\{(x_2 - E\{x_2\})^{k_2}\} \tag{2.196}$$

if $x_1$ and $x_2$ are independent of each other. Equations (2.195) and (2.196) are special cases of the relationship

$$E\{g_1(x_1) g_2(x_2)\} = E\{g_1(x_1)\} E\{g_2(x_2)\} \tag{2.197}$$

for two independent random variables $x_1$ and $x_2$. This equation can be derived from equation (2.166) by a similar procedure to that used to obtain equations (2.195) and (2.196) taking (2.125) into account.

We will now further consider equations (2.195) and (2.196), in particular, for $k_1 = k_2 = 1$. We have

$$E\{x_1 x_2\} = E\{x_1\} E\{x_2\} \tag{2.198}$$

and

$$E\{(x_1 - E\{x_1\})(x_2 - E\{x_2\})\} = E\{x_1 - E\{x_1\}\} E\{x_2 - E\{x_2\}\} \tag{2.199}$$

for independent $x_1$ and $x_2$. However, it has already been shown in Section 2.5.2.1 that

$$E\{x_1 - E\{x_1\}\} = E\{x_2 - E\{x_2\}\} = 0$$

i.e., the second-order central moment $E\{(x_1 - E\{x_1\})(x_2 - E\{x_2\})\}$ of the random variable $\boldsymbol{x} = [x_1, x_2]'$, which is the covariance $\mathrm{cov}(x_1, x_2)$, vanishes if $x_1$ and $x_2$ are independent random variables:

$$\mathrm{cov}(x_1, x_2) = E\{(x_1 - E\{x_1\})(x_2 - E\{x_2\})\} = 0 \tag{2.200}$$

This result could also have been derived from (2.181) and (2.198).

Therefore $\mathrm{cov}(x_1, x_2)$ for the independent random variables $x_1$ and $x_2$ is equal to zero. However, the converse that when $\mathrm{cov}(x_1, x_2)$ vanishes the two random variables $x_1$ and $x_2$ are independent is not necessarily true. There are examples (Cramer, 1965; Heinhold and Gaede, 1979) for which $\mathrm{cov}(x_1, x_2)$ vanishes but the random variables $x_1$ and $x_2$ are not independent. An exception to this is the two-dimensional normal distribution. As was shown in Section 2.4.8, the random variables $x_1$ and $x_2$ are independent and are also normally distributed if $\rho = 0$ (i.e., if $\mu_{11} = 0$). In Section 2.5.2.1 it was also shown that the parameter $\mu_{11}$ is just the covariance $\mathrm{cov}(x_1, x_2)$ for the two-dimensional normal distribution. Therefore it is possible to deduce the independence of the random variables $x_1$ and $x_2$ of the two-dimensional normal distribution from $\mathrm{cov}(x_1, x_2) = 0$. However, we must emphasize that, in general, this deduction is not valid.

We now say that if $\mathrm{cov}(x_1, x_2) = 0$ the random variables $x_1$ and $x_2$ are *uncorrelated*. However, it can be seen from (2.181) $\mathrm{cov}(x_1, x_2) = E\{x_1x_2\} - E\{x_1\}E\{x_2\}$ that for uncorrelated random variables $x_1$ and $x_2$ we must have

$$E\{x_1x_2\} = E\{x_1\}E\{x_2\} \tag{2.201}$$

Using (2.81) it again follows from this that $\mathrm{cov}(x_1, x_2) = 0$. Therefore the two statements $\mathrm{cov}(x_1, x_2) = 0$ and $E\{x_1x_2\} = E\{x_1\}E\{x_2\}$ are equivalent in characterizing the two uncorrelated random variables $x_1$ and $x_2$. That the two random variables $x_1$ and $x_2$ are uncorrelated is a weaker condition than their independence because two independent random variables are always uncorrelated (see (2.198)), whereas the converse of this statement is not correct.

If two random variables $x_1$ and $x_2$ obey the condition

$$E\{x_1x_2\} = 0 \tag{2.202}$$

we say that $x_1$ and $x_2$ are *orthogonal*. Two uncorrelated random variables $x_1$ and $x_2$ are orthogonal if at least one of the two expectations $E\{x_1\}$ or $E\{x_2\}$ is zero.

We assume that the components of an $n$-dimensional random variable $\boldsymbol{x} = [x_1, \ldots, x_n]'$ are independent, so that according to (2.131b) we have

$$f(\xi_1, \ldots, \xi_n) = f_1(\xi_1) \cdots f_n(\xi_n)$$

Then the following equations corresponding to (2.195) and (2.196) are valid for both the moment and the central moment of order $\Sigma_{\nu=1}^{n} k_\nu$ of the random variable $\boldsymbol{x} = [x_1, \ldots, x_n]'$:

$$E\{x_1^{k_1} \cdots x_n^{k_n}\} = E\{x_1^{k_1}\} \cdots E\{x_n^{k_n}\} \tag{2.203}$$

$$\begin{aligned} &E\{(x_1 - E\{x_1\})^{k_1} \cdots (x_n - E\{x_n\})^{k_n}\} \\ &\quad = E\{(x_1 - E\{x_1\})^{k_1}\} \cdots E\{(x_n - E\{x_n\})^{k_n}\} \end{aligned} \tag{2.204}$$

It follows from equation (2.203) that for every second-order moment $E[x_i^1 x_j^1], i \neq j$,

$$E\{x_i x_j\} = E\{x_i\}E\{x_j\} \quad i, j = 1, \ldots, n \quad i \neq j \tag{2.205}$$

and it follows from equation (2.204) that for every second-order central moment $E\{(x_i - E\{x_i\})^1(x_j - E\{x_j\})^1\}$, $i \neq j$,

$$\begin{aligned} E\{(x_i - E\{x_i\})(x_j - E\{x_j\})\} &= E\{x_i - E\{x_i\}\}E\{x_j - E\{x_j\}\} \\ &= \operatorname{cov}(x_i, x_j) = 0 \quad i, j = 1, \ldots, n \quad i \neq j \end{aligned} \tag{2.206}$$

Therefore all possible covariances $\operatorname{cov}(x_i, x_j)$, $i \neq j$, for an $n$-dimensional random variable $\boldsymbol{x} = [x_1, \ldots, x_n]'$ (the $n$ components $x_1, \ldots, x_n$ of which are independent) vanish, i.e., the associated covariance matrix $\mathrm{COV}(\boldsymbol{x}, \boldsymbol{x})$ is a diagonal matrix.

If, for an $n$-dimensional random variable $\boldsymbol{x} = [x_1, \ldots, x_n]'$, all possible covariances $\operatorname{cov}(x_i, x_j)$ (for which equation (2.181) also holds) are equal to zero or, equivalently, if we have for all $i \neq j$

$$E\{x_i x_j\} = E\{x_i\}E\{x_j\}$$

we say that the components $x_1, \ldots, x_n$ of $\boldsymbol{x} = [x_1, \ldots, x_n]'$ are *uncorrelated*. Then, as in the case of the independent components $x_1, \ldots, x_n$, the covariance matrix $\mathrm{COV}(\boldsymbol{x}, \boldsymbol{x})$ is a diagonal matrix.

If for two vector random variables $\boldsymbol{x} = [x_1, \ldots, x_n]'$ and $\boldsymbol{y} = [y_1, \ldots, y_n]'$ all components of $\boldsymbol{x}$ are uncorrelated with all components of $\boldsymbol{y}$, we have

$$E\{x_i y_j\} = E\{x_i\}E\{y_j\} \quad i = 1, \ldots, n \quad i = 1, \ldots, m$$

or, equivalently,

$$\operatorname{cov}(x_i, y_j) = E\{(x_i - E\{x_i\})(y_j - E\{y_j\})\} = 0 \quad i = 1, \ldots, n \quad j = 1, \ldots, m$$

In this case the covariance matrix $\mathrm{COV}(\boldsymbol{x}, \boldsymbol{y})$ is the zero matrix.

### 2.5.3 Conditional Expectations

In this section we consider only a two-dimensional continuous random variable $\boldsymbol{x} = [x_1, x_2]'$ with a density function $f(\xi_1, \xi_2)$ and the two density functions $f_1(\xi_1)$ and $f_2(\xi_2)$ of the marginal distributions.

Following equation (2.152), we obtain the expectation $E\{x_1\}$ of the one-dimensional random variable $x_1$ as

$$E\{x_1\} := \int_{-\infty}^{+\infty} \xi_1 f(\xi_1) \mathrm{d}\xi_1$$

The expectation of the random variable $x_1$ under the condition that the random variable $x_2$ takes the value $\xi_2$ is

$$E\{x_1 | x_2 = \xi_2\} := \int_{-\infty}^{+\infty} \xi_1 f(\xi_1 | \xi_2) \mathrm{d}\xi_1 \tag{2.207}$$

This is the equation defining the *conditional expectation* of the random variable $x_1$, if $x_2 = \xi_2$. The last two equations only differ in that the density function $f(\xi_1)$ appears in the integrand for $E\{x_1\}$ and the conditional density function $f(\xi_1|\xi_2)$ appears in the integrand for $E\{x_1|x_2 = \xi_2\}$. Using equation (2.120) we can rearrange the last equation to give

$$E\{x_1 | x_2 = \xi_2\} := \int_{-\infty}^{+\infty} \xi_1 \frac{f(\xi_1, \xi_2)}{f_2(\xi_2)} \mathrm{d}\xi_1 \tag{2.208}$$

In a corresponding manner, the conditional expectation of the random variable $g_1(x_1)$, if $x_2 = \xi_2$, is defined as

$$E\{g_1(x_1) | x_2 = \xi_2\} := \int_{-\infty}^{+\infty} g_1(\xi_1) \frac{f(\xi_1, \xi_2)}{f_2(\xi_2)} \mathrm{d}\xi_1 \tag{2.209}$$

In this way we have made the same generalization as for the ordinary expectations in which we went from $E\{x\}$ (equation (1.152)) to $E\{g(x)\}$ (equation (2.159)). In a similar manner to equations (2.208) and (2.209), the conditional expectations of the random variables $x_2$ and $g_2(x_2)$, if $x_1 = \xi_1$, are defined as

$$E\{x_2 | x_1 = \xi_1\} := \int_{-\infty}^{+\infty} \xi_2 \frac{f(\xi_1, \xi_2)}{f_1(\xi_1)} \mathrm{d}\xi_2 \tag{2.210}$$

and

$$E\{g_2(x_2)|x_1 = \xi_1\} := \int_{-\infty}^{+\infty} g_2(\xi_2)\frac{f(\xi_1, \xi_2)}{f_1(\xi_1)}\,d\xi_2 \tag{2.211}$$

Of course, we must assume that the relevant integrals exist if the defining equations (2.207)–(2.211) are to be meaningful.

The conditional expectation $E\{x_1|x_2 = \xi_2\}$, which is the mean value of the random variable $x_1$ if $x_2 = \xi_2$, is a "constant" whose numerical value depends on $\xi_2$ (see equation (2.208)). Therefore we have

$$E\{x_1|x_2 = \xi_2\} =: h(\xi_2)$$

However, because $\xi_2$ is a numerical value of the random variable $x_2$, $h(\xi_2)$ must be the numerical value of a random variable:

$$E\{x_1|x_2\} =: h(x_2)$$

Thus the conditional expectation $E\{x_1|x_2\}$ is a random variable which takes the value $E\{x_1|x_2 = \xi_2\} = h(\xi_2)$ when the random variable $x_2$ takes the value $\xi_2$.

We now obtain the (ordinary) expectation or mean value of the random variable $h(x_2)$ using equation (2.159):

$$E\{h(x_2)\} = \int_{-\infty}^{+\infty} h(\xi_2)f_2(\xi_2)d\xi_2 = \int_{-\infty}^{+\infty} E\{x_1|x_2 = \xi_2\}f_2(\xi_2)d\xi_2$$

It follows from this, using equation (2.208), that

$$E\{h(x_2)\} = \int_{-\infty}^{+\infty}\int_{-\infty}^{+\infty} \xi_1 f(\xi_1, \xi_2)d\xi_1 d\xi_2 \tag{2.212}$$

According to (2.171) the expectation $E\{x_1^1 x_2^0\} = E\{x_1\}$ is on the right-hand side of this equation. If we now expand the abbreviation $h(x_2)$ on the left-hand side of (2.212), we obtain the following relationship for the (ordinary) expectation of the conditional expectation $E\{x_1|x_2\}$:

$$E\{E\{x_1|x_2\}\} = E\{x_1\} \tag{2.213}$$

Similarly, we also have

$$E\{E\{x_2|x_1\}\} = E\{x_2\} \tag{2.214}$$

Having established these basic properties of conditional expectations, we will now conclude this section. It should also be noted that for vector random variables $\boldsymbol{x} = [x_1, \ldots, x_n]'$ with $n > 2$ being the number of conditional distribution functions possible, the number of associated conditional density functions and thus also the number of conditional expectations possible rapidly increases. Thus for $\boldsymbol{x} = [x_1, x_2, x_3]'$, for example, no less than six distribution functions $F(\xi_1, \xi_2|\xi_3)$, $F(\xi_1, \xi_3|\xi_2)$, $F(\xi_2, \xi_3|\xi_1)$, $\mathrm{F}(\xi_1|\xi_2, \xi_3)$, $F(\xi_2|\xi_1, \xi_3)$ and $F(\xi_3|\xi_1, \xi_2)$ are possible.

### 2.5.4 Calculation Rules for Expectations

It was stated in Section 2.5.1.6 that expectations can be handled in exactly the same way as the complex functions in the Laplace transformation. We derive the rules necessary for this procedure in this section. These rules provide an important tool which will be used frequently in Chapter 3 and subsequent chapters. Calculation rules for conditional expectations, which are similar to the rules for ordinary expectations, will not be considered here.

#### *2.5.4.1 Calculating Rules for the Expectation with a Random Variable g(x)*

The starting point for deriving the rules for expectations for a one-dimensional random variable $g(x)$ with a density function $f(\xi)$ associated with the random variable $x$ is (2.159):

$$E\{g(x)\} = \int_{-\infty}^{+\infty} g(\xi)f(\xi)\mathrm{d}(\xi)$$

If $g(x) = g_1(x) + g_2(x)$, where $g_1(x)$ and $g_2(x)$ are two uniquely defined but otherwise arbitrary functions, (2.159) becomes

$$\begin{aligned} E\{g_1(x) + g_2(x)\} &= \int_{-\infty}^{+\infty} \{g_1(\xi) + g_2(\xi)\}f(\xi)\mathrm{d}\xi \\ &= \int_{-\infty}^{+\infty} g_1(\xi)f(\xi)\mathrm{d}\xi + \int_{-\infty}^{+\infty} g_2(\xi)f(\xi)\mathrm{d}\xi \end{aligned}$$

The far right-hand side of this equation consists of two expectations $E\{g_1(x)\}$ and $E\{g_2(x)\}$, the existence of which is assumed. Then the expectation $E\{g_1(x) + g_2(x)\}$ must also exist (Fisz, 1988), and we have

$$E\{g_1(x) + g_2(x)\} = E\{g_1(x)\} + E\{g_2(x)\} \tag{2.215}$$

If we apply this formula twice to

$$g(x) = g_1(x) + g_2(x) + g_3(x)$$

we obtain

$$\begin{aligned} E\{g_1(x) + g_2(x) + g_3(x)\} &= E\{g_1(x)\} + E\{g_2(x) + g_3(x)\} \\ &= E\{g_1(x)\} + E\{g_2(x)\} + E\{g_3(x)\} \end{aligned}$$

It is clear from this that (2.215) can be generalized to a finite number of summands:

$$E\left\{\sum_{\nu=1}^{k} g_\nu(x)\right\} = \sum_{\nu=1}^{k} E\{g_\nu(x)\} \tag{2.216}$$

The expectation of the random variable $\lambda g(x)$, where $\lambda$ is a numerical constant, is given by (equation (2.159)),

$$E\{\lambda g(x)\} = \int_{-\infty}^{+\infty} \lambda g(\xi) f(\xi) \mathrm{d}\xi = \lambda \int_{-\infty}^{+\infty} g(\xi) f(\xi) \mathrm{d}\xi$$

Thus, if $E\{g(x)\}$ exists, we have

$$E\{\lambda g(x)\} = \lambda E\{g(x)\} \tag{2.217}$$

Using this result and equation (2.216), we can immediately obtain another rule for combining expectations:

$$E\left\{\sum_{\nu=1}^{k} \lambda_\nu g_\nu(x)\right\} = \sum_{\nu=1}^{k} \lambda_\nu E\{g_\nu(x)\} \tag{2.218}$$

if $\lambda_1, \ldots, \lambda_k$ are arbitrary constants. This result shows that the formation of the expectation is a linear operation.

The expectation of a constant $\lambda$ can also be derived using equation (2.159):

$$E\{\lambda\} = \int_{-\infty}^{+\infty} \lambda f(\xi) \mathrm{d}\xi = \lambda \int_{-\infty}^{+\infty} f(\xi) \mathrm{d}\xi$$

It follows from this and equation (2.90) that

$$E\{\lambda\} = \int_{-\infty}^{+\infty} \lambda f(\xi)\mathrm{d}\xi = \lambda \tag{2.219}$$

This result will now be derived in a different way. The constant $\lambda$ is regarded as a discrete random variable $z$ which can only take the value $z = \zeta^{(1)} = \lambda$ with probability $P(z = \zeta^{(1)} = \lambda) = 1$. According to equation (2.91), the distribution function of the random variable $z$ is

$$F_z(\zeta) = \sum_{i=1}^{1} P(z = \zeta^{(i)})\sigma(\zeta - \zeta^{(i)}) = 1 \times \sigma(\zeta - \lambda)$$

which has the density function (equation (2.92))

$$f_z(\zeta) = \delta(\zeta - \lambda)$$

We now use the expression (equation (2.152))

$$E\{z\} = \int_{-\infty}^{+\infty} \zeta f_z(\zeta)\mathrm{d}\zeta$$

to calculate the expectation of $z = \lambda$. By using the masking property of the $\delta$ function, we obtain

$$E\{z = \lambda\} = \int_{-\infty}^{+\infty} \zeta\delta(\zeta - \lambda)\mathrm{d}\zeta = \lambda \tag{2.220}$$

This is again the result (2.219). It should be noted that here $f_z(\zeta)$ is the density function of $z = \lambda$, whereas the function $f(\xi)$ in equation (2.219) is the density function of $x$ and not of $g(x) = \lambda$.

In the derivation of (2.219) the question of what kind of density function $f(\xi)$ of the random variable $x$ should be used does not arise if, for $g(x) = \lambda$, the "random variable" $g(x)$ is no longer a true random variable. This is because every density function $f(\xi)$ of a one-dimensional random variable $x$ must obey equation (2.90), i.e.,

$$\int_{-\infty}^{+\infty} f(\xi)\mathrm{d}\xi = 1$$

The correctness of the rule

$$E\{\lambda_1 + \lambda_2 g(x)\} = \lambda_1 + \lambda_2 E\{g(x)\} \tag{2.221}$$

where $\lambda_1$ and $\lambda_2$ are two constant parameters, can easily be proved by using equations (2.218) and (2.219).

In Section 2.5.1.5 we evaluated the integral

$$E\{(x - E\{x\})^2\} = \int_{-\infty}^{+\infty} (\xi - E\{x\})^2 f(\xi) \mathrm{d}\xi$$

to obtain the following expression for the variance of the random variable $x$:

$$E\{(x - E\{x\})^2\} = E\{x^2\} - E^2\{x\}$$

However, we can now derive this formula more simply and rapidly by using the calculation rules for expectations given above. We have

$$\begin{aligned} E\{(x - E\{x\})^2\} &= E\{x^2 - 2E\{x\}x + E^2\{x\}\} \\ &= E\{x^2\} - 2E\{x\}E\{x\} + E^2\{x\} = E\{x^2\} - E^2\{x\} \end{aligned}$$

where the mean value $E\{x\}$ is a constant value and not a random variable. The advantage of using the notation $E\{\ \}$ is ultimately based on the fact that the same function $g$ occurs in the argument of the expectation and in the integrand of the associated integral, and that the transformation of $g$ in the integrand can also be carried out in the argument of the expectation. Of course, when these rules are applied, the existence of the individual expectations must be guaranteed. This will always be assumed here and in what follows.

### 2.5.4.2 *Calculating Rules for the Expectation with a Random Variable $g(x)$*

Completely analogous rules to those given in Section 2.5.4.1 can be derived for a one-dimensional random variable $g(\boldsymbol{x})$ which depends on an $n$-dimensional random variable $\boldsymbol{x} = [x_1, \ldots, x_n]'$ with the density function $f(\boldsymbol{\xi})$. In this case we start from equation (2.182c):

$$E\{g(x)\} = \int_{-\infty}^{+\infty} g(\xi) l(\xi) \mathrm{d}\xi$$

and obtain

$$E\left\{\sum_{\nu=1}^{k} g_\nu(\boldsymbol{x})\right\} = \sum_{\nu=1}^{k} E\{g_\nu(\boldsymbol{x})\} \tag{2.222}$$

and

$$E\{\lambda g(\boldsymbol{x})\} = \lambda E\{g(\boldsymbol{x})\} \tag{2.223}$$

or, if both rules are combined,

$$E\left\{\sum_{\nu=1}^{k} \lambda_\nu g_\nu(\boldsymbol{x})\right\} = \sum_{\nu=1}^{k} \lambda_\nu E\{g_\nu(\boldsymbol{x})\} \tag{2.224}$$

The $\lambda_1, \ldots, \lambda_k$ are constant but otherwise arbitrary parameters.

As a special case of equation (2.224) it follows that, for $g_\nu(\boldsymbol{x}) = x_\nu$, $\nu = 1, \ldots, k$,

$$E\left\{\sum_{\nu=1}^{k} \lambda_\nu x_\nu\right\} = \sum_{\nu=1}^{k} \lambda_\nu E\{x_\nu\} \tag{2.225}$$

In this case the expectation $E\{x_\nu\}$ is regarded as the characteristic value of the marginal distribution of the one-dimensional random variable $x_\nu$ with respect to the vector random variable $\boldsymbol{x} = [x_1, \ldots, x_n]'$.

In applying the above rules, the case

$$E\{\lambda_1 + \lambda_2 g_2(\boldsymbol{x})\} = E\{\lambda_1\} + \lambda_2 E\{g_2(\boldsymbol{x})\}$$

may occur. Because $E\{\lambda_1\}$ represents an $n$-fold integral:

$$E\{\lambda_1\} = \int_{-\infty}^{+\infty} \lambda_1 f(\boldsymbol{\xi})\mathrm{d}\boldsymbol{\xi} = \lambda_1 \int_{-\infty}^{+\infty} f(\boldsymbol{\xi})\mathrm{d}\boldsymbol{\xi} \tag{2.226}$$

equation (2.219) cannot be applied indiscriminately to it. However, because

$$\int_{-\infty}^{+\infty} f(\boldsymbol{\xi})\mathrm{d}\boldsymbol{\xi} = 1$$

which is the generalization of equation (2.107), it follows purely formally from equation (2.226) that

$$E\{\lambda_1\} = \int_{-\infty}^{+\infty} \lambda_1 f(\boldsymbol{\xi})\mathrm{d}\boldsymbol{\xi} = \lambda_1 \tag{2.227}$$

Because $g(\boldsymbol{x}) = \lambda_1$ is not a true random variable, we will also consider a second method for deriving equation (2.227), just as we did for (2.219). For this, we use the following theorem (Heinhold and Gaede, 1979): if $f(\boldsymbol{\xi})$ is the density function associated with the vector random variable $\boldsymbol{x}$ and $f_z(\zeta)$ is the density function associated with the scalar random variable $z = g(x)$, then

$$E\{g(\boldsymbol{x})\} = \int_{-\infty}^{+\infty} g(\boldsymbol{\xi})f(\boldsymbol{\xi})\mathrm{d}\boldsymbol{\xi} = E\{z\} = \int_{-\infty}^{+\infty} \zeta f_z(\zeta)\mathrm{d}\zeta$$

if the expectation $E\{g(\boldsymbol{x})\}$ exists.

If we now set $z = g(\boldsymbol{x}) = \lambda_1$ or $g(\boldsymbol{\xi}) = \lambda_1$, because $f_z(\zeta) = \delta(\zeta - \lambda_1)$ (see the derivation of equation (2.220)), this theorem gives

$$\begin{aligned} E\{\lambda_1\} &= \int_{-\infty}^{+\infty} \lambda_1 f(\boldsymbol{\xi})\mathrm{d}\boldsymbol{\xi} \\ &= E\{z = \lambda_1\} = \int_{-\infty}^{+\infty} \zeta\delta(\zeta - \lambda_1)\mathrm{d}\zeta = \lambda_1 \end{aligned}$$

i.e., exactly equation (2.227), which, for $\lambda_1 = \lambda$, can also be written

$$E\{\lambda\} = \int_{-\infty}^{+\infty} \lambda f(\boldsymbol{\xi})\mathrm{d}\boldsymbol{\xi} = \lambda \tag{2.228}$$

Another rule then follows from this, i.e.,

$$E\{\lambda_1 + \lambda_2 g(\boldsymbol{x})\} = \lambda_1 + \lambda_2 E\{g(\boldsymbol{x})\} \tag{2.229}$$

The rules derived in this section can be used, for example, to obtain the covariance property given by equation (2.181) without the explicit use of general calculus:

$$\begin{aligned} \mathrm{cov}(x_1, x_2) &= E\{(x_1 - E\{x_1\})(x_2 - E\{x_2\})\} \\ &= E\{x_1x_2 - E\{x_1\}x_2 - E\{x_2\}x_1 + E\{x_1\}E\{x_2\}\} \\ &= E\{x_1x_2\} - E\{x_1\}E\{x_2\} - E\{x_2\}E\{x_1\} + E\{x_1\}E\{x_2\} \\ &= E\{x_1x_2\} - E\{x_1\}E\{x_2\} \end{aligned}$$

### 2.5.4.3 Calculating Rules for the Expectation with a Random Variable $\boldsymbol{X} = (x_{ij})$

We will consider the $n \times m$ matrix

$$\boldsymbol{X} = \begin{bmatrix} x_{11} & \ldots & x_{1m} \\ \vdots & & \vdots \\ x_{n1} & \ldots & x_{nm} \end{bmatrix} = (x_{ij}) \quad i = 1, \ldots, n \quad j = 1, \ldots, m \qquad (2.230)$$

in which the $nm$ elements are random variables. Let the density function $f(\xi_{11}, \ldots, \xi_{nm})$ be associated with the $nm$-dimensional random variable $\boldsymbol{X}$. The expectation of $\boldsymbol{X}$ is defined as (equation (2.186b)):

$$E\{\boldsymbol{X}\} = E\{(x_{ij})\} := (E\{x_{ij}\}) \quad i = 1, \ldots, n \quad j = 1, \ldots, m$$

or is written explicitly as

$$E\left\{\begin{bmatrix} x_{11} & \ldots & x_{1m} \\ \vdots & & \vdots \\ x_{n1} & \ldots & x_{nm} \end{bmatrix}\right\} := \begin{bmatrix} E\{x_{11}\} & \ldots & E\{x_{1m}\} \\ \vdots & & \vdots \\ E\{x_{n1}\} & \ldots & E\{x_{nm}\} \end{bmatrix}$$

Each element $E\{x_{ij}\}$ represents the characteristic value of the marginal distribution of the random variable $x_{ij}$ with respect to the $nm$-dimensional random variable $\boldsymbol{X}$. The defining equation of the expectation of a matrix $\boldsymbol{X}$ contains the defining equation (2.162b) of the expectation of an $n$-dimensional random vector $\boldsymbol{x}$, i.e.,

$$E\{\boldsymbol{x}\} = E\{(x_i)\} = (E\{x_i\}) \quad i = 1, \ldots, n$$

as a special case.

In subsequent chapters, matrices of the form

$$\boldsymbol{xx}' = \begin{bmatrix} x_1x_1 & \ldots & x_1x_n \\ \vdots & & \vdots \\ x_nx_1 & \ldots & x_nx_n \end{bmatrix}$$

and

$$\boldsymbol{xy}' = \begin{bmatrix} x_1y_1 & \ldots & x_1y_m \\ \vdots & & \vdots \\ x_ny_1 & \ldots & x_ny_m \end{bmatrix}$$

often occur where both $\boldsymbol{x} = [x_1, \ldots, x_n]'$ and $\boldsymbol{y} = [y_1, \ldots, y_m]'$ are vector random variables. Now, to avoid having to write all the rules for $\boldsymbol{xx}'$ and $\boldsymbol{xy}'$, in what follows we let the matrix $\boldsymbol{X}$, (equation (2.230)) stand for $\boldsymbol{xx}'$ and $\boldsymbol{xy}'$ and the matrix $\boldsymbol{Y}$ stand for the dyadic products $\boldsymbol{uu}'$ and $\boldsymbol{uv}'$ of the vector random variables $\boldsymbol{u} = [u_1, \ldots, u_n]'$ and $\boldsymbol{v} = [v_1, \ldots, v_m]'$. We can envisage $\boldsymbol{X}$ as being produced by a simple renaming of the elements of the matrix $\boldsymbol{xx}'$ or the matrix $\boldsymbol{xy}'$, and similarly for the matrix $\boldsymbol{Y}$.

After these preliminary remarks, we will now derive some calculation rules.

(a)

$$E\{\boldsymbol{X}'\} = (E\{\boldsymbol{X}\})' \tag{2.231}$$

*Proof*

$$E\{\boldsymbol{X}'\} = E\{(x_{ij})'\} = E\{(x_{ji})\} = (E\{x_{ji}\}) = (E\{x_{ij}\})' = (E\{\boldsymbol{X}\})'$$

If $\boldsymbol{X}$ is an $n \times 1$ matrix (i.e., a column vector $\boldsymbol{x}$) equation (2.231) becomes

$$E\{\boldsymbol{x}'\} = (E\{\boldsymbol{x}\})' \tag{2.232}$$

(b)

$$E\{\mathrm{tr}\boldsymbol{X}\} = \mathrm{tr}E\{\boldsymbol{X}\} \tag{2.233}$$

Because the trace of a matrix is only defined for square matrices, this equation is also only meaningful for square $\boldsymbol{X}$.

*Proof*

$$E\{\mathrm{tr}\boldsymbol{X}\} = E\left\{\sum_{i=1}^{n} x_{ii}\right\}$$

Using equation (2.225), it follows that

$$E\left\{\sum_{i=1}^{n} x_{ii}\right\} = \sum_{i=1}^{n} E\{x_{ii}\} = \mathrm{tr}E\{\boldsymbol{X}\}$$

(c)

$$E\{\boldsymbol{X} + \boldsymbol{Y}\} = E\{\boldsymbol{X}\} + E\{\boldsymbol{Y}\} \tag{2.234}$$

The expectations $E\{\boldsymbol{X}\}$ and $E\{\boldsymbol{Y}\}$ are characteristic values of marginal distributions if the density function $f(\xi_{11}, \ldots, \xi_{nm}, \nu_{11}, \ldots, \nu_{nm})$ belongs to the $2mn$ random variables in $\boldsymbol{X}$ and $\boldsymbol{Y}$.

*Proof*

$$E\{\boldsymbol{X} + \boldsymbol{Y}\} = E\{(x_{ij}) + (y_{ij})\} = E\{(x_{ij} + y_{ij})\} = (E\{x_{ij} + y_{ij}\})$$

Equation (2.225) can now be applied to this and it follows that

$$(E\{x_{ij} + y_{ij}\}) = (E\{x_{ij}\} + E\{y_{ij}\}) = (E\{x_{ij}\}) + (E\{y_{ij}\}) = E\{X\} + E\{Y\}$$

Equation (2.234) can be generalized for $k$ summands in exactly the same way as was shown for equation (2.215). As a special case of (2.234) we have

$$E\{\boldsymbol{x} + \boldsymbol{y}\} = E\{\boldsymbol{x}\} + E\{\boldsymbol{y}\} \tag{2.235}$$

(d) Let $\boldsymbol{\Lambda}^{(1)} = (\lambda_{ij}^{(1)})$ be a $p \times n$ matrix with elements $\lambda_{ij}^{(1)}$, $i = 1, \ldots, p$, $j = 1, \ldots, n$, that are constant parameters (i.e., parameters that do not depend on chance). Then,

$$E\{\boldsymbol{\Lambda}^{(1)}\boldsymbol{X}\} = \boldsymbol{\Lambda}^{(1)}E\{\boldsymbol{X}\} \tag{2.236}$$

*Proof*

$$E\{\boldsymbol{\Lambda}^{(1)}\boldsymbol{X}\} = E\{(\lambda_{ij}^{(1)})(x_{ij})\} = E\left\{\left(\sum_{\nu=1}^{n} \lambda_{i\nu}^{(1)} x_{\nu j}\right)\right\} = \left(E\left\{\sum_{\nu=1}^{n} \lambda_{i\nu}^{(1)} x_{\nu j}\right\}\right)$$

Again, (2.225) can be applied, and it follows that

$$\left(E\left\{\sum_{\nu=1}^{n} \lambda_{i\nu}^{(1)} x_{\nu j}\right\}\right) = \left(\sum_{\nu=1}^{n} \lambda_{i\nu}^{(1)} E\{x_{\nu j}\}\right) = (\lambda_{ij}^{(1)})(E\{x_{ij}\}) = \boldsymbol{\Lambda}^{(1)}E\{\boldsymbol{X}\}$$

(e) In an analogous manner to (2.236) we have

$$E\{\boldsymbol{X}\boldsymbol{\Lambda}^{(2)}\} = E\{\boldsymbol{X}\}\boldsymbol{\Lambda}^{(2)} \tag{2.237}$$

where $\boldsymbol{\Lambda}^{(2)} = (\lambda_{ij}^{(2)})$ is an $m \times q$ matrix with constant elements $\lambda_{ij}^{(2)}$, $i = 1, \ldots, m$, $j = 1, \ldots, q$. The following special cases of (2.236) and (2.237) are obtained if $\boldsymbol{\Lambda}^{(3)} = (\lambda_{ij}^{(3)})$ is an $n \times q$ matrix with constant elements:

$$E\{\boldsymbol{\Lambda}^{(1)}\boldsymbol{x}\} = \boldsymbol{\Lambda}^{(1)}E\{\boldsymbol{x}\} \tag{2.238}$$

$$E\{\boldsymbol{x}'\boldsymbol{\Lambda}^{(3)}\} = E\{\boldsymbol{x}'\}\boldsymbol{\Lambda}^{(3)} \tag{2.239}$$

(f)
$$E\{\boldsymbol{\Lambda}^{(1)}\boldsymbol{X}\boldsymbol{\Lambda}^{(2)}\} = \boldsymbol{\Lambda}^{(1)}E\{\boldsymbol{X}\}\boldsymbol{\Lambda}^{(2)} \tag{2.240}$$

*Proof*

$$E\{\boldsymbol{\Lambda}^{(1)}\boldsymbol{X}\boldsymbol{\Lambda}^{(2)}\} = \boldsymbol{\Lambda}^{(1)}E\{\boldsymbol{X}\boldsymbol{\Lambda}^{(2)}\} = \boldsymbol{\Lambda}^{(1)}E\{\boldsymbol{X}\}\boldsymbol{\Lambda}^{(2)}$$

(g) We can easily prove the following two rules using (2.234) and (2.236) or (2.237). The first of these rules is

$$E\{\boldsymbol{\Lambda}^{(1)}\boldsymbol{X} + \boldsymbol{\Lambda}^{(4)}\boldsymbol{Y}\} = \boldsymbol{\Lambda}^{(1)}E\{\boldsymbol{X}\} + \boldsymbol{\Lambda}^{(4)}E\{\boldsymbol{Y}\} \tag{2.241}$$

In order to combine with the $n \times m$ matrices $\boldsymbol{X}$ and $\boldsymbol{Y}$, $\boldsymbol{\Lambda}^{(1)}$ and $\boldsymbol{\Lambda}^{(4)}$, whose elements are constant parameters, must both be $p \times n$ matrices. A special case of this ($\boldsymbol{x}$ and $\boldsymbol{y}$ are $n$-dimensional random vectors) is

$$E\{\boldsymbol{\Lambda}^{(1)}\boldsymbol{x} + \boldsymbol{\Lambda}^{(4)}\boldsymbol{y}\} = \boldsymbol{\Lambda}^{(1)}E\{\boldsymbol{x}\} + \boldsymbol{\Lambda}^{(4)}E\{\boldsymbol{y}\} \tag{2.242}$$

However, if $\boldsymbol{\Lambda}^{(2)}$ and $\boldsymbol{\Lambda}^{(5)}$ are two $m \times q$ matrices, we have the second rule:

$$E\{\boldsymbol{X}\boldsymbol{\Lambda}^{(2)} + \boldsymbol{Y}\boldsymbol{\Lambda}^{(5)}\} = E\{\boldsymbol{X}\}\boldsymbol{\Lambda}^{(2)} + E\{\boldsymbol{Y}\}\boldsymbol{\Lambda}^{(5)} \tag{2.243}$$

A special case is

$$E\{\boldsymbol{x}'\boldsymbol{\Lambda}^{(3)} + \boldsymbol{y}'\boldsymbol{\Lambda}^{(6)}\} = E\{\boldsymbol{x}'\}\boldsymbol{\Lambda}^{(3)} + E\{\boldsymbol{y}'\}\boldsymbol{\Lambda}^{(6)} \tag{2.244}$$

where $\boldsymbol{\Lambda}^{(3)}$ and $\boldsymbol{\Lambda}^{(6)}$ are two $n \times q$ matrices with constant elements. Equations (2.241)–(2.244) apply in exactly the same way for more than two summands.

(h) If $\boldsymbol{\Lambda} = (\lambda_{ij})$ is a matrix whose elements $\lambda_{ij}$ are constant parameters, we have

$$E\{\boldsymbol{\Lambda}\} = \boldsymbol{\Lambda} \tag{2.245}$$

*Proof:* It immediately follows from equation (2.227) that

$$E\{\boldsymbol{\Lambda}\} = E\{(\lambda_{ij})\} = (E\{\lambda_{ij}\}) = (\lambda_{ij}) = \boldsymbol{\Lambda}$$

A special case is

$$E\{\boldsymbol{\lambda}\} = \boldsymbol{\lambda} \tag{2.246}$$

where the elements of the vector $\boldsymbol{\lambda}$ are constant parameters.

(i) Using equations (2.234), (2.236) or (2.237), and (2.245), we obtain

$$E\{\boldsymbol{\Lambda}^{(7)} + \boldsymbol{\Lambda}^{(1)}\boldsymbol{X}\} = \boldsymbol{\Lambda}^{(7)} + \boldsymbol{\Lambda}^{(1)}E\{\boldsymbol{X}\} \tag{2.247}$$

and

$$E\{\boldsymbol{\Lambda}^{(3)} + \boldsymbol{X}\boldsymbol{\Lambda}^{(2)}\} = \boldsymbol{\Lambda}^{(3)} + E\{\boldsymbol{X}\}\boldsymbol{\Lambda}^{(2)} \tag{2.248}$$

where $\boldsymbol{\Lambda}^{(1)}$ is a $p \times n$ matrix, $\boldsymbol{\Lambda}^{(2)}$ is a $m \times q$ matrix, $\boldsymbol{\Lambda}^{(3)}$ is an $n \times q$ matrix and $\boldsymbol{\Lambda}^{(7)}$ is a $p \times m$ matrix, all with constant elements.

(j) If two vector random variables $\boldsymbol{x} = [x_1, \ldots, x_n]'$ and $\boldsymbol{y} = [y_1, \ldots, y_m]'$ are uncorrelated; i.e., if

$$E\{x_i y_j\} = E\{x_i\}E\{y_j\} \quad i = 1, \ldots, n, \quad j = 1, \ldots, m$$

it follows that

$$E\{\boldsymbol{x}\boldsymbol{y}'\} = E\{\boldsymbol{x}\}E\{\boldsymbol{y}'\} \tag{2.249}$$

*Proof*

$$E\{\boldsymbol{x}\boldsymbol{y}'\} = \begin{bmatrix} E\{x_1y_1\} & \ldots & E\{x_1y_m\} \\ \vdots & & \vdots \\ E\{x_ny_1\} & \ldots & E\{x_ny_m\} \end{bmatrix}$$

$$= \begin{bmatrix} E\{x_1\}E\{y_1\} & \ldots & E\{x_1\}E\{y_m\} \\ \vdots & & \vdots \\ E\{x_n\}E\{y_1\} & \ldots & E\{x_n\}E\{y_m\} \end{bmatrix} = E\{\boldsymbol{x}\}E\{\boldsymbol{y}'\}$$

Equation (2.249) also holds if $\boldsymbol{x}$ and $\boldsymbol{y}$ are independent of each other (Richter, 1966).

Having established these rules, we can now easily derive relationship (2.191) which is satisfied by the covariance matrix COV($\boldsymbol{x}$, $\boldsymbol{x}$) for an $n$-dimensional random variable $\boldsymbol{x}$. The starting point for this is the defining equation (2.188):

$$\begin{aligned} \mathrm{COV}(\boldsymbol{x}, \boldsymbol{x}) : &= E\{(\boldsymbol{x} - E\{\boldsymbol{x}\})(\boldsymbol{x} - E\{\boldsymbol{x}\})'\} \\ &= E\{\boldsymbol{x}\boldsymbol{x}' - E\{\boldsymbol{x}\}\boldsymbol{x}' - \boldsymbol{x}(\mathrm{E}\{\boldsymbol{x}\})' + E\{\boldsymbol{x}\}(E\{\boldsymbol{x}\})'\} \\ &= E\{\boldsymbol{x}\boldsymbol{x}'\} - \mathrm{E}\{\boldsymbol{x}\}E\{\boldsymbol{x}'\} - E\{\boldsymbol{x}\}(E\{\boldsymbol{x}\})' + E\{\boldsymbol{x}\}(E\{\boldsymbol{x}\})' \end{aligned}$$

Therefore

$$\mathrm{COV}(\boldsymbol{x}, \boldsymbol{x}) = E\{\boldsymbol{x}\boldsymbol{x}'\} - E\{\boldsymbol{x}\}E\{\boldsymbol{x}'\}$$

which is exactly equation (2.191). It can be seen how elegantly and quickly this equation can be derived with the help of the above rules.

In order to illustrate the use of the calculation rules for expectations, the expression

$$E\{(\boldsymbol{Ax} + \boldsymbol{By})(\boldsymbol{Ax} + \boldsymbol{By})'\}$$

will be transformed. $\boldsymbol{A}$ and $\boldsymbol{B}$ are matrices with constant elements which can be combined with $\boldsymbol{x}$ or $\boldsymbol{y}$ with:

$$\begin{aligned}
&E\{(\boldsymbol{Ax} + \boldsymbol{By})(\boldsymbol{Ax} + \boldsymbol{By})'\} \\
&= E\{(\boldsymbol{Ax} + \boldsymbol{By})(\boldsymbol{x}'\boldsymbol{A}' + \boldsymbol{y}'\boldsymbol{B}')\} \\
&= E\{\boldsymbol{Axx}'\boldsymbol{A}' + \boldsymbol{Byx}'\boldsymbol{A}' + \boldsymbol{Axy}'\boldsymbol{B}' + \boldsymbol{Byy}'\boldsymbol{B}'\} \\
&= \boldsymbol{A}E\{\boldsymbol{xx}'\}\boldsymbol{A}' + \boldsymbol{B}E\{\boldsymbol{yx}'\}\boldsymbol{A}' + \boldsymbol{A}E\{\boldsymbol{xy}'\}\boldsymbol{B}' + \boldsymbol{B}E\{\boldsymbol{yy}'\}\boldsymbol{B}'
\end{aligned}$$

#### *2.5.4.4 Calculating Rules for the Expectation with Integrals of Random Variables*

If $x(t)$ is a random variable at every point in time $t$, then in general the stochastic properties of $x(t)$ will depend on time, (i.e., a density function $f(\xi, t)$ is associated with $x(t)$). We will now derive a rule for calculating the expectation of the random variable:

$$g\{x(t)\} = \int_{t_0}^{t} \lambda(\tau)x(\tau)\mathrm{d}\tau$$

where $\lambda(\tau)$ is a function with deterministic function values. To do this, $x$ must be replaced by $x(t)$, $\xi$ by $\xi(t)$ and $f(\xi)$ by $f(\xi, t)$ in equation (2.159):

$$E\{g\{x(t)\}\} = \int_{-\infty}^{+\infty} g\{\xi(t)\}f(\xi, t)\mathrm{d}\xi \tag{2.250}$$

We now substitute the integral expression given above for $g\{x(t)\}$ or $g\{\xi(t)\}$:

$$E\left\{\int_{t_0}^{t} \lambda(\tau)x(\tau)\mathrm{d}\tau\right\} = \int_{-\infty}^{+\infty}\int_{t_0}^{t} \lambda(\tau)\xi(\tau)\mathrm{d}\tau f(\xi, t)\mathrm{d}\xi$$

Assuming that it is permissible to permute the order of the two integrations, we have

$$E\left\{\int_{t_0}^{t} \lambda(\tau)x(\tau)\mathrm{d}\tau\right\} = \int_{t_0}^{t} \lambda(\tau)\left\{\int_{-\infty}^{+\infty} \xi(\tau)f(\xi, t)\mathrm{d}\xi\right\}\mathrm{d}\tau$$

However, according to equation (2.250), the integral in braces on the right-hand side is equal to the expectation of $x(\tau)$. Therefore

$$E\left\{\int_{t_0}^{t} \lambda(\tau)x(\tau)\mathrm{d}\tau\right\} = \int_{t_0}^{t} \lambda(\tau)E\{x(\tau)\}\mathrm{d}\tau \tag{2.251}$$

This result can be generalized further.

Let the density function $f(\boldsymbol{\xi}, t)$ be associated with the $n$-dimensional vector $\boldsymbol{x}(t)$ which is a random vector for every point in time $t$. We now form a new random variable $g\{\boldsymbol{x}(t)\}$ from $\boldsymbol{x}(t)$ through

$$g\{\boldsymbol{x}(t)\} = \int_{t_0}^{t} \boldsymbol{\lambda}'(\tau)\boldsymbol{x}(\tau)\mathrm{d}\tau \tag{2.252}$$

where $\boldsymbol{\lambda}(\tau)$ is an $n$-dimensional vector with deterministic variables as elements. The expectation of this new variable is calculated using equation (2.182c) where we replace $\boldsymbol{x}$ by $\boldsymbol{x}(t)$, $\boldsymbol{\xi}$ by $\boldsymbol{\xi}(t)$ and $f(\boldsymbol{\xi})$ by $f(\boldsymbol{\xi}, t)$:

$$E\{g\{\boldsymbol{x}(t)\}\} = \int_{-\infty}^{+\infty} g\{\boldsymbol{\xi}(t)\}f(\boldsymbol{\xi}, t)\mathrm{d}\boldsymbol{\xi} \tag{2.253}$$

Using equation (2.252) we obtain

$$E\left\{\int_{t_0}^{t} \boldsymbol{\lambda}'(\tau)\boldsymbol{x}(\tau)\mathrm{d}\tau\right\} = \int_{-\infty}^{+\infty} \int_{t_0}^{t} \boldsymbol{\lambda}'(\tau)\boldsymbol{\xi}(\tau)\mathrm{d}\tau f(\boldsymbol{\xi}, t)\mathrm{d}\boldsymbol{\xi}$$

$$= \int_{t_0}^{t} \boldsymbol{\lambda}'(\tau)\left\{\int_{-\infty}^{+\infty} \boldsymbol{\xi}(\tau)f(\boldsymbol{\xi}, t)\mathrm{d}\boldsymbol{\xi}\right\}\mathrm{d}\tau$$

It follows from this, using equation (2.253), that

$$E\left\{\int_{t_0}^{t} \boldsymbol{\lambda}'(\tau)\boldsymbol{x}(\tau)\mathrm{d}\tau\right\} = \int_{t_0}^{t} \boldsymbol{\lambda}'(\tau)E\{\boldsymbol{x}(\tau)\}\mathrm{d}\tau \tag{2.254}$$

Because $\boldsymbol{a}'\boldsymbol{b} = \boldsymbol{b}'\boldsymbol{a}$ ($\boldsymbol{a}$ and $\boldsymbol{b}$ are two $n$-dimensional vectors) we also have

$$E\left\{\int_{t_0}^{t} \boldsymbol{x}'(\tau)\boldsymbol{\lambda}(\tau)\mathrm{d}\tau\right\} = \int_{t_0}^{t} E\{\boldsymbol{x}'(\tau)\}\boldsymbol{\lambda}(\tau)\mathrm{d}\tau \tag{2.255}$$

By using this result, we can now derive the following rule:

$$E\left\{\int_{t_0}^{t} \boldsymbol{\Lambda}(\tau)\boldsymbol{X}(\tau)\mathrm{d}\tau\right\} = \int_{t_0}^{t} \boldsymbol{\Lambda}(\tau)E\{\boldsymbol{X}(\tau)\}\mathrm{d}\tau \tag{2.256}$$

where $\boldsymbol{\Lambda} = (\lambda_{ij}(t))$ is a $p \times n$ matrix with deterministic elements, $\boldsymbol{\lambda}_i'(t) = [\lambda_{i1}(t), \ldots, \lambda_{in}(t)]$ is the $i$th row vector of $\boldsymbol{\Lambda}(t)$, $\boldsymbol{X}(t) = (x_{ij}(t))$ is an $n \times m$ matrix whose $nm$ elements $x_{ij}(t)$ are random variables for every $t$ and $\boldsymbol{x}_j = [x_{1j}(t), \ldots, x_{nj}(t)]'$ is the $j$th column vector of $\boldsymbol{X}(t)$. Let the density function $f(\xi_{11}, \ldots, \xi_{nm}, t)$ be associated with $\boldsymbol{X}(t)$.

*Proof*

$$E\left\{\int_{t_0}^{t} \boldsymbol{\Lambda}(\tau)\boldsymbol{X}(\tau)\mathrm{d}\tau\right\} = E\left\{\int_{t_0}^{t} (\boldsymbol{\lambda}_i'(\tau)\boldsymbol{x}_j(\tau))\mathrm{d}\tau\right\}$$

From the defining equation of the integral of a matrix:

$$\int_{t_0}^{t} \boldsymbol{A}(\tau)\mathrm{d}\tau = \int_{t_0}^{t} (a_{ij}(\tau))\mathrm{d}\tau := \left(\int_{t_0}^{t} a_{ij}(\tau)\mathrm{d}\tau\right)$$

if we also use

$$E\{Y\} = E\{(y_{ij})\} := (E\{y_{ij}\})$$

and equations (2.186b) and (2.254), we have

$$E\left\{\int_{t_0}^{t} \boldsymbol{\Lambda}(\tau)\boldsymbol{X}(\tau)\mathrm{d}\tau\right\} = E\left\{\left(\int_{t_0}^{t} \boldsymbol{\lambda}_i'(\tau)\boldsymbol{x}_j(\tau)\mathrm{d}\tau\right)\right\}$$

$$= \left(E\left\{\int_{t_0}^{t} \boldsymbol{\lambda}_i'(\tau)\boldsymbol{x}_j(\tau)\mathrm{d}\tau\right\}\right) = \left(\int_{t_0}^{t} \boldsymbol{\lambda}_i'(\tau)E\{\boldsymbol{x}_j(\tau)\}\mathrm{d}\tau\right)$$

$$= \int_{t_0}^{t} (\boldsymbol{\lambda}_i'(\tau)E\{\boldsymbol{x}_j(\tau)\})\mathrm{d}\tau = \int_{t_0}^{t} \boldsymbol{\Lambda}(\tau)E\{\boldsymbol{X}(\tau)\}\mathrm{d}\tau$$

Therefore equation (2.256) is proved.

A special case is

$$E\left\{\int_{t_0}^{t} \boldsymbol{\Lambda}(\tau)\boldsymbol{x}(\tau)\mathrm{d}\tau\right\} = \int_{t_0}^{t} \boldsymbol{\Lambda}(\tau)E\{\boldsymbol{x}(\tau)\}\mathrm{d}\tau \tag{2.257}$$

In a corresponding manner to equation (2.256), we also have

$$E\left\{\int_{t_0}^{t} \boldsymbol{X}(\tau)\boldsymbol{\Lambda}(\tau)\mathrm{d}\tau\right\} = \int_{t_0}^{t} E\{\boldsymbol{X}(\tau)\}\boldsymbol{\Lambda}(\tau)\mathrm{d}\tau \tag{2.258}$$

and, as a special case,

$$E\left\{\int_{t_0}^{t} \boldsymbol{x}'(\tau)\boldsymbol{\Lambda}(\tau)\mathrm{d}\tau\right\} = \int_{t_0}^{t} E\{\boldsymbol{x}'(\tau)\}\boldsymbol{\Lambda}(\tau)\mathrm{d}\tau \tag{2.259}$$

Because $\boldsymbol{\Lambda}(\tau)$ can combine with $\boldsymbol{X}(\tau)$ and $\boldsymbol{x}'(\tau)$, it must be an $m \times q$ matrix in equation (2.258) and an $n \times q$ matrix in (2.259) if $\boldsymbol{x}(\tau)$ is $n$-dimensional.

As in Section 2.5.4.3, the matrix $\boldsymbol{X}(t)$ can be envisaged as being formed by renaming the matrices $\boldsymbol{u}(t)\boldsymbol{u}'(t)$ or $\boldsymbol{v}(t)\boldsymbol{u}'(t)$ if $\boldsymbol{u}(t)$ or $\boldsymbol{v}(t)$ are random vectors of corresponding dimensions for every point in time $t$.

## 2.6 RANDOM (STOCHASTIC) PROCESSES

### 2.6.1 The Concept of the Random Process

In the random experiments considered so far, the elementary events have been characterized by scalar or vector numerical values assigned in a fixed manner. However, more complicated random experiments often occur in which every elementary event is assigned not only a single numerical value but a parameter-dependent numerical value (i.e., a function). Here this parameter will be the time $t$. However, in general, the parameter can have any meaning.

Let us define, for example, an elementary event $\omega_i$ as the withdrawal of a resistor with resistance $R$ ohms from the current production batch. A voltage $\xi^{(i)}(t)$ caused by thermal noise can be measured by using this resistor. Neither the magnitude of this voltage at a fixed point in time $t = t_1$ nor the complete variation of the noise voltage can be specified before removing the resistor concerned. Both depend on chance.

However, in a random experiment with its elementary events characterized by the numerical values of a random variable, *all* possible numerical values $\xi^{(1)}, \xi^{(2)}, \ldots$ that the random variable $x$ can take must be used to describe the random experiment. In exactly the same way, if the elementary events are assigned functions, *all* possible functions in this random experiment must be used to describe it. The family (ensemble) of these functions $\xi^{(1)}(t), \xi^{(2)}(t), \ldots$ is called a *random or stochastic* process. It is denoted by $\{x(t)\}$. A single function $\xi^{(i)}(t)$ of the random process $\{x(t)\}$ is called a *sample function* (a sample) or a *realization of the random process*. Of course, this definition is only meaningful if all sample functions $\xi^{(1)}(t), \xi^{(2)}(t), \ldots$ occur under the same conditions (i.e., in the above example all resistor components must have the same resistance $R$ ohms, the same temperature).

Therefore a random process $\{x(t)\}$ represents the generalization of a random variable $x$. Conversely, it is important to note that a random variable $x(t_1)$ which takes the functional values $\xi^{(1)}(t_1)$, $\xi^{(2)}(t_1), \ldots$ of the sample functions $\xi^{(1)}(t)$, $\xi^{(2)}(t)$, $\ldots$, as numerical values can be obtained by taking a snapshot of the random process $\{x(t)\}$ at time $t = t_1$. It is not the occurrence of the sample function $\xi^{(i)}(t)$ that is determined by chance, but the occurrence of the elementary event $\omega_i$. The sample function $\xi^{(i)}(t)$ is assigned to it in a fixed manner and takes the functional value $\xi^{(i)}(t_1)$ at the point $t = t_1$.

If every elementary event $\omega_i$ has not just a single function $\xi^{(i)}(t)$ assigned to it but, for example, $n$ functions $\xi_1^{(i)}(t), \ldots, \xi_n^{(i)}(t)$, then $\xi_k^{(i)}(t)$, $k = 1, \ldots, n$, is the realization of the scalar random process $\{x_k(t)\}$, $k = 1, \ldots, n$, and the vector

$$\boldsymbol{\xi}^{(i)}(t) = [\xi_1^{(i)}(t), \ldots, \xi_n^{(i)}(t)]'$$

of all $n$ sample functions $\xi_i^{(i)}(t), \ldots, \xi_n^{(i)}(t)$ is the realization of the *vector random process:*

$$\{\boldsymbol{x}(t)\} = \{[x_1(t), \ldots, x_n(t)]'\}$$

In this case also a vector random variable

$$\boldsymbol{x}(t_1) = [x_1(t_1), \ldots, x_n(t_1)]'$$

can be obtained from the vector random process $\{\boldsymbol{x}(t)\}$ by taking a snapshot at time $t = t_1$. The vector random variable can take the numerical values:

$$\boldsymbol{\xi}^{(i)}(t_1) = [\xi_1^{(i)}(t_1), \ldots, \xi_n^{(i)}(t_1)]' \quad i = 1, 2, \ldots$$

These relationships are illustrated in Fig. 2.24, which is a generalization of Figure 2.12.

If, as has been assumed so far, every elementary event $\omega_i$ of a random experiment has a function $\xi^{(i)}(t)$ or $\boldsymbol{\xi}^{(i)}(t)$ assigned to it in a fixed manner defined for all points inside a finite or an infinite interval $T$, we have a random process in continuous time. It is denoted by $\{x(t), t \leftarrow T)$ or $\{\boldsymbol{x}(t), t \leftarrow T\}$. However, if $\omega_i$ has a sequence of numbers $\xi^{(i)}(t_\nu)$, $\nu = 1, 2, \ldots$, or $\boldsymbol{\xi}^{(i)}(t_\nu)$, we have a random process in discrete time which is denoted by $\{x(t_\nu), \nu = 1, 2, \ldots\}$ or $\{\boldsymbol{x}(t_\nu), \nu = 1, 2, \ldots\}$. As a numerical sequence $f(t_\nu)$, $\nu = 1, 2, \ldots$, is just a function $f(t)$ with a domain of definition that does not consist of all points on the $t$ axis or inside an interval $T$ but only of the discrete points $t_1, t_2, \ldots$, the random process in discrete time is included as a special case of the random process in continuous time. The distinction between continuous

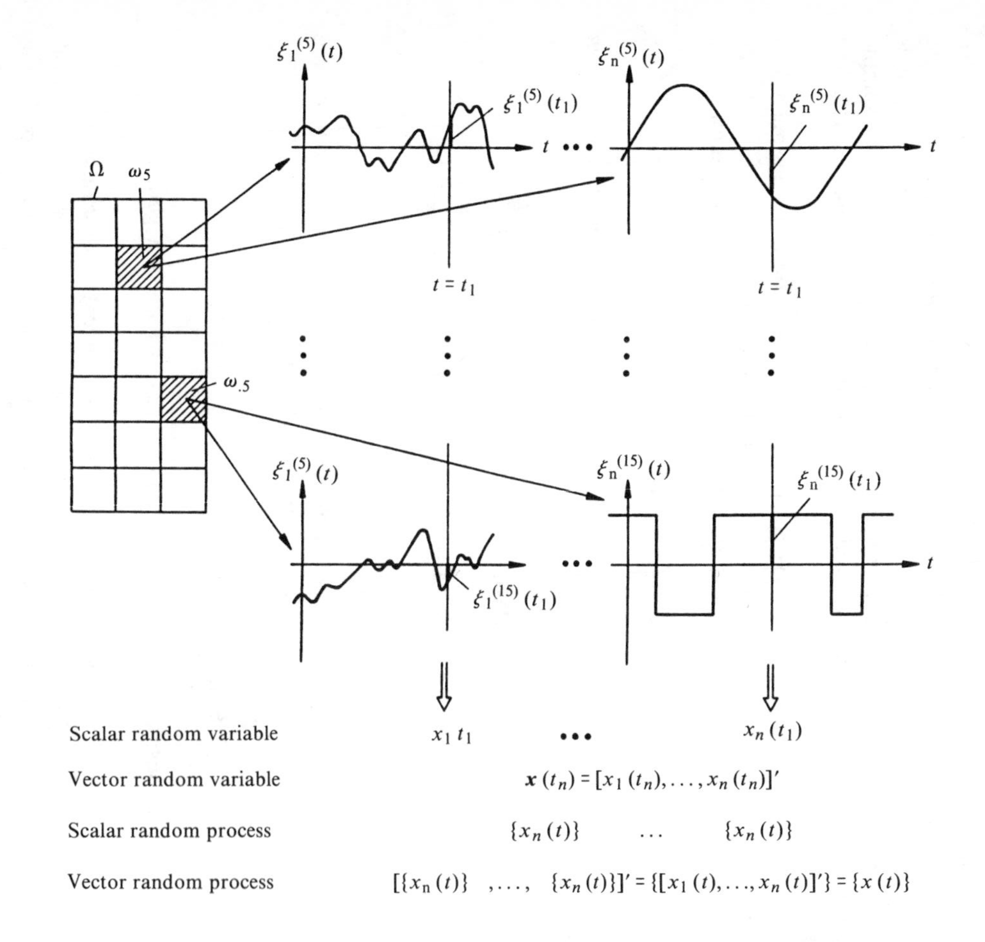


**Figure 2.24** Assignment of $n$ functions $\xi_k^{(i)}(t)$, $k = 1, \ldots, n$, to an elementary event $\omega_i$, $i = 1, 2, \ldots$.

and discrete can be made for the functional values of the samples as well as for their argument $t$.

In the above example of a random process with sample functions that are noise voltages for the same resistors, chance is involved at two points. First, the selection of a particular resistor with resistance $R$ ohms (i.e., the occurrence of a particular elementary event, $\omega_i$, depends on chance). Second, the variation in the noise voltage associated with this resistor is also determined by chance; thus, the function $\xi^{(i)}(t)$ is a random function having a variation that cannot be described by using a formula.

In the definition of a random process, the functions assigned to the elementary events in a fixed manner were not required to be random functions. In other words, even if all these functions are deterministic, their totality is called a random process. The following example is used to illustrate this point. Let four elementary events be possible in a random experiment. These are assigned four deterministic functions in a fixed manner:

$$\xi^{(1)}(t) = 2 \sin\left(t + \frac{\pi}{12}\right) = \psi(\omega_1, t)$$

$$\xi^{(2)}(t) = 0.5 \sin\left(3t + \frac{\pi}{5}\right) = \psi(\omega_2, t)$$

$$\xi^{(3)}(t) = 4 \sin\left(0.2t + \frac{\pi}{3}\right) = \psi(\omega_3, t)$$

$$\xi^{(4)}(t) = -3 \sin\left(12t + \frac{\pi}{9}\right) = \psi(\omega_4, t)$$

These four functions also form a random process $\{x(t)\}$ because every elementary event $\omega_i$ has a function $\xi^{(i)}(t)$ assigned to it in a fixed manner. Although all four sample functions are known, we cannot specify in advance which elementary event and which of the four sample functions will occur as the result of the random experiment. In addition, we cannot predict which of the four known functional values $\xi^{(i)}(t_1)$, $i = 1, 2, 3, 4$, will occur as a result of the experiment at $t = t_1$. Therefore $x(t_1)$ is also a random variable.

Sometimes the sum

$$\{z(t)\} = \{x_1(t)\} + \{x_2(t)\}$$

of two random processes $\{x_1(t)\}$ and $\{x_2(t)\}$ is also used. This consists of all sample functions $\zeta^{(i)}(t)$, where

$$\zeta^{(i)}(t) = \xi_1^{(i)}(t) + \xi_2^{(i)}(t)$$

if $\xi_1^{(i)}(t)$ is a sample function of $\{x_1(t)\}$ and $\xi_2^{(i)}(t)$ is a sample function of $\{x_2(t)\}$. Of course, this definition is only valid if the number of sample functions $\xi_1^{(i)}(t)$ is equal to the number of sample functions $\xi_2^{(i)}(t)$. Similarly, the random variable $z(t_1)$ is the sum of $x_1(t_1)$ and $x_2(t_1)$:

$$z(t_1) = x_1(t_1) + x_2(t_1)$$

Other combination operations are defined for more than one random process in the same manner.

As the conclusion to this section we will describe another way of defining a random process (Gnedenko, 1978). In a random experiment, let $\Omega$ be the set of all elementary events $\omega$ and $t$ be a parameter. A random process $\{x(t)\}$ is then a function of two arguments:

$$\{x(t)\} = \psi(t, \omega) \quad \omega \in \Omega$$

For every value $t_1$ of the parameter $t$, $\psi(t_1, \omega)$ depends only on $\omega$ which represents all possible elementary events whose occurrence or nonoccurrence is determined by chance. Therefore, for a fixed $t = t_1$, $\psi(t_1, \omega)$ is a random variable $x(t_1)$ which can take the numerical values $\psi(t_1, \omega_1)$, $\psi(t_2, \omega_2)$. However, for fixed $\omega = \omega_i$, $\psi(t, \omega_i) = \xi^{(i)}(t)$ is still only a function of $t$ whose variation may of course differ from one elementary event to another. The random process $\{x(t)\}$ can now be regarded as the complete collection of all random variables $x(t_1), x(t_2), \ldots$ or of all sample functions $\xi^{(1)}(t), \xi^{(2)}(t), \ldots$ The collection of all random variables $x(t_1), x(t_2), \ldots$ is expressed by the notation $\{x(t)\}$.

In the next section, we consider the question of how the properties of a random process can be described using the methods of probability theory considered so far. Only random processes whose sample functions can take real values will be allowed.

### 2.6.2 Correlation and Covariance Functions

In describing the properties of a random process with probability theory, we use the fact that, for a fixed point in time $t = t_1$, a random process becomes a random variable. The new viewpoints and concepts arising in this connection will first be discussed for a scalar random process.

### 2.6.2.1 *Autocorrelation and Autocovariance Functions*

For a fixed point in time $t = t_1$, the random process $\{x(t)\}$ becomes a random variable $x(t_1)$. The random variable $x(t_1)$, which is assumed to have originated from the random process $\{x(t)\}$ through a snapshot at $t = t_1$, has a distribution function (see Figure 2.24)

$$F_{\text{I}}(\xi_1, t_1) := P\{x(t_1) < \xi_1\} \tag{2.260}$$

the variation of which, in contrast with the variation of distribution function $F(\xi)$ of a random variable $x$ (equation (2.81)), still depends on the time $t_1$. The subscript I in $F_{\text{I}}$ indicates that $F_{\text{I}}(\xi_1, t_1)$ is the distribution function of a *one*-dimensional random variable $x(t_1)$. $F_{\text{I}}(\xi_1, t_1)$ is also known as the *first distribution function* of the random process $\{x(t)\}$. If the first distribution function depends on $t_1$, the associated *first density function* of the random process $\{x(t)\}$

$$f_{\text{I}}(\xi_1, t_1) := \frac{\partial}{\partial \xi_1} F_{\text{I}}(\xi_1, t_1) \tag{2.261}$$

must also depend on $t_1$. Now that $F_{\text{I}}(\xi_1, t_1)$ and $f_{\text{I}}(\xi_1, t_1)$ have been defined, the "random experiment" associated with the random variable $x(t_1)$ and therefore with the random process $\{x(t)\}$ is completely described, although only for the point in time $t = t_1$.

In addition to the one-dimensional random variable $x(t_1)$, a two-dimensional random variable $[x(t_1), x(t_2)]'$ with the distribution function

$$F_{\text{II}}(\xi_1, t_1, \xi_2, t_2) := P\{[x(t_1) < \xi_1] \cap [x(t_2) < \xi_2]\} \tag{2.262}$$

which is also called the *second distribution function* of the random process $\{x(t)\}$, and the *second density function*

$$f_{\text{II}}(\xi_1, t_1, \xi_2, t_2) := \frac{\partial^2}{\partial \xi_1 \partial \xi_2} F_{\text{II}}(\xi_1, t_1, \xi_2, t_2) \tag{2.263}$$

can be obtained from the random process $\{x(t)\}$ by taking two snapshots at $t = t_1$ and $t = t_2$. The first distribution function $F_{\text{I}}(\xi_1, t_1)$ and the first density function $f_{\text{I}}(\xi_1, t_1)$ only allow the properties of the random process $\{x(t)\}$ to be described point by point for different values of $t$. However, the second distribution function $F_{II}(\xi_1, t_1, \xi_2, t_2)$ and the second density function $f_{II}(\xi_1, t_1, \xi_2, t_2)$ are measures of the internal relationship between the two random variables $x(t_1)$ and $x(t_2)$ (e.g., equation (2.108)) and therefore are also a measure of the internal relationship which exists for $\{x(t)\}$ between the points $t = t_1$ and $t = t_2$.

The third distribution function $F_{III}(\xi_1, t_1, \xi_2, t_2, \xi_3, t_3)$, the fourth distribution function $F_{IV}(\xi_1, t_1, \xi_2, t_2, \xi_3, t_3, \xi_4, t_4)$ *et cetera* give correspondingly more information on the properties of the random process $\{x(t)\}$. We will not consider here whether it is possible to achieve a complete description of the properties of the random process $\{x(t)\}$ in this way or what this complete description would look like. Purely for practical reasons, the exactness of the description of the properties of a random process cannot in general be extended indefinitely. To describe the random process $\{x(t)\}$ with the first distribution function $F_I(\xi_1, t_1)$, this function must be determined by using measurement technology for points lying sufficiently close together, or analytically from the laws governing the sample functions of $\{x(t)\}$. However, we must remember that in general only a small number of the possible sample functions can be determined. The determination of the second distribution function $F_{II}(\xi_1, t_1, \xi_2, t_2)$ for pairs of points $(t_1, t_2)$ lying sufficiently close together and of subsequent distribution functions is obviously even more difficult. For these reasons, we often prefer to determine the associated moments or expectation values rather than describing the properties of a random process by distribution or density functions. However, we again limit ourselves to the first- and second-order moments of the one- or two-dimensional random variables $x(t_1)$ or $[x(t_1), x(t_2)]'$. When these expectations are known, a large number of important practical problems can be solved. However, there are also problems for which this global description of the properties of a random process is inadequate.

The moments or expectations of the random variables $x(t_1)$ or $[x(t_1), x(t_2)]'$ can be formed using the formulas derived in Sections 2.5.1 and 2.5.2 for the random variables $x$ and $[x_1, x_2]'$. Here, however, we also must consider the time dependence of the corresponding density functions. We have

$$E\{x(t_1)\} = \int_{-\infty}^{+\infty} \xi_1 f_I(\xi_1, t_1) d\xi_1 \tag{2.264}$$

$$E\{x^2(t_1)\} = \int_{-\infty}^{+\infty} \xi_1^2 f_I(\xi_1, t_1) d\xi_1 \tag{2.265}$$

$$E\{[x(t_1) - E\{x(t_1)\}]^2\} = \int_{-\infty}^{+\infty} [\xi_1 - E\{x(t_1)\}]^2 f_I(\xi_1, t_1) d\xi_1 \tag{2.266}$$

and so forth. All of these expectations are functions of $t_1$ (i.e., they are functions of time and are not constant numerical values). This is because the mean value $E\{x(t_1)\}$, for example, is an *ensemble mean value*, i.e., a mean value of the functional values $\xi^{(1)}(t_1), \xi^{(2)}(t_1), \ldots$ of *all* sample functions $\xi^{(1)}(t), \xi^{(2)}(t), \ldots$ of $\{x(t)\}$ for fixed $t = t_1$. Of course, this mean value depends on the value of $t_1$. The same argument applies to the other expectations of the random variable $x(t_1)$. A clear distinction must be made between the ensemble mean value $E\{x(t_1)\}$ and the *time mean value* over *all* functional values of a *single* sample function $\xi^{(i)}(t)$ of $\{x(t)\}$.

Like the moments of $x(t_1)$, the moments of the two-dimensional random variable $[x(t_1), x(t_2)]'$, e.g., the second-order ordinary moment

$$E\{x(t_1)x(t_2)\} = \int_{-\infty}^{+\infty}\int_{-\infty}^{+\infty} \xi_1\xi_2 f_{\mathrm{II}}(\xi_1, t_1, \xi_2, t_2)\mathrm{d}\xi_1\mathrm{d}\xi_2 \tag{2.267}$$

the second-order central moment

$$E\{[x(t_1) - E\{x(t_1)\}][x(t_2) - E\{x(t_2)\}]\}$$
$$= \int_{-\infty}^{+\infty}\int_{-\infty}^{+\infty} [\xi_1 - E\{x(t_1)\}][\xi_2 - E\{x(t_2)\}] f_{\mathrm{II}}(\xi_1, t_1, \xi_2, t_2)\mathrm{d}\xi_1\mathrm{d}\xi_2 \tag{2.268}$$

are functions of time. They depend on the two points in time $t_1$ and $t_2$. Since the two second-order moments of the random variable $[x(t_1), x(t_2)]'$ are of great importance in characterizing the random process $\{x(t)\}$, they are given their own names and symbols. The second-order ordinary moment $E\{x(t_1)x(t_2)\}$ is called the *autocorrelation function* $r_{xx}(t_1, t_2)$ of the random process $\{x(t)\}$:†

$$r_{xx}(t_1, t_2) := E\{x(t_1)x(t_2)\} \tag{2.269}$$

The second-order central moment $E\{[x(t_1) - E\{x(t_1)\}][x(t_2) - E\{x(t_2)\}]\}$ is called the *autocovariance function* $c_{xx}(t_1, t_2)$ of the random process $\{x(t)\}$:‡

$$c_{xx}(t_1, t_2) := E\{[x(t_1) - E\{x(t_1)\}][x(t_2) - E\{x(t_2)\}]\} \tag{2.270}$$

†The symbols $R_{xx}(t_1, t_2)$, $R_x(t_1, t_2)$, $\Phi_{xx}(t_1, t_2)$ and $\phi_{xx}(t_1, t_2)$ are also used instead of $r_{xx}(t_1, t_2)$.

‡The symbol $C_{xx}(t_1, t_2)$ is also used instead of $c_{xx}(t_1, t_2)$.

The autocovariance function $c_{xx}(t_1, t_2)$ of the random process $\{x(t)\}$ is identical with the covariance of the two-dimensional random variable $[x(t_1), x(t_2)]'$. By analogy with equation (2.181) there is a relationship between $c_{xx}(t_1, t_2)$ and $r_{xx}(t_1, t_2)$:

$$c_{xx}(t_1, t_2) = r_{xx}(t_1, t_2) - E\{x(t_1)\}E\{x(t_2)\} \tag{2.271}$$

Therefore the autocovariance function coincides with the autocorrelation function if the mean value $E\{x(t_1)\}$ vanishes for all values of $t_1$.

A random process can also be described by conditional distribution or density functions. If, for example, we replace the two random variables $x_1$ and $x_2$ in equation (2.118) by the two random variables $x(t_1)$ and $x(t_2)$ obtained from the random process $\{x(t)\}$ by snapshots at $t = t_1$ and $t = t_2$, the *conditional distribution function* $F(\xi_1, t_1|x(t_2) = \xi_2)$ of the random process $\{x(t)\}$ becomes

$$\begin{aligned} &\lim_{\Delta\xi_2 \to 0} P(x(t_1) < \xi_1 | \xi_2 \leqslant x(t_2) < \xi_2 + \Delta\xi_2) \\ &= \frac{\int_{-\infty}^{\xi_1} f_{\mathrm{II}}(v_1, t_1, \xi_2, t_2)\mathrm{d}v_1}{\int_{-\infty}^{+\infty} f_{\mathrm{II}}(\xi_1, t_1, \xi_2, t_2)\mathrm{d}\xi_1} \\ &=: F(\xi_1, t_1|x(t_2) = \xi_2) \end{aligned} \tag{2.272}$$

This function depends on $\xi_1$, $\xi_2$ and $t_1$, $t_2$. Application of equation (2.115) to the denominator in (2.272) gives the density function $f_{\mathrm{I},2}(\xi_2, t_2)$ of the marginal distribution of the one-dimensional random variable $x(t_2)$ with respect to the two-dimensional random variable $[x(t_1), x(t_2)]'$. The subscript I in $f_{\mathrm{I},2}(\xi_2, t_2)$ means that $f_{\mathrm{I},2}(\xi_2, t_2)$ is the first density function of the random process $\{x(t)\}$:

$$\begin{aligned} F\{\xi_1, t_1|x(t_2) = \xi_2\} &= \frac{\int_{-\infty}^{\xi_1} f_{\mathrm{II}}(v_1, t_1, \xi_2, t_2)\mathrm{d}v_1}{f_{\mathrm{I},2}(\xi_2, t_2)} \\ &= \int_{-\infty}^{\xi_1} \frac{f_{\mathrm{II}}(v_1, t_1, \xi_2, t_2)}{f_{\mathrm{I},2}(\xi_2, t_2)} \mathrm{d}v_1 \end{aligned}$$

The *conditional density function* $f\{\xi_1, t_1|x(t_2) = \xi_2\}$ of the random process $\{x(t)\}$ associated with the conditional distribution function $F\{\xi_1, t_1|x(t_2) = \xi_2\}$ is simply obtained as the integrand in the last expression taken at the upper integration limit (see also equation (2.120)):

$$f\{\xi_1, t_1 | x(t_2) = \xi_2\} := \frac{\partial}{\partial \xi_1} F\{\xi_1, t_1 | x(t_2) = \xi_2\}$$
$$= \frac{f_{\text{II}}(\xi_1, t_1, \xi_2, t_2)}{f_{\text{I},2}(\xi_2, t_2)} \tag{2.273}$$

*2.6.2.2 Cross-Correlation and Cross-Covariance Functions*

So far we have obtained a two-dimensional random variable $[x(t_1), x(t_2)]'$ from *one* scalar random process $\{x(t)\}$ by taking *two* snapshots at $t = t_1$ and $t = t_2$. However, we can also obtain two random variables $x_1(t_1)$ and $x_2(t_2)$, which can also be combined to form a two-dimensional random variable $[x_1(t_1), x_2(t_2)]'$, from two scalar random processes $\{x_1(t)\}$ and $\{x_2(t)\}$ by taking *one snapshot in each case* at $t = t_1$ and $t = t_2$. This two-dimensional random variable has a *second distribution function:*

$$F_{\text{II,d}}(\xi_1, t_1, \xi_2, t_2) := P[\{x_1(t_1) < \xi_1\} \cap \{x_2(t_2) < \xi_2\}] \tag{2.274}$$

and a *second density function:*

$$f_{\text{II,d}}(\xi_1, t_1, \xi_2, t_2) := \frac{\partial^2}{\partial \xi_1 \partial \xi_2} F_{\text{II,d}}(\xi_1, t_1, \xi_2, t_2) \tag{2.275}$$

The subscript d in the distribution function (2.274) and the density function (2.275) of the two-dimensional random variable $[x_1(t_1), x_2(t_2)]'$ is necessary in order to distinguish these two functions from the distribution function (2.262) and the density function (2.263) of a two-dimensional random variable $[x(t_1), x(t_2)]'$ the components of which have been obtained from a *single* random process $\{x(t)\}$.*

The properties of the two random processes $\{x_1(t)\}$ and $\{x_2(t)\}$ can also be described using the moments of the scalar random variables $x_1(t_1)$ and $x_2(t_2)$ and the vector random variable $[x_1(t_1), x_2(t_2)]'$. In particular, the second-order moments of $[x_1(t_1), x_2(t_2)]'$ are the *ordinary moment*

$$E\{x_1(t_1)x_2(t_2)\} = \int_{-\infty}^{+\infty} \int_{-\infty}^{+\infty} \xi_1 \xi_2 f_{\text{II,d}}(\xi_1, t_1, \xi_2, t_2) \mathrm{d}\xi_1 \mathrm{d}\xi_2 \tag{2.276}$$

---

*The notation d indicates that the components in $[x_1(t_1), x_2(t_2)]'$ have been taken from two *different* scalar random processes $\{x_1(t)\}$ and $\{x_2(t)\}$.

and the *central moment*

$$E\{[x_1(t_1) - E\{x_1(t_1)\}][x_2(t_2) - E\{x_2(t_2)\}]\}$$
$$= \int_{-\infty}^{+\infty}\int_{-\infty}^{+\infty} [\xi_1 - E\{x_1(t_1)\}][\xi_2 - E\{x_2(t_2)\}]f_{\mathrm{II,d}}(\xi_1, t_1, \xi_2, t_2)\mathrm{d}\xi_1\mathrm{d}\xi_2 \tag{2.277}$$

The second-order moments measure the statistical relationship that exists between the random variables $x_1(t_1)$ and $x_2(t_2)$, and thus between the random processes $\{x_1(t)\}$ and $\{x_2(t)\}$ at times $t_1$ and $t_2$. Therefore these two moments have a similar importance to the autocorrelation function $r_{xx}(t_1, t_2)$ and the autocovariance function $c_{xx}(t_1, t_2)$ for a single random process $\{x(t)\}$. The second-order ordinary moment $E\{x_1(t_1)x_2(t_2)\}$ is called the *cross-correlation function* $r_{x_1x_2}(t_1, t_2)$ and the second-order central moment $E\{[x_1(t_1) - E\{x_1(t_1)\}][x_2(t_2) - E\{x_2(t_2)\}]\}$ is called the *cross-covariance function* $c_{x_1x_2}(t_1, t_2)$:

$$r_{x_1x_2}(t_1, t_2) := E\{x_1(t_1)x_2(t_2)\} \tag{2.278}$$

$$c_{x_1x_2}(t_1, t_2) := E\{[x_1(t_1) - E\{x_1(t_1)\}][x_2(t_2) - E\{x_2(t_2)\}]\} \tag{2.279}$$

$c_{x_1x_2}(t_1, t_2)$ is the covariance of the random variable $[x_1(t_1), x_2(t_2)]'$. From equation (2.181) we must have

$$c_{x_1x_2}(t_1, t_2) = r_{x_1x_2}(t_1, t_2) - E\{x_1(t_1)\}E\{x_2(t_2)\} \tag{2.280}$$

The cross-covariance function thus becomes the cross-correlation function if the mean value $E\{x_1(t_1)\}$ of the random variable $x_1(t_1)$ is zero for all $t_1$ or if $E\{x_2(t_2)\}$ is zero for all $t_2$.

The concepts and relationships that we have introduced are also valid if the two scalar random processes $\{x_1(t)\}$ and $\{x_2(t)\}$ are combined to form a two-dimensional random process:

$$\{\boldsymbol{x}(t)\} = \{[x_1(t), x_2(t)]'\}$$

The two-dimensional random variables $\boldsymbol{x}(t_1) = [x_1(t_1), x_2(t_1)]'$ and $\boldsymbol{x}(t_2) = [x_1(t_2), x_2(t_2)]'$ can be obtained from $\{\boldsymbol{x}(t)\}$ by taking snapshots at $t = t_1$ and $t = t_2$.

The autocorrelation and cross-correlation functions defined here can be combined to form the *correlation matrix* $\boldsymbol{R}_{xx}(t_1, t_2)$:

$$\boldsymbol{R}_{\boldsymbol{xx}}(t_1, t_2) := E\{\boldsymbol{x}(t_1)\boldsymbol{x}'(t_2)\}$$
$$= \begin{bmatrix} E\{x_1(t_1)x_1(t_2)\} & E\{x_1(t_1)x_2(t_2)\} \\ E\{x_2(t_1)x_1(t_2)\} & E\{x_2(t_1)x_2(t_2)\} \end{bmatrix} \tag{2.281}$$

The main diagonal of this matrix consists of the autocorrelation functions $r_{x_1x_1}(t_1, t_2)$ and $r_{x_2x_2}(t_1, t_2)$ of the components of $\{\boldsymbol{x}(t)\}$. The cross-correlation functions $r_{x_1x_2}(t_1, t_2)$ and $r_{x_2x_1}(t_1, t_2)$ lie outside the main diagonal. Note that the cross-correlation function $r_{x_2x_1}(t_1, t_2) = E\{x_2(t_1)x_1(t_2)\}$ describes the statistical relationship between the random variables $x_1(t_2)$ and $x_2(t_1)$, but not, as in the case of $r_{x_1x_2}(t_1, t_2) = E\{x_1(t_1)x_2(t_2)\}$, the relationship between $x_1(t_1)$ and $x_2(t_2)$. We have

$$r_{x_1x_2}(t_1, t_2) \neq r_{x_2x_1}(t_1, t_2) \quad t_1 \neq t_2 \tag{2.282}$$

However,

$$r_{x_1x_2}(t, t) = r_{x_2x_1}(t, t) \tag{2.283}$$

The correlation matrix is therefore asymmetric for $t_1 \neq t_2$ and symmetric for $t_1 = t_2$.

The *covariance matrix* $\boldsymbol{C}_{\boldsymbol{xx}}(t_1, t_2)$ for a two-dimensional random process $\{\boldsymbol{x}(t)\} = \{[x_1(t), x_2(t)]'\}$ can be obtained in a similar way to the correlation matrix by combining all possible autocovariance and cross-covariance functions:

$$\boldsymbol{C}_{\boldsymbol{xx}}(t_1, t_2) := E\{[\boldsymbol{x}(t_1) - E\{\boldsymbol{x}(t_1)\}][\boldsymbol{x}(t_2) - E\{\boldsymbol{x}(t_2)\}]'\}$$
$$= \begin{bmatrix} E\{[x_1(t_1) - E\{x_1(t_1)\}][x_1(t_2) - E\{x_1(t_2)\}]\} & E\{[x_1(t_1) - E\{x_1(t_1)\}][x_2(t_2) - E\{x_2(t_2)\}]\} \\ E\{[x_2(t_1) - E\{x_2(t_1)\}][x_1(t_2) - E\{x_1(t_2)\}]\} & E\{[x_2(t_1) - E\{x_2(t_1)\}][x_2(t_2) - E\{x_2(t_2)\}]\} \end{bmatrix} \tag{2.284}$$

The main diagonal of $\boldsymbol{C}_{\boldsymbol{xx}}(t_1, t_2)$ contains the autocovariance functions $c_{x_1x_1}(t_1, t_2)$ and $c_{x_2x_2}(t_1, t_2)$ of the two components of $\{\boldsymbol{x}(t)\}$, and the elements outside the main diagonal are the cross-covariance functions $c_{x_1x_2}(t_1, t_2)$ and $c_{x_2x_1}(t_1\ t_2)$, where

$$c_{x_1x_2}(t_1, t_2) \neq c_{x_2x_1}(t_1, t_2) \quad t_1 \neq t_2 \tag{2.285}$$

and

$$c_{x_1x_2}(t, t) = c_{x_2x_1}(t, t) \tag{2.286}$$

Therefore, the covariance matrix $C_{xx}(t_1, t_2)$ is also symmetric for $t_1 = t_2$ and asymmetric for $t_1 \neq t_2$.

In the case of an $n$-dimensional random process,

$$\{\boldsymbol{x}(t)\} = \{[x_1(t), \ldots, x_n(t)]'\}$$

the *correlation matrix* $\boldsymbol{R}_{xx}(t_1, t_2)$, which contains all possible autocorrelation and cross-correlation functions of $\{\boldsymbol{x}(t)\}$, and the *covariance matrix* $\boldsymbol{C}_{xx}(t_1, t_2)$, which contains all possible autocovariance and cross-covariance functions of $\{\boldsymbol{x}(t)\}$, are defined in exactly the same way as for a two-dimensional random process:

$$\boldsymbol{R}_{xx}(t_1, t_2) := E\{\boldsymbol{x}(t_1)\boldsymbol{x}'(t_2)\} = \begin{bmatrix} r_{x_1x_1}(t_1, t_2) \ldots & r_{x_1x_n}(t_1, t_2) \\ \vdots & \vdots \\ r_{x_nx_1}(t_1, t_2) \ldots & r_{x_nx_n}(t_1, t_2) \end{bmatrix} \tag{2.287}$$

$$\boldsymbol{C}_{xx}(t_1, t_2) := E\{[\boldsymbol{x}(t_1) - E\{\boldsymbol{x}(t_1)\}][\boldsymbol{x}(t_2) - E\{\boldsymbol{x}(t_2)\}]'\} = \begin{bmatrix} c_{x_1x_1}(t_1, t_2) \ldots & c_{x_1x_n}(t_1, t_2) \\ \vdots & \vdots \\ c_{x_nx_1}(t_1, t_2) \ldots & c_{x_nx_n}(t_1, t_2) \end{bmatrix} \tag{2.288}$$

The matrices $\boldsymbol{R}_{xx}(t_1, t_2)$ and $\boldsymbol{C}_{xx}(t_1, t_2)$ are asymmetric for $t_1 \neq t_2$ and symmetric for $t_1 = t_2$. By analogy with equations (2.271) and (2.280), $\boldsymbol{C}_{xx}(t_1, t_2)$ and $\boldsymbol{R}_{xx}(t_1, t_2)$ are related as follows:

$$\boldsymbol{C}_{xx}(t_1, t_2) = \boldsymbol{R}_{xx}(t_1, t_2) - E\{\boldsymbol{x}(t_1)\}E\{\boldsymbol{x}'(t_2)\} \tag{2.289}$$

This equation can be derived in exactly the same way as equation (2.191).

The covariance matrix $\boldsymbol{C}_{xx}(t_1, t_2)$ is identical with the correlation matrix $\boldsymbol{R}_{xx}(t_1, t_2)$ if $E\{\boldsymbol{x}(t_1)\} = \boldsymbol{0}$ for all $t_1$. We then have

$$\boldsymbol{C}_{xx}(t_1, t_2) = \boldsymbol{R}_{xx}(t_1, t_2) = E\{\boldsymbol{x}(t_1)\boldsymbol{x}'(t_2)\} \tag{2.290}$$

We have the relationships:

$$\boldsymbol{C}_{xx}(t_1, t_2) = \mathrm{COV}\{\boldsymbol{x}(t_1), \boldsymbol{x}(t_2)\} \tag{2.291}$$

$$\boldsymbol{C}_{xx}(t, t) = \mathrm{COV}\{\boldsymbol{x}(t), \boldsymbol{x}(t)\} \tag{2.292}$$

The covariance matrix $\boldsymbol{C}_{xx}(t_1, t_2)$ characterizes the properties of the vector random process $\{\boldsymbol{x}(t)\}$ through the two vector random variables $\boldsymbol{x}(t_1)$ and $\boldsymbol{x}(t_2)$ obtained from $\{\boldsymbol{x}(t)\}$ by taking snapshots at $t = t_1$ and $t = t_2$. The covariance matrices $\mathrm{COV}(\boldsymbol{x}, \boldsymbol{x})$ and $\mathrm{COV}(\boldsymbol{x}, \boldsymbol{y})$ introduced in Section 2.5.2 describe the probability distributions associated with the vector random variable $\boldsymbol{x}$ or the two vector random variables $\boldsymbol{x}$ and $\boldsymbol{y}$. Because of (2.292), however, the covariance matrix $\boldsymbol{C}_{xx}(t_1, t_2)$ of the vector random process $\{\boldsymbol{x}(t)\}$ must then be positive semidefinite, as was shown in Section 2.5.2.3 for the covariance matrix $\mathrm{COV}(\boldsymbol{x}, \boldsymbol{x})$.

The concepts of the correlation matrix $\boldsymbol{R}_{xx}(t_1, t_2)$ and the covariance matrix $\boldsymbol{C}_{xx}(t_1, t_2)$ can be generalized for *two* vector random processes (Sage and Melsa, 1971):

$$\{\boldsymbol{x}(t)\} = \{[x_1(t), \ldots, x_n(t)]'\} \quad \text{and} \quad \{\boldsymbol{y}(t)\} = \{[y_1(t), \ldots, y_m(t)]'\}$$

where $\boldsymbol{x}(t_1) = [x_1(t_1), \ldots, x_n(t_1)]'$ and $\boldsymbol{y}(t_2) = [y_1(t_2), \ldots, y_m(t_2)]'$ are respectively $n$-dimensional and $m$-dimensional random variables obtained by taking snapshots of $\{\boldsymbol{x}(t)\}$ at $t = t_1$ and $\{\boldsymbol{y}(t)\}$ at $t = t_2$:

$$\boldsymbol{R}_{xy}(t_1, t_2) := E\{\boldsymbol{x}(t_1)\boldsymbol{y}'(t_2)\} \tag{2.293}$$

$$\boldsymbol{C}_{xy}(t_1, t_2) := E\{[\boldsymbol{x}(t_1) - E\{\boldsymbol{x}(t_1)\}][\boldsymbol{y}(t_2) - E\{\boldsymbol{y}(t_2)\}]'\} \tag{2.294}$$

The *correlation matrix* $\boldsymbol{R}_{xy}(t_1, t_2)$ contains all possible cross-correlation functions for the random processes $\{\boldsymbol{x}(t)\}$ and $\{\boldsymbol{y}(t)\}$ but no autocorrelation functions. The *covariance matrix* $\boldsymbol{C}_{xy}(t_1, t_2)$ contains all possible cross-covariance functions for $\{\boldsymbol{x}(t)\}$ and $\{\boldsymbol{y}(t)\}$ but no autocovariance functions. $\boldsymbol{C}_{xy}(t_1, t_2)$ and $\boldsymbol{R}_{xy}(t_1, t_2)$ are related as follows:

$$\boldsymbol{C}_{xy}(t_1, t_2) = \boldsymbol{R}_{xy}(t_1, t_2) - E\{\boldsymbol{x}(t_1)\}E\{\boldsymbol{y}'(t_2)\} \tag{2.295}$$

In addition,

$$\boldsymbol{C}_{xy}(t_1, t_2) = \boldsymbol{R}_{xy}(t_1, t_2) = E\{\boldsymbol{x}(t_1)\boldsymbol{y}'(t_2)\} \tag{2.296}$$

if $E\{\boldsymbol{x}(t_1\} \equiv \boldsymbol{0}$ or $E\{\boldsymbol{y}(t_2)\} \equiv \boldsymbol{0}$.

### *2.6.2.3 Properties of the Autocorrelation and Cross-Correlation Functions*

(a) From the defining equation (2.269) we can see that

$$r_{xx}(t_1, t_2) = r_{xx}(t_2, t_1) \tag{2.297}$$

and

$$r_{xx}(t, t) = E\{x^2(t)\} \tag{2.298}$$

Therefore, if we know the autocorrelation function of the random process $\{x(t)\}$, we also know the second moment of the random variable $x(t)$ which can be obtained from $\{x(t)\}$ by a snapshot.

The situation for the cross-correlation function is not quite as simple. If the factors in the argument of the expectation in equation (2.278) are permuted, it follows that

$$r_{x_1x_2}(t_1, t_2) = r_{x_2x_1}(t_2, t_1) \tag{2.299}$$

As noted in Section 2.6.2.2, the inequality (2.282) holds, i.e.,

$$r_{x_1x_2}(t_1, t_2) \neq r_{x_2x_1}(t_1, t_2) \quad t_1 \neq t_2$$

However, for $t_1 = t_2 = t$, this becomes (2.283):

$$r_{x_1x_2}(t, t) = r_{x_2x_1}(t, t)$$

These two relationships indicate that the correlation matrix $\boldsymbol{R}_{\boldsymbol{xx}}(t_1, t_2)$ is symmetric only for $t_1 = t_2$ and is asymmetric for $t_1 \neq t_2$. It was also shown in Section 2.6.2.2 that the same conditions hold for the covariance matrix $\boldsymbol{C}_{\boldsymbol{xx}}(t_1, t_2)$.

(b) The following equation enables us to estimate the numerical values which the cross-correlation function $r_{x_1x_2}(t_1, t_2)$ can take:

$$|r_{x_1x_2}(t_1, t_2)|^2 \leqslant r_{x_1x_1}(t_1, t_1) r_{x_2x_2}(t_2, t_2) \tag{2.300}$$

*Proof*. The expectation or mean value of the square of a real random variable must always be greater than or equal to zero. For example, we have

$$E\left\{\left[\frac{x_1(t_1)}{\{r_{x_1x_1}(t_1, t_1)\}^{1/2}} \pm \frac{x_2(t_2)}{\{r_{x_2x_2}(t_2, t_2)\}^{1/2}}\right]^2\right\} \geqslant 0$$

For each pair of values $t_1$, $t_2$, the terms $x_1(t_1)$ and $x_2(t_2)$ are two random variables which are obtained by taking snapshots at $t = t_1$ and $t = t_2$ from the components $\{x_1(t)\}$ and $\{x_2(t)\}$ of the two-dimensional random process $\{\boldsymbol{x}(t)\} = \{[x_1(t), x_2(t)]'\}$; $r_{x_1x_1}(t_1, t_1)$ and $r_{x_2x_2}(t_2, t_2)$ are the autocorrelation functions of the components of $\{\boldsymbol{x}(t)\}$ for the same time parameters. They are fixed numerical values for every pair of values $t_1$, $t_2$. We shall now multiply the square in the above expression to give

$$E\left\{\frac{x_1^2(t_1)}{r_{x_1x_1}(t_1, t_1)} \pm 2\frac{x_1(t_1)x_2(t_2)}{\{r_{x_1x_1}(t_1, t_1)r_{x_2x_2}(t_2, t_2)\}^{1/2}} + \frac{x_2^2(t_2)}{r_{x_2x_2}(t_2, t_2)}\right\} \geqslant 0$$

This is the expectation of a one-dimensional random variable that depends on a two-dimensional random variable $[x_1(t_1), x_2(t_2)]'$. If we apply the calculation rule (2.224) to the preceding expression and take into account equations (2.278) and (2.298), it follows that

$$2 \pm 2\frac{r_{x_1x_2}(t_1, t_2)}{\{r_{x_1x_1}(t_1, t_1)r_{x_2x_2}(t_2, t_2)\}^{1/2}} \geqslant 0$$

or

$$\left|\frac{r_{x_1x_2}(t_1, t_2)}{\{r_{x_1x_1}(t_1, t_1)r_{x_2x_2}(t_2, t_2)\}^{1/2}}\right| \leqslant 1$$

or

$$|r_{x_1x_2}(t_1, t_2)| \leqslant |\{r_{x_1x_1}(t_1, t_1)r_{x_2x_2}(t_2, t_2)\}^{1/2}|$$

Because of (2.298), the expression under the square root must always be positive, and thus the modulus sign on the right-hand side of the inequality can be omitted if the positive value of the square root is taken. Therefore we have

$$|r_{x_1x_2}(t_1, t_2)| \leqslant +\{r_{x_1x_1}(t_1, t_1)r_{x_2x_2}(t_2, t_2)\}^{1/2}$$

or

$$|r_{x_1x_2}(t_1, t_2)|^2 \leqslant r_{x_1x_1}(t_1, t_1)r_{x_2x_2}(t_2, t_2)$$

This proves the relationship (2.300) (Davenport and Root 1958). If $x_1 = x_2 = x$, it also provides an estimate of the numerical values of the autocorrelation function $r_{xx}(t_1, t_2)$:

$$|r_{xx}(t_1, t_2)|^2 \leqslant r_{xx}(t_1, t_1)r_{xx}(t_2, t_2) \qquad (2.301)$$

(c) Two random processes $\{x_1(t)\}$ and $\{x_2(t)\}$ are said to be *independent* if the vector random variables $[x_1(t_1), \ldots, x_1(t_n)]'$ and $[x_2(t_1^*), \ldots, x_2(t_m^*)]'$ obtained from them are independent for any $n$ and $m$ and for any points in time $t_1, \ldots, t_n, t_1^*, \ldots, t_m^*$ (Papoulis, 1984). It follows from the independence of two vector random variables (see Section 2.4.7) that every component of one vector must be independent of every component of the other vector. However, because the vector $[x_1(t_1), \ldots, x_1(t_n)]'$ must be independent of the vector $[x_2(t_1^*), \ldots, x_2(t_m^*)]'$ for all points in time $t_1, \ldots, t_n$, $t_1^*, \ldots, t_m^*$, the fact that the random processes $\{x_1(t)\}$ and $\{x_2(t)\}$ are independent means, among other things, that every random variable $x_1(t_1)$ which can be obtained by taking a snapshot of $\{x_1(t)\}$ at any point in time $t = t_1$ is independent of every random variable $x_2(t_1^*)$ which has been obtained by taking a snapshot of $\{x_2(t)\}$ at any point in time $t = t_1^*$.

Random variables $x_1(t_1)$ and $x_2(t_1^*)$ or $x_2(t_2)$ which are independent of each other (for simplicity $t_1^*$ has been replaced by $t_2$) obey the relation (equation (2.198)):

$$E\{x_1(t_1)x_2(t_2)\} = E\{x_1(t_1)\}E\{x_2(t_2)\}$$

However, according to equation (2.278) the expectation $E\{x_1(t_1)x_2(t_2)\}$ is equal to the cross-correlation function $r_{x_1x_2}(t_1, t_2)$, so that we must have

$$r_{x_1x_2}(t_1, t_2) = E\{x_1(t_1)x_2(t_2)\} = E\{x_1(t_1)\}\mathrm{E}\{x_2(t_2)\} \tag{2.302}$$

It follows from this and equation (2.280) that the cross-covariance function $c_{x_1x_2}(t_1, t_2)$ is given by

$$c_{x_1x_2}(t_1, t_2) = E\{[x_1(t_1) - E\{x_1(t_1)\}][x_2(t_2) - E\{x_2(t_2)\}]\} = 0 \tag{2.303}$$

Equations (2.302) and (2.303) are satisfied for the independent random processes $\{x_1(t)\}$ and $\{x_2(t)\}$ for every time pair $t_1$, $t_2$.

However, for random processes $\{x_1(t)\}$ and $\{x_2(t)\}$, which are independent of each other, not only the random variables $x_1(t_1)$ and $x_2(t_2)$ but also the random variables $x_2(t_1)$ and $x_1(t_2)$ are independent of each other; thus we also have

$$r_{x_2x_1}(t_1, t_2) = E\{x_2(t_1)x_1(t_2)\} = E\{x_2(t_1)\}E\{x_1(t_2)\} \tag{2.304}$$

and

$$c_{x_2x_1}(t_1, t_2) = E\{[x_2(t_1) - E\{x_2(t_1)\}][x_1(t_2) - E\{x_1(t_2)\}]\} = 0 \tag{2.305}$$

for all time pairs $t_1$, $t_2$. Therefore the cross-covariance functions $c_{x_1x_2}(t_1, t_2)$ and $c_{x_2x_1}(t_1, t_2)$ for independent random processes $\{x_1(t)\}$ and $\{x_2(t)\}$ vanish for all $t_1$, $t_2$, whereas the cross-correlation functions $r_{x_1x_2}(t_1, t_2)$ and $r_{x_2x_1}(t_1, t_2)$ are only identically zero if in addition $E\{x_1(t_1)\} \equiv 0$ or $E\{x_2(t_2)\} \equiv 0$.

Two random processes $\{x_1(t)\}$ and $\{x_2(t)\}$ are uncorrelated if, for every pair of values $t_1$, $t_2$,

$$c_{x_1x_2}(t_1, t_2) = c_{x_2x_1}(t_1, t_2) = 0 \tag{2.306}$$

or, which is equivalent because of equation (2.280),

$$r_{x_1x_2}(t_1, t_2) = E\{x_1(t_1)\}E\{x_2(t_2)\} \tag{2.307a}$$

and

$$r_{x_2x_1}(t_1, t_2) = E\{x_2(t_1)\}E\{x_1(t_2)\} \tag{2.307b}$$

Again, as in Section 2.5.2.4, the uncorrelatedness of two random processes is a weaker property than their independence. Two independent random processes are always uncorrelated, whereas the converse is not generally true.

If we combine two independent random processes $\{x_1(t)\}$ and $\{x_2(t)\}$ to form a two-dimensional random process $\{\boldsymbol{x}(t)\} = \{[x_1(t), x_2(t)]'\}$, the elements in the covariance matrix $\boldsymbol{C}_{\boldsymbol{xx}}(t_1, t_2)$ associated with $\{\boldsymbol{x}(t)\}$ according to equation (2.284) vanish, except for the elements in the main diagonal, according to relationships (2.303) and (2.305). $\boldsymbol{C}_{\boldsymbol{xx}}(t_1, t_2)$ is then a diagonal matrix.

If in an $n$-dimensional random process $\{\boldsymbol{x}(t)\} = \{[x_1(t), \ldots, x_n(t)]'\}$ the components $\{x_1(t)\}, \ldots, \{x_n(t)\}$ are pairwise independent, the covariance matrix $\boldsymbol{C}_{\boldsymbol{xx}}(t_1, t_2)$ (equation (2.288)) of the $n$-dimensional random process $\{\boldsymbol{x}(t)\}$ is also a diagonal matrix. However, if every component of the $n$-dimensional random process $\{\boldsymbol{x}(t)\} = \{[x_1(t), \ldots, x_n(t)]'\}$ is independent of every component of the $m$-dimensional random process $\{\boldsymbol{y}(t)\} = \{[y_1(t), \ldots, y_m(t)]'\}$, the covariance matrix $\boldsymbol{C}_{\boldsymbol{xy}}(t_1, t_2)$ (equation (2.294)) is the null matrix because its elements are all cross-covariance functions which, according to the assumption, all vanish:

$$\boldsymbol{C}_{\boldsymbol{xy}}(t_1, t_2) \equiv \boldsymbol{0} \tag{2.308}$$

Because of equation (2.295) it follows from (2.308) that the correlation matrix $\boldsymbol{R}_{\boldsymbol{xy}}(t_1, t_2)$ of two vector random processes $\{\boldsymbol{x}(t)\}$ and $\{\boldsymbol{y}(t)\}$ defined by (2.293) is given by

$$\boldsymbol{R}_{\boldsymbol{xy}}(t_1, t_2) \equiv E\{\boldsymbol{x}(t_1)\}\mathrm{E}\{\boldsymbol{y}'(t_2)\} \tag{2.309}$$

This equation applies only to two vector random processes $\{\boldsymbol{x}(t)\}$ and $\{\boldsymbol{y}(t)\}$ which are such that every component of $\{\boldsymbol{x}(t)\}$ is independent of every component of $\{\boldsymbol{y}(t)\}$.

If either equation (2.308) or (2.309) holds for two vector random processes $\{\boldsymbol{x}(t)\}$ and $\{\boldsymbol{y}(t)\}$, we say that these two processes are *uncorrelated* for all $t_1$, $t_2$ (Sage and Melsa, 1971), where, in general, the independence of $\{\boldsymbol{x}(t)\}$ and $\{\boldsymbol{y}(t)\}$ or the independence of their components cannot be deduced from this.

(d) The concept of the sum of two random processes was explained in Section 2.6.1. We shall now determine the autocorrelation function of the sum or difference process:

$$\{z(t)\} = \{x_1(t)\} \pm \{x_2(t)\}$$

according to equation (2.269):

$$r_{zz}(t_1, t_2) = E\{z(t_1)z(t_2)\}$$

Because (see Section 2.6.1)

$$z(t_1) = x_1(t_1) \pm x_2(t_1) \quad z(t_2) = x_1(t_2) \pm x_2(t_2)$$

it follows that the autocorrelation function $r_{zz}(t_1, t_2)$ is

$$\begin{aligned} r_{zz}(t_1, t_2) &= E\{\{x_1(t_1) \pm x_2(t_1)\}\{x_1(t_2) \pm x_2(t_2)\}\} \\ &= \mathrm{E}\{x_1(t_1)x_1(t_2) \pm x_2(t_1)x_1(t_2) \pm x_1(t_1)x_2(t_2) + x_2(t_1)x_2(t_2)\} \end{aligned}$$

Using the calculation rule (2.224) and equations (2.269) and (2.278) we can rewrite the above equation as

$$\begin{aligned} r_{x_1 \pm x_2, x_1 \pm x_2}(t_1, t_2) = r_{x_1x_1}(t_1, t_2) \pm r_{x_2x_1}(t_1, t_2) \pm r_{x_1x_2}(t_1, t_2) \\ + r_{x_2x_2}(t_1, t_2) \end{aligned} \tag{2.310}$$

If the two random processes $\{x_1(t)\}$ and $\{x_2(t)\}$ are independent or uncorrelated and if we also have $E\{x_1(t_1)\} \equiv 0$ or $E\{x_2(t_2)\} \equiv 0$, then because of (2.302) and (2.304) or, for uncorrelated processes, because of (2.307a) and (2.307b) the two cross-correlation functions $r_{x_2x_1}(t_1, t_2)$ and $r_{x_1x_2}(t_1, t_2)$ vanish and equation (2.310) can be expressed in the simple form

$$r_{x_1 \pm x_2, x_1 \pm x_2}(t_1, t_2) = r_{x_1x_1}(t_1, t_2) + r_{x_2x_2}(t_1, t_2) \tag{2.311}$$

We can calculate the cross-correlation function, the autocovariance function and the cross-covariance function of sum processes in the same way, even if the sum process contains more than two summands.

### 2.6.3 Stationary and Ergodic Random Processes

Random processes can be classified as either stationary or nonstationary. In Section 2.6.3.1 we will explain what a stationary random process is and derive some of its properties. Then, in Section 2.6.3.2 we will briefly explain a special stationary process—the *ergodic process*.

### 2.6.3.1 *Stationary Random Processes*

A random process $\{x(t)\}$ is called *stationary* in the narrow (or strict) sense if for $n = 1, 2, \ldots$, for any point in time $t_1, \ldots, t_n$ and for any $\Delta t$, the distribution functions associated with the $n$-dimensional random variable $[x(t_1), \ldots, x(t_n)]'$ and the $n$-dimensional random variable $[x(t_1 + \Delta t), \ldots, x(t_n + \Delta t)]'$ are the same, such that

$$F_n(\xi_1, t_1, \ldots, \xi_n, t_n) = F_n(\xi_1, t_1 + \Delta t, \ldots, \xi_n, t_n + \Delta t) \qquad (2.312)$$

Thus, in the case of a *stationary* random process, the statistical properties of an $n$-dimensional random variable, $n = 1, 2, \ldots$, which has been obtained by taking $n$ snapshots of $\{x(t)\}$, do not change if the complete grid of $n$ snapshots is shifted along the $t$ axis by any $\Delta t$ without altering the interval between the individual snapshots. However, for this to be valid, all components of the vector random variable $[x(t_1 + \Delta t), \ldots, x(t_n + \Delta t)]'$ must remain within the domain of definition of the sample functions $\xi^{(i)}(t)$ of the random process (see Section 2.6.1).

For a vector random process $\{\boldsymbol{x}(t)\} = \{[x_1(t), \ldots, x_n(t)]'\}$ to be stationary, it is not sufficient for its components $\{x_1(t)\}, \ldots, \{x_n(t)\}$ to be stationary random processes. The joint distribution functions of vector random variables whose elements have been produced by snapshots of different components of $\{\boldsymbol{x}(t)\}$ must also be invariant with respect to a shift in time in the above sense (Papoulis, 1984).

We will now investigate the consequences of the defining equation (2.312) for the characteristics of a random process.

For the first distribution function of a stationary random process $\{x(t)\}$ we have

$$F_{\mathrm{I}}(\xi_1, t_1) = F_{\mathrm{I}}(\xi_1, t_1 + \Delta t)$$

This holds for any $\Delta t$, including $\Delta t = -t_1$, so that

$$F_{\mathrm{I}}(\xi_1, t_1) = F_{\mathrm{I}}(\xi_1, 0)$$

Because this equation must also be true for any $t_1$, the first distribution function of the stationary random process is independent of $t_1$. Therefore,

$$F_{\mathrm{I}}(\xi_1, t_1) = F_{\mathrm{I}}(\xi_1) \qquad (2.313)$$

Thus the first density function of the stationary random process $\{x(t)\}$ must also be independent of time:

$$f_{\mathrm{I}}(\xi_1, t_1) = f_{\mathrm{I}}(\xi_1) \qquad (2.314)$$

For $\Delta t = -t_1$, the second distribution function of the stationary random process $\{x(t)\}$ becomes

$$F_{II}(\xi_1, t_1, \xi_2, t_2) = F_{II}(\xi_1, t_1 + \Delta t, \xi_2, t_2 + \Delta t) = F_{II}(\xi_1, 0, \xi_2, t_2 - t_1)$$

We substitute

$$t_2 - t_1 = \tau$$

We then have

$$F_{II}(\xi_1, t_1, \xi_2, t_2) = F_{II}(\xi_1, \xi_2, \tau) \tag{2.315}$$

and hence the second density function is given by

$$f_{II}(\xi_1, t_1, \xi_2, t_2) = f_{II}(\xi_1, \xi_2, \tau) \tag{2.316}$$

Now, if the first density function of the stationary random process $\{x(t)\}$ is independent of time, this must also apply to the various expectations of the random variable $x(t_1)$. It then follows from equations (2.264)–(2.266) that

$$E\{x(t_1)\} = E\{x\} \tag{2.317}$$

$$E\{x^2(t_1)\} = E\{x^2\} \tag{2.318}$$

$$E\{[x(t_1) - E\{x(t_1)\}]^2\} = E\{(x - E\{x\})^2\} \tag{2.319}$$

*et cetera*. Because

$$f_{II}(\xi_1, t_1, \xi_2, t_2) = f_{II}(\xi_1, \xi_2, \tau)$$

the autocorrelation function given by equation (2.267) or (2.269) for a stationary random process $\{x(t)\}$ no longer depends on the individual points in time $t_1$ and $t_2$ but only on the difference $t_2 - t_1 = \tau$:

$$r_{xx}(\tau) = \int_{-\infty}^{+\infty} \int_{-\infty}^{+\infty} \xi_1 \xi_2 f_{II}(\xi_1, \xi_2, \tau) \mathrm{d}\xi_1 \mathrm{d}\xi_2$$

If we substitute $t_2 = t_1 + \tau$ in (2.269) and replace $t_1$ by $t$, we also have

$$r_{xx}(\tau) = E\{x(t)x(t + \tau)\} \tag{2.320}$$

Using the same transformations, we obtain the autocovariance function of the stationary random process $\{x(t)\}$ (equations (2.268) and (2.270)) as

$$c_{xx}(\tau) = E\{[x(t) - E\{x\}][x(t + \tau) - \mathrm{E}\{x\}]\} \tag{2.321}$$

If the two-dimensional random process $\{\boldsymbol{x}(t)\} = \{[x_1(t), x_2(t)]'\}$ is stationary (its two components $\{x_1(t)\}$ and $\{x_2(t)\}$ are also stationary), the second distribution function is given by

$$F_{\mathrm{II,d}}(\xi_1, t_1, \xi_2, t_2) = F_{\mathrm{II,d}}(\xi_1, \xi_2, \tau) \tag{2.322}$$

and the second density function is given by

$$f_{\mathrm{II,d}}(\xi_1, t_1, \xi_2, t_2) = f_{\mathrm{II,d}}(\xi_1, \xi_2, \tau) \tag{2.323}$$

Then, according to (2.276) and (2.278), the cross-correlation function must be given by

$$r_{x_1x_2}(\tau) = E\{x_1(t)x_2(t + \tau)\} \tag{2.324}$$

and, according to equations (2.277) and (2.279), the cross-covariance function must be given by

$$c_{x_1x_2}(\tau) = E\{[x_1(t) - E\{x_1\}][x_2(t + \tau) - E\{x_2\}]\} \tag{2.325}$$

If we reverse the substitutions $t_2 - t_1 = \tau$ and $t_1 = t$ in equation (2.324), we obtain, using relationship (2.278),

$$r_{x_1x_2}(t_2 - t_1) = E\{x_1(t_1)x_2(t_2)\} = r_{x_1x_2}(t_1, t_2) \tag{2.326}$$

If we permute $x_1$ with $x_2$ and $t_1$ with $t_2$, this equation becomes

$$r_{x_2x_1}(t_1 - t_2) = E\{x_2(t_2)x_1(t_1)\} = r_{x_2x_1}(t_2, t_1)$$

The preceding two expressions are equal (see equation (2.299)), and hence

$$r_{x_1x_2}(t_2 - t_1) = r_{x_2x_1}(t_1 - t_2)$$

or

$$r_{x_1x_2}(\tau) = r_{x_2x_1}(-\tau) \tag{2.327}$$

By setting $x_1 = x_2 = x$, we obtain a corresponding relationship for the autocorrelation function:

$$r_{xx}(\tau) = r_{xx}(-\tau) \tag{2.328}$$

By setting $x_1 = x_2 = x$ and $t_1 = t_2 = t$ in equation (2.326) we obtain

$$r_{xx}(t, t) = r_{xx}(0)$$

or, from equations (2.298) and (2.317),

$$r_{xx}(t, t) = r_{xx}(0) = E\{x^2\} \tag{2.329}$$

Using equations (2.326) and (2.329), we can rewrite (2.300) as

$$|r_{x_1x_2}(\tau)|^2 \leqslant r_{x_1x_1}(0) r_{x_2x_2}(0) = E\{x_1^2\}E\{x_2^2\} \tag{2.330}$$

or, for $x_1 = x_2 = x$,

$$|r_{xx}(\tau)| \leqslant r_{xx}(0) = E\{x^2\} \tag{2.331}$$

Thus the autocorrelation function $r_{xx}(\tau)$ for a stationary random process $\{x(t)\}$ is an even function (equation (2.328)) which has its maximum, which is always positive, at $\tau = 0$ (equation (2.331)). However, because of the permutation of the subscripts in equation (2.327), the cross-correlation function $r_{x_1x_2}(\tau)$ is neither even nor odd.

It is obvious that in practice it is not feasible to use equation (2.312) to determine whether a random process $\{x(t)\}$ is stationary. As we have seen, this equation must hold for $n = 1, 2, \ldots,$ for any $t_1, \ldots, t_n$ and any $\Delta t$. Therefore simpler definitions of a "stationary" random process have been introduced.

If equation (2.312) is only satisfied for $n = 1, 2, \ldots, k$, the associated random process is said to be stationary of order $k$. Every random process that is stationary of order $k$ is also stationary of order $n < k$.

A random process $\{x(t)\}$ that is stationary of second order has a first density function $f_{\mathrm{I}}(\xi_1)$ which is independent of time and a second density function $f_{\mathrm{II}}(\xi_1, \xi_2, \tau)$ which depends only on $t_2 - t_1 = \tau$. As we have seen, it follows from this that the expectations $E\{x(t_1)\}$ and $E\{x^2(t_1)\}$ are constants and the autocorrelation function $r_{xx}(t_1, t_2)$ depends only on $\tau$ (equation (2.320)). Then, from equation (2.271), the autocovariance function $c_{xx}(t_1, t_2)$ also depends only on $t_2 - t_1 = \tau$.

If it is known that, for a random process $\{x(t)\}$, the expectations $E\{x(t_1)\}$ and $E\{x^2(t_1)\}$ exist and are independent of time and the autocorrelation function $r_{xx}(t_1, t_2)$ exists and depends only on $t_2 - t_1 = \tau$, from which it follows that the autocovariance function $c_{xx}(t_1, t_2)$ also depends only on $\tau$, the random process $\{x(t)\}$ is said to be stationary in the broad sense (alternatively, weakly stationary or covariance stationary). Every random process that is stationary or stationary of second order is also stationary in the broad sense. In general, however, the converse of this statement is not true.

A two-dimensional random process $\{\boldsymbol{x}(t)\} = \{[x_1(t), x_2(t)]'\}$ is stationary in the broad sense if every component $\{x_1(t)\}$ and $\{x_2(t)\}$ is stationary in the broad sense and if the cross-correlation function $r_{x_1x_2}(t_1, t_2)$ exists and depends only on $t_2 - t_1 = \tau$. It then follows from equation (2.280) that the cross-covariance function $c_{x_1x_2}(t_1, t_2)$ also exists and depends only on $t_2 - t_1 = \tau$ (Papoulis, 1984).

The above definitions can be transferred to $n$-dimensional random processes $\{\boldsymbol{x}(t)\}$ without any difficulty. The same applies to the derived results which can be transferred in an analogous manner to the correlation matrix $\boldsymbol{R}_{xx}(t_1, t_2)$ (equation (2.287)), the covariance matrix $\boldsymbol{C}_{xx}(t_1, t_2)$ (equation (2.288)), the correlation matrix $\boldsymbol{R}_{xy}(t_1, t_2)$ (equation (2.293)) and the covariance matrix $\boldsymbol{C}_{xy}(t_1, t_2)$ (equation (2.294)).

### *2.6.3.2 Ergodic Random Processes*

In what follows we consider only stationary random processes whose sample functions take real numerical values.

If we want to calculate the expectations $E\{x\}$, $E\{x^2\}$, $E\{x(t)x(t + \tau)\}$ *et cetera* for a random process $\{x(t)\}$, we need to know the density functions $f_{\mathrm{I}}(\xi_1)$, $f_{\mathrm{II}}(\xi_1, \xi_2, \tau)$ *et cetera*. Otherwise we must remember that the expectations were introduced as ensemble mean values and we can then calculate, for example, $E\{x\}$ as the mean value of as many functional values $\xi^{(i)}(t_1)$, $i = 1, 2, \ldots$, as possible of the sample functions $\xi^{(i)}(t)$, $i = 1, 2, \ldots$, for an arbitrary but fixed point in time $t = t_1$. The value for $E\{x\}$ calculated in this way becomes increasingly accurate as the total number of possible sample functions of $\{x(t)\}$ available for evaluation increases. However, if only a few sample functions of the random process $\{x(t)\}$ are available (e.g., in the form of measurements) the second method for determining the statistical characteristics of $\{x(t)\}$ cannot be used. The question now arises as to whether these characteristics can be calculated from a single sample function $\xi^{(i)}(t)$ of $\{x(t)\}$. In a corresponding manner to the ensemble mean value $E\{x\}$, a time average over all functional values of one sample function $\xi^{(i)}(t)$ can be introduced:

$$\overline{x(t)} := \lim_{T \to \infty} \frac{1}{2T} \int_{-T}^{T} \xi^{(i)}(t) \mathrm{d}t \tag{2.332}$$

In exactly the same way, the autocorrelation function $r_{xx}(\tau)$, which was introduced in equation (2.320) $\nu_{xx}(\tau) = E\{x(t)xt + \tau)\}$ as the ensemble mean value of the products $\xi^{(i)}(t)\xi^{(i)}(t + \tau)$, $i = 1, 2, \ldots$, of all sample functions of $\{x(t)\}$, corresponds to a time average

$$\overline{x(t)x(t+\tau)} := \lim_{T\to\infty} \frac{1}{2T} \int_{-T}^{T} \xi^{(i)}(t)\xi^{(i)}(t+\tau)\mathrm{d}t \tag{2.333}$$

We can define the corresponding time averages for all other expectations of $\{x(t)\}$ in exactly the same way. In general, these time averages will not coincide with the associated ensemble mean values. However, there are special cases where there is agreement.

A stationary random process $\{x(t)\}$ for which the ensemble mean values coincide with a probability of unity with the corresponding time averages of any sample function $\xi^{(i)}(t)$, $i = 1, 2, \ldots$, is known as an ergodic random process. The qualifying statement "with a probability of unity" means that of all the sample functions of an ergodic process, those having time averages which do not coincide with the corresponding ensemble mean values are allowed only if the probability that they occur is zero.

We will not consider here what the conditions are for the existence of, for example, the limiting value of the integral in (2.332) where the integrand is a sample function of a random process, an integral which, in addition, takes the same value as for every other sample function of the same random process with a probability of unity. Neither will we consider what criteria a random process must satisfy in order to be ergodic. We restrict ourselves to introducing the concept of the ergodic random process.

In conclusion, we emphasize again that the definition of an ergodic random process being linked to a stationary random process does not mean that every stationary random process is also an ergodic process.

### 2.6.4 Normal Processes (Gaussian Processes), White Noise and Markov Processes

#### *2.6.4.1 Normal (Gaussian) Processes*

A random process $\{(x(t)\}$ is called a *normal process* or a *Gaussian process* if the vector random variable $[x(t_1^*), \ldots, x(t_n^*)]'$, the components of which are produced by taking $n$ snapshots of $\{x(t)\}$ at times $t = t_1^*, \ldots, t = t_n^*$, is normally distributed for any $n$ and for any points in time $t_1^*, \ldots, t_n^*$. Thus if the $n$th density function

$$f_n(\xi_1, t_1^*, \ldots, \xi_n, t_n^*) = \frac{1}{(2\pi)^{n/2}|\pm\mathbf{N}|^{1/2}} \exp\left\{-\frac{1}{2}(\boldsymbol{\xi}-\boldsymbol{m})'\boldsymbol{N}^{-1}(\boldsymbol{\xi}-\boldsymbol{m})\right\} \tag{2.334}$$

is associated with $[x(t_1^*), \ldots, x(t_n^*)]'$, according to (2.144), where

$$\boldsymbol{\xi} = [\xi_1, \ldots, \xi_n]'$$

$$\boldsymbol{m} = E\{[x(t_1^*), \ldots, x(t_n^*)]'\} = [E\{x(t_1^*)\}, \ldots, E\{x(t_n^*)\}]' \tag{2.335}$$

and

$$\boldsymbol{N} = E\left\{\begin{bmatrix} x(t_1^*) - E\{x(t_1^*)\} \\ \vdots \\ x(t_n^*) - E\{x(t_n^*)\} \end{bmatrix} \begin{bmatrix} x(t_1^*) - E\{x(t_1^*)\} \\ \vdots \\ x(t_n^*) - E\{x(t_n^*)\} \end{bmatrix}^{\mathrm{T}}\right\}$$

$$= \begin{bmatrix} c_{xx}(t_1^*, t_1^*) & \ldots c_{xx}(t_1^*, t_n^*) \\ \vdots & \vdots \\ c_{xx}(t_n^*, t_1^*) \ldots & c_{xx}(t_n^*, t_n^*) \end{bmatrix} \tag{2.336}$$

$\boldsymbol{N}$ is the covariance matrix for the vector random variable $[x(t_1^*), \ldots, x(t_n^*)]'$. The elements of $\boldsymbol{N}$ are formed from the autocovariance function

$$c_{xx}(t_1, t_2) = E\{[x(t_1) - E\{x(t_1)\}][x(t_2) - E\{x(t_2)\}]\}$$

of the random process $\{x(t)\}$. The two equations (2.335) and (2.336) were derived in Section 2.5.2.1 for $n = 2$. However, they are also valid for $n > 2$ (Fisz, 1988). In contrast with Section 2.4.8, the elements of the vector $\boldsymbol{m}$ and the covariance matrix $\boldsymbol{N}$ are still time dependent here.

The density functions given by equation (2.334) for $n = 1, 2, \ldots$, which describe the statistical properties of the normal process, are uniquely defined for any $n$ by the function $E\{x(t_1)\}$ that for different values of $t_1$ gives the elements of $\boldsymbol{m}$ and the autocovariance function $c_{xx}(t_1, t_2)$ that for different pairs of values $t_1$, $t_2$ gives the elements of $\boldsymbol{N}$ . Because of (2.271)

$$c_{xx}(t_1, t_2) = r_{xx}(t_1, t_2) - E\{x(t_1)\}E\{x(t_2)\}$$

the density functions associated with the normal process $\{x(t)\}$ are uniquely defined for all $n$ by the function $E\{x(t_1)\}$ and the autocorrelation function $r_{xx}(t_1, t_2)$.

If the normal process $\{x(t)\}$ is stationary, we have $E\{x(t_1)\} = E\{x\}$ and $c_{xx}(t_1, t_2) = c_{xx}(t_2 - t_1)$; thus,

$$\boldsymbol{m} = E\{x\}[1, \ldots, 1]' \tag{2.337}$$

and

$$N = \begin{bmatrix} c_{xx}(0) & \dots c_{xx}(t_n^* - t_1^*) \\ \vdots & \vdots \\ c_{xx}(t_1^* - t_n^*) & \dots c_{xx}(0) \end{bmatrix} \tag{2.338}$$

If the normal process is stationary in the broad sense, we have $E\{x(t_1)\} = E\{x\}$ and $r_{xx}(t_1, t_2) = r_{xx}(t_2 - t_1)$ or $c_{xx}(t_1, t_2) = c_{xx}(t_2 - t_1)$. Hence equations (2.337) and (2.338) are satisfied; i.e., the $n$th density function no longer depends explicitly on the times $t_1^*, \dots, t_n^*$, but only on the time differences $t_2^* - t_1^*, \dots, t_n^* - t_1^*$. Therefore the normal process is also stationary in the narrow sense. However, we should again remember that in general it is not permissible to infer "stationary in the narrow sense" from "stationary in the broad sense."

A two-dimensional random process $\{\boldsymbol{x}(t)\} = \{[x_1(t), x_2(t)]'\}$ is known as a normal process if the vector random variable $[x_1(t_1^*), \dots, x_1(t_n^*), x_2(t_1^{**}, \dots, x_2(t_m^{**})]'$, the elements of which have been obtained by taking snapshots of $\{x_1(t)\}$ at times $t = t_1^*, \dots, t = t_n^*$, and of $\{x_2(t)\}$ at times $t = t_1^{**} \dots, t = t_m^{**}$, is normally distributed for any $n$ and $m$ and for any points in time $t_1^*, \dots, t_n^*, t_1^{**} \dots, t_m^{**}$. This definition applies in an analogous manner to an $n$-dimensional normal process (Papoulis, 1984). However, whereas the covariance matrix $N$ of the vector random variable $[x(t_1^*), \dots, x(t_n^*)]'$ of a one-dimensional normal process contains only autocovariance functions as elements, the covariance matrix $N$ of the vector random variable $[x_1(t_1^*), \dots, x_1(t_n^*), x_2(t_1^{**}), \dots, x_2(t_m^{**})]'\}$ of a two-dimensional normal process contains cross-covariance functions in addition to autocovariance functions.

### 2.6.4.2 *White Noise*

If the relationship

$$E\{[x(k) - E\{x(k)\}][x(\kappa) - E\{x(\kappa)\}]\} = q(k)\delta_{k\kappa} \quad k, \kappa = 0, 1, \dots, \tag{2.339}$$

applies to the autocovariance function of a scalar normal process in discrete-time $\{x(k), k = 0, 1, \dots\}$, whose sample functions $\xi^{(i)}(k)$ are therefore only defined for discrete points in time $k$, the random process $\{x(k), k = 0, 1, \dots\}$ is called Gaussian white noise in discrete time or, more concisely, white noise in discrete time; $\delta_{k\kappa}$ is the Kronecker symbol (Kronecker delta):

$$\delta_{k\kappa} = \begin{cases} 1 & \text{for} \quad k = \kappa \\ 0 & \text{for} \quad k \neq \kappa \end{cases} \tag{2.340}$$

According to equation (2.177), the expectation in (2.339) is the covariance cov$\{x(k), x(\kappa)\}$ of the two random variables $x(k)$ and $x(\kappa)$. Equation (2.339) therefore implies that every two random variables $x(k)$ and $x(\kappa)$ which have been obtained at different points in time from $\{x(k), k = 0, 1, \ldots\}$ are uncorrelated and, because a normal process is being considered, are also independent. It also follows from this that more than two random variables are independent if they are obtained at different points in time from $\{x(k), k = 0, 1, \ldots\}$ (Fisz, 1988).

A scalar normal random process in continuous time $\{x(t)\}$ is called Gaussian white noise in continuous time or, more concisely, white noise in continuous time if its autocovariance function obeys the relationship

$$E\{[x(t_1) - E\{x(t_1)\}][x(t_2) - E\{x(t_2)\}]\} = q(t_1)\delta(t_1 - t_2) \qquad (2.341)$$

where $\delta(t_1 - t_2)$ is the Dirac delta function. Therefore, in this case, the two random variables $x(t_1)$ and $x(t_2)$ of $\{x(t)\}$, which are arbitrarily close together, are independent of each other. If, in addition, $\{x(t)\}$ is also a stationary random process with $E\{x(t_1)\} \equiv 0$, $q(t_1)$ is a constant and, for $t_1 = t$ and $t_2 = t + \tau$, equation (2.341) becomes

$$E\{x(t)x(t + \tau)\} = q\delta(\tau) \qquad (2.342)$$

A random process in continuous time with such an autocovariance function (autocorrelation function) has sample functions whose frequency components cover the entire frequency spectrum and all have a finite power volume of the same magnitude (Laning and Battin, 1956). These sample functions must therefore have an infinite power, and so it is not feasible to implement such a random process. However, because of the special advantages in terms of mathematical operations, white noise is always a useful approximation of an actually occurring random process if the variation of the spectral power density is constant over a region which is substantially broader than the frequency range over which the dynamic system associated with this random process can appreciably react to input signals. The adjective "white" for characterizing the properties of such a random process was chosen by analogy with white light, which also contains all visible frequencies at the same intensity. Quite generally, white noise in continuous time can be understood as a useful approximation of an actual random process.

The concept of white noise, which has so far been restricted to scalar random processes, can be transferred to vector random processes. The normal process $\{\boldsymbol{x}(k), k = 0, 1, \ldots\}$ is called vector Gaussian white noise (white Gaussian process, white random process) in discrete time if its covariance matrix satisfies

$$E\{[\boldsymbol{x}(k) - E\{\boldsymbol{x}(k)\}][\boldsymbol{x}(\kappa) - \mathrm{E}\{\boldsymbol{x}(\kappa)\}]'\} = \boldsymbol{Q}(k)\delta_{k\kappa} \quad k, \kappa = 0, 1, \ldots \qquad (2.343)$$

Similarly, a normal process $\{\boldsymbol{x}(t)\}$ in continuous time with the covariance matrix

$$E\{[\boldsymbol{x}(t_1) - E\{\boldsymbol{x}(t_1)\}][\boldsymbol{x}(t_2) - E\{\boldsymbol{x}(t_2)\}]'\} = \boldsymbol{Q}(t_1)\delta(t_1 - t_2) \qquad (2.344)$$

is called vector Gaussian white noise in continuous time. In view of the arguments presented in Section 2.6.2.2, matrices $\boldsymbol{Q}(k)$ and $\boldsymbol{Q}(t_1)$ must be symmetric and positive semidefinite. If the components of the Gaussian white noise $\{\boldsymbol{x}(k),\ k = 0.1\ldots\}$ or $\{\boldsymbol{x}(t)\}$ are independent, then $\boldsymbol{Q}(k)$ and $\boldsymbol{Q}(t_1)$ are diagonal matrices.

### 2.6.4.3 Markov Processes

A random process $\{x(t)\}$ in discrete or continuous time is called a *Markov process* if for any points in time $t_1 < t_2 < \ldots < t_n$, $n = 2, 3, \ldots$, and any real numerical values $\xi_1, \ldots, \xi_n$,

$$\begin{aligned} &P[x(t_n) < \xi_n | \{x(t_{n-1}) = \xi_{n-1}\} \cap \ldots \cap \{x(t_1) = \xi_1\}] \\ &\quad = P\{x(t_n) < \xi_n | x(t_{n-1}) = \xi_{n-1}\} \end{aligned} \qquad (2.345)$$

The random variables $x(t_1), \ldots, x(t_n)$ are obtained by taking $n$ snapshots of the random process $\{x(t)\}$ at times $t = t_1, \ldots, t = t_n$. According to this, in a Markov process $\{x(t)\}$ the state of the random process at times $t_{n-2}, \ldots, t_1$ has no effect on the conditional probability

$$P[x(t_n) < \xi_n | \{x(t_{n-1}) = \xi_{n-1})\} \cap \ldots \cap (x(t_n) = \{x(t_1) = \xi_1\}]$$

which depends only on the state of the random process at the immediately preceding point in time $t_{n-1}$.

Property (2.345) can also be expressed using the conditional distribution function, following the notation given in Section 2.6.2.1:

$$F[\xi_n, t_n | \{x(t_{n-1}) = \xi_{n-1}\} \cap \ldots \cap \{x(t_1) = \xi_1\}]\ F\{\xi_n, t_n | x(t_{n-1}) = \xi_{n-1}\} \qquad (2.346)$$

Alternatively, if the Markov process is a random process in continuous time, it can be expressed by using the conditional density function:

$$f[\xi_n, t_n | \{x(t_{n-1}) = \xi_{n-1}\} \cap \ldots \cap \{x(t_1) = \xi_1\}] = f\{\xi_n, t_n | x(t_{n-1}) = \xi_{n-1}\} \qquad (2.347)$$

By rewriting equation (2.273), which applies to every random process in continuous time, and omitting the subscripts of the density functions $f$, we obtain

$$f(\xi_1, t_1, \xi_2, t_2) = f(\xi_1, t_1 | x(t_2) = \xi_2) f(\xi_2, t_2)$$

This relationship can be generalized to (Jazwinski, 1970; Papoulis, 1984):

$$f(\xi_1, t_1, \ldots, \xi_n, t_n) = f[\xi_1, t_1|\{x(t_2) = \xi_2\}\cap \ldots \cap\{x(t_n) = \xi_n\}] \times f(\xi_2, t_2, \ldots, \xi_n, t_n)$$

or, if we relabel the subscripts,

$$f(\xi_n, t_n, \ldots, \xi_1, t_1) = f[\xi_n, t_n|\{x(t_{n-1}) = \xi_{n-1}\} \cap \ldots \cap \{x(t_1) = \xi_1\}] \times f(\xi_{n-1}, t_{n-1}, \ldots, \xi_1, t_1) \quad (2.348)$$

If we write this equation for a Markov process, because of property (2.347), it becomes

$$f(\xi_n, t_n, \ldots, \xi_1, t_1) = f\{\xi_n, t_n|x(t_{n-1}) = \xi_{n-1}\} \times f(\xi_{n-1}, t_{n-1}, \ldots, \xi_1, t_1)$$

If the density function $f(\xi_{n-1}, t_{n-1}, \ldots, \xi_1, t_1)$ in the above equation is transformed using equation (2.348) and the Markov property (2.347), we finally obtain the following expression for the $n$th density function $f(\xi_n, t_n, \ldots, \xi_1, t_1)$ of a Markov process $\{x(t)\}$:

$$f(\xi_n, t_n, \ldots, \xi_1, t_1) = f\{\xi_n, t_n|x(t_{n-1}) = \xi_{n-1}\} \times \ldots \times f\{\xi_2, t_2|x(t_1) = \xi_1\}f(\xi_1, t_1) \quad (2.349)$$

The $n$th density function of a Markov process $\{x(t)\}$ is therefore fully described by the first density function $f(\xi_1, t_1)$ and the *conditional transition density functions* $f\{\xi_n, t_n|x(t_{n-1}) = \xi_{n-1}\}, \ldots, f\{\xi_2, t_2|x(t_1) = \xi_1\}$ of the random process $\{x(t)\}$ which are second-order density functions. We again emphasize that, although the same symbol $f$ is used for the different density functions in equation (2.349), we cannot infer that these functions differ only in their arguments.

## REFERENCES

Cramer, H. (1965). *The Elements of Probability Theory and Some of Its Applications,* (8th printing) Wiley, New York.

Davenport, W.B., Jr., and Root, W.L. (1958). *An Introduction to the Theory of Random Signals and Noise,* McGraw-Hill, New York.

Fisz, M. (1988). *Wahrscheinlichkeitsrechnung und Mathematische Statistik* (11th edn), VEB Deutscher Verlag der Wissenschaften, Berlin.

Gerthsen, C., Kneser, H.O. and Vogel, H. (1986). *Physik* (15th edn), Springer, Berlin.

Gnedenko, B.W. (1978). *Lehrbuch der Wahrscheinlichkeitsrechnung* (7th ed.), Akademie-Verlag, Berlin.

Heinhold, J. and Gaede, K.-W. (1979). *Ingenieur-Statistik* (4th ed.), Oldenbourg, Munich.

Jazwinski, A.H. (1970). *Stochastic Processes and Filtering Theory*, Academic Press, New York.

Kreyszig, E. (1979). *Statistische Methoden und ihre Anwendungen* (7th ed.), Vandenhoeck and Ruprecht, Göttingen.

Laning, J.H., Jr., and Battin, R.H. (1956). *Random Processes in Automatic Control*, McGraw-Hill, New York.

Meschkowski, H. (1976). *Mathematisches Begriffswörterbuch* (4th ed.), Bibliographical Institute, Mannheim.

Papoulis, A. (1984). *Probability, Random Variables and Stochastic Processes* (2nd ed.), McGraw-Hill, New York.

Richter, H. (1966). *Wahrscheinlichkeitstheorie* (2nd ed.), Springer, Berlin.

Sage, A.P. and Melsa, J.L. (1971). *Estimation Theory with Applications to Communications and Control*, McGraw-Hill, New York.

Wunsch, G. (1972). *Systemanalyse*, Vol. 2, *Statistische Systemanalyse* (3rd ed.), Hüthig, Heidelberg.

Wunsch, G. and Schreiber, H. (1984). *Stochastische Systeme*, Springer, Vienna.

## BIBLIOGRAPHY

Bauer, H. (1978). *Wahrscheinlichkeitstheorie und Grundzuge der Masstheorie* (3rd ed.), De Gruyter, Berlin.

Bendat, J.S. and Piersol, A.G. (1986). *Random Data: Analysis and Measurement Procedures* (2nd ed.), Wiley, New York.

Cramér, H. (1966). *Mathematical Methods of Statistics*, (11th printing) Princeton University Press, Princeton, NJ.

Cramér, H. and Leadbetter, M.R. (1967). *Stationary and Related Stochastic Processes*, Wiley, New York.

Doob, J.L. (1953). *Stochastic Processes*, Wiley, New York.

Elliot, R.J. (1982). *Stochastic Calculus and Applications*, Springer, Berlin.

Gikhman, I.I. and Skorokhod, A.V. (1969). *Introduction to the Theory of Random Processes*, Saunders, Philadelphia, PA.

Gray, R.M. (1988). *Probability, Random Processes and Ergodic Properties*, Springer, Berlin.

Karlin, S. (1969). *A First Course in Stochastic Processes*, Academic Press, New York.

Krickeberg, K. (1963). *Wahrscheinlichkeitstheorie*, Teubner, Stuttgart.

Larson, H.J. and Shubert, B.O. (1979). *Probabilistic Models in Engineering Sciences*, Vol. I, *Random Variables and Stochastic Processes*, and Vol. II, *Random Noise, Signals and Dynamic Systems*, Wiley, New York.

Melsa, J.L. and Sage, A.P. (1973). *An Introduction to Probability and Stochastic Processes*, Prentice-Hall, Englewood Cliffs, NJ.

von Mises, R. (1972). *Wahrscheinlichkeit Statistik Wahrheit* (4th ed.), Springer, Vienna.

Morgenstern, D. (1968). *Einfuhrung in die Wahrscheinlichkeitsrechnung und Mathematische Statistik* (2nd ed.), Springer, Berlin.

Muller, P.H. (ed.) (1983). *Lexikon der Stochastik* (4th ed.), Akademie-Verlag, Berlin.

Parzen, E. (1965). *Stochastic Processes*, Holden-Day, San Francisco, CA.

Rényi, A. (1977). *Wahrscheinlichkeitsrechnung*, (5th ed.), VEB Deutscher Verlag der Wissenschaften, Berlin.

Schmetterer, L. (1966). *Einfuhrung in die Mathematische Statistik* (2nd ed.), Springer, Vienna.

Winkler, G. (1977). *Stochastische Systeme*, Akademische Verlagsgesellschaft, Wiesbaden.

Wong, E. and Hajek, B. (1985). *Stochastic Processes in Engineering Systems*, Springer, Berlin.

# Chapter 3
# Response of Linear Systems to Stochastic Processes

If a sample function of a random process (the input random process) is switched to the input of a dynamic system, the corresponding output function will also be the sample function of a random process (the output random process). It is assumed that the output function can be measured or calculated from the input function and the system equations. However, as in the characterization of the input random process, where not only one but all possible sample functions were used, all possible sample functions (output functions of the dynamic system for a given input random process) will be needed to characterize the output random process. In this chapter we will not determine the output function of the dynamic system if the input function is the sample function of a random process, but on the basis of the statistical properties of the input random process we will calculate the corresponding statistical properties of the output random process by making use of the properties of the dynamic system. These statistical properties may be the various distribution or density functions introduced in Section 2.6.2 or the associated moments or expectations. This general formulation of the problem will be subject to the two following restrictions in the discussion presented in this chapter.

(1) Only linear dynamic systems which satisfy the state equations

$$\dot{\boldsymbol{x}}(t) = \boldsymbol{A}(t)\boldsymbol{x}(t) + \boldsymbol{B}(t)\boldsymbol{u}(t) \tag{3.1}$$

$$\boldsymbol{y}(t) = \boldsymbol{C}(t)\boldsymbol{x}(t) + \boldsymbol{D}(t)\boldsymbol{u}(t) \tag{3.2}$$

will be considered. In the above equations $\boldsymbol{u}(t)$ is the $p \times T$ input vector, $\boldsymbol{y}(t)$ is the $m \times T$ output vector and $\boldsymbol{x}(t)$ is the $n \times T$ state vector, and the matrices $\boldsymbol{A}(t)$, $\boldsymbol{B}(t)$, $\boldsymbol{C}(t)$ and $\boldsymbol{D}(t)$ have the appropriate dimensions. It is assumed that their elements are known. In Section 1.2.1, the relationship

$$\boldsymbol{x}(t) = \boldsymbol{\phi}(t,t_0)\boldsymbol{x}(t_0) + \int_{t_0}^{t} \boldsymbol{\phi}(t,\tau)\boldsymbol{B}(\tau)\boldsymbol{u}(\tau)\mathrm{d}\tau \tag{3.3}$$

was given as the solution of the state differential equation (3.1). In this case the $n \times n$ matrix $\boldsymbol{\phi}$ is the transition matrix of the linear dynamical system whose properties (see Section 1.2.2)

$$\boldsymbol{\phi}(t_0,t_0) = \boldsymbol{I} \tag{3.4}$$

$$\boldsymbol{\phi}(t_2,t_1)\boldsymbol{\phi}(t_1,t_0) = \boldsymbol{\phi}(t_2,t_0) \tag{3.5}$$

$$\frac{\partial}{\partial t}\boldsymbol{\phi}(t,t_0) = \boldsymbol{A}(t)\boldsymbol{\phi}(t,t_0) \tag{3.6}$$

will be used frequently in what follows.

(2) The random vector $\boldsymbol{x}(t_0)$ and the random processes $\{\boldsymbol{u}(t)\}$, $\{\boldsymbol{y}(t)\}$ and $\{\boldsymbol{x}(t)\}$ will be described using only their mean values and covariance matrices (i.e., the first- and second-order moments). Although this means that a complete statistical description of these random processes cannot be given, characterization using the first- and second-order moments is adequate in many applications. The situation is different for a normal process (Gaussian process). This is completely statistically described by the first and second moments (see Section 2.6.4). Furthermore, by a linear transformation a normal process becomes another normal process (Melsa and Sage, 1973). Therefore if the input random process is a normal process and the initial vector condition $\boldsymbol{x}(t_0)$ of a linear dynamic system for example with equations (3.1) and (3.2), is normally distributed, the associated output random process must also be a normal process. Hence, if the first and second moments of the output random process can be calculated, it is completely statistically defined.

We shall now look more closely at equation (3.3). This equation describes the relationship between $\boldsymbol{u}(t)$ and the state vector $\boldsymbol{x}(t)$; $\boldsymbol{y}(t)$ can then be calculated from $\boldsymbol{u}(t)$ and $\boldsymbol{x}(t)$ using (3.2); $\boldsymbol{u}(t)$, which is a vector sample function of the input random process $\{\boldsymbol{u}(t)\}$, occurs in the integrand of the integral in (3.3). Therefore this integral can be regarded as a stochastic integral and does not have to be evaluated using deterministic methods. However, in what follows the various vector sample functions $\boldsymbol{x}(t)$ will not be calculated as such. Only the expectations (vector mean value, covariance matrix) of the corresponding random process $\{\boldsymbol{x}(t)\}$, which, in turn, are defined as integral operations, will be determined. In addition, only linear dynamic systems will be considered.

Because of the special properties of the stochastic integrals, if we limit ourselves to linear dynamic systems and formally calculate the stochastic integrals occurring in the expressions for the required expectations in the same way as in deterministic methods, we obtain the same results as if we had calculated these stochastic integrals by using Ito calculus, which is required in the general case. Therefore in what follows (for linear dynamical systems) we shall proceed with the calculation of expectations using the integral in (3.3) in the same way as is customary with deterministic methods. This simplification cannot be used for nonlinear dynamic systems. The problems involved here are examined in greater detail in Chapter 6, Section 6.8.

In Section 3.1 we examine linear dynamic systems in continuous time when $\{\boldsymbol{u}(t)\}$ is also a random process in continuous time. Then, in Section 3.2, we derive completely analogous results for linear dynamic systems in discrete time and input random processes in discrete time $\{\boldsymbol{u}(t_\nu),\ \nu = 0,1,\ldots\}$. In both cases first the moments for $\{\boldsymbol{x}(t)\}$ and from these the expectations of the output random process $\{\boldsymbol{y}(t)\}$ are determined. The calculation rules for expectations derived in Chapter 2, Section 2.5.4, will be used frequently.

## 3.1 STOCHASTIC PROCESSES AND LINEAR SYSTEMS IN CONTINUOUS TIME

In order to calculate the mean value and the covariance matrix of the output random process, the mean value

$$E\{\boldsymbol{x}(t_0)\} = \boldsymbol{m}_x(t_0) \tag{3.7}$$

and the covariance matrix

$$E\{[\boldsymbol{x}(t_0) - \boldsymbol{m}_x(t_0)][\boldsymbol{x}(t_0) - \boldsymbol{m}_x(t_0)]'\} = \boldsymbol{C}_{xx}(t_0,t_0) \tag{3.8}$$

(see equations (2.189), (2.288) and (2.292)) of the vector random variable $\boldsymbol{x}(t_0)$, which is the initial value of the state differential equation (3.1), as well as the mean value

$$E\{\boldsymbol{u}(t)\} = \boldsymbol{m}_u(t) \tag{3.9}$$

and the covariance matrix

$$E\{[\boldsymbol{u}(t_1) - \boldsymbol{m}_u(t_1)][\boldsymbol{u}(t_2) - \boldsymbol{m}_u(t_2)]'\} = \boldsymbol{C}_{uu}(t_1,t_2) \tag{3.10}$$

of the input random process $\{\boldsymbol{u}(t)\}$ must be known numerically.

First we calculate the mean value

$$\boldsymbol{m}_x(t) = E\{\boldsymbol{x}(t)\} \tag{3.11}$$

To do so, we substitute (3.3) into (3.11):

$$\boldsymbol{m}_x(t) = E\left\{\boldsymbol{\Phi}(t,t_0)\boldsymbol{x}(t_0) + \int_{t_0}^{t} \boldsymbol{\Phi}(t,\tau)\boldsymbol{B}(\tau)\boldsymbol{u}(\tau)\mathrm{d}\tau\right\}$$

We then use equations (2.235), (2.238) and (2.256). The transition matrix $\boldsymbol{\Phi}$ and the matrix $\boldsymbol{B}$ have deterministic elements, so that we obtain

$$\boldsymbol{m}_x(t) = \boldsymbol{\Phi}(t,t_0)\boldsymbol{m}_x(t_0) + \int_{t_0}^{t} \boldsymbol{\Phi}(t,\tau)\boldsymbol{B}(\tau)\boldsymbol{m}_u(\tau)\mathrm{d}\tau \tag{3.12}$$

The form of this result should be compared with the solution (3.3) of the state differential equation (3.1).

The numerical evaluation of this integral representation of $\boldsymbol{m}_x(t)$ presents difficulties because of the presence of the integral and because it is necessary to know the transition matrix. Therefore a differential representation for $\boldsymbol{m}_x(t)$, which is more suitable than equation (3.12) for computational evaluation, is also derived.

To do so, (3.12) is expressed in terms of $t$ using the Leibniz formula for the differentiation of a definite integral:

$$\frac{\partial}{\partial t}\int_{a(t)}^{b(t)} f(t,\tau)\mathrm{d}\tau = \int_{a(t)}^{b(t)} \frac{\partial}{\partial t} f(t,\tau)\mathrm{d}\tau + f\{t,b(t)\}\frac{\mathrm{d}}{\mathrm{d}t}b(t) - f\{t,a(t)\}\frac{\mathrm{d}}{\mathrm{d}t}a(t) \tag{3.13}$$

This also applies if the integrand is a matrix and is assumed to be applicable here (Rothe, 1962):

$$\dot{\boldsymbol{m}}_x(t) = \frac{\partial}{\partial t}\boldsymbol{\Phi}(t,t_0)\boldsymbol{m}_x(t_0) + \int_{t_0}^{t}\frac{\partial}{\partial t}\boldsymbol{\Phi}(t,\tau)\boldsymbol{B}(\tau)\boldsymbol{m}_u(\tau)\mathrm{d}\tau + \boldsymbol{\Phi}(t,t)\boldsymbol{B}(t)\boldsymbol{m}_u(t)$$

Using (3.4) and (3.6), we can also write this as

$$\dot{\boldsymbol{m}}_x(t) = \boldsymbol{A}(t)\left\{\boldsymbol{\Phi}(t,t_0)\boldsymbol{m}_x(t_0) + \int_{t_0}^{t}\boldsymbol{\Phi}(t,\tau)\boldsymbol{B}(\tau)\boldsymbol{m}_u(\tau)\mathrm{d}\tau\right\} + \boldsymbol{B}(t)\boldsymbol{m}_u(t)$$

However, the expression in braces in this equation is equal to $\boldsymbol{m}_x(t)$ according to (3.12). We thus obtain the differential equation

$$\dot{\boldsymbol{m}}_x(t) = \boldsymbol{A}(t)\boldsymbol{m}_x(t) + \boldsymbol{B}(t)\boldsymbol{m}_u(t) \tag{3.14}$$

for $\boldsymbol{m}_x(t)$ which has a completely analogous structure to that of the state differential equation (3.1). The mean value $\boldsymbol{m}_x(t_0)$ is the associated initial value.

Next we calculate the covariance matrix

$$\boldsymbol{C}_{xx}(t_1,t_2) = \mathrm{E}\{[\boldsymbol{x}(t_1) - \boldsymbol{m}_x(t_1)][\boldsymbol{x}(t_2) - \boldsymbol{m}_x(t_2)]'\} \tag{3.15}$$

of the random process $\{\boldsymbol{x}(t)\}$. In order to clarify the expressions occurring here, the covariance matrix $\boldsymbol{C}_{xx}(t_1,t_2)$ is calculated using the correlation matrix (2.287):

$$\boldsymbol{R}_{xx}(t_1,t_2) = E\{\boldsymbol{x}(t_1)\boldsymbol{x}'(t_2)\}$$

Using (2.289), and taking equations (2.232) and (3.11) into account, we obtain the relationship

$$\boldsymbol{C}_{xx}(t_1,t_2) = \boldsymbol{R}_{xx}(t_1,t_2) - \boldsymbol{m}_x(t_1)\boldsymbol{m}_x'(t_2) \tag{3.16}$$

between $\boldsymbol{C}_{xx}(t_1,t_2)$ and $\boldsymbol{R}_{xx}(t_1,t_2)$. We now substitute equation (3.3) in the expression for $\boldsymbol{R}_{xx}(t_1,t_2)$ and multiply out the term in braces. Then, remembering that the transpose of the sum of matrices is equal to the sum of the transposed matrices and using (A.21) and the relationship

$$\left[\int_{t_0}^{t_1} \boldsymbol{A}(t)\mathrm{d}t\right]' = \int_{t_0}^{t_1} \boldsymbol{A}'(t)\mathrm{d}t \tag{3.17}$$

which can easily be derived from the definitions in the Appendix, we obtain

$$\begin{aligned}
\boldsymbol{R}_{xx}(t_1,t_2) = E\Big\{ & \boldsymbol{\Phi}(t_1,t_0)\boldsymbol{x}(t_0)\boldsymbol{x}'(t_0)\boldsymbol{\Phi}'(t_2,t_0) \\
& + \boldsymbol{\Phi}(t_1,t_0)\boldsymbol{x}(t_0)\int_{t_0}^{t_2} \boldsymbol{u}'(\tau)\boldsymbol{B}'(\tau)\boldsymbol{\Phi}'(t_2,\tau)\mathrm{d}\tau \\
& + \int_{t_0}^{t_1} \boldsymbol{\Phi}(t_1,\tau)\boldsymbol{B}(\tau)\boldsymbol{u}(\tau)\mathrm{d}\tau\boldsymbol{x}'(t_0)\boldsymbol{\Phi}'(t_2,t_0) \\
& + \int_{t_0}^{t_1} \boldsymbol{\Phi}(t_1,\tau)\boldsymbol{B}(\tau)\boldsymbol{u}(\tau)\mathrm{d}\tau \int_{t_0}^{t_2} \boldsymbol{u}'(\tau)\boldsymbol{B}'(\tau)\boldsymbol{\Phi}'(t_2,\tau)\mathrm{d}\tau \Big\}
\end{aligned}$$

Using equations (2.234), (2.236), (2.237), (2.240) and (2.256) and labeling the integration variables in the last summand with subscripts 1 and 2 to distinguish them, we can rewrite this equation as

$$
\begin{aligned}
\boldsymbol{R}_{xx}(t_1,t_2) = {} & \boldsymbol{\Phi}(t_1,t_0)E\{\boldsymbol{x}(t_0)\boldsymbol{x}'(t_0)\}\boldsymbol{\Phi}'(t_2,t_0) \\
& + \boldsymbol{\Phi}(t_1,t_0)\int_{t_0}^{t_2} E\{\boldsymbol{x}(t_0)\boldsymbol{u}'(\tau)\}\boldsymbol{B}'(\tau)\boldsymbol{\Phi}'(t_2,\tau)\mathrm{d}\tau \\
& + \int_{t_0}^{t_1} \boldsymbol{\Phi}(t_1,\tau)\boldsymbol{B}(\tau)E\{\boldsymbol{u}(\tau)\boldsymbol{x}'(t_0)\}\mathrm{d}\tau\,\boldsymbol{\Phi}'(t_2,t_0) \\
& + \int_{\tau_1=t_0}^{t_1}\int_{\tau_2=t_0}^{t_2} \boldsymbol{\Phi}(t_1,\tau_1)\boldsymbol{B}(\tau_1)E\{\boldsymbol{u}(\tau_1)\boldsymbol{u}'(\tau_2)\}\boldsymbol{B}'(\tau_2)\boldsymbol{\Phi}'(t_2,\tau_2)\mathrm{d}\tau_1\mathrm{d}\tau_2
\end{aligned}
$$

or, using (2.287) and (2.293), as

$$
\begin{aligned}
\boldsymbol{R}_{xx}(t_1,t_2) = {} & \boldsymbol{\Phi}(t_1,t_0)\boldsymbol{R}_{xx}(t_0,t_0)\boldsymbol{\Phi}'(t_2,t_0) \\
& + \boldsymbol{\Phi}(t_1,t_0)\int_{t_0}^{t_2} \boldsymbol{R}_{xu}(t_0,\tau)\boldsymbol{B}'(\tau)\boldsymbol{\Phi}'(t_2,\tau)\mathrm{d}\tau \\
& + \int_{t_0}^{t_1} \boldsymbol{\Phi}(t_1,\tau)\boldsymbol{B}(\tau)\boldsymbol{R}_{ux}(\tau,t_0)\mathrm{d}\tau\,\boldsymbol{\Phi}'(t_2,t_0) \\
& + \int_{\tau_1=t_0}^{t_1}\int_{\tau_2=t_0}^{t_2} \boldsymbol{\Phi}(t_1,\tau_1)\boldsymbol{B}(\tau_1)\boldsymbol{R}_{uu}(\tau_1,\tau_2)\boldsymbol{B}'(\tau_2)\boldsymbol{\Phi}'(t_2,\tau_2)\mathrm{d}\tau_1\mathrm{d}\tau_2
\end{aligned}
\tag{3.18}
$$

The product $\boldsymbol{m}_x(t_1)\boldsymbol{m}_x'(t_2)$ must also be rearranged. Using (3.12), we obtain

$$
\begin{aligned}
\boldsymbol{m}_x(t_1)\boldsymbol{m}_x'(t_2) = {} & \boldsymbol{\Phi}(t_1,t_0)\boldsymbol{m}_x(t_0)\boldsymbol{m}_x'(t_0)\boldsymbol{\Phi}'(t_2,t_0) \\
& + \boldsymbol{\Phi}(t_1,t_0)\int_{t_0}^{t_2} \boldsymbol{m}_x(t_0)\boldsymbol{m}_u'(\tau)\boldsymbol{B}'(\tau)\boldsymbol{\Phi}'(t_2,\tau)\mathrm{d}\tau \\
& + \int_{t_0}^{t_1} \boldsymbol{\Phi}(t_1,\tau)\boldsymbol{B}(\tau)\boldsymbol{m}_u(\tau)\boldsymbol{m}_x'(t_0)\mathrm{d}\tau\,\boldsymbol{\Phi}'(t_2,t_0) \\
& + \int_{\tau_1=t_0}^{t_1}\int_{\tau_2=t_0}^{t_2} \boldsymbol{\Phi}(t_1,\tau_1)\boldsymbol{B}(\tau_1)\boldsymbol{m}_u(\tau_1)\boldsymbol{m}_u'(\tau_2)\boldsymbol{B}'(\tau_2)\boldsymbol{\Phi}'(t_2,\tau_2)\mathrm{d}\tau_1\mathrm{d}\tau_2
\end{aligned}
\tag{3.19}
$$

Again, the integration variables are labeled $\tau_1$ and $\tau_2$ in the double integral.

If we now substitute equations (3.18) and (3.19) into (3.16), we obtain, by using (2.289) and (2.295), the required covariance matrix $\boldsymbol{C}_{xx}(t_1,t_2)$:

$$C_{xx}(t_1,t_2) = \boldsymbol{\Phi}(t_1,t_0)C_{xx}(t_0,t_0)\boldsymbol{\Phi}'(t_2,t_0)$$
$$+ \boldsymbol{\Phi}(t_1,t_0)\int_{t_0}^{t_2} C_{xu}(t_0,\tau)B'(\tau)\boldsymbol{\Phi}'(t_2,\tau)d\tau$$
$$+ \int_{t_0}^{t_1} \boldsymbol{\Phi}(t_1,\tau)B(\tau)C_{ux}(\tau,t_0)d\tau\,\boldsymbol{\Phi}'(t_2,t_0)$$
$$+ \int_{\tau_1=t_0}^{t_1}\int_{\tau_2=t_0}^{t_2} \boldsymbol{\Phi}(t_1,\tau_1)B(\tau_1)C_{uu}(\tau_1,\tau_2)B'(\tau_2)\boldsymbol{\Phi}'(t_2,\tau_2)d\tau_1 d\tau_2 \quad (3.20)$$

If the vector random variable $\boldsymbol{x}(t_0)$ is not correlated with the random process $\{\boldsymbol{u}(\tau)\}$ for $\tau > t_0$, we have (see Section 2.6.2.3)

$$C_{xu}(t_0,\tau) = \boldsymbol{0} \quad \text{for} \quad \tau > t_0$$

and

$$C_{ux}(\tau,t_0) = \boldsymbol{0} \quad \text{for} \quad \tau > t_0$$

(the last equation is obtained from the penultimate equation by transposition). Then equation (3.20) can be simplified to

$$C_{xx}(t_1,t_2) = \boldsymbol{\Phi}(t_1,t_0)C_{xx}(t_0,t_0)\boldsymbol{\Phi}'(t_2,t_0)$$
$$+ \int_{t_1=t_0}^{\tau_1}\int_{\tau_2=t_0}^{t_2} \boldsymbol{\Phi}(t_1,\tau_1)B(\tau_1)C_{uu}(\tau_1,\tau_2)B'(\tau_2)\boldsymbol{\Phi}'(t_2,\tau_2)d\tau_1 d\tau_2 \quad (3.21)$$

As in the evaluation of $\boldsymbol{m}_x(t)$, it is necessary to derive a differential equation for the covariance matrix $C_{xx}(t_1,t_2)$ in addition to equation (3.21). To do so, we must place another restriction on the input random process $\{\boldsymbol{u}(t)\}$: we assume that $\{\boldsymbol{u}(t)\}$ is a vector white noise with the covariance matrix

$$C_{uu}(t_1,t_2) = Q_u(t_1)\delta_D(t_1 - t_2) \quad (3.22)$$

In this case, the Dirac function must be defined somewhat differently than in Section 2.4.3, for example, because of the difficulties noted earlier in this chapter. This is indicated by the subscript D. However, the definition used earlier is contained in the definition here (Sage and Melsa, 1971, p. 50; Melsa and Sage, 1973, p. 252). The Dirac function $\delta_D(t_1 - t_2)$ is introduced through its masking property

$$\int_a^b f(\tau)\delta_D(\tau - t)d\tau = \begin{cases} 0 & \text{for} \quad t < a \text{ or } t > b \\ f(a)/2 & \text{for} \quad t = a \\ f(b)/2 & \text{for} \quad t = b \\ f(\tau) & \text{for} \quad a < t < b \end{cases} \tag{3.23}$$

$f(\tau)$ must be continuous over the closed interval $[a,b]$.

Using assumption (3.22), we shall now look at (3.21). So as not to lose the case $\tau_1 = \tau_2$ in the double integral, the integration must first be performed over the larger of the two intervals. We then obtain, for $t_2 > t_1$,

$$\begin{aligned} \boldsymbol{C}_{xx}(t_1,t_2) &= \boldsymbol{\Phi}(t_1,t_0)\boldsymbol{C}_{xx}(t_0,t_0)\boldsymbol{\Phi}'(t_2,t_0) \\ &+ \int_{\tau_1=t_0}^{\tau_1}\int_{\tau_2=t_0}^{t_2} \boldsymbol{\Phi}(t_1,\tau_1)\boldsymbol{B}(\tau_1)\boldsymbol{Q}_u(\tau_1)\ \boldsymbol{B}'(\tau_2)\boldsymbol{\Phi}'(t_2,\tau_2)\delta_D(\tau_1 - \tau_2)d\tau_2 d\tau_1 \end{aligned}$$

or

$$\begin{aligned} \boldsymbol{C}_{xx}(t_1,t_2) &= \boldsymbol{\Phi}(t_1,t_0)\boldsymbol{C}_{xx}(t_0,t_0)\boldsymbol{\Phi}'(t_2,t_0) \\ &+ \int_{\tau_1=t_0}^{t_1} \boldsymbol{\Phi}(t_1,\tau_1)\boldsymbol{B}(\tau_1)\boldsymbol{Q}_u(\tau_1)\boldsymbol{B}'(\tau_1)\boldsymbol{\Phi}'(t_2,\tau_1)d\tau_1 \end{aligned} \tag{3.24}$$

The corresponding result for $t_1 > t_2$ is the same as (3.24) except that the integration variable is $\tau_2$ and the upper integration limit is $t_2$. The two cases can be combined to give

$$\begin{aligned} \boldsymbol{C}_{xx}(t_1,t_2) &= \boldsymbol{\Phi}(t_1,t_0)\boldsymbol{C}_{xx}(t_0,t_0)\boldsymbol{\Phi}'(t_2,t_0) \\ &+ \int_{t_0}^{\min(t_1,t_2)} \boldsymbol{\Phi}(t_1,\tau)\boldsymbol{B}(\tau)\boldsymbol{Q}_u(\tau)\boldsymbol{B}'(\tau)\boldsymbol{\Phi}'(t_2,\tau)d\tau \end{aligned} \tag{3.25}$$

where $\min(t_1,t_2)$ means the smaller of the two numerical values $t_1$ and $t_2$ (Sage and Melsa, 1971; Melsa and Sage, 1973).

We now set $t_1 = t_2 = t$ which, as will be shown later, does not impose any additional restrictions. For $t_1 = t_2 = t$, equation (3.25) becomes

$$\begin{aligned} \boldsymbol{C}_{xx}(t,t) &= \boldsymbol{\Phi}(t,t_0)\boldsymbol{C}_{xx}(t_0,t_0)\boldsymbol{\Phi}'(t,t_0) \\ &+ \int_{t_0}^{t} \boldsymbol{\Phi}(t,\tau)\boldsymbol{B}(\tau)\boldsymbol{Q}_u(\tau)\boldsymbol{B}'(\tau)\boldsymbol{\Phi}'(t,\tau)d\tau \end{aligned} \tag{3.26}$$

This expression will now be differentiated with respect to $t$ by using the rule given by equation (3.13) and taking (A.60) into account:

$$\dot{\boldsymbol{C}}_{xx}(t,t) = \left[\frac{\partial}{\partial t}\boldsymbol{\Phi}(t,t_0)\right]\boldsymbol{C}_{xx}(t_0,t_0)\boldsymbol{\Phi}'(t,t_0)$$

$$+ \boldsymbol{\Phi}(t,t_0)\boldsymbol{C}_{xx}(t_0,t_0)\frac{\partial}{\partial t}\boldsymbol{\Phi}'(t,t_0)$$

$$+ \int_{t_0}^{t}\left\{\frac{\partial}{\partial t}\boldsymbol{\Phi}(t,\tau)\right\}\boldsymbol{B}(\tau)\boldsymbol{Q}_u(\tau)\boldsymbol{B}'(\tau)\boldsymbol{\Phi}'(t,\tau)\mathrm{d}\tau$$

$$+ \int_{t_0}^{t}\boldsymbol{\Phi}(t,\tau)\boldsymbol{B}(\tau)\boldsymbol{Q}_u(\tau)\boldsymbol{B}'(\tau)\frac{\partial}{\partial t}\boldsymbol{\Phi}'(t,\tau)\mathrm{d}\tau$$

$$+ \boldsymbol{\Phi}(t,t)\boldsymbol{B}(t)\boldsymbol{Q}_u(t)\boldsymbol{B}'(t)\boldsymbol{\Phi}'(t,t) \qquad (3.27)$$

Using (3.4), (3.6) and (A.21) together with the relationship

$$\frac{\mathrm{d}}{\mathrm{d}t}\boldsymbol{A}'(t) = \left\{\frac{\mathrm{d}}{\mathrm{d}t}\boldsymbol{A}(t)\right\}' \qquad (3.28)$$

which can easily be verified from the definitions in the appendix, we obtain from (3.27)

$$\boldsymbol{C}_{xx}(t,t) = \boldsymbol{A}(t)\left\{\boldsymbol{\Phi}(t,t_0)\boldsymbol{C}_{xx}(t_0,t_0)\boldsymbol{\Phi}'(t,t_0)\right.$$

$$\left.+ \int_{t_0}^{t}\boldsymbol{\Phi}(t,\tau)\boldsymbol{B}(\tau)\boldsymbol{Q}_u(\tau)\boldsymbol{B}'(\tau)\boldsymbol{\Phi}'(t,\tau)\mathrm{d}\tau\right\}$$

$$+ \left\{\boldsymbol{\Phi}(t,t_0)\boldsymbol{C}_{xx}(t_0,t_0)\boldsymbol{\Phi}'(t,t_0)\right.$$

$$\left.+ \int_{t_0}^{t}\boldsymbol{\Phi}(t,\tau)\boldsymbol{B}(\tau)\boldsymbol{Q}_u(\tau)\boldsymbol{B}'(\tau)\boldsymbol{\Phi}'(t,\tau)\mathrm{d}\tau\right\}\boldsymbol{A}'(t)$$

$$+ \boldsymbol{B}(t)\boldsymbol{Q}_u(t)\boldsymbol{B}'(t)$$

Then, using equation (3.26), we obtain from the above equation a differential equation for the covariance matrix $\boldsymbol{C}_{xx}(t,t)$:

$$\dot{\boldsymbol{C}}_{xx}(t,t) = \boldsymbol{A}(t)\boldsymbol{C}_{xx}(t,t) + \boldsymbol{C}_{xx}(t,t)\boldsymbol{A}'(t) + \boldsymbol{B}(t)\boldsymbol{Q}_u(t)\boldsymbol{B}'(t) \tag{3.29}$$

The covariance matrix $\boldsymbol{C}_{xx}(t_0,t_0)$ (equation (3.8)) is the associated initial value.

Because of its advantages with respect to numerical evaluation, the differential equation (3.29) is preferred to an integral representation for $\boldsymbol{C}_{xx}(t,t)$, e.g., (3.26). However, (3.29) only allows us to calculate $\boldsymbol{C}_{xx}(t,t)$ and not $\boldsymbol{C}_{xx}(t_1,t_2)$ as is the case with equation (3.25), for example.

If we write down equation (3.26) for $t = t_1$ and multiply it on the right-hand side by $\boldsymbol{\Phi}'(t_2,t_1)$, we obtain

$$\begin{aligned}\boldsymbol{C}_{xx}(t_1,t_1)\boldsymbol{\Phi}'(t_2,t_1) = {} & \boldsymbol{\Phi}(t_1,t_0)\boldsymbol{C}_{xx}(t_0,t_0)\boldsymbol{\Phi}'(t_1,t_0)\boldsymbol{\Phi}'(t_2,t_1) \\ & + \int_{t_0}^{t_1} \boldsymbol{\Phi}(t_1,\tau)\boldsymbol{B}(\tau)\boldsymbol{Q}_u(\tau)\boldsymbol{B}'(\tau)\boldsymbol{\Phi}'(t_1,\tau)\mathrm{d}\tau\,\boldsymbol{\Phi}'(t_2,t_1)\end{aligned}$$

Then, using equations (3.5) and (A.21), we obtain

$$\begin{aligned}\boldsymbol{C}_{xx}(t_1,t_1)\boldsymbol{\Phi}'(t_2,t_1) = {} & \boldsymbol{\Phi}(t_1,t_0)\boldsymbol{C}_{xx}(t_0,t_0)\boldsymbol{\Phi}'(t_2,t_0) \\ & + \int_{t_0}^{t_1} \boldsymbol{\Phi}(t_1,\tau)\boldsymbol{B}(\tau)\boldsymbol{Q}_u(\tau)\boldsymbol{B}'(\tau)\boldsymbol{\Phi}'(t_2,\tau)\mathrm{d}\tau\end{aligned}$$

The right-hand side of this equation is identical with (3.25) for $t_1 \leq t_2$. We therefore have

$$\boldsymbol{C}_{xx}(t_1,t_2) = \boldsymbol{C}_{xx}(t_1,t_1)\boldsymbol{\Phi}'(t_2,t_1) \quad \text{for} \quad t_1 \leq t_2$$

In the same way we obtain

$$\boldsymbol{C}_{xx}(t_1,t_2) = \boldsymbol{\Phi}(t_1,t_2)\boldsymbol{C}_{xx}(t_2,t_2) \quad \text{for} \quad t_2 \leq t_1$$

By combining these two equations to give

$$\boldsymbol{C}_{xx}(t_1,t_2) = \begin{cases}\boldsymbol{C}_{xx}(t_1,t_1)\boldsymbol{\Phi}'(t_2,t_1) & \text{for} \quad t_1 \leq t_2 \\ \boldsymbol{\Phi}(t_1,t_2)\boldsymbol{C}_{xx}(t_2,t_2) & \text{for} \quad t_1 \geq t_2\end{cases} \tag{3.30}$$

we obtain a possible method of calculating $\boldsymbol{C}_{xx}(t_1,t_2)$ which exploits the numerical advantages of (3.29). At this point, we emphasize again that (3.25), (3.29) and (3.30) are only valid if $\{\boldsymbol{u}(t)\}$ is a Gaussian white noise and is not correlated with the random vector $\boldsymbol{x}(t_0)$.

We will now calculate the covariance matrix from the given data

$$C_{xu}(t_1,t_2) = E\{[x(t_1) - m_x(t_1)][u(t_2) - m_u(t_2)]'\} \tag{3.31}$$

(see (2.294)) which, for example, occur in (3.20). However, we will not perform this calculation as explicitly as has been done earlier in this section where the operation of the calculation rules for expectations was demonstrated and practiced.

In conjunction with $C_{xu}(t_1,t_2)$, the relationship

$$C'_{xu}(t_1,t_2) = C_{ux}(t_2,t_1) \tag{3.32}$$

is also of interest. It can easily be verified by using the definition equation (3.31) and taking (2.231) and (A.21) into account. To make the derivation of $C_{xu}(t_1,t_2)$ easier to follow, we also obtain the covariance matrix from the correlation matrix $R_{xu}(t_1,t_2)$. We have

$$C_{xu}(t_1,t_2) = R_{xu}(t_1,t_2) - m_x(t_1)m'_u(t_2) \tag{3.33}$$

according to equation (2.295), and

$$R_{xu}(t_1,t_2) = E\{x(t_1)u'(t_2)\} \tag{3.34}$$

according to (2.293). From equation (3.3), $R_{xu}(t_1,t_2)$ is given by

$$R_{xu}(t_1,t_2) = \Phi(t_1,t_0)E\{x(t_0)u'(t_2)\} + \int_{t_0}^{t_1} \Phi(t_1,\tau)B(\tau)E\{u(\tau)u'(t_2)\}d\tau$$

which, using (2.287) and (3.34), becomes

$$R_{xu}(t_1,t_2) = \Phi(t_1,t_0)R_{xu}(t_0,t_2) + \int_{t_0}^{t_1} \Phi(t_1,\tau)B(\tau)R_{uu}(\tau,t_2)d\tau \tag{3.35}$$

From equation (3.12), the product $m_x(t_1)m'_u(t_2)$ is given by

$$m_x(t_1)m'_u(t_2) = \Phi(t_1,t_0)m_x(t_0)m'_u(t_2) + \int_{t_0}^{t_1} \Phi(t_1,\tau)B(\tau)m_u(\tau)m'_u(t_2)d\tau \tag{3.36}$$

The two equations (3.35) and (3.36) are now substituted into (3.33), taking (2.289) into account. We obtain

$$\boldsymbol{C}_{xu}(t_1,t_2) = \boldsymbol{\Phi}(t_1,t_0)\boldsymbol{C}_{xu}(t_0,t_2) + \int_{t_0}^{t_1} \boldsymbol{\Phi}(t_1,\tau)\boldsymbol{B}(\tau)\boldsymbol{C}_{uu}(\tau,t_2)\mathrm{d}\tau \tag{3.37}$$

This result can be considerably simplified if, as has been assumed above, $\boldsymbol{x}(t_0)$ and $\{\boldsymbol{u}(t_2)\}$ are uncorrelated for $t_2 > t_0$, and therefore $\boldsymbol{C}_{xu}(t_0,t_2) = \boldsymbol{0}$ for $t_2 > t_0$. Then we have

$$\boldsymbol{C}_{xu}(t_1,t_2) = \int_{t_0}^{t_1} \boldsymbol{\Phi}(t_1,\tau)\boldsymbol{B}(\tau)\boldsymbol{C}_{uu}(\tau,t_2)\mathrm{d}\tau \tag{3.38}$$

If $\{\boldsymbol{u}(t)\}$ is assumed to be a white noise, this expression can be simplified further. First, using (3.22), equation (3.38) becomes

$$\boldsymbol{C}_{xu}(t_1,t_2) = \int_{t_0}^{t_1} \boldsymbol{\Phi}(t_1,\tau)\boldsymbol{B}(\tau)\boldsymbol{Q}_u(\tau)\delta_{\mathrm{D}}(\tau - t_2)\mathrm{d}\tau$$

If the masking property of the Dirac function (equation (3.23)) is used here, we obtain for $t_1 > t_0$

$$\boldsymbol{C}_{xu}(t_1,t_2) = \begin{cases} \boldsymbol{0} & \text{for} \quad t_0 < t_1 < t_2 \\ \dfrac{\boldsymbol{B}(t_1)\boldsymbol{Q}_u(t_1)}{2} & \text{for} \quad t_0 < t_1 = t_2 \\ \boldsymbol{\Phi}(t_1,t_2)\boldsymbol{B}(t_2)\boldsymbol{Q}_u(t_2) & \text{for} \quad t_0 < t_2 < t_1 \end{cases} \tag{3.39}$$

Because of the precondition $t_2 > t_0$, the cases $t_2 = t_0$ and $t_2 < t_0$ do not appear in the equation for the covariance matrix $\boldsymbol{C}_{xu}(t_1,t_2)$ which exhibits a discontinuity at $t_1 = t_2$.

So far, we have determined only the mean value $\boldsymbol{m}_x(t)$ and the covariance matrices $\boldsymbol{C}_{xx}(t_1,t_2)$ and $\boldsymbol{C}_{xu}(t_1,t_2)$ for the random process $\{\boldsymbol{x}(t)\}$. However, the actual output variable of the dynamic system under consideration is $\boldsymbol{y}(t)$ or the output random process $\{\boldsymbol{y}(t)\}$. Therefore we will now use equation (3.2) to determine the mean value $\boldsymbol{m}_y(t)$ and the covariance matrix $\boldsymbol{C}_{yy}(t_1,t_2)$ of the output random process $\{\boldsymbol{y}(t)\}$ from the expectations of $\{\boldsymbol{x}(t)\}$ and $\{\boldsymbol{u}(t)\}$. Because the calculations for this proceed in a similar manner to those already described in this section, we will only give the results here.

The mean value $\boldsymbol{m}_y(t)$ of the output random process $\{\boldsymbol{y}(t)\}$ is obtained as

$$\boldsymbol{m}_y(t) = E\{\boldsymbol{y}(t)\} = \boldsymbol{C}(t)\boldsymbol{m}_x(t) + \boldsymbol{D}(t)\boldsymbol{m}_u(t) \tag{3.40}$$

In this way $\boldsymbol{m}_y(t)$ can be calculated without any difficulty from the given mean value $\boldsymbol{m}_u(t)$ and from $\boldsymbol{m}_x(t)$ according to equation (3.14). The covariance matrix

$$\boldsymbol{C}_{yy}(t_1,t_2) = E\{[\boldsymbol{y}(t_1) - \boldsymbol{m}_y(t_1)][\boldsymbol{y}(t_2) - \boldsymbol{m}_y(t_2)]'\} \tag{3.41}$$

is given by the relationship

$$\begin{aligned}\boldsymbol{C}_{yy}(t_1,t_2) = {} & \boldsymbol{C}(t_1)\boldsymbol{C}_{xx}(t_1,t_2)\boldsymbol{C}'(t_2) \\ & + \boldsymbol{D}(t_1)\boldsymbol{C}_{ux}(t_1,t_2)\boldsymbol{C}'(t_2) \\ & + \boldsymbol{C}(t_1)\boldsymbol{C}_{xu}(t_1,t_2)\boldsymbol{D}'(t_2) \\ & + \boldsymbol{D}(t_1)\boldsymbol{C}_{uu}(t_1,t_2)\boldsymbol{D}'(t_2)\end{aligned} \tag{3.42}$$

A careful distinction must be made here between the covariance matrices and the measurement matrix $\boldsymbol{C}(t)$ of the linear dynamic system, both of which are denoted by $\boldsymbol{C}$.

The covariance matrix $\boldsymbol{C}_{yy}(t_1,t_2)$ can be calculated from the given covariance matrix $\boldsymbol{C}_{uu}(t_1,t_2)$ and from the covariance matrices $\boldsymbol{C}_{xx}(t_1,t_2)$ and $\boldsymbol{C}_{xu}(t_1,t_2)$ by using equations (3.20) and (3.37) respectively in the general case or by using equations (3.30) and (3.39) respectively in the special case. In addition, we also need the covariance matrix $\boldsymbol{C}_{ux}(t_1,t_2)$ for which we have not yet obtained an expression. However, this expression can be derived in a completely corresponding manner to that for $\boldsymbol{C}_{xu}(t_1,t_2)$.

In addition to the formulas derived in this section, the corresponding results for linear time-invariant systems with a stationary input random process are given by Sage and Melsa (1971) and Melsa and Sage (1973).

## 3.2 STOCHASTIC PROCESSES AND LINEAR SYSTEMS IN DISCRETE TIME

We start from the state equations (1.39) and (1.40) derived in Section 1.3 for a linear dynamic system in discrete time:

$$\boldsymbol{x}(t_{k+1}) = \boldsymbol{A}(k)\boldsymbol{x}(t_k) + \boldsymbol{B}(k)\boldsymbol{u}(t_k) \tag{3.43}$$

$$\boldsymbol{y}(t_k) = \boldsymbol{C}(k)\boldsymbol{x}(t_k) + \boldsymbol{D}(k)\boldsymbol{u}(t_k) \tag{3.44}$$

The notation used can be simplified if it is assumed that the sampling times $t_0$, $t_1$, . . . are equally spaced; thus,

$$t_{k+1} - t_k = T \qquad k = 0,1,\ldots \tag{3.45}$$

where $T$ is the sampling period. If it is also assumed that $t_0 = 0$, we can set

$$t_k = kT \qquad k = 0,1,\ldots \tag{3.46}$$

and

$$\boldsymbol{x}(t_k) = \boldsymbol{x}(kT) \qquad k = 0,1,\ldots$$

and use the abbreviated notation $\boldsymbol{x}(k)$ for $\boldsymbol{x}(kT)$. The abbreviations $\boldsymbol{x}[k]$ or $\boldsymbol{x}_k$ are also frequently used instead of $\boldsymbol{x}(k)$. The same applies to $\boldsymbol{u}(t_k)$ and $\boldsymbol{y}(t_k)$.

Using this notation, we can now write the state equations as

$$\boldsymbol{x}(k + 1) = \boldsymbol{A}(k)\boldsymbol{x}(k) + \boldsymbol{B}(k)\boldsymbol{u}(k) \tag{3.47}$$

$$\boldsymbol{y}(k) = \boldsymbol{C}(k)\boldsymbol{x}(k) + \boldsymbol{D}(k)\boldsymbol{u}(k) \tag{3.48}$$

The abbreviated notation is also used for the properties of the initial state $\boldsymbol{x}(t_0) = \boldsymbol{x}(0)$, which are assumed to be known,

$$E\{\boldsymbol{x}(0)\} = \boldsymbol{m}_x(0) \tag{3.49}$$

$$E\{[\boldsymbol{x}(0) - \boldsymbol{m}_x(0)][\boldsymbol{x}(0) - \boldsymbol{m}_x(0)]'\} = \boldsymbol{C}_{xx}(0,0) \tag{3.50}$$

and for the known properties of the input random process $\{\boldsymbol{u}(k)\}$:

$$E\{\boldsymbol{u}(k)\} = \boldsymbol{m}_u(k) \tag{3.51}$$

$$E\{[\boldsymbol{u}(k) - \boldsymbol{m}_u(k)][\boldsymbol{u}(j) - \boldsymbol{m}_u(j)]'\} = \boldsymbol{C}_{uu}(k,j) \tag{3.52}$$

Because the formulas and the derivations of the mean values and the covariance matrices of the random processes $\{\boldsymbol{x}(k)\}$ and $\{\boldsymbol{y}(k)\}$ differ only slightly from the corresponding formulas and derivations for linear systems in continuous time, we will only make brief references to the derivations here. Cross references will be made to analogous procedures.

If the expectation is formed from the solution (1.43) of equation (3.47) written in the abbreviated notation

$$\boldsymbol{x}(k) = \boldsymbol{\Phi}(k,0)\boldsymbol{x}(0) + \sum_{\nu=0}^{k-1} \boldsymbol{\Phi}(k,\nu + 1)\boldsymbol{B}(\nu)\boldsymbol{u}(\nu) \tag{3.53}$$

we already have the first equation for the mean value of $\{\boldsymbol{x}(k)\}$:

$$\boldsymbol{m}_x(k) = E\{\boldsymbol{x}(k)\} \tag{3.54}$$

Using equations (3.49) and (3.51), we can rewrite this as

$$\boldsymbol{m}_x(k) = \boldsymbol{\Phi}(k,0)\boldsymbol{m}_x(0) + \sum_{\nu=0}^{k-1} \boldsymbol{\Phi}(k,\nu + 1)\boldsymbol{B}(\nu)\boldsymbol{m}_u(\nu) \tag{3.55}$$

Another equation for $\boldsymbol{m}_x(k)$ can be obtained from (3.47) if the expectation is formed on both sides:

$$\boldsymbol{m}_x(k + 1) = \boldsymbol{A}(k)\boldsymbol{m}_x(k) + \boldsymbol{B}(k)\boldsymbol{m}_u(k) \tag{3.56}$$

Equation (3.55) is the analog of (3.12), and (3.56) is the analog of (3.14). In this connection, we emphasize again that the matrices $\boldsymbol{A}(t)$ and $\boldsymbol{B}(t)$ for linear systems in continuous time do not coincide with the matrices $\boldsymbol{A}(k)$ and $\boldsymbol{B}(k)$ for linear systems in discrete time (see the remarks at the end of Section 1.3).

We can obtain a first equation for the covariance matrix

$$\boldsymbol{C}_{xx}(k,j) = E\{[\boldsymbol{x}(k) - \boldsymbol{m}_x(k)][\boldsymbol{x}(j) - \boldsymbol{m}_x(j)]'\} \tag{3.57}$$

by substituting (3.53) and (3.55) into (3.57):

$$\begin{aligned}\boldsymbol{C}_{xx}(k,j) = {} & \boldsymbol{\Phi}(k,0)\boldsymbol{C}_{xx}(0,0)\boldsymbol{\Phi}'(j,0) \\ & + \boldsymbol{\Phi}(k,0) \sum_{\mu=0}^{j-1} \boldsymbol{C}_{xu}(0,\mu)\boldsymbol{B}'(\mu)\boldsymbol{\Phi}'(j,\mu + 1) \\ & + \sum_{\nu=0}^{k-1} \boldsymbol{\Phi}(k,\nu + 1)\boldsymbol{B}(\nu)\boldsymbol{C}_{ux}(\nu,0)\boldsymbol{\Phi}'(j,0) \\ & + \sum_{\nu=0}^{k-1} \sum_{\mu=0}^{j-1} \boldsymbol{\Phi}(k,\nu + 1)\boldsymbol{B}(\nu)\boldsymbol{C}_{uu}(\nu,\mu)\boldsymbol{B}'(\mu)\boldsymbol{\Phi}'(j,\mu + 1)\end{aligned} \tag{3.58}$$

In the same way we obtain for the covariance matrix

$$\boldsymbol{C}_{xu}(k,j) = E\{[\boldsymbol{x}(k) - \boldsymbol{m}_x(k)][\boldsymbol{u}(j) - \boldsymbol{m}_u(j)]'\} \tag{3.59}$$

the expression

$$C_{xu}(k,j) = \boldsymbol{\Phi}(k,0)C_{xu}(0,j) + \sum_{\nu=0}^{k-1} \boldsymbol{\Phi}(k,\nu+1)B(\nu)C_{uu}(\nu,j) \tag{3.60}$$

If $\boldsymbol{x}(0)$ is not correlated with $\{\boldsymbol{u}(k)\}$, in effect,

$$C_{xu}(0,k) \equiv \boldsymbol{0}$$

and

$$C_{ux}(k,0) \equiv \boldsymbol{0}$$

equation (3.58) becomes

$$C_{xx}(k,j) = \boldsymbol{\Phi}(k,0)C_{xx}(0,0)\boldsymbol{\Phi}'(j,0) + \sum_{\nu=0}^{k-1}\sum_{\mu=0}^{j-1} \boldsymbol{\Phi}(k,\nu+1)B(\nu)C_{uu}(\nu,\mu)B'(\mu)\boldsymbol{\Phi}'(j,\mu+1) \tag{3.61}$$

and (3.60) becomes

$$C_{xu}(k,j) = \sum_{\nu=0}^{k-1} \boldsymbol{\Phi}(k,\nu+1)B(\nu)C_{uu}(\nu,j) \tag{3.62}$$

Because covariance matrices $C_{xx}(0,0)$ and $C_{uu}(k,j)$ are known, $C_{xx}(k,j)$ and $C_{xu}(k,j)$ can be calculated from these equations. Equations (3.58) and (3.61) correspond to (3.20) and (3.21) respectively, and equations (3.60) and (3.62) correspond to (3.37) and (3.38) respectively.

These expressions can be simplified further if the input random process $\{\boldsymbol{u}(k)\}$ is a vector white noise in discrete time with the covariance matrix

$$C_{uu}(k,j) = Q_u(k)\delta_{k,j} \tag{3.63}$$

where $\delta_{k,j}$ is the Kronecker symbol with the property

$$\delta_{k,j} = \begin{cases} 1 & \text{for} \quad k = j \\ 0 & \text{for} \quad k \neq j \end{cases}$$

In this way we obtain (Sage and Melsa, 1971; Melsa and Sage, 1973):

$$C_{xx}(k,j) = \boldsymbol{\Phi}(k,0)C_{xx}(0,0)\boldsymbol{\Phi}'(j,0) + \sum_{\nu=0}^{\min(k-1,j-1)} \boldsymbol{\Phi}(k,\nu+1)B(\nu)Q_u(\nu)B'(\nu)\boldsymbol{\Phi}'(j,\nu+1) \tag{3.64}$$

and

$$C_{xu}(k,j) = \begin{cases} 0 & \text{for } j > k - 1 \\ \Phi(k,j+1)B(j)Q_u(j) & \text{for } j \leq k - 1 \end{cases} \tag{3.65}$$

Equations (3.64) and (3.65) correspond to (3.25) and (3.39) for linear systems in continuous time.

We will now derive another equation for the covariance matrix $C_{xx}(k,j)$. If we substitute (3.47) and (3.56) in

$$C_{xx}(k+1,k+1) = E\{[x(k+1) - m_x(k+1)][x(k+1) - m_x(k+1)]'\}$$

we obtain

$$\begin{aligned} C_{xx}(k+1,k+1) &= A(k)C_{xx}(k,k)A'(k) + A(k)C_{xu}(k,k)B'(k) \\ &\quad + B(k)C_{ux}(k,k)A'(k) + B(k)C_{uu}(k,k)B'(k) \end{aligned} \tag{3.66}$$

If (3.65)—in which it is assumed that $x(0)$ and $\{u(k)\}$ are uncorrelated and $\{u(k)\}$ is white noise—is substituted in equation (3.66), it is simplified to

$$C_{xx}(k+1,k+1) = A(k)C_{xx}(k,k)A'(k) + B(k)C_{uu}(k,k)B'(k) \tag{3.67}$$

if, in addition, the following relationship is taken into account:

$$C'_{xu}(k,k) = C_{ux}(k,k)$$

Equation (3.67), which is the analog of equation (3.29), provides another method for calculating $C_{xx}(k,k)$ in addition to (3.64).

As in the case of linear systems in continuous time, we will also attempt to express $C_{xx}(k,j)$ in terms of $C_{xx}(k,k)$ or $C_{xx}(j,j)$. This can be done by writing down (3.64) for $j = k$ and multiplying it on the right-hand side by $\Phi'(j,k)$. In this way we obtain the equation

$$\begin{aligned} C_{xx}(k,k)\Phi'(j,k) &= \Phi(k,0)C_{xx}(0,0)\Phi'(j,0) \\ &\quad + \sum_{\nu=0}^{k-1} \Phi(k,\nu+1)B(\nu)Q_u(\nu)B'(\nu)\Phi'(j,\nu+1) \end{aligned}$$

the right-hand side of which coincides with $C_{xx}(k,j)$ according to (3.64) for $k \leq j$. We therefore have

$$C_{xx}(k,j) = C_{xx}(k,k)\boldsymbol{\Phi}'(j,k) \quad \text{for} \quad k \leq j \tag{3.68}$$

In the same way we can derive the equation

$$C_{xx}(k,j) = \boldsymbol{\Phi}(k,j)C_{xx}(j,j) \quad \text{for} \quad j \leq k \tag{3.69}$$

If equations (3.68) and (3.69) are combined, we obtain the expression

$$C_{xx}(k,j) = \begin{cases} C_{xx}(k,k)\boldsymbol{\Phi}'(j,k) & \text{for} \quad k \leq j \\ \boldsymbol{\Phi}(k,j)C_{xx}(j,j) & \text{for} \quad k \geq j \end{cases} \tag{3.70}$$

which agrees with (3.30) if we remember the abbreviated notation for the time parameter which was introduced at the beginning of this section.

The output equation (3.48) for linear systems in discrete time only differs from the output equation (3.2) for linear systems in continuous time by the argument. This also applies to the equations defining the mean values and covariance matrices for the random processes $\{u(k)\}$ or $\{u(t)\}$ and $\{x(k)\}$ or $\{x(t)\}$. Therefore the equations for the mean value

$$m_y(k) = E\{y(k)\} \tag{3.71}$$

and the covariance matrix

$$C_{yy}(k,j) = E\{[y(k) - m_y(k)][y(j) - m_y(j)]'\} \tag{3.72}$$

of the output random process $\{y(k)\}$ must be obtained from the corresponding results of Section 3.1 by a simple renaming of the arguments. If we replace $t$ by $k$ in equation (3.40), it follows that

$$m_y(k) = C(k)m_x(k) + D(k)m_u(k) \tag{3.73}$$

If we replace $t_1$ by $k$ and $t_2$ by $j$ in equation (3.42), it follows that

$$\begin{aligned} C_{yy}(k,j) = {} & C(k)C_{xx}(k,j)C'(j) \\ & + D(k)C_{ux}(k,j)C'(j) \\ & + C(k)C_{xu}(k,j)D'(j) \\ & + D(k)C_{uu}(k,j)D'(j) \end{aligned} \tag{3.74}$$

Here, also, the covariance matrices and the measurement matrix $C$ must be carefully distinguished. The remarks made after equation (3.42) apply here in exactly the same way.

The corresponding equations for a linear time-invariant dynamic system in discrete time with a stationary input random process are given by Sage and Melsa (1971) and Melsa and Sage (1973).

## REFERENCES

Liebelt, P.B. (1967). *An Introduction to Optimal Estimation,* Addison-Wesley, Reading, MA.

Melsa, J.L. and Sage, A.P. (1973). *An Introduction to Probability and Stochastic Processes,* Prentice-Hall, Englewood Cliffs, NJ.

Rothe, R. (1962). *Höhöre Mathematik,* Part II (13th ed.), Teubner, Stuttgart.

Sage, A.P. and Melsa, J.L. (1971). *Estimation Theory with Applications to Communications and Control,* McGraw-Hill, New York.

# Chapter 4
# Observation of the State Vector

## 4.1 INTRODUCTION

From time immemorial, scientists and engineers have been interested in acquiring information on the physical and technological environment. To describe the essential characteristics of a given process, the underlying laws must be established by experiments and observations, and then expressed in the form of mathematical models. A considerable advance was made in this field with the introduction of **dynamic models,** which began with the discovery of Newton's well known laws. Usually, the goal of model construction is to obtain quantitative information on the behavior of the system under study, either to describe a process taking place in the present or to predict future events and experimental results.

Subsequently mathematical models of the physical environment became increasingly refined, and the corresponding requirements for the accuracy of the measurement, estimation and prediction procedures for the variables concerned became increasingly stringent. For example, when the first planetoid Ceres was discovered in 1801 and was then lost in the Sun's radiation after only a fortieth of its orbit had been observed, astronomers tried in vain to locate it on the other side of the Sun. However, with the help of his least-squares method, developed in 1795, C. F. Gauss was able to determine the orbit of Ceres so exactly that the planetoid was found again (Gauss, 1809).

The least-squares method, which in the simplest special case gives the algebraic mean value, is the classical method for the systematic compensation of random measurement errors. By 1821, Gauss had also provided a recursive variant of this method which makes it possible to correct a previously calculated estimate after an additional measurement value has been obtained without having to repeat the entire calculation from the beginning (Gauss, 1873). This concept was re-examined by Plackett (1950) and generalized to include several simultaneous extra measurements.

In this chapter we formulate the observation problem in the context of a control problem. As is usual in this case, we assume that the state vector of the controlled process is inaccessible and must be estimated from measurements of the output variables. Disturbance variables at the input to the process are not considered in the observation problem; measurement errors are included in the output variables but are not specified. Therefore this problem is a precursor and special case of the filter problem.

The treatment is done in state space throughout, where the concepts of observability, duality and controllability developed by Kalman (1960a, 1961) are readily defined. The most important results of this chapter are the observation laws, which are presented in various forms and constitute the solution of the observation problem. Both continuous and discrete time are taken into account. It is shown that the estimation of the state vector by block processing of cumulative measurement variables fits into the framework of the Gauss least-squares method and that the Kalman observability matrix is a Gaussian normal matrix corresponding to the problem.

The cumulative observation laws for continuous and discrete time are in the form of integrals or sums. These can be suitably transformed to systems of differential or difference equations which represent observation laws for the recursive processing of measurements arriving sequentially. There was a sudden increase in interest in this form of observation and estimation at the end of the 1950s. Demands were being made by the new field of astronautics, and the powerful digital computers which were then emerging offered the possibility of performing recursive estimation in real time. There was also a change in the concept of observers and filters: they were no longer regarded as frequency responses or transfer functions as in the time of Wiener, Bode and Shannon, but as computing algorithms for the real time calculation of Gaussian, Gauss–Markov or minimum variance estimates, conditional expectations or even complete conditional distributions.

The recursive observation and filter laws presented in the form of differential or difference equations can be directly implemented in terms of hardware or software. They can also be applied to time-varying systems and finite observation intervals. The large number of applications in observation and filter technology (e.g., in the determination of the trajectories of aircraft, spacecraft and submarines) provide evidence for the practical importance of this theory. The first implementations are described by Smith *et al.* (1962) and Battin (1964).

The filter problem is a generalization of the observation problem which is treated as a preparatory step in this chapter. In filtering, the stochastic disturbance variables and measurement errors are taken into account explicitly. The filter has the same structure as the corresponding observer. The difference is that the gain factors of the filter are optimal with respect to the given statistical properties of the stochastic disturbance and measurement noises, whereas the observer's gain factors can be chosen by using other criteria. We return to the filter problem in Chapter 5.

## 4.2 THE OBSERVATION PROBLEM

The need for estimation and filtering methods became particularly pressing when the modern theory of optimum control was developed. The control laws which are obtained by applying the calculus of variations (Carathéodory, 1935), dynamic programming (Bellman and Dreyfus, 1962) and the maximum principle (Pontryagin *et al.*, 1964) normally require the feedback of the entire state vector of the controlled process. The classical feedback of a single scalar output variable is no longer sufficient. Instead, we face the problem of determining the state vector of the process from the measured output variable or variables. If the accompanying disturbance variables and measurement noises are so small that they do not have a major effect on the design, the problem is known as an observation problem; otherwise, it is known as a filter problem. In this section we present a mathematical formulation and discussion of the observation problem.

Consider a dynamic system which is expressed in the form

$$\dot{\boldsymbol{x}}(t) = \boldsymbol{A}(t)\boldsymbol{x}(t) + \boldsymbol{B}(t)\boldsymbol{u}(t) \qquad t_0 \le t \tag{4.1a}$$

$$\boldsymbol{y}(t) = \boldsymbol{C}(t)\boldsymbol{x}(t) \tag{4.1b}$$

where $\boldsymbol{x}$ is the $n$-dimensional unknown state vector, $\boldsymbol{u}$ is the $p$-dimensional known control vector and $\boldsymbol{y}$ is the vector of the $m$ measured output variables. $\boldsymbol{A}$, $\boldsymbol{B}$ and $\boldsymbol{C}$ are given matrices of corresponding dimensions (Figure 4.1, upper part). We require an estimate for $\boldsymbol{x}(t_0)$ or $\boldsymbol{x}(t)$.

*Note*. In principle, knowledge of $\boldsymbol{x}(t_0)$ is equivalent to knowledge of $\boldsymbol{x}(t)$ in the observation problem because one state can be calculated from the other by forward or backward integration of the differential equation (4.1a).

The solution of the observation problem is trivial when the measurement matrix $\boldsymbol{C}(t)$ is square and regular because then we obviously have

$$\boldsymbol{x}(t) = \boldsymbol{C}^{-1}(t)\boldsymbol{y}(t) \tag{4.2}$$

This case occurs if, for example, enough sensors can be installed to obtain $n$ linearly independent simultaneous measurements involving all the state variables of the system and systematic and random measurement errors can be neglected. Apart from the fact that in practice every additional sensor involves extra costs, a system configuration yielding a square regular measurement matrix is not feasible in most cases. Therefore other observation procedures must be found for systems with fewer output variables than state variables.

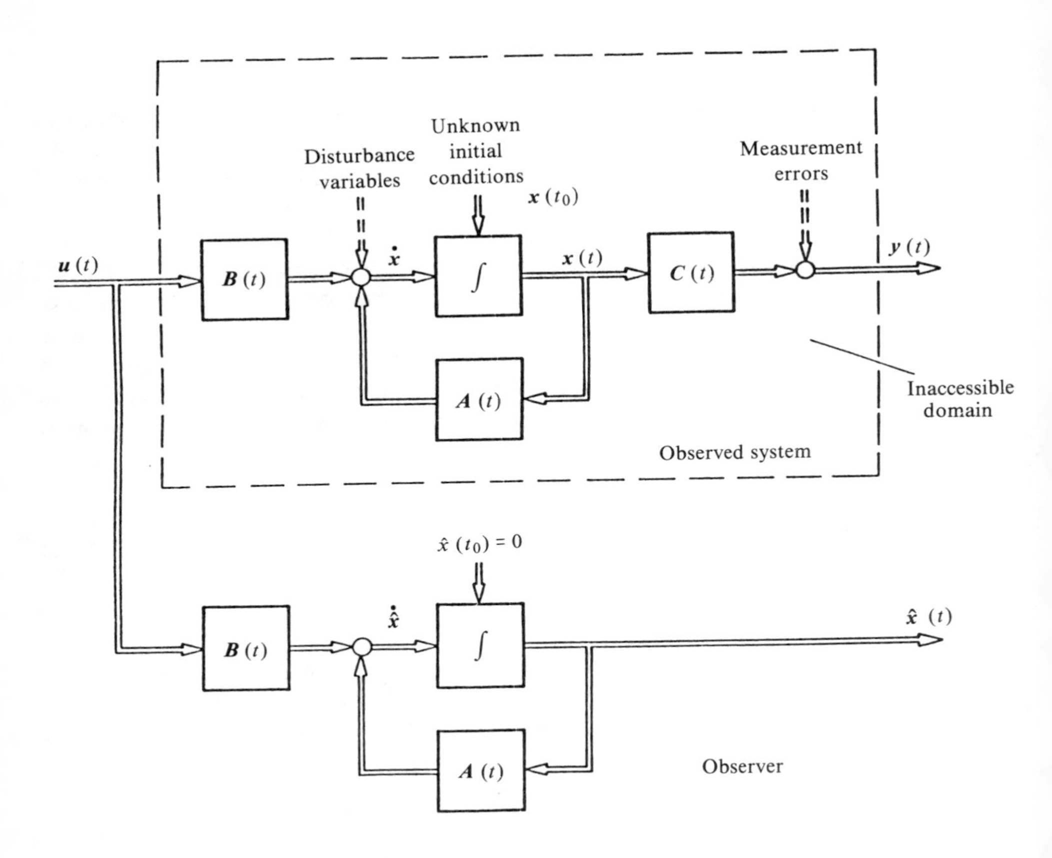


**Figure 4.1** Observed system with the observer in the form of an uncoupled model.

An obvious approach is to differentiate the output variable $\boldsymbol{y}(t)$ sufficiently often and to augment the matrix $\boldsymbol{C}$ by adding corresponding rows until a regular matrix is obtained. In order to illustrate the essential features of this approach, it is sufficient to consider a time-invariant single-input, single-output system of $n$th order:

$$\dot{\boldsymbol{x}}(t) = \boldsymbol{A}\boldsymbol{x}(t) + \boldsymbol{b}u(t) \tag{4.3a}$$

$$y(t) = \boldsymbol{c}'\boldsymbol{x}(t) \tag{4.3b}$$

Equation (4.3b) is now repeatedly differentiated with respect to $t$, where $\dot{\boldsymbol{x}}$ is substituted each time according to (4.3a):

$$
\begin{aligned}
y &= & &= c'x & & \\
\dot{y} &= c'\dot{x} & &= c'Ax & &+ c'bu \\
\ddot{y} &= c'A\dot{x} + c'b\dot{u} & &= c'A^2x & &+ c'Abu + c'b\dot{u} \\
&\vdots & & & & \\
\overset{(n-1)}{y} &= \ldots & &= c'A^{n-1}x & &+ \sum_{i=0}^{n-2} c'A^{n-2-i}b\overset{(i)}{u}
\end{aligned}
\tag{4.3c}
$$

*Note.* This procedure can also be used for the general time-variant system (4.1). Then the product rule of differential calculus must be applied. Of course, it is necessary to assume that the matrices $A(t)$, $B(t)$ and $C(t)$ are sufficiently differentiable. In addition to the terms in (4.3c), the result will contain derivatives of $A$, $B$ and $C$ (see (4.25)).

The row vectors involved in the scalar products with $x$ in equation system (4.3c) are stacked to form an $n \times n$ matrix, which is postmultiplied by $x$. All other terms in (4.3c) are known quantities, which are combined on the other side of the equation:

$$
\begin{bmatrix} c' \\ c'A \\ \vdots \\ c'A^{n-1} \end{bmatrix} x(t) = \begin{bmatrix} y(t) \\ \dot{y}(t) - c'bu(t) \\ \vdots \\ \overset{(n-1)}{y}(t) - \sum_{i=0}^{n-2} c'A^{n-2-i}b\overset{(i)}{u}(t) \end{bmatrix}
\tag{4.3d}
$$

If the matrix on the left of $x(t)$ is regular, this equation can be solved for $x(t)$. Otherwise, it is inherently impossible to calculate $x$. This will be shown in Section 4.3 on observability, where it is also shown that it is impossible to obtain additional linearly independent equations for $x$ by further differentiation (i.e., by adding the rows $c'A^n$, $c'A^{n+1}$ *et cetera*) [Cayley–Hamilton theorem]. Apart from this, there is little to criticize in this method from the purely theoretical point of view. However, in practice it is not feasible for higher-order systems ($n \geq 3$). The neglected measurement noise would be amplified to considerable levels by the repeated differentiation. Furthermore, discontinuities in the control variable $u$ and its derivatives would rapidly lead to saturation of the differentiating elements. Finally, the neglected disturbance variables on the input side would contribute their derivatives to the right-hand side of (4.3d) in a similar manner to $u$, resulting in a further deterioration in the result.

As the next approach to solving the observation problem, we will attempt to use a model for the observed process. It is well known that the solution of the differential equation (4.1a) has the general form

$$\boldsymbol{x}(t) = \boldsymbol{\phi}(t,t_0)\boldsymbol{x}(t_0) + \int_{t_0}^{t} \boldsymbol{\phi}(t,\tau)\boldsymbol{B}(\tau)\boldsymbol{u}(\tau)d\tau \tag{4.4}$$

where $\boldsymbol{\phi}(t,t_0)$ is the transition matrix associated with $\boldsymbol{A}(t)$ (see Section 1.2). The solution consists of the sum of a free component and a forced component. If the system is sufficiently stable, the free solution $\boldsymbol{\phi}(t,t_0)\boldsymbol{x}(t_0)$ rapidly decays and the state $\boldsymbol{x}(t)$ is then only determined by the forced solution:

$$\int_{t_0}^{t} \boldsymbol{\phi}(t,\tau)\boldsymbol{B}(\tau)\boldsymbol{u}(\tau)\mathrm{d}\tau$$

This component of the solution can be generated, for example, by a model of the observed process, which we start with zero initial conditions and supply with the same control input as the real process. According to this, the observer has the form

$$\frac{\mathrm{d}}{\mathrm{d}t}\hat{\boldsymbol{x}}(t) = \boldsymbol{A}(t)\hat{\boldsymbol{x}}(t) + \boldsymbol{B}(t)\boldsymbol{u}(t) \qquad \hat{\boldsymbol{x}}(t_0) = \boldsymbol{0} \tag{4.5}$$

where $\hat{\boldsymbol{x}}(t)$ is the estimate for $\boldsymbol{x}(t)$ (Figure 4.1). Obviously, no use is made of the output variable $\boldsymbol{y}(t)$.

This observation method has the advantage that the problems associated with differentiating the measurement errors, control inputs and disturbances are avoided (only integration is done). The disadvantage is that the time behavior of the estimation error

$$\tilde{\boldsymbol{x}}(t) := \boldsymbol{x}(t) - \hat{\boldsymbol{x}}(t) \tag{4.6}$$

depends exclusively on the dynamics of the observed system and cannot be influenced. This can be seen very easily if the differential equation for the estimation error is formed by subtracting (4.5) from (4.1a):

$$\frac{\mathrm{d}}{\mathrm{d}t}\tilde{\boldsymbol{x}}(t) = \boldsymbol{A}(t)\tilde{\boldsymbol{x}}(t)$$

However, this disadvantage can be overcome by adding the output equation

$$\hat{\boldsymbol{y}}(t) = \boldsymbol{C}(t)\hat{\boldsymbol{x}}(t) \tag{4.7}$$

to the model, where $\hat{\boldsymbol{y}}(t)$ can be interpreted as the estimated output variable. We then compare $\hat{\boldsymbol{y}}(t)$ with the actual measurement variable $\boldsymbol{y}(t)$ and feed the difference back to improve the estimate $\hat{\boldsymbol{x}}(t)$. In this way the observer takes the form

$$\frac{\mathrm{d}}{\mathrm{d}t}\hat{\boldsymbol{x}}(t) = \boldsymbol{A}(t)\hat{\boldsymbol{x}}(t) + \boldsymbol{B}(t)\boldsymbol{u}(t) + \boldsymbol{K}(t)\{\boldsymbol{y}(t) - \boldsymbol{C}(t)\hat{\boldsymbol{x}}(t)\} \tag{4.8}$$

where $\boldsymbol{K}(t)$ is a freely selectable $n \times m$ gain matrix (Figure 4.2). In order to investigate the behavior of the estimation error, we again form the difference between equations (4.1a) and (4.8), and apply (4.1b). We obtain

$$\begin{aligned}\frac{\mathrm{d}}{\mathrm{d}t}\tilde{\boldsymbol{x}}(t) &= \boldsymbol{A}(t)\tilde{\boldsymbol{x}}(t) - \boldsymbol{K}(t)\{\boldsymbol{C}(t)\boldsymbol{x}(t) - \boldsymbol{C}(t)\hat{\boldsymbol{x}}(t)\}\\ &= \{\boldsymbol{A}(t) - \boldsymbol{K}(t)\boldsymbol{C}(t)\}\tilde{\boldsymbol{x}}(t)\end{aligned} \tag{4.9}$$

*Note*. The differential equations (4.8) and (4.9) for $\hat{\boldsymbol{x}}$ and $\tilde{\boldsymbol{x}}$ have the same dynamic matrix $\boldsymbol{A} - \boldsymbol{KC}$.

We now have the means of adjusting the dynamics of the observer by a suitable choice of the gain matrix $\boldsymbol{K}(t)$ so that the observation error $\tilde{\boldsymbol{x}}(t)$ decays sufficiently rapidly. High values for $\boldsymbol{K}$ are preferable for this, but we will have to compromise for reasons of stability and noise sensitivity (bandwidth). In any case, the design of the observer is now reduced to the specification of $\boldsymbol{K}(t)$: we *only* need to pick suitable values for the gain factors $k_{ij}(t)$.

Unfortunately, the observer for an $n$th-order system with $m$ output variables needs $n \times m$ gain factors, and it is by no means trivial to adjust them adequately if not optimally for all $t$. For simple systems, it is sometimes possible to manage with a trial-and-error method by using computer simulation. However, it is clear that systematic methods are needed to synthesize $\boldsymbol{K}(t)$. Anticipating the next chapter, we can already say here that the observer in equation (4.8) has exactly the same structure as the Kalman–Bucy filter. Their method is tailored for a computer-aided synthesis of $\boldsymbol{K}(t)$, however, the statistical properties of the disturbance and measurement noise are required. An elementary method which manages without this requirement is presented in Section 4.3. Apart from this, there are some special methods for time-invariant systems, and two of these are briefly described here.

We consider a time-invariant $n$th-order system with a single output variable and we choose the observer canonical form as the state representation. For this canonical form, the last column of the $\boldsymbol{A}$ matrix consists of the coefficients $\alpha_i$ of the characteristic equation multiplied by $-1$ (see Section 1.4.2). Otherwise, only values of zero and unity occur in $\boldsymbol{A}$ and $\boldsymbol{c}'$:

$$\dot{\boldsymbol{x}}(t) = \begin{bmatrix} 0 & & & & -\alpha_0 \\ 1 & & & & -\alpha_1 \\ & 1 & & & -\alpha_2 \\ & & \ddots & & \vdots \\ 0 & & & 1 & -\alpha_{n-1} \end{bmatrix} \boldsymbol{x}(t) + \boldsymbol{B}\boldsymbol{u}(t)$$

$$y(t) = [\,0 \quad \ldots \quad 0 \quad 1\,]\ \boldsymbol{x}(t)$$

In this case as well, the observer takes the structure of equation (4.8), where $\boldsymbol{A}$, $\boldsymbol{B}$, and $\boldsymbol{C}$ are substituted from the above canonical form. The gain matrix $\boldsymbol{K}$ is now of the type $n \times 1$, i.e., it is an $n$-dimensional column vector. The dynamic matrix $\boldsymbol{A} - \boldsymbol{KC}$ in the differential equation (4.9) of the estimation error is now

$$\begin{bmatrix} 0 & & -\alpha_0 \\ 1 & & -\alpha_1 \\ & \ddots & \vdots \\ 0 & 1 & -\alpha_{n-1} \end{bmatrix} - \begin{bmatrix} k_1 \\ k_2 \\ \vdots \\ k_n \end{bmatrix} [0 \ldots 0\ 1] = \begin{bmatrix} 0 & & -\alpha_0 - k_1 \\ 1 & & -\alpha_1 - k_2 \\ & \ddots & \vdots \\ 0 & 1, & -\alpha_{n-1} - k_n \end{bmatrix}$$

The special forms of $\boldsymbol{A}$ and $\boldsymbol{C}$ produce a change in only the final column of $\boldsymbol{A}$. Thus the differential equation of the error is again in the observer canonical form. The final column now contains the negative coefficients of the characteristic equation for $\tilde{\boldsymbol{x}}$ or $\hat{\boldsymbol{x}}$. We can now prescribe a certain characteristic equation or pole (eigenvalue) configuration for the observer or for the error differential equation. It is then very easy to calculate the gain factors $k_1, \ldots, k_n$ by simply comparing coefficients. (Exactly dual conditions are obtained for the design of a controller using the control canonical form (Morgan, 1966).)

*Example 4.1.* An aircraft continuously acquires sufficiently exact position values from its standard navigation aids (e.g., VOR/DME or precision landing system). However, in addition it requires speed data relative to the ground. In order to avoid the installation of an expensive speed sensor (doppler radar, inertial system), the speed is to be estimated by an observer. For simplicity, we consider only one geometric dimension. In order to model the system in the observer canonical form, we define the position state variable as $x_2$ and the speed state variable as $x_1$:

$$\begin{bmatrix} \dot{x}_1 \\ \dot{x}_2 \end{bmatrix} = \begin{bmatrix} 0 & 0 \\ 1 & 0 \end{bmatrix} \begin{bmatrix} x_1 \\ x_2 \end{bmatrix} + \begin{bmatrix} 1 \\ 0 \end{bmatrix} u$$

$$y = [0 \quad 1] \begin{bmatrix} x_1 \\ x_2 \end{bmatrix}$$

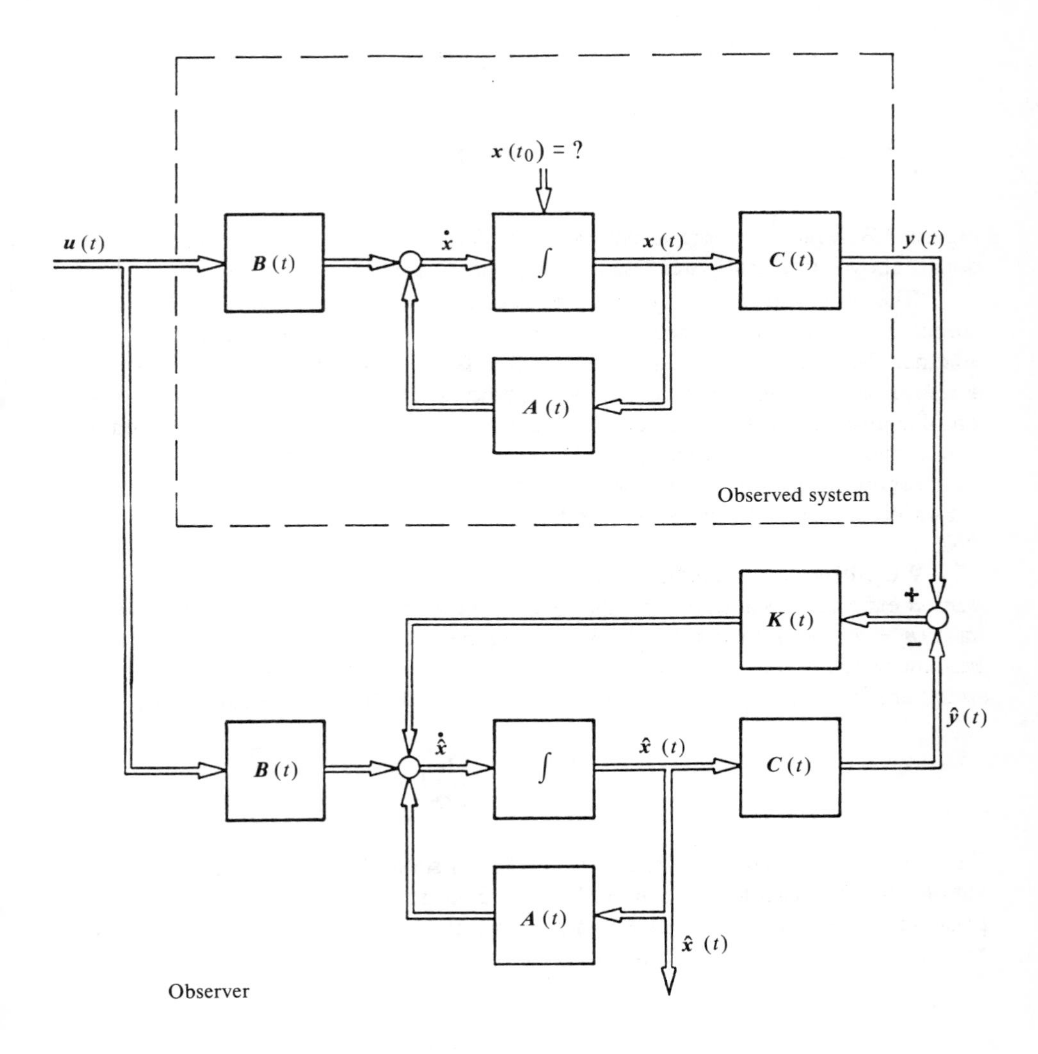


**Figure 4.2** Observed system with the observer in the form of a model with feedback of the output differences between system and observer.

A simple accelerometer is installed to measure the acceleration $u$. We want the observer to have its poles at $s = -1 \pm j$. Accordingly, its characteristic equation is $s^2 + 2s + 2 = 0$. Comparison of coefficients gives $0 - k_1 = -2$ and $0 - k_2 = -2$; therefore $k_1 = k_2 = 2$. The resulting observer has the form

$$\frac{\mathrm{d}}{\mathrm{d}t}\begin{bmatrix}\hat{x}_1\\ \hat{x}_2\end{bmatrix} = \begin{bmatrix}0 & 0\\ 1 & 0\end{bmatrix}\begin{bmatrix}\hat{x}_1\\ \hat{x}_2\end{bmatrix} + \begin{bmatrix}1\\ 0\end{bmatrix}u + \begin{bmatrix}2\\ 2\end{bmatrix}(y - \hat{x}_2)$$
$$= \begin{bmatrix}0 & -2\\ 1 & -2\end{bmatrix}\begin{bmatrix}\hat{x}_1\\ \hat{x}_2\end{bmatrix} + \begin{bmatrix}1\\ 0\end{bmatrix}u + \begin{bmatrix}2\\ 2\end{bmatrix}y$$

Figure 4.3 shows the corresponding circuit diagram. In addition to the sensors mentioned above, two integrators and an inverting amplifier are necessary.

The observation of a system with the aid of an exact copy of its mathematical model is conceptually extremely rewarding. Nevertheless, the question arises as to whether the relevant hardware is excessive. Is it, for example, really necessary for a fourth-order system with three measurement variables to implement four differential equations for the observer? After all, only one more linearly independent equation is required for a purely algebraic determination of $\boldsymbol{x}$! The general answer is that an observer of order $n - m$ is sufficient to determine the state vector of an observable $n$th-order system with $m$ linearly independent measurement variables (Luenberger, 1966).

We will follow Luenberger and consider a system of the form (4.1) with *constant* coefficients. Starting from the measurement matrix $\boldsymbol{C}$ of rank $m$ we add a constant $(n - m) \times n$ matrix $\boldsymbol{D}$ still to be determined. Its rows must be linearly independent with respect to the rows of $\boldsymbol{C}$. We denote by $z$ the vector which is obtained after transformation of the state $\boldsymbol{x}(t)$ with the matrix $\boldsymbol{D}$. We therefore have

$$\begin{bmatrix}\boldsymbol{y}(t)\\ z(t)\end{bmatrix} = \begin{bmatrix}\boldsymbol{C}\\ \boldsymbol{D}\end{bmatrix}\boldsymbol{x}(t) \tag{4.10}$$

The combined matrix consisting of $\boldsymbol{C}$ and $\boldsymbol{D}$ is square and nonsingular. The observation problem can now be solved by inversion of this combined matrix if it is possible to generate an estimate $\hat{z}$ of $z$ so that the error $\tilde{z} = z - \hat{z}$ tends to zero. For this, we modify the model procedure and postulate

$$\frac{\mathrm{d}}{\mathrm{d}t}\hat{z}(t) = \boldsymbol{F}\hat{z}(t) + \boldsymbol{G}\boldsymbol{u}(t) + \boldsymbol{H}\boldsymbol{y}(t) \tag{4.11}$$

In order to investigate the time behavior of the error $\tilde{z}$, we form the corresponding differential equation

$$\frac{\mathrm{d}}{\mathrm{d}t}\tilde{z}(t) = \frac{\mathrm{d}}{\mathrm{d}t}(z - \hat{z}) = \boldsymbol{D}\dot{\boldsymbol{x}}(t) - \frac{\mathrm{d}}{\mathrm{d}t}\hat{z}(t)$$

The differential equations (4.1a) and (4.11) are substituted for $\dot{\boldsymbol{x}}$ and $d\hat{z}/\mathrm{d}t$:

$$\frac{d}{dt}\tilde{z}(t) = \boldsymbol{DAx} - \boldsymbol{F}\hat{z} + (\boldsymbol{DB} - \boldsymbol{G})\boldsymbol{u} - \boldsymbol{Hy}$$

Substitution of $\hat{z} = \boldsymbol{Dx} - \tilde{z}$ and $\boldsymbol{y} = \boldsymbol{Cx}$ and rearrangement gives

$$\frac{d}{dt}\tilde{z}(t) = \boldsymbol{F}\tilde{z}(t) + (\boldsymbol{DA} - \boldsymbol{FD} - \boldsymbol{HC})\boldsymbol{x}(t) + (\boldsymbol{DB} - \boldsymbol{G})\boldsymbol{u}(t)$$

In order to let the error $\tilde{z}$ decay independently of $\boldsymbol{x}$ and $\boldsymbol{u}$, the observer matrices must be chosen as follows: $\boldsymbol{F}$ must be asymptotically stable (eigenvalues with negative real part)

$$\boldsymbol{G} = \boldsymbol{DB} \tag{4.12}$$

and $\boldsymbol{H}$ must satisfy the equation

$$\boldsymbol{DA} - \boldsymbol{FD} = \boldsymbol{HC} \tag{4.13}$$

Then we will have $d\tilde{z}/dt = \boldsymbol{F}\tilde{z}$ or $\tilde{z}(t) = \exp(\boldsymbol{F}t)\,\tilde{z}(0)$, or

$$\hat{z}(t) = \boldsymbol{Dx}(t) - \exp(\boldsymbol{F}t)\,\tilde{z}(0)$$

so that $\hat{z}(t)$ can be used as a good approximation for $z(t)$ in equation (4.10). The design of the observer now essentially consists of solving equation (4.13) where the stability of $\boldsymbol{F}$ and the linear independence of the matrix $\boldsymbol{D}$ with respect to $\boldsymbol{C}$ must be provided as a secondary condition. The solution exists if $\boldsymbol{A}$ and $\boldsymbol{F}$ have unequal eigenvalues (for more details see Luenberger (1966)).

*Example 4.2.* When the method explained above is used, the order of the observer in Example 4.1 can be reduced from 2 to 1. We set $\boldsymbol{D} = [d_1, d_2]$ and choose $F = -1$ and $H = +1$. With these assumptions and the values for $\boldsymbol{A}$ and $\boldsymbol{C}$ from Example 4.1, (4.13) is now

$$[d_2 \quad 0] + [d_1 \quad d_2] = [0 \quad 1]$$

The solution is obviously $\boldsymbol{D} = [-1, +1]$. Equation (4.12) produces $G = -1$. According to (4.11) and (4.10), the observer now has the form

$$\frac{d}{dt}\hat{z}(t) = -\hat{z}(t) - u(t) + y(t)$$

$$\hat{x}_1(t) = y(t) - \hat{z}(t)$$

$$\hat{x}_2(t) = y(t)$$

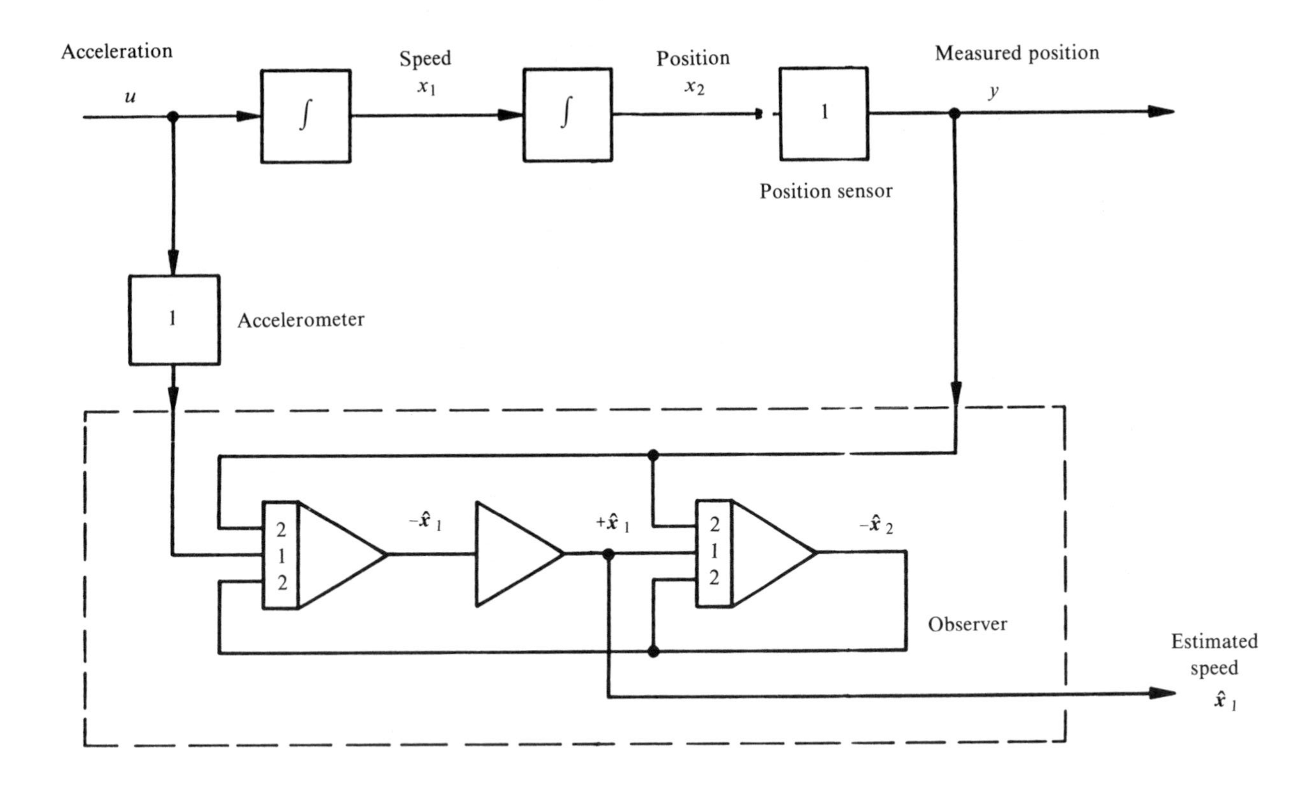


**Figure 4.3** Circuit diagram for the observation of speed based on position and acceleration measurements.

Figure 4.4 shows the corresponding circuit diagram. The number of integrators has been reduced from two to one.

We have given a brief introduction to the observation problem in this section. We have deliberately adopted an elementary approach. Nevertheless, we have already touched upon some important concepts, problems, relationships and approaches which we will meet frequently in a more generalized form. The presentation has been restricted to continuous time. The observation problem in discrete time has a very similar structure and will be comprehensively discussed in Section 4.6.

## 4.3 OBSERVABILITY AND OBSERVATION IN CONTINUOUS TIME

The concepts of observability and controllability are of fundamental importance in modern system theory. It is absolutely essential that a system has these properties if its state is to be successfully observed or controlled. A particular need for these criteria arose with the advent of controlled processes with multiple inputs and multiple outputs. It is far more difficult to recognize degenerate structures in these systems than in conventional single-input, single-output systems. The concept of controllability originated in principle from the calculus of variations (Carathéodory, 1935) and was subsequently used in the theory of optimal control (Pontryagin *et al.*, 1964). The concept of observability is related to a special normal matrix of the Gaussian least-squares method and was first used in the theory of optimal filtering. The development and formalization of the two concepts in the context of control theory, the demonstration of their duality and their common acceptance in control engineering are generally ascribed to Kalman (Kalman, 1960, 1961).

We first examine observability as a continuation of the previous section. There, the observation problem consisted in estimating the state vector $\boldsymbol{x}(t)$ of the dynamic system (4.1) on the basis of measurements of the output vector $\boldsymbol{y}(t)$ and the input vector $\boldsymbol{u}(t)$. In examining (4.4), which represents the general solution of the differential equation (4.1a), we saw that the input variable $\boldsymbol{u}(t)$ generates the forced solution component and that this is added to the free solution component produced by the initial condition $\boldsymbol{x}(t_0)$. Note that this is a consequence of the superposition principle for linear systems. Therefore the effect of $\boldsymbol{u}(t)$ on $\boldsymbol{x}(t)$ can readily be calculated separately. The intrinsic difficulty of the observation problem is to determine the remaining free solution component of $\boldsymbol{x}(t)$. To obtain the corresponding free component in $\boldsymbol{y}(t)$ the forced part of $\boldsymbol{x}(t)$ is premultiplied by $\boldsymbol{C}(t)$ and this product is subtracted from the measured output variable.

In what follows, we assume that the measurement variable $\boldsymbol{y}(t)$ has already had the forced components removed in the manner described above. Thus we will consider the free system of $n$th order:

$$\dot{\boldsymbol{x}}(t) = \boldsymbol{A}(t)\boldsymbol{x}(t) \quad \boldsymbol{x}(t_0) = ? \tag{4.14a}$$

$$\boldsymbol{y}(t) = \boldsymbol{C}(t)\boldsymbol{x}(t) \tag{4.14b}$$

Let the matrices $\boldsymbol{A}$ and $\boldsymbol{C}$ be given for all $t_0 \leq t \leq t_1$. Let the output variable $\boldsymbol{y}(t)$ be $m$ dimensional and known in the observation interval $[t_0, t_1]$.

*Definition 4.1*

(i) A state $\boldsymbol{x}(t_1)$ of system (4.14) is said to be *observable in the interval* $[t_0,t_1]$ if it can be uniquely determined on the basis of knowing $\boldsymbol{y}(t)$ in the interval $[t_0, t_1]$.

(ii) If and only if every state is observable in this sense, the system (4.14) is said to be *completely observable in the interval* $[t_0, t_1]$.

(iii) If and only if there is a $t_0$ for every $t_1$ so that (ii) is satisfied, system (4.14) is said to be *completely observable* as such.

In system design, only complete observability (i.e., properties (ii) and (iii)) is really relevant. We are now interested in useful criteria for complete observability and in manageable observation laws. The following treatment may appear somewhat formal at first glance. However, the practical aspects will become clear when it is related to the least-squares method (Section 4.6).

We start by constructing a first observation law as follows. The solution of the free differential equation (4.14a) with the transition matrix $\boldsymbol{\phi}$ is

$$\boldsymbol{x}(t_1) = \boldsymbol{\phi}(t_1,t)\boldsymbol{x}(t) \qquad \text{for} \quad t_0 \leq t \leq t_1$$

Solving for $\boldsymbol{x}(t)$ and substituting in (4.14b) gives

$$\boldsymbol{y}(t) = \boldsymbol{C}(t)\boldsymbol{\phi}(t,t_1)\boldsymbol{x}(t_1) \tag{4.15}$$

Both sides are premultiplied by $\boldsymbol{\phi}'\boldsymbol{C}'$:

$$\boldsymbol{\phi}'(t,t_1)\boldsymbol{C}'(t)\boldsymbol{y}(t) = \boldsymbol{\phi}'(t,t_1)\boldsymbol{C}'(t)\boldsymbol{C}(t)\boldsymbol{\phi}(t,t_1)\boldsymbol{x}(t_1)$$

We replace $t$ by $\tau$ and integrate over $\tau$ from $t_0$ to $t_1$. On the right-hand side, $\boldsymbol{x}(t_1)$ can be extracted from the integral as a constant and we abbreviate the remaining integral as $\boldsymbol{M}$:

$$\boldsymbol{M}(t_1,t_0) := \int_{t_0}^{t_1} \boldsymbol{\phi}'(\tau,t_1)\boldsymbol{C}'(\tau)\boldsymbol{C}(\tau)\boldsymbol{\phi}(\tau,t_1)\mathrm{d}\tau \tag{4.16}$$

We therefore have

$$\int_{t_0}^{t_1} \boldsymbol{\phi}'(\tau,t_1)\boldsymbol{C}'(\tau)\boldsymbol{y}(\tau)\mathrm{d}\tau = \boldsymbol{M}(t_1,t_0)\boldsymbol{x}(t_1) \tag{4.17}$$

Obviously, the left-hand side of this equation represents a weighted integration of the measurement variable $\boldsymbol{y}$ and can always be evaluated as such. In order to solve equation (4.17) for the unknown $\boldsymbol{x}(t_1)$, we have to invert $\boldsymbol{M}$ and premultiply both sides by $\boldsymbol{M}^{-1}$ or, more practically, use the Gauss or Cholesky algorithm with $\boldsymbol{M}$ as the coefficient matrix (Zurmuehl and Falk, 1984). A unique solution for $\boldsymbol{x}(t_1)$ exists in such a case if and only if $\boldsymbol{M}(t_1,t_0)$ is nonsingular.

*Notes*

(i) The matrix $\boldsymbol{M}$ defined by (4.16) is called the *observability matrix of the first kind*. It is $n \times n$ square and symmetric.

(ii) We now premultiply both sides of (4.17) by $\boldsymbol{x}'(t_1)$. On the left-hand side of the equation, we bring this vector under the integral and finally use (4.15) in its transposed form. We thus obtain an expression for the *measurement energy:*

$$\int_{t_0}^{t_1} \boldsymbol{y}'(\tau)\boldsymbol{y}(\tau)\mathrm{d}\tau = \boldsymbol{x}'(t_1)\boldsymbol{M}(t_1,t_0)\boldsymbol{x}(t_1) \tag{4.18}$$

(iii) The left-hand side of this equation is the integral of a sum of squares which can never be negative. Thus $\boldsymbol{M}$ is always positive semidefinite (i.e., $\boldsymbol{x}'\boldsymbol{M}\boldsymbol{x} \geq 0$ for all $\boldsymbol{x}$).

(iv) Later, we will derive equivalent differential equations for (4.16) and (4.17), which are easier to handle than the integrals.

*Theorem 4.1* For complete observability of the system (4.14) in the interval $[t_0,t_1]$, it is necessary and sufficient that the observability matrix $\boldsymbol{M}(t_1,t_0)$ in equation (4.16) is nonsingular.*

*Proof.* The sufficiency of this condition has already been established in discussing the observation law (4.17). The necessity of the condition is proved indirectly. If $\boldsymbol{M}(t_1,t_0)$ is singular, then there exists at least one $\boldsymbol{x}(t_1) \neq \boldsymbol{0}$ so that $\boldsymbol{M}(t_1,t_0)\boldsymbol{x}(t_1) = \boldsymbol{0}$. The right-hand side of equation (4.18) vanishes for this $\boldsymbol{x}(t_1)$. It follows from

*A symmetric positive semidefinite matrix which is nonsingular is positive definite; all of its eigenvalues are positive.

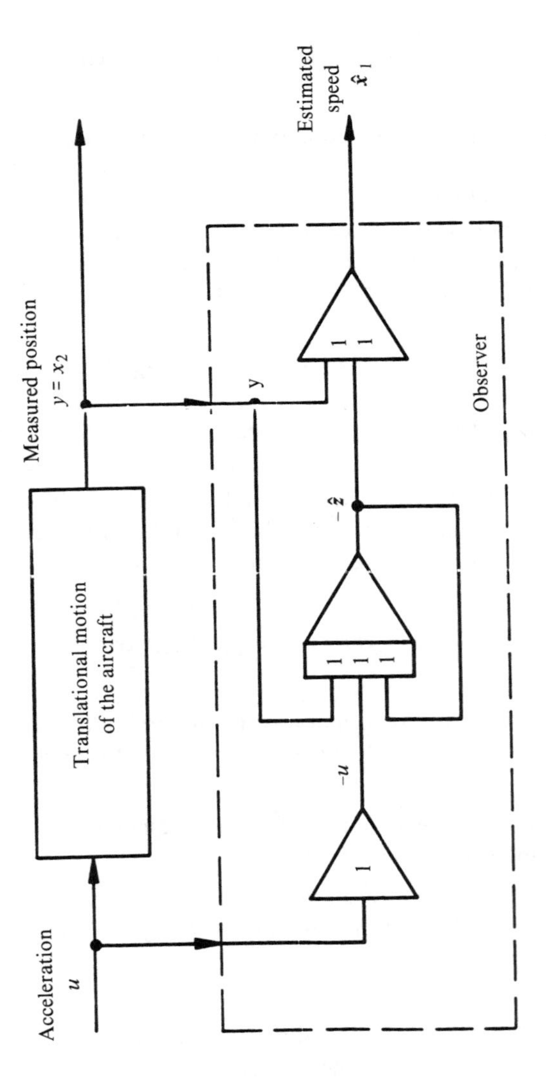


**Figure 4.4** Circuit diagram for a Luenberger observer.

this that, on the left-hand side, $y(\tau)$ is identically zero over the entire interval $[t_0,t_1]$ because it is continuous. However, the same output function is also obtained for $x(t) \equiv 0$ so that, while observing $y$, at least two different states cannot be distinguished.

*Lemma 4.1.* The system is completely observable as such if and only if a $t_0 < t_1$ can be given for every $t_1$ so that $M(t_1,t_0)$ is nonsingular.

We can further state that if $M(t_1,t_0)$ is positive definite for a certain $t_1$ then $M(t,t_0)$ is also positive definite for all $t > t_1$ since the integrand in (4.16) is positive semidefinite.

If the state $x(t_1)$ of the free system is exactly known at just one single point in time $t_1$, then the state of the system can be calculated from this for all times $t$ larger or smaller than $t_1$. This can be done by integrating the free differential equation or by multiplication by the transition matrix. This statement has, however, mainly theoretical value because in practice observed systems are neither completely free of disturbances nor can the effect of known input variables be specified with absolute exactness.

The observation law (4.17) requires the integration of the measurement variable $y(t)$. We have already seen in Section 4.2 that the observation problem, at least from the theoretical point of view, can also be solved by repeated differentiation of the output variable. We return to this concept here because it enables us to arrive at a significantly simpler criterion for complete observability in the case of time-invariant systems. We therefore now let system (4.14) have constant parameters $A$ and $C$, and we form

$$\begin{aligned}
y(t) &= & &= Cx(t) \\
\dot{y}(t) &= C\dot{x}(t) & &= CAx(t) \\
\ddot{y}(t) &= CA\dot{x}(t) & &= CA^2x(t) \\
&\vdots & & \\
\overset{(n-1)}{y}(t) &= & &= CA^{n-1}x(t)
\end{aligned} \tag{4.19}$$

There is no need to go on collecting higher-order derivatives. According to the Cayley–Hamilton theorem (A.41), every $n \times n$ matrix $A$ satisfies its own characteristic equation. Thus $A^n$ can be expressed as a linear combination of the previous powers:

$$A^n = -\alpha_0 I - \alpha_1 A - \ldots -\alpha_{n-1}A^{n-1} \tag{4.20}$$

(The $\alpha_i$ are the coefficients of the characteristic equation.) By premultiplication of equation (4.20) by $C$, we obtain

$$CA^n = -\alpha_0 C - \alpha_1 CA - \ldots - \alpha_{n-1} CA^{n-1} \tag{4.21}$$

It follows from this after postmultiplication by $\boldsymbol{x}(t)$ that

$$\overset{(n)}{y}(t) = -\alpha_0 y(t) - \alpha_1 \dot{y}(t) - \ldots - \alpha_{n-1} \overset{(n-1)}{y}(t)$$

By repeated postmultiplication of (4.21) by $A$ and substitution of equation (4.20) in the last summand each time, the matrices $CA^{n+1}$, $CA^{n+2}$, ... can be written as a linear combination of the matrices $C, \ldots, CA^{n-1}$. In a corresponding manner, the derivatives $\overset{(n+1)}{y}$, $\overset{(n+2)}{y}$, ... and $\overset{(n)}{y}$ are just linear combinations of $y, \dot{y}, \ldots, \overset{(n-1)}{y}$. The derivatives of $n$th and higher order therefore do not contribute any additional information.

We now combine the $n$ matrices $C$ up to $CA^{n-1}$ to form the $nm \times n$ matrix $\bar{M}$:

$$\bar{M} := \begin{bmatrix} C \\ CA \\ CA^2 \\ \vdots \\ CA^{n-1} \end{bmatrix} \tag{4.22}$$

In this way the equation system (4.19) can be written

$$\begin{bmatrix} y(t) \\ \dot{y}(t) \\ \vdots \\ \overset{(n-1)}{y}(t) \end{bmatrix} = \bar{M}x(t) \tag{4.23}$$

This is an observation law in the form of a system of $nm$ linear equations for the $n$ unknown state variables. The required state $x(t)$ can be uniquely determined from it if and only if the coefficient matrix $\bar{M}$ has rank $n$ (i.e., if it has $n$ linearly independent rows).

The matrix $\bar{M}$ defined by equation (4.22) is called the *observability matrix of the second kind*. It is much easier to obtain than the observability matrix of the first kind, given by (4.16), because neither the transition matrix nor an integration is required. It is sufficient to postmultiply $C$ or $CA^i$ by $A$ each time. However, in contrast with $M$, $\bar{M}$ is no longer symmetric and, for $n > 1$, is also no longer square.

*Theorem 4.2.* For a system (4.14) with time-invariant parameters $A$ and $C$ to be completely observable, it is necessary and sufficient for the observability matrix $\bar{M}$ in equation (4.22) to have rank $n$ (where $n$ is the order of the system or of the matrix $A$).

*Proof.* The sufficiency of this condition has already been established by the observation law (4.23). The necessity is again proved indirectly. If the rank of $\bar{\boldsymbol{M}}$ is less than $n$, then for arbitrary $t_1$ there is at least one $\boldsymbol{x}(t_1) \neq \boldsymbol{0}$ such that $\bar{\boldsymbol{M}}\boldsymbol{x}(t_1) = \boldsymbol{0}$. For this $\boldsymbol{x}(t_1)$, $\boldsymbol{y}$ and *all* its derivatives vanish at time $t_1$ so that $\boldsymbol{y}(t)$ is identically zero. However, the same output variable is obtained for $\boldsymbol{x}(t) \equiv \boldsymbol{0}$. Therefore at every point in time there are at least two unequal states which cannot be distinguished by observing $\boldsymbol{y}$.

Theorem 4.2 is quite easy to handle. Consequently, time-invariant systems in particular can and should be rapidly checked for observability before the observer (or filter) is designed.

The absolute observation interval $[t_0,t_1]$ is no longer important for time-invariant systems. In everything that happens in these systems it is only the time difference that matters, so that the beginning of the interval can always be set to zero. In addition, the following theorem states that the length of the interval can be made arbitrarily small. (This fact is not surprising in view of the observability law (4.23).)

*Theorem 4.3.* For complete observability of a system (4.14) with time-invariant parameters $\boldsymbol{A}$ and $\boldsymbol{C}$, it is necessary and sufficient that the observability matrix $\boldsymbol{M}$ $(\varepsilon,0)$ from equation (4.16) is nonsingular for any $\varepsilon > 0$.

*Proof of sufficiency.* If $M(\varepsilon,0)$ is nonsingular, then owing to the time invariance of the observed system $\boldsymbol{M}(t_1,t_1 - \varepsilon)$ is also nonsingular for every $t_1$. According to Lemma 4.1, the system is therefore completely observable. The proof of the necessity is given by Athans and Falb (1966) or Bucy and Joseph (1968) and others (see Kalman (1960) for the dual case).

We now return to free systems with *time-varying* parameters. For constant systems, repeated differentiation of the measurement variable $\boldsymbol{y}(t)$ produced the very simple observability matrix of the second kind. This can also be performed for systems with time-varying parameters provided that the matrices $\boldsymbol{A}$ and $\boldsymbol{C}$ are differentiable sufficiently often. In order to reduce the amount of notation here we use the operator:

$$L_A\{.\} := .\ \boldsymbol{A}(t) + \frac{\mathrm{d}}{\mathrm{d}t}\,. \tag{4.24}$$

When we apply this operator to a matrix we must proceed as follows. Postmultiply this matrix by $\boldsymbol{A}$ and also differentiate it once with respect to $t$ and add the two results together. For example, for $\boldsymbol{y}(t) = \boldsymbol{C}(t)\ \boldsymbol{x}(t)$, we have

$$\begin{aligned}\dot{\boldsymbol{y}}(t) &= \boldsymbol{C}(t)\dot{\boldsymbol{x}}(t) + \dot{\boldsymbol{C}}(t)\boldsymbol{x}(t)\\ &= [\boldsymbol{C}(t)\boldsymbol{A}(t) + \dot{\boldsymbol{C}}(t)]\boldsymbol{x}(t)\end{aligned}$$

$$= L_A\{C(t)\}x(t)$$

Thus for time-varying systems the observability matrix of the second kind takes the generalized form†

$$\bar{M}(t) := \begin{bmatrix} C(t) \\ L_A\{C(t)\} \\ L_A^2\{C(t)\} \\ \vdots \\ L_A^{n-1}\{C(t)\} \end{bmatrix} \tag{4.25}$$

Here, it is *sufficient* for complete observability if $\bar{M}(t)$ has rank $n$. For more details, in particular for the necessary condition, see Silverman (1966a,b) and Bucy and Joseph (1968).

The matrix $\bar{M}(t)$ could be used in a generalized observation law corresponding to equation (4.23), but in the context of filtering theory the general observation law (4.16) and (4.17) established earlier is much more interesting. However, in that form it is still not practical because it requires the evaluation of two complicated integrals. We will therefore derive an equivalent system of differential equations which is much more useful for numerical calculations. We do this in four main steps.

(i) In equation (4.16), the fixed end time $t_1$ is replaced by the running time $t$ and we proceed to differentiate both sides with respect to $t$. Here it should be noted that the integrand itself now also depends on $t$. The easiest way to remove this difficulty is to premultiply both sides by $\phi'(t,t_0)$ and postmultiply both sides by $\phi(t,t_0)$ before differentiation. These factors can be taken under the integral because they do not depend on $\tau$. We therefore have

$$\phi'(t,t_0)M(t,t_0)\phi(t,t_0) = \int_{t_0}^{t} \phi'(\tau,t_0)C'(\tau)C(\tau)\phi(\tau,t_0)\mathrm{d}\tau$$

We can now differentiate this equation with respect to $t$. On the left-hand side we use the product rule and the property of the transition matrix that $\mathrm{d}\phi(t,t_0)/\mathrm{d}t = A(t)\phi(t,t_0)$. We obtain

$$\phi'(t,t_0)\{A'(t)M(t,t_0) + \frac{\mathrm{d}}{\mathrm{d}t}M(t,t_0) + M(t,t_0)A(t)\}\phi(t,t_0)$$
$$= \phi'(t,t_0)C'(t)C(t)\phi(t,t_0)$$

†We have $L_Ak+1\{.\} := L_A\{L_A^k\{.\}\}$, $k = 1,2,3,\ldots$

Because the transition matrix is always nonsingular, the contents of the braces must be equal to $\boldsymbol{C}'\boldsymbol{C}$. By rearranging the terms, we obtain the desired differential equation for $\boldsymbol{M}$:

$$\frac{\mathrm{d}}{\mathrm{d}t}\boldsymbol{M}(t,t_0) = -\boldsymbol{A}'(t)\boldsymbol{M}(t,t_0) - \boldsymbol{M}(t,t_0)\boldsymbol{A}(t) + \boldsymbol{C}'(t)\boldsymbol{C}(t) \tag{4.26}$$

According to equation (4.16), the initial condition $\boldsymbol{M}(t_0,t_0)$ is equal to $\boldsymbol{0}$. This bilinear matrix differential equation is equivalent to the integral form (4.16) for the observability matrix of the first kind. It has the advantage that the transition matrix is no longer needed.

(ii) In equation (4.17), $t_1$ is also replaced by the running time $t$. Before differentiating with respect to $t$, we premultiply both sides by $\boldsymbol{\phi}'(t,t_0)$ for the same reason as above:

$$\boldsymbol{\phi}'(t,t_0)\boldsymbol{M}(t,t_0)\boldsymbol{x}(t) = \int_{t_0}^{t} \boldsymbol{\phi}'(\tau,t_0)\boldsymbol{C}'(\tau)\boldsymbol{y}(\tau)\mathrm{d}\tau$$

In differentiating this equation with respect to $t$, we follow the same procedure as above but treat the product $\boldsymbol{Mx} := \boldsymbol{z}$ as an entity in itself. We obtain

$$\dot{\boldsymbol{z}}(t) = -\boldsymbol{A}'(t)\boldsymbol{z}(t) + \boldsymbol{C}'(t)\boldsymbol{y}(t) \tag{4.27}$$

In an equivalent manner to the observation law (4.17) this is a differential equation for the vector

$$\boldsymbol{z}(t) := \boldsymbol{M}(t,t_0)\boldsymbol{x}(t) \tag{4.28a}$$

with the initial condition $\boldsymbol{z}(t_0) = \boldsymbol{0}$. As soon as $\boldsymbol{M}(t,t_0)$ has become nonsingular as a solution of the differential equation (4.26), we have complete observability. Then, according to (4.28a), $\boldsymbol{x}(t)$ is uniquely defined by

$$\boldsymbol{x}(t) = \boldsymbol{M}^{-1}(t,t_0)\boldsymbol{z}(t) \tag{4.28b}$$

(iii) In continuing the observation process, $\boldsymbol{M}$ will change but will remain nonsingular. Instead of inverting $\boldsymbol{M}$ over and over again, in view of solution (4.28b), it would be very useful to obtain a differential equation for $\boldsymbol{M}^{-1}$ itself directly. For this purpose, we examine the product

$$\boldsymbol{M}^{-1}(t,t_0)\boldsymbol{M}(t,t_0) = \boldsymbol{I}$$

We differentiate this equation with respect to $t$ using the product rule and apply the differential equation (4.26) for $\dot{\boldsymbol{M}}$. We then postmultiply both sides by $\boldsymbol{M}^{-1}$, rearrange and obtain

$$\frac{\mathrm{d}}{\mathrm{d}t}\boldsymbol{M}^{-1}(t,t_0) = \boldsymbol{A}(t)\boldsymbol{M}^{-1}(t,t_0) + \boldsymbol{M}^{-1}(t,t_0)\boldsymbol{A}'(t) - \boldsymbol{M}^{-1}(t,t_0)\boldsymbol{C}'(t)\boldsymbol{C}(t)\boldsymbol{M}^{-1}(t,t_0) \tag{4.29}$$

This is a nonlinear matrix differential equation of the Riccati type for $\boldsymbol{M}^{-1}(t,t_0)$. It is a follow-on solution to the bilinear differential equation (4.26) for $\boldsymbol{M}(t,t_0)$ and is valid as soon as complete observability has occurred. The initial condition for (4.29) is obtained by inversion of the final value of (4.26).

(iv) When $\boldsymbol{M}$ has become nonsingular, the differential equation (4.27) can also be transformed so that it no longer provides $z(t)$ but $\boldsymbol{x}(t)$ itself. At this point we call attention to the fact that when solving an observation law for $\boldsymbol{x}(t)$ we are actually only dealing with an estimate of $\boldsymbol{x}(t)$. This distinction is relevant if we consider the disturbance and measurement noise which has been neglected so far. To emphasize the distinction we introduce the notation:

$$\hat{\boldsymbol{x}}(t) := \text{estimate of } \boldsymbol{x}(t)$$

We now apply the product rule to the left-hand side of (4.27) and combine terms with $\hat{\boldsymbol{x}}(t)$ to obtain

$$\boldsymbol{M}(t,t_0)\frac{\mathrm{d}}{\mathrm{d}t}\hat{\boldsymbol{x}}(t) = \left\{-\frac{\mathrm{d}}{\mathrm{d}t}\boldsymbol{M}(t,t_0) - \boldsymbol{A}'(t)\boldsymbol{M}(t,t_0)\right\}\hat{\boldsymbol{x}}(t) + \boldsymbol{C}'(t)\boldsymbol{y}(t)$$

The derivative of $\boldsymbol{M}$ in braces is now substituted according to equation (4.26). The remaining terms are recombined and premultiplied by $\boldsymbol{M}^{-1}$:

$$\frac{\mathrm{d}}{\mathrm{d}t}\hat{\boldsymbol{x}}(t) = \boldsymbol{A}(t)\hat{\boldsymbol{x}}(t) + \boldsymbol{M}^{-1}(t,t_0)\boldsymbol{C}'(t)\{\boldsymbol{y}(t) - \boldsymbol{C}(t)\hat{\boldsymbol{x}}(t)\} \tag{4.30}$$

The initial condition is given by the solution (4.28b). This observation law has the form which has by now become familiar: it consists of a model of the process with weighted feedback of the difference between the output variables of the process and the model (cf. equation (4.8)). The gain matrix for the feedback is now

$$\boldsymbol{K}(t) := \boldsymbol{M}^{-1}(t,t_0)\boldsymbol{C}'(t) \tag{4.31}$$

We no longer have to rely on trial and error or special procedures for determining $\boldsymbol{K}(t)$. The gain matrix of the observer is essentially calculated by integration of the bilinear differential equation (4.26), followed by inversion of $\boldsymbol{M}$ and subsequent integration of the matrix Riccati differential equation (4.29). To calculate the estimate of the state, equations (4.27), (4.28b) and (4.30) are solved in this order. If an estimate of $\boldsymbol{x}$ is required in the initial phase, when $\boldsymbol{M}$ is still singular, Penrose's pseudo-inverse can be used in equation (4.28b) instead of the true inverse (Penrose, 1955; see also Kalman, 1960a; Zadeh and Desoer, 1963; Kalman and Englar, 1965).

We will see later that, using this purely deterministic approach, we have already come fairly close to the Kalman–Bucy filter. This is why we have gone into so much detail at this point.

## 4.4 THE DUALITY PRINCIPLE

Observation and filtering on the one hand and actuation and control on the other are interrelated in a remarkable manner. This is shown, for example, in the mirror-image correspondence of the observer canonical form and the control canonical form (see Section 1.4, equations (1.57)–(1.60) and (1.65)–(1.68)). In this context, the adjoint system should also be mentioned: it has long been used in the theory of differential equations and optimum control as well as in simulation applications (Laning and Battin, 1956; Zadeh and Desoer, 1963). However, the full extent of this reciprocal relationship was not recognized until about 1960 when Kalman formulated the duality principle (Kalman, 1960a,b; 1961; Kalman and Bucy, 1968). The impetus for this was the mirror-image similarity of the equations which he developed for linear optimal observation or filtering with respect to the equations of linear optimal control that were already known.

The duality principle can be expressed in a loose form as follows. Let

$$\dot{\boldsymbol{x}}(t) = \boldsymbol{A}(t)\boldsymbol{x}(t) + \boldsymbol{B}(t)\boldsymbol{u}(t)$$
$$\boldsymbol{y}(t) = \boldsymbol{C}(t)\boldsymbol{x}(t)$$

be the original system (equation (4.1)). We then obtain the dual system by using the following measures:

(i) time reversal;
(ii) interchanging the input and output matrices; (4.32)
(iii) transposition of the matrices $\boldsymbol{A}$, $\boldsymbol{B}$ and $\boldsymbol{C}$.

Let $\boldsymbol{\xi}$, $\boldsymbol{\nu}$ and $\boldsymbol{\eta}$ be the state, input variable and output variable of the dual system respectively. According to (4.32) this now has a form in which the triplet $\boldsymbol{A}$, $\boldsymbol{B}$, $\boldsymbol{C}$ appears reflected about the main diagonal of $\boldsymbol{A}$:

$$-\dot{\boldsymbol{\xi}}(t) = \boldsymbol{A}'(t)\boldsymbol{\xi}(t) + \boldsymbol{C}'(t)\boldsymbol{\nu}(t) \tag{4.33a}$$

$$\boldsymbol{\eta}(t) = \boldsymbol{B}'(t)\boldsymbol{\xi}(t) \tag{4.33b}$$

Observation and filtering of the original system (4.1) correspond to actuation and control of the dual system (4.33) and *vice versa*.

This duality has a considerable scope of application from both the theoretical and the practical point of view: concepts, definitions, criteria and miscellaneous theorems can be directly "translated" formally from the observation domain to the control domain and *vice versa* by using duality. Duality is also an effective aid in executing any of the necessary proofs. In this way the development of observation and filtering theory, which is less mature than optimal control theory, was considerably accelerated. The practical importance of duality is that the same computer programs can be used to synthesize optimal filters and optimal controllers. This is particularly convenient in the case of the program for the matrix Riccati equation.

The reflection law also applies to the transition matrices of dual systems. Let the adjoint system (4.33a) have the transition matrix $\boldsymbol{\psi}$. Thus, we write

$$\frac{\mathrm{d}}{\mathrm{d}t}\boldsymbol{\psi}(t, t_0) = -\boldsymbol{A}'(t)\boldsymbol{\psi}(t, t_0) \tag{4.34}$$

We bring everything over to the left-hand side and premultiply by the transposed transition matrix $\boldsymbol{\phi}'$ of the original system (4.1a):

$$\boldsymbol{\phi}'(t, t_0)\frac{\mathrm{d}}{\mathrm{d}t}\boldsymbol{\psi}(t, t_0) + \boldsymbol{\phi}'(t, t_0)\boldsymbol{A}'(t)\boldsymbol{\psi}(t, t_0) = \boldsymbol{0}$$

$$\boldsymbol{\phi}'(t, t_0)\frac{\mathrm{d}}{\mathrm{d}t}\boldsymbol{\psi}(t, t_0) + \frac{\mathrm{d}}{\mathrm{d}t}\boldsymbol{\phi}'(t, t_0)\boldsymbol{\psi}(t, t_0) = \boldsymbol{0}$$

We now integrate with respect to $t$ by reversing the product rule of differentiation (note that $\boldsymbol{\phi}'$ is the transpose of $\boldsymbol{\phi}$). The result is

$$\boldsymbol{\phi}'(t, t_0)\boldsymbol{\psi}(t, t_0) = \boldsymbol{I} \tag{4.35}$$

The integration constant on the right-hand side is the unit matrix because both $\boldsymbol{\phi}'$ and $\boldsymbol{\psi}$ are equal to $\boldsymbol{I}$ for $t = t_0$. Using the property (1.14) of the transition matrix and equation (A.32), it follows from (4.35) that the transition matrix of the adjoint system is indeed dual to that of the original system:

$$\boldsymbol{\psi}(t, t_0) = \boldsymbol{\phi}'(t_0, t) \tag{4.36}$$

A further consequence of (4.35) is that the scalar product of the states of two dual systems is constant over time:

$$\boldsymbol{x}'(t)\boldsymbol{\xi}(t) = \boldsymbol{x}'(t_0)\boldsymbol{\phi}'(t, t_0)\boldsymbol{\psi}(t, t_0)\boldsymbol{\xi}(t_0) = \boldsymbol{x}'(t_0)\boldsymbol{\xi}(t_0) \tag{4.37}$$

In what follows, we will repeatedly return to duality. In particular, extensive use will be made of it in the next section.

## 4.5 CONTROLLABILITY IN CONTINUOUS TIME

Whereas the question of observability is of direct and central importance for every observation and filtering problem, the concept of controllability is only of marginal importance in this respect. There are three reasons why controllability is nevertheless considered in this book. Because of its duality to the observation problem, the control problem can be briefly outlined without too much extra effort. Second, complete controllability is required as a precondition in certain theorems of filtering theory. Third, every control engineer is interested in the closed control loop (i.e., the combination of observation and actuation to feedback control). If a process possesses both complete observability and complete controllability, we can be sure that the feedback control problem has been properly formulated and can be solved in principle.

It is sufficient here to introduce the most important concepts and criteria briefly without proof. The dual proofs and further inferences can be found in the cited literature (or performed by the reader as an exercise). Because in the case of actuation it is not the output variable $\boldsymbol{y}$ but the control variable $\boldsymbol{u}$ which matters, we consider the $n$th-order inhomogeneous system without the output equation:

$$\dot{\boldsymbol{x}}(t) = \boldsymbol{A}(t)\boldsymbol{x}(t) + \boldsymbol{B}(t)\boldsymbol{u}(t) \tag{4.38}$$

Let the matrices $\boldsymbol{A}$ and $\boldsymbol{B}$ be given for all $t_0 \leq t \leq t_1$. The input variable $\boldsymbol{u}(t)$ is $p$ dimensional as before. Let the initial state $\boldsymbol{x}(t_0)$ be known and a desired end state $\boldsymbol{x}(t_1)$ be given. The (open-loop) control problem consists in choosing the actuating variable $\boldsymbol{u}(t)$ in the control interval $[t_0, t_1]$ so that the initial state is transferred to the final state.

*Definition 4.2*

(i) A state $\boldsymbol{x}(t_0)$ of the system (4.38) is said to be *controllable in the interval* $[t_0, t_1]$ if a control function $\boldsymbol{u}(t)$ exists in the interval $[t_0, t_1]$ so that $\boldsymbol{x}(t_1) = \boldsymbol{0}$.

(ii) If and only if every state $\boldsymbol{x}(t_0)$ is controllable in this sense, the system (4.38) is said to be *completely controllable in the interval* $[t_0, t_1]$.

(iii) If and only if there is a $t_1$ for every $t_0$ so that (ii) is satisfied, the system (4.38) is said to be *completely controllable* as such.

In this definition the final state is set equal to zero as usual. Provided that there is complete controllability in the sense of (ii) and (iii), this does not imply any restriction with regard to the general control problem. To verify this we note that the transition from $\boldsymbol{x}(t_0)$ to $\boldsymbol{x}(t_1)$ satisfies the general solution equation of system (4.38):

$$\boldsymbol{x}(t_1) = \boldsymbol{\phi}(t_1, t_0)\boldsymbol{x}(t_0) + \int_{t_0}^{t_1} \boldsymbol{\phi}(t_1, \tau)\boldsymbol{B}(\tau)\boldsymbol{u}(\tau)\mathrm{d}\tau \tag{4.39}$$

or

$$\int_{t_0}^{t_1} \boldsymbol{\phi}(t_1, \tau)\boldsymbol{B}(\tau)\boldsymbol{u}(\tau)\mathrm{d}\tau = \boldsymbol{x}(t_1) - \boldsymbol{\phi}(t_1, t_0)\boldsymbol{x}(t_0) \tag{4.40}$$

Complete controllability means that this equation has a solution in terms of $\boldsymbol{u}(\tau)$ if $\boldsymbol{x}(t_1) = \boldsymbol{0}$ while $\boldsymbol{x}(t_0)$ is arbitrary. We now set the remaining right-hand side of equation (4.40) equal to $z$, i.e., $-\boldsymbol{\phi}(t_1, t_0)\boldsymbol{x}(t_0) := z$. Because $\boldsymbol{\phi}$ is nonsingular, we can find an $\boldsymbol{x}(t_0)$ for any given $z$. Therefore a solution must also exist for any $z$. Since $z$ can be chosen arbitrarily, a solution must also exist for $z^* = z + \boldsymbol{x}(t_1)$ with $\boldsymbol{x}(t_1) \neq \boldsymbol{0}$. Therefore, if a system is completely controllable, its state can be forced by suitable choice of $\boldsymbol{u}$ to move from any arbitrary point in state space to any other arbitrary point.

In what follows, we shall always set $\boldsymbol{x}(t_1) = \boldsymbol{0}$ for simplicity. Doing this in (4.40), premultiplying the remaining part of this equation by $\boldsymbol{\phi}(t_0, t_1)$ and placing this factor under the integral will immediately give

$$\int_{t_0}^{t_1} \boldsymbol{\phi}(t_0, \tau)\boldsymbol{B}(\tau)\boldsymbol{u}(\tau)\mathrm{d}\tau = -\boldsymbol{x}(t_0) \tag{4.41}$$

This relationship should be compared with the dual equation (4.17). We form the expression that is dual to (4.16) and obtain the symmetric *controllability matrix of the first kind:*

$$W(t_1, t_0) := \int_{t_0}^{t_1} \boldsymbol{\phi}(t_0, \tau)\boldsymbol{B}(\tau)\boldsymbol{B}'(\tau)\boldsymbol{\phi}'(t_0, \tau)\mathrm{d}\tau \tag{4.42}$$

If $\boldsymbol{W}$ is nonsingular, the control problem can obviously be solved by using the following control function:

$$\boldsymbol{u}(\tau) = -\boldsymbol{B}'(\tau)\boldsymbol{\phi}'(t_0, \tau)\boldsymbol{W}^{-1}(t_1, t_0)\boldsymbol{x}(t_0) \quad t_0 \leqslant \tau \leqslant t_1 \tag{4.43}$$

This is easy to verify by substituting this control law in (4.41) and referring to (4.42).

If we premultiply both sides of (4.41) by $-\boldsymbol{x}'(t_0)\boldsymbol{W}^{-1}(t_1, t_0)$ and simplify the left-hand side using the transposed equation (4.43), we obtain the *actuation energy* of the law (4.43):

$$\int_{t_0}^{t_1} \boldsymbol{u}'(\tau)\boldsymbol{u}(\tau)\mathrm{d}\tau = \boldsymbol{x}'(t_0)\boldsymbol{W}^{-1}(t_1, t_0)\boldsymbol{x}(t_0) \tag{4.44}$$

This expression is dual to the measurement energy (4.18).

*Theorem 4.4.* For complete controllability of the system (4.38) in the interval $[t_0, t_1]$ it is necessary and sufficient that the controllability matrix $\boldsymbol{W}(t_1, t_0)$ in (4.42) is nonsingular.

*Proof.* The proof of the sufficiency has already been given above. As far as the necessity of the condition is concerned, a dual argument as used in Theorem 4.1 applies (Kalman, 1960a; Zadeh and Desoer, 1963; Athans and Falb, 1966). For time-invariant systems, the *controllability matrix of the second kind* is (compare (4.22)):

$$\bar{\boldsymbol{W}} := [\boldsymbol{B}, \boldsymbol{AB}, \boldsymbol{A}^2\boldsymbol{B}, \ldots, \boldsymbol{A}^{n-1}\boldsymbol{B}] \tag{4.45}$$

*Theorem 4.5.* For complete controllability of a system (4.38) with time-invariant parameters $\boldsymbol{A}$ and $\boldsymbol{B}$, it is necessary and sufficient that the controllability matrix $\bar{\boldsymbol{W}}$ in (4.45) has rank $n$ ($n$ is the order of the system or of the matrix $\boldsymbol{A}$).

For the proof, see Kalman (1960a), Zadeh and Desoer (1963) and Athans and Falb (1966) among others.

An alternative to this theorem is to verify the nonsingularity of $\boldsymbol{W}(\varepsilon, 0)$. Here, the length $\varepsilon$ of the control interval can be chosen to be arbitrarily small. However, in practice, for $\varepsilon \rightarrow 0$, the actuation energy according to (4.44) tends to infinity because

$$\lim_{\varepsilon \rightarrow 0} \boldsymbol{W}(\varepsilon, 0) = \boldsymbol{0}$$

In Section 4.3 we derived an equivalent matrix differential equation for the observability matrix of the first kind as well as for its inverse. In order to obtain the corresponding differential equations for the controllability matrix, we perform the steps here in a dual manner. We proceed as follows.

(i) In (4.42) the fixed initial time $t_0$ is replaced by the running time $t$. In order to make the integrand independent of $t$ once more, both sides are premultiplied by $\boldsymbol{\phi}(t_1, t)$ and postmultiplied by $\boldsymbol{\phi}'(t_1, t)$:

$$\boldsymbol{\phi}(t_1, t)\boldsymbol{W}(t_1, t)\boldsymbol{\phi}'(t_1, t) = \int_t^{t_1} \boldsymbol{\phi}(t_1, \tau)\boldsymbol{B}(\tau)\boldsymbol{B}'(\tau)\boldsymbol{\phi}'(t_1, \tau)\mathrm{d}\tau$$

Both sides are now differentiated with respect to $t$. The derivative of the transition matrix with respect to its second argument now appears on the left-hand side. This derivative can be calculated in the following way. We have

$$\boldsymbol{\phi}(t_1, t)\boldsymbol{\phi}(t, t_1) = \boldsymbol{I}$$

Differentiation of both sides with respect to $t$ gives

$$\frac{\mathrm{d}}{\mathrm{d}t}\boldsymbol{\phi}(t_1, t)\boldsymbol{\phi}(t, t_1) + \boldsymbol{\phi}(t_1, t)\boldsymbol{A}(t)\boldsymbol{\phi}(t, t_1) = \boldsymbol{0}$$

Therefore we have

$$\frac{\mathrm{d}}{\mathrm{d}t}\boldsymbol{\phi}(t_1, t) = -\boldsymbol{\phi}(t_1, t)\boldsymbol{A}(t) \tag{4.46}$$

In this way and after some rearrangement, the result of differentiating (4.45a) becomes

$$\frac{\mathrm{d}}{\mathrm{d}t}\boldsymbol{W}(t_1, t) = \boldsymbol{A}(t)\boldsymbol{W}(t_1, t) + \boldsymbol{W}(t_1, t)\boldsymbol{A}'(t) - \boldsymbol{B}(t)\boldsymbol{B}'(t) \tag{4.47}$$

In this case the boundary condition is known at the end of the interval: $\boldsymbol{W}(t_1, t_1) = \boldsymbol{0}$. The bilinear matrix differential equation (4.47) must therefore be integrated backwards.

(ii) The open-loop control law (4.43) can be written as a closed-loop control law with feedback of the instantaneous state if $\tau$ and $t_0$ are replaced by the running time $t$:

$$\boldsymbol{u}(t) = -\boldsymbol{B}'(t)\boldsymbol{W}^{-1}(t_1, t)\boldsymbol{x}(t) \tag{4.48}$$

(iii) Here we require the matrix $\boldsymbol{W}^{-1}$. In order to obtain it, we differentiate the product $\boldsymbol{W}\boldsymbol{W}^{-1} = \boldsymbol{I}$ with respect to $t$ and substitute (4.47). Then we premultiply the result by $\boldsymbol{W}^{-1}$ and obtain a nonlinear matrix differential equation of the Riccati type:

$$-\frac{\mathrm{d}}{\mathrm{d}t}\boldsymbol{W}^{-1}(t_1, t) = \boldsymbol{A}'(t)\boldsymbol{W}^{-1}(t_1, t) + \boldsymbol{W}^{-1}(t_1, t)\boldsymbol{A}(t) - \boldsymbol{W}^{-1}(t_1, t)\boldsymbol{B}(t)\boldsymbol{B}'(t)\boldsymbol{W}^{-1}(t_1, t) \tag{4.49}$$

This differential equation is dual to the Riccati differential equation (4.29) for the case of observation. More details are given in Section 6.1.

## 4.6 OBSERVATION IN DISCRETE TIME

Observation in discrete time means that the output variable of the system under consideration is no longer measured continuously but is sampled at discrete points in time. The sampling times $t_k$ need not be equally spaced.

We will see that the observation problem in discrete time has many parallels to the continuous-time case. To help visualize these parallel features, we choose the notation such that corresponding variables in discrete and continuous time have the same symbols. There is a risk of confusion but it is only significant when continuous- and discrete-time models are considered at the same time. In such cases we will identify one set of variables by bars. In this section, where the continuous variables are of secondary importance, they will be identified by a bar when there is any ambiguity. The discrete-time case is mathematically simpler both for observation and subsequently for filtering because differentials are replaced by differences and integrals by sums.

We begin the formulation of the observation problem in discrete time by constructing a discrete model of an observed system. As a typical example, we choose a process which is originally given in continuous time. We therefore examine the system

$$\dot{\boldsymbol{x}}(t) = \bar{\boldsymbol{A}}(t)\boldsymbol{x}(t) + \bar{\boldsymbol{B}}(t)\boldsymbol{u}(t) \tag{4.50a}$$

$$\boldsymbol{y}(t) = \boldsymbol{C}(t)\boldsymbol{x}(t) \tag{4.50b}$$

The state $\boldsymbol{x}(t)$ is $n$ dimensional as before, and the control variable $\boldsymbol{u}(t)$ is again a $p$-vector. The $m$-dimensional measurement variable is sampled at discrete points in time $t_k$. The $t_k$ need not be equally spaced.

The relationship between the values of the state $\boldsymbol{x}$ at two successive sampling points is described by the general solution formula for the differential equation (4.50a) ($\boldsymbol{\phi}$ is the transition matrix to $\bar{\boldsymbol{A}}$):

$$x(t_{k+1}) = \phi(t_{k+1}, t_k)x(t_k) + \int_{t_k}^{t_{k+1}} \phi(t_{k+1}, \tau)\bar{B}(\tau)u(\tau)\mathrm{d}\tau \tag{4.50c}$$

According to (4.50b), the relationship between the sampled measurement variable and the state is

$$y(t_k) = C(t_k)x(t_k) \tag{4.50d}$$

Two methods are customary for implementing the actuating variable of sampled systems (Figure 4.5): impulse-shaped actuator signals

$$u(t) = \bar{u}(t_k)\delta(t - t_k - \Delta t)$$

and piecewise constant actuator signals

$$u(t) = u(t_k) \quad \text{for} \quad t_k \leqslant t < t_{k+1}$$

For example, impulse control is used for course corrections of spacecraft and step function control is generated by sample-and-hold units.

To simplify the notation, we adopt the following conventions.

(i) The points in time $t_k$ are numbered throughout using only the integers $k$.

(ii) According to this

$$x(t_k) \equiv x(k) \quad \text{and} \quad y(t_k) \equiv y(k) \tag{4.51a}$$

$$\phi(t_{k+1}, t_k) \equiv \phi(k + 1, k) \quad \text{and} \quad C(t_k) \equiv C(k) \tag{4.51b}$$

(iii) We introduce the abbreviation

$$\phi(k + 1, k) := A(k) \tag{4.52}$$

where $A(k)$ is a transition matrix and is always nonsingular.

(iv) For an impulse-shaped actuating signal, $\bar{u}(t_k) := u(k)$ is taken out on the right from the integral (4.50c) and the masking property of the $\delta$ function is then used to evaluate the integral. The result is

$$\phi(t_{k+1}, t_k)\bar{B}(t_k)\bar{u}(t_k) := B(k)u(k) \tag{4.53}$$

(v) For a piecewise constant control variable, $u(t_k) = u(k)$ is removed from the integral in (4.50c) and we make the following definition:

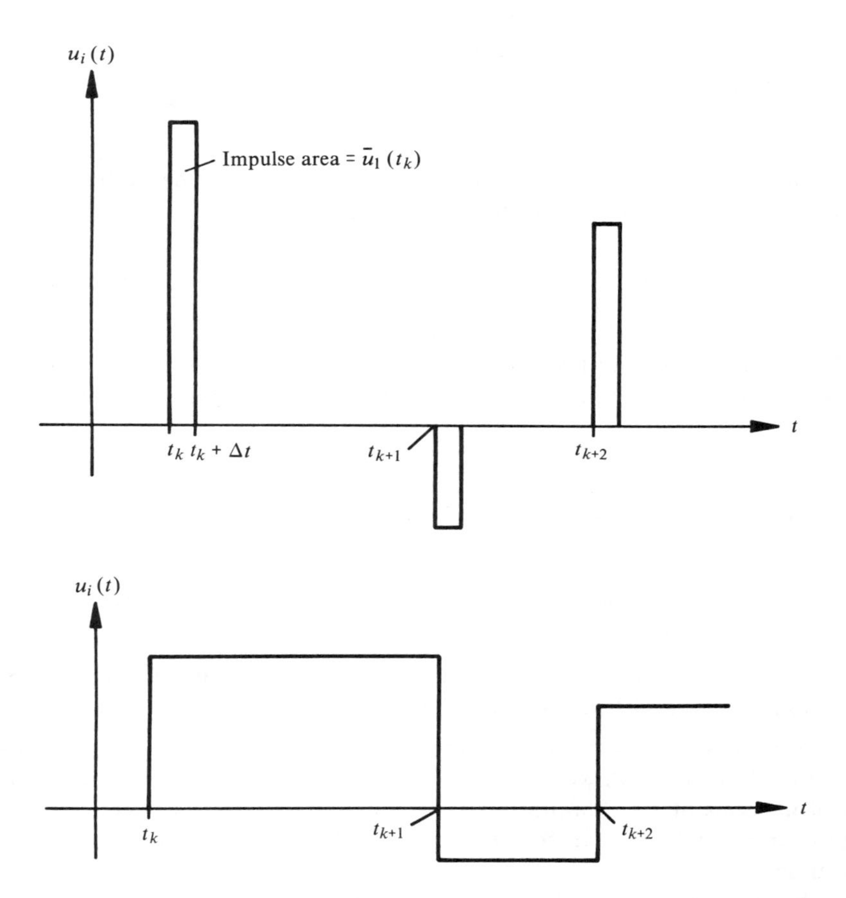


**Figure 4.5** Impulse-shaped and piecewise constant actuator signals.

$$\int_{t_k}^{t_{k+1}} \boldsymbol{\phi}(t_{k+1}, \tau)\bar{\boldsymbol{B}}(\tau)\mathrm{d}\tau := \boldsymbol{B}(k) \tag{4.54}$$

*Note*. The matrix $\boldsymbol{A}(k)$ or the two different matrices $\boldsymbol{B}(k)$ which have just been defined are not equal to $\bar{\boldsymbol{A}}(t_k)$ or $\bar{\boldsymbol{B}}(t_k)$. Neither is $\boldsymbol{u}(k)$ equal to $\boldsymbol{u}(t_k)$ in the case of an impulse-shaped control variable. There is no risk of confusion for the other variables.

According to (4.50c) and (4.50d) and with the conventions (4.51)–(4.54), the continuous model for the system (4.50a) and (4.50b) turns into the discrete version:

$$x(k+1) = A(k)x(k) + B(k)u(k) \tag{4.55a}$$

$$y(k) = C(k)x(k) \tag{4.55b}$$

The state $x$ is $n$ dimensional as before, the control vector $u$ has $p$ components and the measurement vector $y$ is $m$ dimensional.

We now consider the relationship

$$\phi(t_{k+1}, t_{k_0}) = \phi(t_{k+1}, t_k)\phi(t_k, t_{k_0})$$

If (4.51b) and (4.52) are taken into account, this can be written as the matrix difference equation for the transition matrix of the discrete system (4.55a):

$$\phi(k+1, k_0) = A(k)\phi(k, k_0) \quad \phi(k_0, k_0) = I \tag{4.55c}$$

The general solution of the vector difference equation (4.55a) has the form

$$x(k) = \phi(k, k_0)x(k_0) + \sum_{\kappa=k_0}^{k-1} \phi(k, \kappa+1)B(\kappa)u(\kappa) \quad k_0 < k \tag{4.55d}$$

The proof of this can be found inductively by recursive application of (4.55a) or *vice versa* by using (4.55d) on both sides of (4.55a).

*Example 4.3.* The measurement of the position of the aircraft which was performed continuously in Example 4.1 will now be carried out at discrete points in time at intervals of 1 s. Starting from the continuous form of the system $\dot{x} = \bar{A}x(t) + \bar{b}u(t)$ in Example 4.1, we construct the model of the observation situation in discrete time.

The transition matrix of this time-invariant system is

$$\begin{aligned} A(k) &:= \phi(t_{k+1}, t_k) = \exp\{\bar{A} \times (t_{k+1} - t_k)\} \\ &= I + \bar{A} \times (t_{k+1} - t_k) + \bar{A}^2 \frac{(t_{k+1} - t_k)^2}{2!} + \cdots \\ &= \begin{bmatrix} 1 & 0 \\ 0 & 1 \end{bmatrix} + \begin{bmatrix} 0 & 0 \\ 1 & 0 \end{bmatrix} (t_{k+1} - t_k) \end{aligned}$$

$\bar{A}^2$ and the higher powers of $\bar{A}$ vanish. Because $t_{k+1} - t_k = 1$ s for all $k$ (equidistant sampling), we also obtain a constant $A$ matrix in discrete time:

$$A = \begin{bmatrix} 1 & 0 \\ 1 & 1 \end{bmatrix}$$

We calculate the new control matrix according to (4.54):

$$\boldsymbol{b} = \int_{t_k}^{t_{k+1}} \begin{bmatrix} 1 & 0 \\ t_{k+1} - \tau & 1 \end{bmatrix} \begin{bmatrix} 1 \\ 0 \end{bmatrix} \mathrm{d}\tau$$

$$= \int_{t_k}^{t_{k+1}} \begin{bmatrix} 1 \\ t_{k+1} - \tau \end{bmatrix} \mathrm{d}\tau = \begin{bmatrix} 1 \\ 0.5 \end{bmatrix}$$

The model for the discrete-time position measurement is therefore

$$\begin{bmatrix} x_1(k+1) \\ x_2(k+1) \end{bmatrix} = \begin{bmatrix} 1 & 0 \\ 1 & 1 \end{bmatrix} \begin{bmatrix} x_1(k) \\ x_2(k) \end{bmatrix} + \begin{bmatrix} 1 \\ 0.5 \end{bmatrix} u(k)$$

$$y(k) = [0 \quad 1]\boldsymbol{x}(k)$$

Here, as before, $x_1$ is the speed and $x_2$ is the position. A minor drawback is that we have assumed that the acceleration is constant between two sampling points. It should be noticed in passing that the discrete model, in contrast with the original model, is no longer in the observer canonical form.

Leaving this example, we now turn to the general problem of observing the discrete-time system (4.55). Let the sequence of measurement vectors $\boldsymbol{y}(k)$ be given for all $k$ in the interval $[k_0, k_1]$. We wish to know the state $\boldsymbol{x}$ at the final point in time $k_1$.

Because of the principle of superposition, which is expressed in the general solution equation (4.55d), the effect of the known control sequence $\boldsymbol{u}(k_0)$, $\boldsymbol{u}(k_0 + 1)$, ... $\boldsymbol{u}(k - 1)$ on $\boldsymbol{x}(k)$ and $\boldsymbol{y}(k)$ can be calculated separately and subtracted from the measurement variables. Again, it is therefore sufficient to consider only the free system

$$\boldsymbol{x}(k+1) = \boldsymbol{A}(k)\boldsymbol{x}(k) \quad \boldsymbol{x}(k_0) = ? \tag{4.56a}$$

$$\boldsymbol{y}(k) = \boldsymbol{C}(k)\boldsymbol{x}(k) \tag{4.56b}$$

We are given the measurement values $\boldsymbol{y}(k_0)$, $\boldsymbol{y}(k_0 + 1)$, . . ., $\boldsymbol{y}(k_1)$ and we wish to find $\boldsymbol{x}(k_1)$. We eliminate the state vectors $\boldsymbol{x}(k_0)$ *et cetera* by expressing them in terms of the desired state $\boldsymbol{x}(k_1)$ with the help of the homogeneous part of the solution equation (4.55d):

$$\boldsymbol{x}(k_1) = \boldsymbol{\phi}(k_1, k)\boldsymbol{x}(k) \quad \text{or} \quad \boldsymbol{x}(k) = \boldsymbol{\phi}(k, k_1)\boldsymbol{x}(k_1)$$

In this way we obtain

$$
\begin{aligned}
\boldsymbol{y}(k_0) &= \boldsymbol{C}(k_0)\boldsymbol{x}(k_0) &&= \boldsymbol{C}(k_0)\boldsymbol{\phi}(k_0, k_1)\boldsymbol{x}(k_1) \\
\boldsymbol{y}(k_0 + 1) &= \boldsymbol{C}(k_0 + 1)\boldsymbol{x}(k_0 + 1) &&= \boldsymbol{C}(k_0 + 1)\boldsymbol{\phi}(k_0 + 1, k_1)\boldsymbol{x}(k_1) \\
&\vdots && \vdots \\
\boldsymbol{y}(k_1) &= \boldsymbol{C}(k_1)\boldsymbol{x}(k_1) &&= \boldsymbol{C}(k_1)\boldsymbol{\phi}(k_1, k_1)\boldsymbol{x}(k_1)
\end{aligned}
$$

The measurement vectors are combined to form a vector $\boldsymbol{z}$. On the right-hand side, we form a corresponding matrix $\boldsymbol{D}$:

$$
\boldsymbol{z}(k_1, k_0) := \begin{bmatrix} \boldsymbol{y}(k_0) \\ \boldsymbol{y}(k_0 + 1) \\ \vdots \\ \boldsymbol{y}(k_1) \end{bmatrix}
\qquad
\boldsymbol{D}(k_1, k_0) := \begin{bmatrix} \boldsymbol{C}(k_0)\boldsymbol{\phi}(k_0, k_1) \\ \boldsymbol{C}(k_0 + 1)\boldsymbol{\phi}(k_0 + 1, k_1) \\ \vdots \\ \boldsymbol{C}(k_1)\boldsymbol{\phi}(k_1, k_1) \end{bmatrix}
\tag{4.57}
$$

With these definitions, the above system of equations becomes the matrix equation:

$$
\boldsymbol{z}(k_1, k_0) = \boldsymbol{D}(k_1, k_0)\boldsymbol{x}(k_1) \tag{4.58}
$$

The vector $\boldsymbol{z}$ has the dimension $(k_1 - k_0 + 1)m$, and in what follows we assume that this value is greater than or equal to the order $n$ of the system. Both $\boldsymbol{z}$ and $\boldsymbol{D}$ are known quantities. If the dimension of $\boldsymbol{z}$ is exactly equal to $n$, equation (4.58) represents a system of $n$ linear algebraic equations for the $n$ unknown state variables $x_i(k_1)$. If, in addition, $\boldsymbol{D}$ is nonsingular, the system of equations can be uniquely solved for $\boldsymbol{x}(k_1)$ as usual.

If more than $n$ scalar measurements have been accumulated, the excess equations can, of course, simply be omitted. However, this is not usually done. Instead, all measurement values which have been obtained are kept. In this way it is possible to compensate for the measurement errors neglected so far. By exploiting the redundant information contained in the excess measurements, it is possible to increase the accuracy of the solution for $\boldsymbol{x}$. The measurement errors also have the effect that not all the equations of the system (4.58) are mutually compatible. In general this means that there will be no value of $\boldsymbol{x}$ which exactly satisfies all the individual equations. We must therefore content ourselves with an estimate $\hat{\boldsymbol{x}}$ for $\boldsymbol{x}$ which satisfies the system (4.58) to the best possible approximation. The oldest, but still effective, method for solving this problem is the least-squares method developed by Gauss in 1801. We will look at this method somewhat more closely in what follows.

*The least-squares method.* We are given a linear algebraic system of equations in which the number of equations exceeds the number of unknowns. Let the system of equations have the form

$$z = \boldsymbol{D}\boldsymbol{x} \tag{4.59}$$

where $\boldsymbol{x}$ is the vector of the $n$ unknown variables and $z$ is an $r$-vector of given numbers (measurements) with $r > n$. The $r \times n$ matrix $\boldsymbol{D}$ consists of known elements. To compensate for the inaccuracies contained in $z$, the estimate $\hat{\boldsymbol{x}}$ of $\boldsymbol{x}$ is to be determined so that the error vector

$$\tilde{z} := z - \boldsymbol{D}\hat{\boldsymbol{x}} \tag{4.60}$$

becomes as "small" as possible. The Euclidean norm $\|\tilde{z}\|_2$ is chosen as a scale for the magnitude of $\tilde{z}$ (see equation (A.54)). The square of this norm is the sum of the squares of the error components:

$$\|\tilde{z}\|_2^2 = \tilde{z}'\tilde{z} = \tilde{z}_1^2 + \tilde{z}_2^2 + \cdots + \tilde{z}_r^2$$

This expression should assume a minimum by proper choice of the $\hat{x}_i$. The condition for this is that all its partial derivatives with respect to the $\hat{x}_i$ vanish. In other words, its gradient with respect to $\hat{\boldsymbol{x}}$ must be zero:

$$\frac{\partial}{\partial \hat{\boldsymbol{x}}} (\tilde{z}'\tilde{z}) = \boldsymbol{0}'$$

$$\frac{\partial}{\partial \hat{\boldsymbol{x}}} ([z - \boldsymbol{D}\hat{\boldsymbol{x}}]'[z - \boldsymbol{D}\hat{\boldsymbol{x}}]) = \boldsymbol{0}'$$

$$\frac{\partial}{\partial \hat{\boldsymbol{x}}} (z'z - 2z'\boldsymbol{D}\hat{\boldsymbol{x}} + \hat{\boldsymbol{x}}'\boldsymbol{D}'\boldsymbol{D}\hat{\boldsymbol{x}}) = \boldsymbol{0}'$$

The first term in parentheses is independent of $\hat{\boldsymbol{x}}$, the second is a linear form and the third is a quadratic form in $\hat{\boldsymbol{x}}$. We apply the appropriate rules (A.62) and (A.63) for the evaluation of gradients and obtain

$$-2z'\boldsymbol{D} + 2\hat{\boldsymbol{x}}'\boldsymbol{D}'\boldsymbol{D} = \boldsymbol{0}'$$

By transposing, we obtain from this an $n$th order system of equations for $\hat{\boldsymbol{x}}$:

$$\boldsymbol{D}'z = \boldsymbol{D}'\boldsymbol{D}\hat{\boldsymbol{x}} \tag{4.61}$$

This result is easy to remember: we only need to premultiply the original system of equations (4.59) by $\boldsymbol{D}'$ (Gaussian transformation) and replace $\boldsymbol{x}$ by $\hat{\boldsymbol{x}}$. The matrix $\boldsymbol{D}'\boldsymbol{D}$ is called the Gaussian normal matrix (Zurmuehl and Falk, 1984). It is always symmetric and positive semidefinite. If it is positive definite (nonsingular), equation (4.61) can be solved for the required $\hat{\boldsymbol{x}}$ with one of the usual methods, for example by using the Gauss or Cholesky algorithm (Zurmuehl and Falk, 1984). This completes our outline of the fundamentals of the least-squares method; a more thorough discussion is given in the next chapter.

We now apply equation (4.61) to the accumulated observations of our discrete process. For this, we must replace $\boldsymbol{x}(k_1)$ in equation (4.58) by $\hat{\boldsymbol{x}}(k_1)$ and premultiply the equation by $\boldsymbol{D}'(k_1, k_0)$. In full notation, this means (using (4.57))

$$
[\boldsymbol{\phi}'(k_0, k_1)\boldsymbol{C}'(k_0), \ldots, \boldsymbol{\phi}'(k_1, k_1)\boldsymbol{C}'(k_1)] \begin{bmatrix} \boldsymbol{y}(k_0) \\ \vdots \\ \boldsymbol{y}(k_1) \end{bmatrix}
$$

$$
= [\boldsymbol{\phi}'(k_0, k_1)\boldsymbol{C}'(k_0), \ldots, \boldsymbol{\phi}'(k_1, k_1)\boldsymbol{C}'(k_1)] \begin{bmatrix} \boldsymbol{C}(k_0)\boldsymbol{\phi}(k_0, k_1) \\ \vdots \\ \boldsymbol{C}(k_1)\boldsymbol{\phi}(k_1, k_1) \end{bmatrix} \hat{\boldsymbol{x}}(k_1) \quad (4.62)
$$

The normal matrix $\boldsymbol{D}'(k_1, k_0)\boldsymbol{D}(k_1, k_0)$ on the right-hand side of (4.62) can obviously be written in the form of the following sum denoted by $\boldsymbol{M}(k_1, k_0)$:

$$
\boldsymbol{M}(k_1, k_0) := \sum_{\kappa=k_0}^{k_1} \boldsymbol{\phi}'(\kappa, k_1)\boldsymbol{C}'(\kappa)\boldsymbol{C}(\kappa)\boldsymbol{\phi}(\kappa, k_1) \quad (4.63)
$$

This is the *observability matrix of the first kind* for discrete time. A glance at equation (4.16) shows complete analogy with the continuous-time case.

The estimation equation (4.62) now takes the form

$$
\sum_{\kappa=k_0}^{k_1} \boldsymbol{\phi}'(\kappa, k_1)\boldsymbol{C}'(\kappa)\boldsymbol{y}(\kappa) = \boldsymbol{M}(k_1, k_0)\hat{\boldsymbol{x}}(k_1) \quad (4.64)
$$

This observation law corresponds exactly to equation (4.17). The estimate $\hat{\boldsymbol{x}}(k_1)$ can be uniquely determined if and only if the observability matrix $\boldsymbol{M}(k_1, k_0)$ is nonsingular.

Because of the analogy between the continuous- and discrete-time cases, almost all considerations, in particular the definition of observability and the associated criteria, apply in a completely equivalent manner. Therefore we can save repetition and restrict ourselves to a quite concise summary.

The observability matrix of the second kind for discrete time-invariant systems can be derived from equations (4.57) and (4.58). To do so, we set $k_0 = 0$, $k_1 = n - 1$ and $\boldsymbol{x}(k_1) = \boldsymbol{\phi}(k_1, k_0)\boldsymbol{x}(0)$ in these equations. In full notation, we obtain

$$\begin{bmatrix} \boldsymbol{y}(0) \\ \boldsymbol{y}(1) \\ \vdots \\ \boldsymbol{y}(n-1) \end{bmatrix} = \begin{bmatrix} \boldsymbol{C} \\ \boldsymbol{CA} \\ \vdots \\ \boldsymbol{CA}^{n-1} \end{bmatrix} \boldsymbol{x}(0)$$

The *observability matrix of the second kind* is defined by

$$\bar{\boldsymbol{M}} := \begin{bmatrix} \boldsymbol{C} \\ \boldsymbol{CA} \\ \vdots \\ \boldsymbol{CA}^{n-1} \end{bmatrix} \tag{4.65}$$

A system of the form (4.56) with constant parameters $\boldsymbol{A}$ and $\boldsymbol{C}$ is completely observable if and only if $\bar{\boldsymbol{M}}$ has rank $n$. The rank of $\bar{\boldsymbol{M}}$ cannot be increased by adding further submatrices of the form $\boldsymbol{CA}^n$, $\boldsymbol{CA}^{n+1}$ *et cetera*. This follows from the Cayley–Hamilton theorem just as in Section 4.3.

One of the most noticeable differences between the discrete- and continuous-time cases is the minimum observation time which is necessary for the determination of $\boldsymbol{x}$. In continuous time, the relevant observation interval $[t_0, t_1]$ can become arbitrarily small under certain circumstances. For example, this is always true when the system has constant parameters and is completely observable (see Theorem 4.3). However, in discrete time between one and $n$ sampling times are necessary for calculating $\boldsymbol{x}$ depending on the nature of $\boldsymbol{C}$ and $\boldsymbol{A}$, assuming complete observability at all times. A single sampling point in time is sufficient if $\boldsymbol{C}$ is square and regular; $n$ sampling points are necessary if the observed system has only a single output variable.

We now return to the general observation law (4.64). In a similar manner to the continuous-time case we ask whether this sum equation can be transformed to an equivalent system of difference equations. Such a recursive version of the observation law would be particularly suitable for a situation involving successive observations. In order to realize this, assume that the estimate $\hat{\boldsymbol{x}}(k)$ has already been calculated at time $k$. We assume further that $m$ new observations arrive at $k + 1$, which collectively form the vector $\boldsymbol{y}(k + 1)$. It would now require considerable computing effort to evaluate equations (4.63) and (4.64) all over again. This applies even more if new measurements are constantly being received. Fortunately, there is a recursive solution which makes it possible to update the old estimate by small corrections. In its original form it was given by Gauss in 1821 (Gauss, 1873). His solution is valid for a single additional observation, i.e., $y$ is a scalar. Plackett (1950) extended this method to an $m$-tuple of simultaneous additional measurements, i.e., for the case where $\boldsymbol{y}$ is an $m$-component vector.

To obtain the desired recursive observation law in our case, we transform equations (4.63) and (4.64) in a similar manner as was done for continuous time in Section 4.3. We will encounter a similar procedure in the next chapter when we consider the generalized least-squares method. The necessary steps are as follows.

(i) In equation (4.63), the fixed end time $k_1$ is replaced by the running time $k + 1$. Then the final term is separated from the sum:

$$\boldsymbol{M}(k+1, k_0) = \sum_{\kappa=k_0}^{k} \boldsymbol{\phi}'(\kappa, k+1)\boldsymbol{C}'(\kappa)\boldsymbol{C}(\kappa)\boldsymbol{\phi}(\kappa, k+1) + \boldsymbol{C}'(k+1)\boldsymbol{C}(k+1)$$

In the residual sum, the matrix $\boldsymbol{\phi}(\kappa, k+1)$ is replaced by $\boldsymbol{\phi}(\kappa, k)\boldsymbol{\phi}(k, k+1)$ and then $\boldsymbol{\phi}(k, k+1) = \boldsymbol{A}^{-1}(k)$ is removed to the right. The same takes place in a transposed manner with the matrix $\boldsymbol{\phi}'(\kappa, k+1)$ in the residual sum on the left-hand side:

$$\boldsymbol{M}(k+1, k_0) = \boldsymbol{A}^{-1}(k)' \sum_{\kappa=k_0}^{k} \boldsymbol{\phi}'(\kappa, k)\boldsymbol{C}'(\kappa)\boldsymbol{C}(\kappa)\boldsymbol{\phi}(\kappa, k)\boldsymbol{A}^{-1}(k)$$
$$+ \boldsymbol{C}'(k+1)\boldsymbol{C}(k+1)$$

The sum which is left is equal to the old $\boldsymbol{M}(k, k_0)$; i.e.,

$$\boldsymbol{M}(k+1, k_0) = \boldsymbol{A}^{-1}(k)'\boldsymbol{M}(k, k_0)\boldsymbol{A}^{-1}(k) + \boldsymbol{C}'(k+1)\boldsymbol{C}(k+1) \quad (4.66)$$

This is a matrix difference equation for the observability matrix $\boldsymbol{M}$. The initial condition is $\boldsymbol{M}(k_0, k_0) = \boldsymbol{C}'(k_0)\boldsymbol{C}(k_0)$ (see equation (4.63)). The difference equation (4.66) is the discrete-time counterpart of the bilinear matrix differential equation (4.26).

(ii) In equation (4.64), $k_1$ is also replaced by $k + 1$ and the final term is separated from the sum. $\boldsymbol{A}^{-1}(k)'$ is removed from the residual sum to the left as above:

$$\boldsymbol{A}^{-1}(k)' \sum_{\kappa=k_0}^{k} \boldsymbol{\phi}'(\kappa, k)\boldsymbol{C}'(\kappa)\boldsymbol{y}(\kappa) + \boldsymbol{C}'(k+1)\boldsymbol{y}(k+1) = \boldsymbol{M}(k+1, k_0)\hat{\boldsymbol{x}}(k+1)$$

The sum which now remains is equal to the old product $\boldsymbol{M}(k, k_0)\hat{\boldsymbol{x}}(k)$ so that we have

$$\boldsymbol{M}(k+1, k_0)\hat{\boldsymbol{x}}(k+1) = \boldsymbol{A}^{-1}(k)'\boldsymbol{M}(k, k_0)\hat{\boldsymbol{x}}(k) + \boldsymbol{C}'(k+1)\boldsymbol{y}(k+1) \quad (4.67)$$

This difference equation for $\boldsymbol{M}\hat{\boldsymbol{x}} := z$ corresponds to the differential equation (4.27). According to (4.64), it has the initial condition $\boldsymbol{M}(k_0, k_0)\hat{\boldsymbol{x}}(k_0) = \boldsymbol{C}'(k_0)\boldsymbol{y}(k_0)$. As soon as $\boldsymbol{M}$ has become nonsingular, we have complete observability and can determine $\hat{\boldsymbol{x}}$ from $\hat{z}$.

(iii) For the following intervals, difference equations are sought for $\hat{\boldsymbol{x}}$ itself and for the required inverse $\boldsymbol{M}^{-1}$. To do this, we introduce the extrapolated estimate $\boldsymbol{x}^*$ as an auxiliary variable:

$$\boldsymbol{x}^*(k + 1) := \boldsymbol{A}(k)\hat{\boldsymbol{x}}(k) \tag{4.I}$$

We replace $\hat{\boldsymbol{x}}(k)$ in (4.67) by $\boldsymbol{A}^{-1}(k)\boldsymbol{x}^*(k + 1)$ and apply equation (4.66):

$$\begin{aligned}\boldsymbol{M}(k + 1, k_0)\hat{\boldsymbol{x}}(k + 1) = {} & \{\boldsymbol{M}(k + 1, k_0) - \boldsymbol{C}'(k + 1)\boldsymbol{C}(k + 1)\}\boldsymbol{x}^*(k + 1) \\ & + \boldsymbol{C}'(k + 1)\boldsymbol{y}(k + 1)\end{aligned}$$

We then premultiply by $\boldsymbol{M}^{-1}(k + 1, k_0)$, rearrange and obtain the desired result:

$$\begin{aligned}\hat{\boldsymbol{x}}(k + 1) = {} & \boldsymbol{x}^*(k + 1) + \boldsymbol{M}^{-1}(k + 1, k_0)\boldsymbol{C}'(k + 1) \\ & \cdot \{\boldsymbol{y}(k + 1) - \boldsymbol{C}(k + 1)\boldsymbol{x}^*(k + 1)\}\end{aligned} \tag{4.II}$$

This equation, together with equation (4.I), forms the observation law in recursive form (compare differential equation (4.30)). The recursive law has the familiar structure of a model of the process with weighted feedback of the output differences (Figure 4.6). The gain matrix for the feedback path is obviously

$$\boldsymbol{K}(k + 1) := \boldsymbol{M}^{-1}(k + 1, k_0)\boldsymbol{C}'(k + 1) \tag{4.68}$$

This is the only quantity which is still missing in the recursive observation algorithm. It will be calculated in the next and final step.

(iv) Starting from equation (4.66), we again introduce an auxiliary variable, the practical importance of which will become clear in the next chapter:

$$\boldsymbol{P}^*(k + 1) := \boldsymbol{A}(k)\boldsymbol{M}^{-1}(k, k_0)\boldsymbol{A}'(k) \tag{4.III}$$

We postmultiply both sides of (4.66) by $\boldsymbol{P}^*(k + 1)$. In doing so, we apply the right-hand side of (4.III) to the middle term:

$$\boldsymbol{M}(k + 1, k_0)\boldsymbol{P}^*(k + 1) = I_n + \boldsymbol{C}'(k + 1)\boldsymbol{C}(k + 1)\boldsymbol{P}^*(k + 1)$$

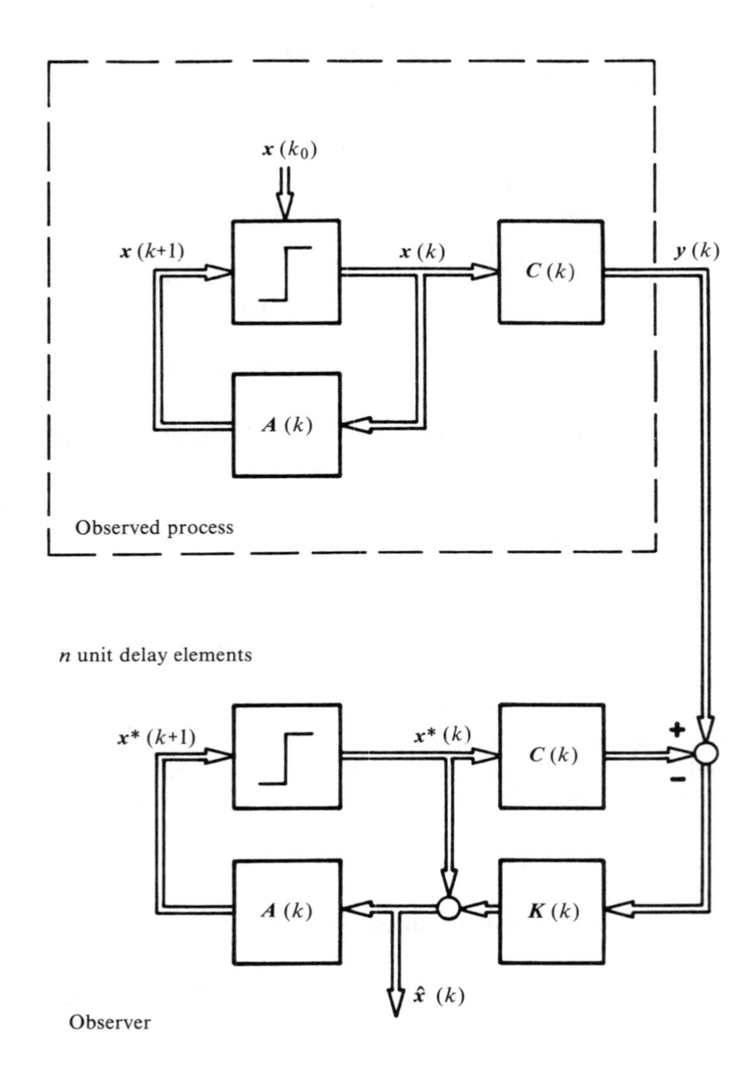


**Figure 4.6** Observed process and observer in discrete time. (The area enclosed by the broken lines is inaccessible.)

Both sides are premultiplied by $\boldsymbol{M}^{-1}(k+1, k_0)$:

$$\boldsymbol{P}^*(k+1) = \boldsymbol{M}^{-1}(k+1, k_0) + \boldsymbol{M}^{-1}(k+1, k_0)\boldsymbol{C}'(k+1)\boldsymbol{C}(k+1)\boldsymbol{P}^*(k+1) \quad (4.69)$$

We now postmultiply by $\boldsymbol{C}'(k+1)$ and solve for $\boldsymbol{M}^{-1}\boldsymbol{C}' = \boldsymbol{K}$:

$$\boldsymbol{K}(k+1) = \boldsymbol{P}^*(k+1)\boldsymbol{C}'(k+1)\{\boldsymbol{C}(k+1)\boldsymbol{P}^*(k+1)\boldsymbol{C}'(k+1) + \boldsymbol{I}\}^{-1} \quad (4.\text{IV})$$

This is an expression for the desired gain matrix of the observer. If we finally replace $\boldsymbol{M}^{-1}\boldsymbol{C}'$ in equation (4.69) by $\boldsymbol{K}$, we can solve for $\boldsymbol{M}^{-1}$:

$$\boldsymbol{M}^{-1}(k+1, k_0) = \boldsymbol{P}^*(k+1) - \boldsymbol{K}(k+1)\boldsymbol{C}(k+1)\boldsymbol{P}^*(k+1) \quad (4.\text{V})$$

In this way we have reached our goal of calculating $\boldsymbol{M}^{-1}$ recursively. If $\boldsymbol{M}^{-1}(k, k_0)$ is known for a certain $k$, the matrix $\boldsymbol{P}^*$ can be calculated for the next sampling point by using (4.III). In this way, all matrices on the right-hand side of (4.IV) are known so that we can also determine $\boldsymbol{K}$ at the next sampling point. The inverse in braces has the order $m$, which is the number of simultaneous measurements $y_i$. If $m < n$, the evaluation of this inverse matrix is simpler than the inversion of $\boldsymbol{M}$ itself which has order $n$. Finally, the new value of the inverse matrix of $\boldsymbol{M}$ is obtained very simply from (4.V). The system of difference equations (4.III), (4.IV) and (4.V) for $\boldsymbol{M}^{-1}$ corresponds to the matrix Riccati differential equation (4.29) for continuous time. Equations (4.I) and (4.II) constitute the observer.

*Summary.* The algorithm for the recursive Gaussian observation of the discrete-time free system

$$\boldsymbol{x}(k+1) = \boldsymbol{A}(k)\boldsymbol{x}(k)$$
$$\boldsymbol{y}(k) = \boldsymbol{C}(k)x(k)$$

can be summarized as follows.

(i) *Observer* (Figure 4.6):

$$\boldsymbol{x}^*(k+1) = \boldsymbol{A}(k)\hat{\boldsymbol{x}}(k) \quad (4.70\text{a})$$

$$\hat{\boldsymbol{x}}(k) = \boldsymbol{x}^*(k) + \boldsymbol{K}(k)\{\boldsymbol{y}(k) - \boldsymbol{C}(k)\boldsymbol{x}^*(k)\} \quad (4.70\text{b})$$

(ii) *Gain matrix and inverse of the observability matrix:*

$$P^*(k+1) = \boldsymbol{A}(k)\boldsymbol{M}^{-1}(k, k_0)\boldsymbol{A}'(k) \quad (4.71\text{a})$$

$$\boldsymbol{K}(k) = \boldsymbol{P}^*(k)\boldsymbol{C}'(k)\{\boldsymbol{C}(k)\boldsymbol{P}^*(k)\boldsymbol{C}'(k) + \boldsymbol{I}\}^{-1} \quad (4.71\text{b})$$

$$\boldsymbol{M}^{-1}(k, k_0) = \boldsymbol{P}^*(k) - \boldsymbol{K}(k)\boldsymbol{C}(k)\boldsymbol{P}^*(k) \quad (4.71\text{c})$$

(iii) *Initial conditions*. The observation begins at time $k_0$ when the first measurements $\boldsymbol{y}(k_0)$ arrive. As long as $\boldsymbol{M}(k, k_0)$ is still singular, equations (4.66) and (4.67) are calculated by using the initial conditions given there. As soon as $\boldsymbol{M}(k, k_0)$ has become nonsingular, the inverse $\boldsymbol{M}^{-1}(k, k_0)$ is formed and substituted in equation (4.71a) as an initial value. The estimate $\hat{\boldsymbol{x}}(k)$ is calculated at the same time and forms the initial condition for equation (4.70a).

Measurable and nonmeasurable stochastic input variables will be considered in the next chapter. Examples are calculated in Section 5.3.

## REFERENCES


Athans, M. and Falb, P.L. (1966). *Optimal Control,* McGraw-Hill, New York.

Battin, R.H. (1964). *Astronautical Guidance,* McGraw-Hill, New York.

Bellman, R. and Dreyfus, S.E. (1962). *Applied Dynamic Programming,* Princeton University Press, Princeton, NJ.

Bucy R.S. and Joseph, P.D. (1968). *Filtering for Stochastic Processes with Applications to Guidance,* Interscience, New York.

Carathéodory, C.C. (1935). *Variationsrechnung und partielle Differentialgleichungen erster Ordnung,* Teubner, Leipzig.

Gauss, C.F. (1963). *Theory of the Motion of the Heavenly Bodies Moving about the Sun in Conic Sections* (1809), reprinted by Dover Publications, New York.

Gauss, C.F. (1873). *Theoria Combinationis Observationum Erroribus Minimus Obnoxiae, 1821*. In *Collected Works,* Vol. 4, Göttingen.

Kalman, R.E. (1960a). Contributions to the theory of optimal control, *Bol. Soc. Math. Mex.*, pp. 102–119.

Kalman, R.E. (1960b). A new approach to linear filtering and prediction problems, *Trans. ASME, Ser. D, Basic Eng.*, **82,** 35–45.

Kalman, R.E. (1961). On the general theory of control systems. In Coales, J.F., *et al.* (eds), *Automatic and Remote Control, Proc. 1st Int. Congr. on Automatic Control (IFAC), Moscow 1960,* Vol. I, pp. 481–491, Oldenbourg, Munich, and Butterworth, London.

Kalman, R.E. and Bucy, R.S. (1961). New results in linear filtering and prediction theory, *Trans. ASME, Ser. D, Basic Eng.*, **83,** 95–108.

Kalman, R.E. and Englar, T.S. (1965). A user's manual for the automatic synthesis program (ASP-C), *Tech. Rep. NASA-CR-475,* Ames Research Center, Moffet Field, CA.

Laning, J.H. and Battin, R.H. (1956). *Random Processes in Automatic Control,* McGraw-Hill, New York.

Luenberger, D.G. (1966). Observers for multivariable systems, *IEEE Trans. Autom. Control,* **11,** 190–197.

Morgan, B.S. (1966). Sensitivity analysis and synthesis of multivariable systems, *IEEE Trans. Autom. Control,* **11,** 506–512.

Penrose, R. (1955). A generalized inverse for matrices, *Proc. Cambridge Philos. Soc.*, **51,** 406–413.

Plackett, R.L. (1950). Some theorems in least squares, *Biometrika,* **37,** 149–157.

Pontryagin, L.S., Boltyanskii, V.G., Gamkrelidze, R.V. and Misčenko, E.F. (1964). *Mathematische Theorie optimaler Prozesse,* Oldenbourg, Munich.

Silverman, L.M. (1966a). Transformation of time-variable systems to canonical (phase-variable) form, *IEEE Trans. Autom. Control,* **11,** 300–303.

Silverman, L.M. (1966b). Representation and realization of time-variable linear systems, *Tech. Rep. 94,* Office of Naval Research, Columbia University, New York.

Smith, G.L., Schmidt, S.F. and McGee, L.A. (1962). Application of statistical filter theory to the optimal estimation of position and velocity on-board a circumlunar vehicle, *NASA Tech. Rep. R-135*.

Zadeh, L.A. and Desoer, C.A. (1963). *Linear System Theory,* McGraw-Hill, New York.

Zurmuehl, R. and Falk, S. (1984). *Matrizen und ihre Anwendungen,* Part 1 (5th edn), Springer, Berlin.

## BIBLIOGRAPHY

Barnett, S. (1971). *Matrices in Control Theory,* Van Nostrand-Reinhold, New York.

Eykhoff, P. (1974). *System Identification; Parameter and State Estimation,* Wiley, London.

Kailath, T. (ed.) (1977). *Linear Least-squares Estimation,* Dowden, Hutchinson & Ross, Stroudsburg, PA.

Schneeweiss, W.G. (1974). *Zufallsprozesse in dynamischen Systemen,* Springer, Berlin.

Winkler, G. (1977). *Stochastische Systeme,* Akademische Verlagsgesellschaft, Wiesbaden.

# *Chapter 5*
# *Linear Optimal Filtering*

We continue with our considerations of the problem of estimating the state vector of a linear system on the basis of observations of the output variables. In the previous chapter the observation problem was formulated and solved in a purely deterministic manner. In this chapter it is treated by explicitly including stochastic disturbance variables and measurement errors. Here, the unknown state is regarded as a vector random variable or as a vector random process in discrete or continuous time. If the observation problem is placed in a stochastic framework in this way, it is known as a filter problem. To this extent, this chapter is a generalization of the previous chapter going from the deterministic observation situation to a stochastic observation situation.

Mathematically, the filter problem is treated by using the tools of probability theory and the theory of stochastic processes introduced in Chapters 2 and 3 of this book.

## 5.1 ORIGIN OF FILTER THEORY

The starting point of modern filter theory is commonly considered to be the independent development of the methods of Kolmogorov (1940) and Wiener (1949) in about 1940 (the impetus for Wiener's studies was the measurement of flight paths with radar and his work was only published after 1945). Kolmogorov and Wiener dealt with scalar signals and noise in the form of stationary processes. They considered infinitely long observation intervals and time-invariant filters. The performance criterion was the mean square error. It was required to find the filter weighting function or weighting sequence which made the criterion as small as possible. For continuous time, the resulting necessary and sufficient condition for the optimal weighting function was the well-known Wiener–Hopf integral equation. The counterpart in discrete time is a corresponding sum equation for the optimal weighting sequence. We will become acquainted with both these equations in generalized form.

Wiener also gave an elegant method for solving the Wiener–Hopf integral equation. It consists in transforming the equation to the frequency domain, decomposing the resulting power densities in a special way (*spectral factorization*) and combining selected factors to obtain the filter in the form of a frequency response.

The decomposition of the spectral density of a random process into two factors with mirror symmetry involves the construction of the frequency response of a hypothetical dynamic system which generates the given process from white noise. This interpretation was given by Bode and Shannon (1950) on the basis of the Wiener solution. Even today, this so-called *shaping-filter* concept is of central importance in filter theory.

Very soon, attempts were made to lift the rather restrictive assumptions of the Wiener–Kolmogorov theory in order to extend its area of application. Booton (1952) generalized the original Wiener–Hopf integral equation to nonstationary processes and time-varying filters. However, this was only a solution in the mathematical sense. Technically, the result was not useful since, in contrast with the Wiener method, it did not provide a practical method for either evaluating the integral equation or implementing time-varying weighting functions.

Around 1955, Follin and others investigated the filter problem for finite observation time (Follin and Carlton, 1956; Hanson, 1957). They found that the parameters of the corresponding filter vary with time even for stationary processes and satisfy certain differential equations. It was also shown that the solution of this system of differential equations converges to the set of parameters of the Wiener filter for $t \rightarrow \infty$. Furthermore, Bucy (1959) showed that the method of calculating the optimal filter parameters by differential equations can also be applied to nonstationary processes.

Similar investigations were carried out for discrete time by Swerling (1959). Kalman combined his theory of observation for linear systems in state space with the concept of orthogonal projections of random variables (Doob, 1953). In 1960, he published a recursive algorithm in the form of a system of difference equations for the filter and its gain factors (Kalman, 1960). This result applies to nonstationary vector processes in discrete time, time-varying filters and arbitrary observation intervals. It attracted considerable attention because of its general validity, mathematical elegance and widespread technical application. Together with Bucy, Kalman derived the corresponding solution for continuous time in 1961 (Kalman and Bucy, 1961). The filter problem was formulated in state space representation, and then a matrix-valued time-varying version of the Wiener–Hopf integral equation was derived. Finally, this was transformed to an equivalent system of differential equations for the required optimal filter parameters.

Since these pioneering studies a large number of publications have appeared which deal with the Kalman–Bucy filter method in one form or another. What is striking here is the large number of different ways that the solution equations can be derived. With the passage of time, ever more relationships with other estimation methods were discovered (e.g., with Bayesian estimation), with the maximum likelihood method and with the method of minimum variance and the least-squares method (see for example the textbooks by Bucy and Joseph (1968), Jazwinski (1970) and, in particular, Sage and Melsa (1971)).

Here, we shall take the simplest approach and consider the filter problem as the generalization of the Gaussian least-squares method (Wiener, 1949). By the beginning of this century, the Gaussian estimation had already been refined by the introduction of weighting factors in order to achieve a better trade-off between measurements with unequal error dispersions (Gauss–Markov estimation). Starting from this elementary and classical method and with the help of Plackett's recursion method (Plackett, 1950) we can derive the Kalman filter algorithm in a few steps (Brammer, 1971). However, a straightforward derivation requires a sufficient number of measurements since the recursive Gauss–Markov estimation can only be started when more measurements than unknowns have been accumulated. We have already come across this restriction when dealing with recursive observation in Chapter 4. It can now be removed if a certain statistical *a priori* knowledge of the initial state is given. It is then possible to make an estimation at the initial sampling points. For this, however, we need a further generalization of the method of minimum variance.

Therefore we begin with the introduction and discussion of this method, then use it to solve the discrete-time filter problem and finally go over to the continuous-time case.

## 5.2 THE METHOD OF MINIMUM VARIANCE

We are given a linear system of $r$ algebraic equations representing the relationships between measurements and unknowns. The number of measurements may be smaller than, equal to or larger than the number of unknowns. Each measurement has observation errors superimposed additively on it. In vector matrix form, the system of equations is

$$z = \boldsymbol{D}\boldsymbol{x} + \boldsymbol{s} \tag{5.1}$$

where $\boldsymbol{x}$ is the vector of the $n$ required unknowns, $z$ is the vector of the $r$ given measurements and $\boldsymbol{D}$ is a known $r \times n$ matrix. The components of the $r$-vector $\boldsymbol{s}$ are the unknown measurement errors.

The unknown variables $\boldsymbol{x}$ and $\boldsymbol{s}$ are considered as stochastic variables which are characterized by certain statistical parameters. Let the required $\boldsymbol{x}$ be a random vector with zero expectation and a covariance matrix $\boldsymbol{P}$:

$$E\{\boldsymbol{x}\} = \boldsymbol{0} \tag{5.2a}$$

$$E\{\boldsymbol{x}\boldsymbol{x}'\} = \boldsymbol{P} \tag{5.2b}$$

Let the unknown measurement error $\boldsymbol{s}$ also have zero expectation and a covariance matrix $\boldsymbol{S}$:

$$E\{\boldsymbol{s}\} = \boldsymbol{0} \tag{5.2c}$$

$$E\{\boldsymbol{s}\boldsymbol{s}'\} = \boldsymbol{S} \tag{5.2d}$$

Let $\boldsymbol{x}$ and $\boldsymbol{s}$ be uncorrelated:

$$E\{\boldsymbol{x}\boldsymbol{s}'\} = \boldsymbol{0} \tag{5.2e}$$

We first assume that there are more equations than unknowns ($r > n$) and analyze the properties of the least-squares estimate. According to equation (4.61), the Gaussian estimate has the form

$$\hat{\boldsymbol{x}} = (\boldsymbol{D}'\boldsymbol{D})^{-1}\boldsymbol{D}'\boldsymbol{z} \tag{5.3}$$

This estimate is obviously a linear algebraic function of the observations $\boldsymbol{z}$. Substitution of equation (5.1) shows how the estimate $\hat{\boldsymbol{x}}$ is related to the true value $\boldsymbol{x}$ and the observation errors $\boldsymbol{s}$:

$$\hat{\boldsymbol{x}} = \boldsymbol{x} + (\boldsymbol{D}'\boldsymbol{D})^{-1}\boldsymbol{D}'\boldsymbol{s} \tag{5.4}$$

If we take the expectation of both sides of this expression, the last summand vanishes because $E\{\boldsymbol{s}\} = \boldsymbol{0}$ and what remains is

$$E\{\hat{\boldsymbol{x}}\} = E\{\boldsymbol{x}\}$$

The Gaussian estimate is therefore unbiased.

The estimation error is defined as the difference between the estimate and the true value:

$$\tilde{\boldsymbol{x}} := \boldsymbol{x} - \hat{\boldsymbol{x}} \tag{5.5}$$

It is directly obtained from equation (5.4):

$$\tilde{\boldsymbol{x}} = -(\boldsymbol{D}'\boldsymbol{D})^{-1}\boldsymbol{D}'\boldsymbol{s} \tag{5.6}$$

The expectation of the estimation error obviously vanishes. To obtain its covariance matrix we use equation (5.2d) and obtain the expression

$$E\{\tilde{\boldsymbol{x}}\tilde{\boldsymbol{x}}'\} = (\boldsymbol{D}'\boldsymbol{D})^{-1}\boldsymbol{D}'\boldsymbol{S}\boldsymbol{D}(\boldsymbol{D}'\boldsymbol{D})^{-1} \tag{5.7}$$

If $\boldsymbol{S}$ is equal to the unit matrix, in other words, if all observations are normalized (variances $E\{s_j^2\} = 1$) and are mutually uncorrelated (covariances $E\{s_j s_k\} = 0$ for $j \neq k$), equation (5.7) simplifies to

$$E\{\tilde{\boldsymbol{x}}\tilde{\boldsymbol{x}}'\} = (\boldsymbol{D}'\boldsymbol{D})^{-1} \tag{5.8}$$

Thus, for the special case $\boldsymbol{S} = \boldsymbol{I}$, we see that the covariance matrix of the estimation error is equal to the inverse of the Gaussian normal matrix. Referring to the observability matrix $\boldsymbol{M}$ in equation (4.63) which was equal to the normal matrix $\boldsymbol{D}'\boldsymbol{D}$, we see that the inverse of $\boldsymbol{M}$ is the covariance matrix of the Gaussian estimation error.

We now consider the trace of the covariance matrix of $\tilde{\boldsymbol{x}}$:

$$\text{tr}E\{\tilde{\boldsymbol{x}}\tilde{\boldsymbol{x}}'\} = \sum_{i=1}^{n} E\{\tilde{x}_i^2\} \tag{5.9a}$$

$$= E\left\{\sum_{i=1}^{n} \tilde{x}_i^2\right\} = E\{\tilde{\boldsymbol{x}}'\tilde{\boldsymbol{x}}\} \tag{5.9b}$$

The right-hand side of equation (5.9a) consists of the sum of the error variances. As equation (5.9b) shows, it is equal to the expectation of the square of the Euclidean length of $\tilde{\boldsymbol{x}}$. We now ask whether the Gaussian estimate which minimizes the square of the norm $\|\boldsymbol{z} - \boldsymbol{D}\hat{\boldsymbol{x}}\|_2$, also provides the minimum of $E\{\|\boldsymbol{x} - \hat{\boldsymbol{x}}\|_2^2\}$. In other words, in requiring the least sum of the squares $\tilde{z}_j^2$ do we also obtain the minimum sum of the variances $E\{\tilde{x}_i^2\}$? In general, the answer is no because the minimum variance estimate and the Gaussian estimate are different as will be shown below.

We return to the original formulation of the problem which was defined by equations (5.1) and (5.2). We again assume that $r$ does not have to be greater than or equal to $n$, but can also be smaller than $n$. We require an estimate $\hat{\boldsymbol{x}}$ of $\boldsymbol{x}$ with the following properties:

(i) linearity in terms of the measurements;
(ii) zero bias;
(iii) minimum variances of the estimate errors $\tilde{x}_i$.

The restriction to linearity dictates the following relationship between $\hat{\boldsymbol{x}}$ and $\boldsymbol{z}$:

$$\hat{\boldsymbol{x}} = \boldsymbol{g} + \boldsymbol{G}\boldsymbol{z} \tag{5.10}$$

The $n$-vector $\boldsymbol{g}$ and the $n \times r$ weighting matrix $\boldsymbol{G}$ are real but otherwise still arbitrary. The requirement of zero bias means that the following must hold:

$$E\{\hat{\boldsymbol{x}}\} = \boldsymbol{g} + \boldsymbol{G}E\{\boldsymbol{z}\} = E\{\boldsymbol{x}\}$$

It follows from this that $\boldsymbol{g}$ must vanish because both $\boldsymbol{z}$ and $\boldsymbol{x}$ have zero expectation. Relationship (2.10) for $\hat{\boldsymbol{x}}$ therefore takes the simpler form

$$\hat{\boldsymbol{x}} = \boldsymbol{G}\boldsymbol{z} \tag{5.11}$$

The matrix $\boldsymbol{G}$ will be determined by the requirement that all the variances of the estimation errors $\tilde{x}_i := x_i - \hat{x}_i$ must be a minimum. According to (5.11), every estimation component $\hat{x}_i$ depends on $\boldsymbol{z}$ and on the $i$th row of the matrix $\boldsymbol{G}$. We denote this row vector by $\boldsymbol{g}^i$. Now, we can state that $\hat{x}_i = \boldsymbol{g}^i\boldsymbol{z}$. The requirement for minimum variances therefore means that

$$E\{\tilde{x}_i^2\} \rightarrow \min_{\boldsymbol{g}^i} \quad \text{for} \quad i = 1 \ldots n$$

We have

$$\begin{aligned} E\{\tilde{x}_i^2\} &= E\{(x_i - \hat{x}_i)^2\} \\ &= E\{(x_i - \boldsymbol{g}^i\boldsymbol{z})^2\} \\ &= E\{x_i^2 - 2x_i\boldsymbol{g}^i\boldsymbol{z} + (\boldsymbol{g}^i\boldsymbol{z})^2\} \\ &= E\{x_i^2\} - 2E\{x_i\boldsymbol{z}'\}(\boldsymbol{g}^i)' + \boldsymbol{g}^iE\{\boldsymbol{z}\boldsymbol{z}'\}(\boldsymbol{g}^i)' \end{aligned} \tag{5.12}$$

Therefore the variance of the $i$th estimate error is the sum of a component independent of $\boldsymbol{g}^i$, a linear form in $\boldsymbol{g}^i$ and a quadratic form in $\boldsymbol{g}^i$(note that $(\boldsymbol{g}^i)'$ is a column vector). For an extreme value of this expression, it is necessary that all its partial derivatives with respect to the elements $g_{ik}$ vanish. In other words the gradient with respect to $(\boldsymbol{g}^i)'$ must be zero, i.e.,

$$\frac{\partial}{\partial(\boldsymbol{g}^i)'} E\{\tilde{x}_i^2\} = \boldsymbol{0}'$$

The rules for forming the gradient are given in Section A.9, equations (A.61)–(A.63). Applying these rules to the right-hand side of (5.12) gives

$$-2E\{x_i\boldsymbol{z}'\} + 2\boldsymbol{g}^iE\{\boldsymbol{z}\boldsymbol{z}'\} = \boldsymbol{0}'$$

or

$$E\{x_i\boldsymbol{z}'\} - \boldsymbol{g}^iE\{\boldsymbol{z}\boldsymbol{z}'\} = \boldsymbol{0}' \tag{5.13}$$

This condition applies to all $i = 1, \ldots, n$. Every term is a row vector with $r$ components. If the individual rows are placed underneath each other for $i = 1, 2, \ldots, n$, we obtain the result in compact form:

$$E\{\boldsymbol{x}\boldsymbol{z}'\} - \boldsymbol{G}E\{\boldsymbol{z}\boldsymbol{z}'\} = \boldsymbol{0} \tag{5.14}$$

This is a linear system of algebraic equations for the required weighting matrix $\boldsymbol{G}$. It is the necessary condition for the situation that every individual error variance $E\{\tilde{x}_i^2\}$ takes an extreme value. For this to be a minimum, it is sufficient that the second partial derivatives of $E\{\tilde{x}_i^2\}$ with respect to $g_{ik}$ form a positive definite matrix. In other words the Hessian matrix with respect to $(\boldsymbol{g}^i)'$ must be positive definite for all $i$; i.e.,

$$\frac{\partial^2}{\partial\{(\boldsymbol{g}^i)'\}^2} E\{\tilde{x}_i^2\} \quad \text{is positive definite.}$$

Partial differentiation of the left-hand side of (5.13) with respect to $(\boldsymbol{g}^i)'$ provides the desired Hessian matrix. Alternatively, equation (A.65) can be applied directly to (5.12). Thus the above condition becomes

$$E\{\boldsymbol{z}\boldsymbol{z}'\} \quad \text{is positive definite.} \tag{5.15}$$

This condition is sufficient for the extreme values of the variances determined by (5.14) to be a minimum. At the same time, condition (5.15) is necessary and sufficient for the existence of a unique solution to (5.14) in terms of $\boldsymbol{G}$.

Equation (5.14) can be regarded as a precursor of the Wiener–Hopf equation for discrete observations (compare (5.81b)). However, it can be brought into a more compact form. By combining the expectations and factoring out $\boldsymbol{z}'$, it becomes

$$E\{(\boldsymbol{x} - \boldsymbol{G}\boldsymbol{z})\boldsymbol{z}'\} = \boldsymbol{0}$$

or, with $\boldsymbol{x} - \boldsymbol{G}\boldsymbol{z} = \boldsymbol{x} - \hat{\boldsymbol{x}} = \tilde{\boldsymbol{x}}$,

$$E\{\tilde{\boldsymbol{x}}\boldsymbol{z}'\} = \boldsymbol{0} \tag{5.16}$$

This result is very interesting. It means that all the cross-covariances between the estimation errors on the one hand and the observations on the other hand must vanish. In such a case, the $\tilde{x}_i$ are said to be *orthogonal* with respect to the $z_j$. Because $\tilde{\boldsymbol{x}}$ and $\boldsymbol{z}$ have zero expectation, we can also say that $\tilde{\boldsymbol{x}}$ and $\boldsymbol{z}$ are *uncorrelated*. If we postmultiply (5.16) by $\boldsymbol{G}'$ we also obtain

$$E\{\tilde{\boldsymbol{x}}\hat{\boldsymbol{x}}'\} = \boldsymbol{0}$$

This means that the estimate error must not only be uncorrelated with the observations but also with the estimate itself. We use this relationship to rearrange the covariance matrix of the estimate error:

$$\begin{aligned} E\{\tilde{\boldsymbol{x}}\tilde{\boldsymbol{x}}'\} &= E\{\tilde{\boldsymbol{x}}(\boldsymbol{x} - \hat{\boldsymbol{x}})'\} \\ &= E\{\tilde{\boldsymbol{x}}\boldsymbol{x}'\} && (5.17a) \\ &= E\{(\boldsymbol{x} - \boldsymbol{G}\boldsymbol{z})\boldsymbol{x}'\} \\ &= E\{\boldsymbol{x}\boldsymbol{x}'\} - \boldsymbol{G}E\{\boldsymbol{z}\boldsymbol{x}'\} && (5.17b) \end{aligned}$$

The minimum sum of the error variances is given by the trace of one of the expressions in equation (5.17) (compare (5.9a)). Because $\boldsymbol{G}$ is optimal, each of the summands $E\{\tilde{x}_i^2\}$ is a minimum so that their sum is also a minimum.

In order finally to determine the optimal $\boldsymbol{G}$ from equation (5.14), both covariance matrices must be evaluated. With relationships (5.2b), (5.2d) and (5.2e), we obtain

$$E\{\boldsymbol{x}\boldsymbol{z}'\} = E\{\boldsymbol{x}(\boldsymbol{x}'\boldsymbol{D}' + \boldsymbol{s}')\} = \boldsymbol{P}\boldsymbol{D}' \qquad (5.18a)$$

$$E\{\boldsymbol{z}\boldsymbol{z}'\} = E\{(\boldsymbol{D}\boldsymbol{x} + \boldsymbol{s})(\boldsymbol{x}'\boldsymbol{D}' + \boldsymbol{s}')\} = \boldsymbol{D}\boldsymbol{P}\boldsymbol{D}' + \boldsymbol{S} \qquad (5.18b)$$

In this way, equation (5.14) becomes

$$\boldsymbol{G}(\boldsymbol{D}\boldsymbol{P}\boldsymbol{D}' + \boldsymbol{S}) = \boldsymbol{P}\boldsymbol{D}' \qquad (5.19)$$

The parentheses enclose an $r \times r$ matrix. If the number $r$ of observations is smaller than the number $n$ of unknowns $x_i$, (5.19) is the most suitable form for calculating $\boldsymbol{G}$. In this case we obtain

$$\boldsymbol{G} = \boldsymbol{P}\boldsymbol{D}'(\boldsymbol{D}\boldsymbol{P}\boldsymbol{D}' + \boldsymbol{S})^{-1} \qquad (5.20)$$

Substitution in (5.11) gives the *first form of the linear unbiased minimum variance estimate*:

$$\hat{\boldsymbol{x}} = \boldsymbol{P}\boldsymbol{D}'(\boldsymbol{D}\boldsymbol{P}\boldsymbol{D}' + \boldsymbol{S})^{-1}\boldsymbol{z} \qquad (5.21)$$

The covariance matrix of the corresponding estimate error is given by equation (5.17b) together with the relationships (5.2b) and (5.18a):

$$E\{\tilde{\boldsymbol{x}}\tilde{\boldsymbol{x}}'\} = \boldsymbol{P} - \boldsymbol{G}\boldsymbol{D}\boldsymbol{P} \qquad (5.22)$$

Substitution of equation (5.20) gives

$$E\{\tilde{x}\tilde{x}'\} = P - PD'(DPD' + S)^{-1}DP \tag{5.23}$$

The trace of this expression is the minimum value of the sum of the error variances.

An interesting special case of the above estimate occurs when we make the following assumptions:

(i) $P = I$, i.e., the unknowns $x_i$, are normalized and mutually uncorrelated.

(ii) $S = 0$, i.e., the observations do not contain any measurement errors.

(iii) $D$ has rank $r$ for $r \leqslant n$.

Equation (5.21) then becomes

$$\hat{x} = D'(DD')^{-1}z \tag{5.24}$$

In comparison, according to equation (5.3), the Gaussian least-squares estimate has the form

$$\hat{x} = (D'D)^{-1}D'z$$

Remember that $D$ is an $r \times n$ matrix in both cases and note that $DD'$ is $r \times r$ (singular for $r > n$) and $D'D$ is $n \times n$ (singular for $r < n$).

When the special assumptions (i)–(iii) stated above are applied to (5.23), the covariance matrix of the estimation error becomes

$$E\{\tilde{x}\tilde{x}\} = I - D'(DD')^{-1}D \tag{5.25}$$

We now return to equation (5.19). There is a second form for this condition which is more suitable for calculating $G$ as soon as the number of observations exceeds the number of unknowns ($r > n$). We first note two remarkable symmetry properties. The first two terms in (5.22) are symmetric. Therefore the last term must also be symmetric:

$$GDP = PD'G' \tag{5.26}$$

Premultiplication of (5.19) by $D$ gives

$$DGDPD' + DGS = DPD'$$

The first term of this is symmetric because of (5.26); in the last term, the symmetry is obvious. It follows from this that the middle term must also be symmetric:

$$DGS = SG'D' \tag{5.27}$$

In order to be able to bring equation (5.19) into the desired second form, it must be assumed that the covariance matrices $\boldsymbol{P}$ and $\boldsymbol{S}$ are nonsingular. We multiply out the left-hand side of (5.19) and use the symmetry property (5.26) in the first summand:

$$\boldsymbol{PD'G'D'} + \boldsymbol{GS} = \boldsymbol{PD'}$$

The product $\boldsymbol{G'D'}$ in the first summand is substituted by using (5.27):

$$\boldsymbol{PD'S}^{-1}\boldsymbol{DGS} + \boldsymbol{GS} = \boldsymbol{PD'}$$

Finally, both sides are premultiplied by $\boldsymbol{P}^{-1}$ and postmultiplied by $\boldsymbol{S}^{-1}$. The required result is

$$(\boldsymbol{D'S}^{-1}\boldsymbol{D} + \boldsymbol{P}^{-1})\boldsymbol{G} = \boldsymbol{D'S}^{-1} \tag{5.28}$$

The parentheses represent a matrix of dimensions $n \times n$. This solution form is equivalent to (5.19). However, it is easier to evaluate if $n \ll r$ and $\boldsymbol{S}$ is diagonal. With the $\boldsymbol{G}$ from equation (5.28), we obtain the *second form of the linear unbiased minimum variance estimate*:

$$\hat{\boldsymbol{x}} = (\boldsymbol{D'S}^{-1}\boldsymbol{D} + \boldsymbol{P}^{-1})^{-1}\boldsymbol{D'S}^{-1}\boldsymbol{z} \tag{5.29}$$

The error covariance matrix in equation (5.22) can be rearranged here as follows:

$$\begin{aligned} E\{\tilde{\boldsymbol{x}}\tilde{\boldsymbol{x}}'\} &= (\boldsymbol{I} - \boldsymbol{GD})\boldsymbol{P} \\ &= \{\boldsymbol{I} - (\boldsymbol{D'S}^{-1}\boldsymbol{D} + \boldsymbol{P}^{-1})^{-1}\boldsymbol{D'S}^{-1}\boldsymbol{D}\}\boldsymbol{P} \\ &= (\boldsymbol{D'S}^{-1}\boldsymbol{D} + \boldsymbol{P}^{-1})^{-1}\{(\boldsymbol{D'S}^{-1}\boldsymbol{D} + \boldsymbol{P}^{-1}) - \boldsymbol{D'S}^{-1}\boldsymbol{D}\}\boldsymbol{P} \\ &= (\boldsymbol{D'S}^{-1}\boldsymbol{D} + \boldsymbol{P}^{-1})^{-1} \end{aligned} \tag{5.30}$$

The best known special case of the estimation (5.29) occurs when the following conditions are satisfied.

(i) $\boldsymbol{P} \rightarrow \infty$ or $\boldsymbol{P}^{-1} = \boldsymbol{0}$: this means that the dispersions of the unknowns $x_i$ increase beyond all bounds; i.e., there is no *a priori* knowledge of the $x_i$.

(ii) Rank $\boldsymbol{D} = n$, i.e., there are at least $n$ linearly independent measurements available.

Under these assumptions, we obtain the *Gauss–Markov estimate*:

$$\hat{\boldsymbol{x}} = (\boldsymbol{D}'\boldsymbol{S}^{-1}\boldsymbol{D})^{-1}\boldsymbol{D}'\boldsymbol{S}^{-1}\boldsymbol{z} \tag{5.31}$$

According to (5.30) the associated estimate error has the covariance matrix

$$E\{\tilde{\boldsymbol{x}}\tilde{\boldsymbol{x}}'\} = (\boldsymbol{D}'\boldsymbol{S}^{-1}\boldsymbol{D})^{-1} \tag{5.32}$$

The elements in the main diagonal of this matrix represent the minimum values of the error variances which can be obtained using the given *a priori* knowledge. If, in addition to the above conditions, $\boldsymbol{S}$ happens to be equal to the unit matrix, we finally come back to the Gaussian estimation (equations (5.3) and (5.8)).

We can now conclude that the requirement for the least sum of squares $\tilde{z}_j^2$ produces the minimum sum of variances $E\{\tilde{x}_i^2\}$ if and only if the observation errors $\boldsymbol{s}$ are normalized and uncorrelated and there is no *a priori* statistical knowledge of the unknowns $\boldsymbol{x}$.

As has just been shown, the Gauss–Markov estimation is a generalization of the least-squares method. The Gauss–Markov law can also be found in a simpler manner by using a heuristic approach. For this, we start from the classical least-squares solution and assume that $\boldsymbol{S}$ is known and diagonal with distinct main elements $s_{jj}$. In other words, the observation errors are mutually uncorrelated but have unequal variances. It is immediately plausible that it must be worthwhile to adjust the measurements to the same quality first before the Gaussian least-squares formula is applied. For this purpose, we divide every measurement equation by the square root of the variance of its measurement error. This can be interpreted as a premultiplication of equation (5.1) by a matrix $\boldsymbol{S}^{-1/2}$. This matrix is diagonal and its main elements are $1/s_{jj}^{1/2}$. We therefore transform equation (5.1) as follows:

$$\boldsymbol{S}^{-1/2}\boldsymbol{z} = \boldsymbol{S}^{-1/2}(\boldsymbol{D}\boldsymbol{x} + \boldsymbol{s})$$

In this way, we have achieved our aim that the transformed observation errors $\boldsymbol{S}^{-1/2}\boldsymbol{s}$ have the covariance matrix $\boldsymbol{I}$. We now continue to generalize by assuming that $\boldsymbol{S}$ is nondiagonal (i.e., the observation errors may be mutually correlated). In this case it is certainly expedient to transform the original measurement equation (5.1) so that this correlation is removed. For this, we conceptually decompose the matrix $\boldsymbol{S}$ into its eigenvector matrix $\boldsymbol{V}$ and its Jordan canonical form $\boldsymbol{J}$ (see (A.46)):

$$\boldsymbol{S} = \boldsymbol{V}\boldsymbol{J}\boldsymbol{V}^{-1}$$

For real symmetric matrices such as $\boldsymbol{S}$, it is known that the Jordan canonical form is diagonal. In our case $\boldsymbol{S}$ is also positive definite so that the main elements of $\boldsymbol{J}$ are all positive. The square roots of these are used to form the diagonal matrix $\boldsymbol{J}^{1/2}$. The eigenvector matrix of real symmetric matrices is orthonormal, i.e., $\boldsymbol{V}'\boldsymbol{V} = \boldsymbol{I}$. It follows from this that $\boldsymbol{V}^{-1} = \boldsymbol{V}'$. We therefore have

$$\boldsymbol{S} = \boldsymbol{V}\boldsymbol{J}^{1/2}\boldsymbol{J}^{1/2}\boldsymbol{V}'$$

This decomposition of $\boldsymbol{S}$ can be interpreted as taking its square root in the nondiagonal case. The measurement equation (5.1) is now premultiplied by the inverse "square root" of $\boldsymbol{S}$:

$$(\boldsymbol{V}\boldsymbol{J}^{1/2})^{-1}\boldsymbol{z} = (\boldsymbol{V}\boldsymbol{J}^{1/2})^{-1}(\boldsymbol{D}\boldsymbol{x} + \boldsymbol{s}) \tag{5.33}$$

The transformed observation errors have the covariance matrix

$$\boldsymbol{J}^{-1/2}\boldsymbol{V}^{-1}E\{\boldsymbol{s}\boldsymbol{s}'\}(\boldsymbol{V}^{-1})'\boldsymbol{J}^{-1/2} = \boldsymbol{J}^{-1/2}\boldsymbol{V}^{-1}\{\boldsymbol{V}\boldsymbol{J}\boldsymbol{V}^{-1}\}\boldsymbol{V}\boldsymbol{J}^{-1/2} = \boldsymbol{I}$$

As required, the transformed measurement equation (5.33) has normalized and uncorrelated "observation errors." At this point we apply formula (5.3) for the least-squares estimate. Of course the transformed values of $\boldsymbol{z}$ and $\boldsymbol{D}$ from (5.33) must be used:

$$\begin{aligned}\hat{\boldsymbol{x}} &= \{\boldsymbol{D}'(\boldsymbol{V}^{-1})'\boldsymbol{J}^{-1/2}\cdot\boldsymbol{J}^{-1/2}\boldsymbol{V}^{-1}\boldsymbol{D}\}^{-1}\boldsymbol{D}'(\boldsymbol{V}^{-1})'\boldsymbol{J}^{-1/2}\cdot\boldsymbol{J}^{-1/2}\boldsymbol{V}^{-1}\boldsymbol{z}\\ &= \{\boldsymbol{D}'(\boldsymbol{V}\boldsymbol{J}\boldsymbol{V}^{-1})^{-1}\boldsymbol{D}\}^{-1}\boldsymbol{D}'(\boldsymbol{V}\boldsymbol{J}\boldsymbol{V}^{-1})^{-1}\boldsymbol{z}\\ &= (\boldsymbol{D}'\boldsymbol{S}^{-1}\boldsymbol{D})^{-1}\boldsymbol{D}'\boldsymbol{S}^{-1}\boldsymbol{z}\end{aligned}$$

This is exactly the same result as we derived before for the Gauss–Markov estimate (see (5.31)).

The Gauss–Markov estimation law can also be obtained by using a second but essentially identical heuristic approach. According to this philosophy, the error squares are balanced by a weighting matrix which should be chosen equal to $\boldsymbol{S}^{-1}$. According to this, instead of the straightforward norm $\|\boldsymbol{z} - \boldsymbol{D}\hat{\boldsymbol{x}}\|_2^2$, the following generalized norm should be minimized:

$$\|\boldsymbol{z} - \boldsymbol{D}\hat{\boldsymbol{x}}\|_{S^{-1}}^2 = (\boldsymbol{z} - \boldsymbol{D}\hat{\boldsymbol{x}})'\boldsymbol{S}^{-1}(\boldsymbol{z} - \boldsymbol{D}\hat{\boldsymbol{x}}) \tag{5.34}$$

We can multiply out, form the gradient with respect to $\hat{\boldsymbol{x}}$ and set it to zero in a similar manner as for the least-squares method in Section 4.6. In this way, we arrive at the Gauss–Markov solution (5.31) without any difficulty.

*Note.* The Gaussian solution (4.61) was derived by using purely deterministic methods. Its mirror image (5.24) can also be derived deterministically. In this case we have to assume that the equation $\boldsymbol{z} = \boldsymbol{D}\boldsymbol{x}$ has fewer independent rows than unknowns ($r < n$). There are then infinitely many solutions for $\boldsymbol{x}$. We wish to select one of these as an estimate which has the smallest Euclidean norm. We therefore minimize $\hat{\boldsymbol{x}}'\hat{\boldsymbol{x}}$ subject to the constraint $\boldsymbol{z} - \boldsymbol{D}\hat{\boldsymbol{x}} = \boldsymbol{0}$. This is done most rapidly using Lagrange multipliers (here, $\boldsymbol{\lambda}$ is an $r$-vector):

$$\hat{\boldsymbol{x}}'\hat{\boldsymbol{x}} + 2\boldsymbol{\lambda}'(\boldsymbol{z} - \boldsymbol{D}\hat{\boldsymbol{x}}) \rightarrow \min_{\hat{\boldsymbol{x}}}$$

Differentiation with respect to $\hat{\boldsymbol{x}}$, setting the partial derivatives to zero and transposing gives

$$\hat{\boldsymbol{x}} = \boldsymbol{D}'\boldsymbol{\lambda}$$

From this it follows that

$$\boldsymbol{D}\hat{\boldsymbol{x}} = \boldsymbol{D}\boldsymbol{D}'\boldsymbol{\lambda} = \boldsymbol{z}$$

With the help of this equation, $\boldsymbol{\lambda}$ is expressed in terms of $\boldsymbol{z}$ and substituted in the preceding equation:

$$\hat{\boldsymbol{x}} = \boldsymbol{D}'(\boldsymbol{D}\boldsymbol{D}')^{-1}\boldsymbol{z}$$

This result is identical with the special minimum variance estimate of equation (5.24).

## 5.3 THE KALMAN OPTIMUM FILTER (DISCRETE TIME)

The discrete-time filter problem can be solved without any extra difficulty using the procedures of the previous section. We begin by defining the problem in its usual form. For the case of more observations than state variables, it is first shown somewhat heuristically but in a very elementary fashion that the Kalman filter algorithm corresponds to a recursive Gauss–Markov estimation. Then we derive this algorithm in a second rigorous manner as a recursive estimate of minimum variance. This second approach is more general and includes the first observation moments when the total number of measurements is smaller than the number of state variables ($r < n$).

### 5.3.1 Statement of the Problem

As with the deterministic observation in Section 4.6, we consider a linear system in discrete time. We assume that the effect of the known input variables has already been eliminated (this requirement will subsequently be dropped). However, stochastic disturbance variables at the input of the system and stochastic measurement errors at the output are now explicitly taken into consideration. The mathematical model of the observed system can be expressed as follows (see also Figure 5.1, upper part):

$$\boldsymbol{x}(k+1) = \boldsymbol{A}(k)\boldsymbol{x}(k) + \boldsymbol{v}(k) \quad k \geqslant k_0 \tag{5.35a}$$

$$\boldsymbol{y}(k) = \boldsymbol{C}(k)\boldsymbol{x}(k) + \boldsymbol{w}(k) \tag{5.35b}$$

The inaccessible state vector $\boldsymbol{x}(k)$ is $n$ dimensional. Let its initial value $\boldsymbol{x}(k_0)$ be a random vector with zero expectation (this requirement will subsequently be dropped) and a given covariance matrix $\boldsymbol{P}(k_0)$. In its general formulation, this covariance matrix is defined as follows:

$$E\{[\boldsymbol{x}(k_0) - E\{\boldsymbol{x}(k_0)\}][\boldsymbol{x}(k_0) - E\{\boldsymbol{x}(k_0)\}]'\} = \boldsymbol{P}(k_0) \tag{5.36a}$$

The measurement variable $\boldsymbol{y}(k)$ is $m$ dimensional, and the system parameters $\boldsymbol{A}(k)$ and $\boldsymbol{C}(k)$ are known matrices of suitable type. The $n$-component disturbance variable $\boldsymbol{v}(k)$ and the $m$-component measurement error $\boldsymbol{w}(k)$ are white random vector processes in discrete time with given symmetric positive semidefinite covariance matrices (the requirement for whiteness can be dropped):

$$E\{\boldsymbol{v}(k)\} \equiv \boldsymbol{0} \quad E\{\boldsymbol{v}(k)\boldsymbol{v}'(\kappa)\} = \boldsymbol{Q}(k)\delta_{k\kappa} \tag{5.36b}$$

$$E\{\boldsymbol{w}(k)\} \equiv \boldsymbol{0} \quad E\{\boldsymbol{w}(k)\boldsymbol{w}'(\kappa)\} = \boldsymbol{R}(k)\delta_{k\kappa} \tag{5.36c}$$

where $\delta_{k\kappa}$ is the Kronecker delta ($\delta_{k\kappa} = 1$ for $k = \kappa$ and $\delta_{k\kappa} = 0$ for $k \neq \kappa$). Let the initial state, the disturbance process and the measurement process be uncorrelated with each other:

$$E\{\boldsymbol{x}(k_0)\boldsymbol{v}'(k)\} \equiv \boldsymbol{0} \tag{5.36d}$$

$$E\{\boldsymbol{x}(k_0)\boldsymbol{w}'(k)\} \equiv \boldsymbol{0} \tag{5.36e}$$

$$E\{\boldsymbol{v}(k)\boldsymbol{w}'(\kappa)\} \equiv \boldsymbol{0} \tag{5.36f}$$

(the last requirement can be dropped). We seek a linear unbiased estimate of $\boldsymbol{x}(k)$ based on the sample sequence $\boldsymbol{y}(k_0)$, $\boldsymbol{y}(k_0 + 1), \ldots, \boldsymbol{y}(k)$. This estimate is again denoted by $\hat{\boldsymbol{x}}(k)$; as before, the estimation error is defined as the difference between $\boldsymbol{x}$ and $\hat{\boldsymbol{x}}$:

$$\tilde{\boldsymbol{x}}(k) := \boldsymbol{x}(k) - \hat{\boldsymbol{x}}(k) \tag{5.37}$$

For unbiased estimation, its covariance matrix is defined by

$$\tilde{P}(k) := E\{\tilde{x}(k)\tilde{x}'(k)\} \tag{5.38}$$

*Remark:* In the minimum variance procedure it is required that the random variables involved have zero expectation. Accordingly, we start here by assuming that the expectation of the initial state $x(k_0)$ of the model (5.35a) vanishes. The sequence of disturbance variables $v(k_0)$, $v(k_0 + 1), \ldots$ is also centered with respect to zero. Hence the sequence $x(k_0 + 1)$, $x(k_0 + 2), \ldots$ is unbiased for all measurement times. Furthermore, according to (5.35b) and in view of the fact that all the $w(k)$ have an expectation of zero, we conclude that the expectation of the sequence $y(k_0)$, $y(k_0 + 1), \ldots$ also vanishes. In other words the above assumptions are such that the random processes $\{v(k)\}$, $\{w(k)\}$, $\{x(k)\}$ and $\{y(k)\}$ have a zero expectation for all $k$. As far as the noises $v(k)$ and $w(k)$ are concerned, this condition will always apply in what follows. However, the condition that $x(k)$ and $y(k)$ are unbiased will be lifted in Section 5.3.5.

### 5.3.2 Recursive Gauss–Markov Estimation

We refer to the above formulation of the problem. In addition, we assume that there is no *a priori* knowledge of the initial state of the system ($P(k_0) \to \infty$) and that the covariance matrix $R(k)$ of the measurement errors is nonsingular. Beginning at time $k_0$, we collect individual measurement values $y_j(k)$ during the first sample times until $n$ or more than $n$ are available in total. Only then can the Gauss–Markov estimate be determined for the first time. We will not do this here explicitly but we will briefly explain the procedure.

In a similar manner to Section 4.6, equation (4.57), the collected observations are combined to form a system of equations. Again, all past state values are expressed by the current state $x(k)$. Now, however, the extra components originating from the disturbance variable $v(\kappa)$ have to be considered. In this way we obtain the same matrix $D$ as in Section 4.6. What is new is the covariance matrix $S$. It can be expressed in terms of $Q(\kappa)$ and $R(\kappa)$ with $k_0 \leqslant \kappa \leqslant k$. When the values for $D$ and $S$ have been determined in this way, the estimate $\hat{x}(k)$ can be determined from the Gauss–Markov formula (5.31) for the first time. At the same time, the first value of $\tilde{P}(k)$ can be calculated by using equation (5.32).

With continuing observation a new set of measurements $y$ is added at every subsequent sampling time. However, as in Section 4.6, we want to avoid repeating the entire laborious calculation procedure in each interval. Instead, each new estimate $\hat{x}$ and new variance $\tilde{P}$ should be generated from the preceding values by making a relatively simple correction at every sampling time. To find the required recursive solution, we start by assuming that $\hat{x}$ and $\tilde{P}$ are known at time $k$. We then extrapolate $\hat{x}$ using the model (5.35a). There is no contribution from $v(k)$ because it is unknown, unbiased and white so that we obtain

$$x^*(k + 1) = A(k)\hat{x}(k) \tag{5.I}$$

This is the first of five equations of the filter algorithm (compare (4.70a)). Initially the filter equations are designated by Roman numerals.

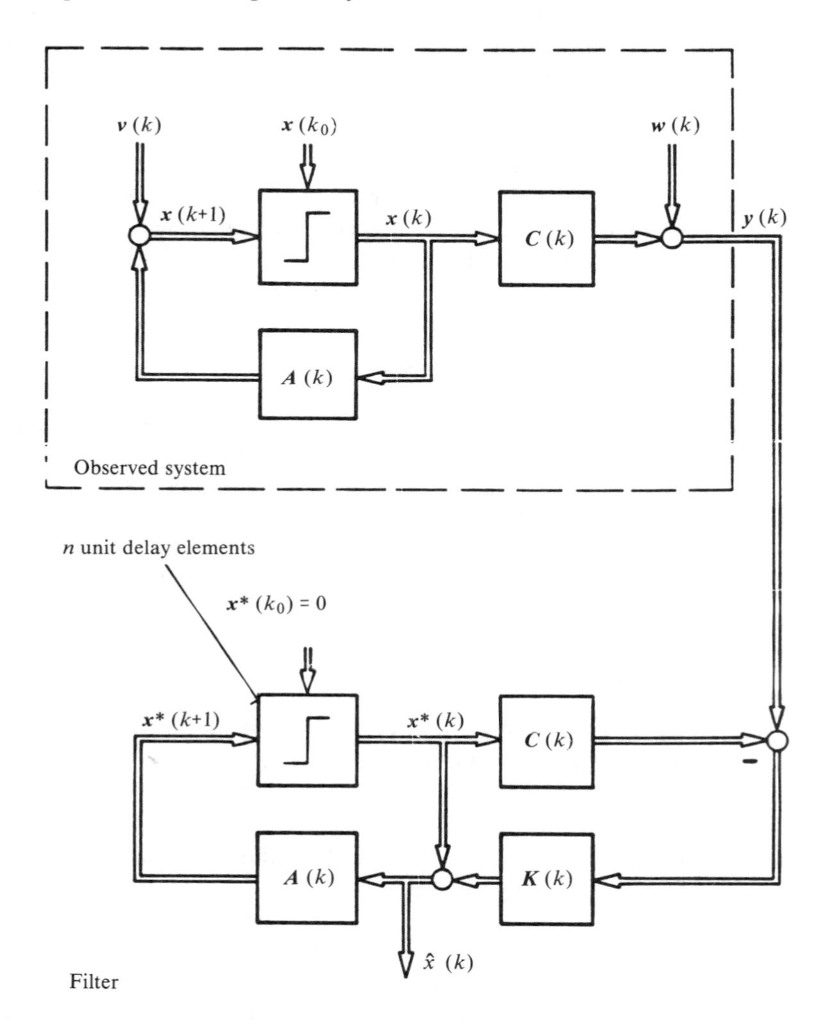


**Figure 5.1** Observed system and Kalman filter in discrete time (the area enclosed by the broken line is inaccessible).

It can be said that the extrapolated $x^*(k + 1)$ contains the summary of all accumulated knowledge of the $x(k + 1)$ to come. The relationship between these two variables becomes clear when $\hat{\boldsymbol{x}}(k)$ in equation (5.I) is replaced by $\boldsymbol{x}(k) - \tilde{\boldsymbol{x}}(k)$ and $A(k)x(k)$ is then substituted by using (5.35a):

$$\boldsymbol{x}^*(k+1) = \boldsymbol{x}(k+1) - \boldsymbol{A}(k)\tilde{\boldsymbol{x}}(k) - \boldsymbol{v}(k)$$

The required $\boldsymbol{x}(k+1)$ is on the right-hand side with the "errors" $A\tilde{\boldsymbol{x}}$ and $\boldsymbol{v}$ added to it. The covariance matrix of the sum of these errors is denoted by $\boldsymbol{P}^*(k+1)$. We then have

$$\boldsymbol{P}^*(k+1) = E\{[\boldsymbol{A}(k)\tilde{\boldsymbol{x}}(k) + \boldsymbol{v}(k)][\tilde{\boldsymbol{x}}'(k)A'(k) + \boldsymbol{v}'(k)]\}$$

According to (5.35a), the state $\boldsymbol{x}(k)$ only depends on $\boldsymbol{x}(k_0)$ and the values $\boldsymbol{v}(k_0), \ldots, \boldsymbol{v}(k-1)$. Therefore, because of (5.36b) and (5.36d), $\boldsymbol{x}(k)$ and $\boldsymbol{v}(k)$ are uncorrelated. The estimate $\hat{\boldsymbol{x}}(k)$ is a function of $\boldsymbol{y}(k_0), \ldots, \boldsymbol{y}(k)$ and so only depends on $\boldsymbol{x}(k_0), \ldots, \boldsymbol{x}(k)$ and the sample sequence $\boldsymbol{w}(k_0), \ldots, \boldsymbol{w}(k)$. All of these sequences have no correlation with the value $\boldsymbol{v}(k)$, i.e., $\hat{\boldsymbol{x}}(k)$ and $\boldsymbol{v}(k)$ are also uncorrelated. Therefore, the cross-covariance between $\tilde{\boldsymbol{x}}(k)$ and $\boldsymbol{v}(k)$ vanishes in the above equation for $\boldsymbol{P}^*(k+1)$ and we are left with

$$\boldsymbol{P}^*(k+1) = \boldsymbol{A}(k)\tilde{\boldsymbol{P}}(k)\boldsymbol{A}'(k) + \boldsymbol{Q}(k) \tag{5.II}$$

This is the second equation of the recursive filter algorithm (compare (4.71a)). When we look at the definition of the matrix $\boldsymbol{P}^*(k+1)$, we realize that it is the covariance matrix of the difference $\boldsymbol{x}(k+1) - \boldsymbol{x}^*(k+1)$, i.e., the extrapolation error.

The above equation for $\boldsymbol{x}^*(k+1)$ can be interpreted as a first set of observations for the required $\boldsymbol{x}(k+1)$ which we already have at our disposal. Appropriate additional observations arrive in the form of the measurement equation (5.35b) taken at time $k+1$. These two sets of observations are combined to form an integrated system:

$$\begin{bmatrix} \boldsymbol{x}^*(k+1) \\ \boldsymbol{y}(k+1) \end{bmatrix} = \begin{bmatrix} \boldsymbol{I} \\ \boldsymbol{C}(k+1) \end{bmatrix} \boldsymbol{x}(k+1) + \begin{bmatrix} -\boldsymbol{A}(k)\tilde{\boldsymbol{x}}(k) - \boldsymbol{v}(k) \\ \boldsymbol{w}(k+1) \end{bmatrix}$$

or

$$\boldsymbol{z} = \boldsymbol{D}\boldsymbol{x}(k+1) + \boldsymbol{s}$$

where $\boldsymbol{z}$, $\boldsymbol{D}$ and $\boldsymbol{s}$ are defined by direct comparison. The covariance matrix of the top part of $\boldsymbol{s}$ has already been declared as $\boldsymbol{P}^*(k+1)$ and calculated above. The covariance matrix of the bottom part, i.e., $\boldsymbol{w}(k+1)$, is $\boldsymbol{R}(k+1)$. The cross-covariances between $\boldsymbol{w}(k+1)$ on the one hand and $\tilde{\boldsymbol{x}}(k)$ or $\boldsymbol{v}(k)$ on the other hand vanish. We therefore have

$$E\{\boldsymbol{ss}'\} = \boldsymbol{S} = \begin{bmatrix} \boldsymbol{P}^*(k+1) & \boldsymbol{0} \\ \boldsymbol{0} & \boldsymbol{R}(k+1) \end{bmatrix}$$

The Gauss–Markov estimation for $\boldsymbol{x}(k+1)$ is now performed by using the appropriate versions of equations (5.31) and (5.32). We have

$$\hat{\boldsymbol{x}}(k+1) = \tilde{\boldsymbol{P}}(k+1)\boldsymbol{D}'\boldsymbol{S}^{-1}\boldsymbol{z}$$

with

$$\tilde{\boldsymbol{P}}(k+1) = (\boldsymbol{D}'\boldsymbol{S}^{-1}\boldsymbol{D})^{-1}$$

The following now applies:

$$\begin{aligned}
\boldsymbol{D}'\boldsymbol{S}^{-1} &= [\boldsymbol{I}, \boldsymbol{C}'(k+1)] \begin{bmatrix} (\boldsymbol{P}^*)^{-1}(k+1) & \boldsymbol{0} \\ \boldsymbol{0} & \boldsymbol{R}^{-1}(k+1) \end{bmatrix} \\
&= [(\boldsymbol{P}^*)^{-1}(\boldsymbol{k}+1), \boldsymbol{C}'(k+1)\boldsymbol{R}^{-1}(k+1)] \\
\boldsymbol{D}'\boldsymbol{S}^{-1}\boldsymbol{D} &= [(\boldsymbol{P}^*)^{-1}(k+1), \boldsymbol{C}'(k+1)\boldsymbol{R}^{-1}(k+1)] \begin{bmatrix} \boldsymbol{I} \\ \boldsymbol{C}(k+1) \end{bmatrix} \\
\boldsymbol{D}'\boldsymbol{S}^{-1}\boldsymbol{z} &= [(P^*)^{-1}(k+1), \boldsymbol{C}'(k+1)\boldsymbol{R}^{-1}(k+1)] \begin{bmatrix} \boldsymbol{x}^*(k+1) \\ \boldsymbol{y}(k+1) \end{bmatrix}
\end{aligned}$$

Substitution of this expression for $\boldsymbol{D}'\boldsymbol{S}^{-1}\boldsymbol{z}$ in the above estimation equation gives

$$\hat{\boldsymbol{x}}(k+1) = \tilde{\boldsymbol{P}}(k+1)\{(\boldsymbol{P}^*)^{-1}(k+1)\boldsymbol{x}^*(k+1) + \boldsymbol{C}'(k+1)\boldsymbol{R}^{-1}(k+1)\boldsymbol{y}(k+1)\} \tag{5.39a}$$

Evaluation of the expression $\boldsymbol{D}'\boldsymbol{S}^{-1}\boldsymbol{D}$ leads to

$$\tilde{\boldsymbol{P}}^{-1}(k+1) = (P^*)^{-1}(k+1) + \boldsymbol{C}'(k+1)\boldsymbol{R}^{-1}(k+1)\boldsymbol{C}(k+1) \tag{5.39b}$$

Because the arguments in these two intermediate results are everywhere equal to $k + 1$, they can be formally replaced by $k$, where $k$ now represents the incremented number of the next sampling time. The combined coefficient of $y(k + 1)$ in (5.39a) is called the gain matrix $\boldsymbol{K}$ (compare (4.68)):

$$\boldsymbol{K}(k) := \tilde{\boldsymbol{P}}(k)\boldsymbol{C}'(k)\boldsymbol{R}^{-1}(k) \tag{5.40}$$

Equation (5.39b) is premultiplied by $\tilde{\boldsymbol{P}}$. The two intermediate results (5.39) now take the form

$$\hat{x}(k) = \tilde{P}(k)(P^*)^{-1}(k)x^*(k) + K(k)y(k)$$
$$I = \tilde{P}(k)(P^*)^{-1}(k) + K(k)C(k)$$

The last equation is solved for $\tilde{P}(P^*)^{-1}$ and substituted in the equation above (compare (4.70b)):

$$\hat{x}(k) = x^*(k) + K(k)\{y(k) - C(k)x^*(k)\} \tag{5.III}$$

In addition, this equation is postmultiplied by $P^*$ (compare (4.71c)):

$$P^*(k) = \tilde{P}(k) + K(k)C(k)P^*(k) \tag{5.IV}$$

Finally, we postmultiply by $C'$, expand the first term of the right-hand side with $R^{-1}R$, take (5.40) into account and solve for $K$ (compare (4.71b)):

$$K(k) = P^*(k)C'(k)\{C(k)P^*(k)C'(k) + R(k)\}^{-1} \tag{5.V}$$

In this way we have reached our goal: the required recursion algorithm consists of equations (5.I) and (5.III) for the filter as such and (5.II), (5.IV) and (5.V) for the error covariance matrices of the estimates $x^*$ and $\hat{x}$ and the gain matrix $K$ of the filter. The results are identical with the Kalman filter algorithm.

*Summary.* The algorithm for recursive Gauss–Markov filtering of the discrete-time stochastically disturbed system

$$x(k + 1) = A(k)x(k) + v(k)$$
$$y(k) = C(k)x(k) + w(k)$$

is as follows: for the filter (Figure 5.1)

$$x^*(k + 1) = A(k)\hat{x}(k) \tag{5.41a}$$

$$\hat{x}(k) = x^*(k) + K(k)\{y(k) - C(k)x^*(k)\} \tag{5.41b}$$

and for the error covariance matrices and gain matrix

$$P^*(k + 1) = A(k)\tilde{P}(k)A'(k) + Q(k) \tag{5.42a}$$

$$K(k) = P^*(k)C'(k)\{C(k)P^*(k)C'(k) + R(k)\}^{-1} \tag{5.42b}$$

$$\tilde{P}(k) = P^*(k) - K(k)C(k)P^*(k) \tag{5.42c}$$

*Initial conditions.* The first measurements arrive at the sampling time $k_0$. We wait for a sufficient number of sampling times to allow the total number of individual measurement values $y_j(\kappa)$ to be greater than or equal to the number $n$ of state variables. Then the first estimate $\hat{\boldsymbol{x}}(k)$ and the first value of the covariance matrix $\tilde{\boldsymbol{P}}(k)$ are determined by batch processing using the Gauss–Markov rules (5.31) and (5.32). These values form the initial conditions for the recursion formulas (5.41a) and (5.42a).

*A priori knowledge of the initial state.* There is no bias, i.e., $E\{\boldsymbol{x}(k_0)\} = \boldsymbol{0}$, and infinite dispersion, i.e., $\boldsymbol{P}^{-1}(k_0) = \boldsymbol{0}$.

A comparison of these results with equations (4.70) and (4.71) shows that recursive Gauss–Markov filtering represents a generalization of recursive Gaussian observation of the case

$$\boldsymbol{Q}(k) = \boldsymbol{0} \quad \boldsymbol{R}(k) = \boldsymbol{I}$$

to arbitrary time-varying values of $\boldsymbol{Q}$ and $\boldsymbol{R}$. The actual observation or filter equations are identical in both cases (see also Figures 4.6 and 5.1). However, in general, the gain matrix has different values for the observer and the filter.

The numerical evaluation of the filter algorithm can be performed in the order (5.41a), (5.42a), increase count of $k$ by one (5.42b), (5.41b) and (5.42c). In the first two steps, the existing estimate and its error covariance matrix are extrapolated into the next interval. In the third step, the new optimal gain matrix $\boldsymbol{K}$ is calculated and the extrapolated estimate is improved with the new $\boldsymbol{y}$ data. The last step produces the covariance matrix of the updated estimate.

Note that the system of variance equations (5.42) is completely independent of the current values of the observations $\boldsymbol{y}$ and the estimates $\hat{\boldsymbol{x}}$. Provided that $\boldsymbol{A}$, $\boldsymbol{C}$, $\boldsymbol{R}$ and $\boldsymbol{Q}$ are available in good time, this part of the algorithm can be evaluated before the actual filter equations (5.41), in particular before the filter is put into operation. Actually, this part of the algorithm is simply the computer-aided synthesis of the optimum filter, in particular its gain. The elements in the principal diagonal of $\tilde{\boldsymbol{P}}(k)$ give a measure of the quality of the filter. If the sequence of gain matrices $\boldsymbol{K}(k)$ is calculated off-line, it can be stored until called up at time $k$ in subsequent operation of the filter. In stationary observation situations (i.e., for constant parameters $\boldsymbol{A}$, $\boldsymbol{C}$, $\boldsymbol{Q}$ and $\boldsymbol{R}$), the values for the $k_{ij}$ usually settle down to steady state values after a certain number of intervals (5–20). The memory requirements for storing the $k_{ij}$ are then quite modest.

The steady state solution for $\boldsymbol{K}$ corresponds to the Wiener–Kolmogorov filter. Some examples are given in the next section.

### 5.3.3 Recursive Minimum Variance Estimation

In Section 5.3.2 the Kalman filter algorithm was derived as a recursive version of the corresponding Gauss–Markov estimation. This is conceptually the simplest approach. However, it has the disadvantage that the first estimate and the subsequent recursion can only be started when there are $n$ or more individual observations $y_j(\kappa)$ available.

We have seen that the minimum variance procedure makes it possible to perform an estimation for $\boldsymbol{x}(k_0)$ immediately after obtaining the first observation $\boldsymbol{y}(k_0)$. However, a condition for this is that, in contrast with the above, the covariance matrix of $\boldsymbol{x}(k_0)$ is *finite* and known. This means that the corresponding amount of *a priori* information on the initial state must be available.

In what follows, we begin by determining the minimum variance estimate $\hat{\boldsymbol{x}}(k_0)$ at time $k_0$. For all further sample times a recursion method is again used, whereby on arrival of new measurements a given estimate is improved in such a way that the updated estimate again exhibits the smallest possible variance. It will be shown that exactly the same algorithm as above is produced. The difference mainly lies in the altered initial conditions and in the earlier starting time. From the mathematical point of view, the minimum variance method has two further advantages.

(i) The covariance matrix $\boldsymbol{R}(k)$ of the measurement error no longer has to be assumed to be nonsingular (note that $\boldsymbol{R}^{-1}$ itself does not appear in the filter algorithm (5.41) and (5.42).

(ii) It can rigorously be proved that the extrapolation (5.I) is optimum.

At the first sample time, the following measurements arrive:

$$\boldsymbol{y}(k_0) = \boldsymbol{C}(k_0)\boldsymbol{x}(k_0) + \boldsymbol{w}(k_0)$$

where

$$E\{\boldsymbol{x}(k_0)\boldsymbol{x}'(k_0)\} = \boldsymbol{P}(k_0)$$

We compare this with equations (5.1) and (5.2b) and substitute the appropriate quantities into (5.21) for the linear unbiased minimum variance estimate. We obtain directly

$$\hat{\boldsymbol{x}}(k_0) = \boldsymbol{K}(k_0)\boldsymbol{y}(k_0) \tag{5.43a}$$

where

$$\boldsymbol{K}(k_0) = \boldsymbol{P}(k_0)\boldsymbol{C}'(k_0)\{\boldsymbol{C}(k_0)\boldsymbol{P}(k_0)\boldsymbol{C}'(k_0) + \boldsymbol{R}(k_0)\}^{-1} \tag{5.43b}$$

It follows immediately from equation (5.22) that

$$\tilde{\boldsymbol{P}}(k_0) = \boldsymbol{P}(k_0) - \boldsymbol{K}(k_0)\boldsymbol{C}(k_0)\boldsymbol{P}(k_0) \tag{5.43c}$$

This initial solution readily fits into the framework of the filter equations (5.41b), (5.42b) and (5.42c) if we set $\boldsymbol{x}^*(k_0) = \boldsymbol{0}$ and $\boldsymbol{P}^*(k_0) = \boldsymbol{P}(k_0)$.

We then generate the solution for all subsequent sample times by induction from $k$ to $k + 1$. For this, the simplest approach is to use the matrix Wiener–Hopf equation for discrete time. It is a necessary and sufficient condition for the minimum variance estimation. In what follows, we will use it in the form of equation (5.16). We will subsequently also need equation (5.17a). For easy reference, these two equations are given again here:

$$E\{\tilde{\boldsymbol{x}}\boldsymbol{z}'\} = \boldsymbol{0} \tag{5.44a}$$

$$E\{\tilde{\boldsymbol{x}}\tilde{\boldsymbol{x}}'\} = E\{\tilde{\boldsymbol{x}}\boldsymbol{x}'\} \tag{5.44b}$$

We now assume that the observations $\boldsymbol{y}(k_0), \boldsymbol{y}(k_0 + 1), \ldots, \boldsymbol{y}(k)$ have already arrived and have been processed to give the estimate $\hat{\boldsymbol{x}}(k)$. Let this estimate be optimum in the sense of minimum variance. This means that it satisfies the corresponding form of condition (5.44a):

$$E\{[\boldsymbol{x}(k) - \hat{\boldsymbol{x}}(k)][\boldsymbol{y}'(k_0), \boldsymbol{y}'(k_0 + 1) \ldots \boldsymbol{y}'(k)]\} = \boldsymbol{0} \tag{5.45a}$$

Note that this equation is valid because of our assumption. Next, we seek the optimum extrapolated estimate $\boldsymbol{x}^*(k + 1)$. Now, the following form of condition (5.44a) must hold:

$$E\{[\boldsymbol{x}(k + 1) - \boldsymbol{x}^*(k + 1)][\boldsymbol{y}'(k_0), \boldsymbol{y}'(k_0 + 1) \ldots \boldsymbol{y}'(k)]\} = \boldsymbol{0} \tag{5.45b}$$

After obtaining the new measurements $\boldsymbol{y}(k + 1)$, the next estimate $\hat{\boldsymbol{x}}(k + 1)$ can be formed. In this case, condition (5.44a) takes the form

$$E\{[\boldsymbol{x}(k + 1) - \hat{\boldsymbol{x}}(k + 1)][\boldsymbol{y}'(k_0), \boldsymbol{y}'(k_0 + 1) \ldots \boldsymbol{y}'(k) | \ \boldsymbol{y}'(k + 1)]\} = \boldsymbol{0} \tag{5.45c}$$

In order to satisfy condition (5.45b), we first compare it with (5.45a). The two equations differ only in the first factor. This similarity suggests that $\boldsymbol{x}(k + 1)$ in (5.45b) can be replaced by $\boldsymbol{A}(k)\boldsymbol{x}(k) + \boldsymbol{v}(k)$. As $\boldsymbol{v}(k)$ is not correlated with $\boldsymbol{y}(k_0), \ldots, \boldsymbol{y}(k)$ (see Section 5.3.2), it can be ignored here so that we are left with

$$E\{[\boldsymbol{A}(k)\boldsymbol{x}(k) - \boldsymbol{x}^*(k + 1)][\boldsymbol{y}'(k_0), \boldsymbol{y}'(k_0 + 1) \ldots \boldsymbol{y}'(k)]\} = \boldsymbol{0}$$

This prescription is obviously satisfied by making the selection

$$\boldsymbol{x}^*(k + 1) = \boldsymbol{A}(k)\hat{\boldsymbol{x}}(k) \tag{5.I}$$

because we can now remove $\boldsymbol{A}(k)$ on the left from the expectation and are left with statement (5.45a) which has been assumed to be valid.

The error in the extrapolation is

$$\begin{aligned} \boldsymbol{x}(k+1) - \boldsymbol{x}^*(k+1) &= \boldsymbol{A}(k)\boldsymbol{x}(k) + \boldsymbol{v}(k) - \boldsymbol{A}(k)\hat{\boldsymbol{x}}(k) \\ &= \boldsymbol{A}(k)\tilde{\boldsymbol{x}}(k) + \boldsymbol{v}(k) \end{aligned} \tag{5.46}$$

Its covariance matrix is given by

$$\begin{aligned} \boldsymbol{P}^*(k+1) &:= E\{[\boldsymbol{x}(k+1) - \boldsymbol{x}^*(k+1)][\boldsymbol{x}(k+1) - \boldsymbol{x}^*(k+1)]'\} \\ &= E\{[\boldsymbol{A}(k)\tilde{\boldsymbol{x}}(k) + \boldsymbol{v}(k)][\tilde{\boldsymbol{x}}'(k)\boldsymbol{A}'(k) + \boldsymbol{v}'(k)]\} \end{aligned}$$

The disturbance $\boldsymbol{v}(k)$ is not correlated with the estimation error $\tilde{\boldsymbol{x}}(k)$ (see Section 5.3.2). We are therefore left with

$$P^*(k+1) = \boldsymbol{A}(k)\tilde{P}(k)\boldsymbol{A}'(k) + \boldsymbol{Q}(k) \tag{5.II}$$

This completes the extrapolation.

We now turn to condition (5.45c). Here, it is a question of finding a suitable expression for $\hat{\boldsymbol{x}}(k+1)$ which could satisfy this prescription. In the light of the observation and Gauss–Markov algorithms we try the following hypothesis:

$$\hat{\boldsymbol{x}}(k+1) = \boldsymbol{x}^*(k+1) + \boldsymbol{K}(k+1)\{\boldsymbol{y}(k+1) - \boldsymbol{C}(k+1)\boldsymbol{x}^*(k+1)\} \tag{5.III}$$

We replace $\boldsymbol{y}$ by $\boldsymbol{Cx} + \boldsymbol{w}$ so that the new estimate error takes the form

$$\begin{aligned} \boldsymbol{x}(k+1) - \hat{\boldsymbol{x}}(k+1) = [\boldsymbol{I} - \boldsymbol{K}(k+1)\boldsymbol{C}(k+1)][\boldsymbol{x}(k+1) \\ - \boldsymbol{x}^*(k+1)] - \boldsymbol{K}(k+1)\boldsymbol{w}(k+1) \end{aligned} \tag{5.47}$$

This expression is substituted in equation (5.45c). We first consider the part of the expectation which is formed with the old $\boldsymbol{y}$ values to the left of the broken line in (5.45c). We know meanwhile that the new measurement error $\boldsymbol{w}(k+1)$ is not correlated with the old observations $\boldsymbol{y}(k_0), \ldots, \boldsymbol{y}(k)$. Neither is $\boldsymbol{x}(k+1) - \boldsymbol{x}^*(k+1)$ because (5.45b) is valid in the meantime. It follows from this that hypothesis (5.III) satisfies the left-hand part of condition (5.45c). The remaining part of prescription (5.45c) enables the gain matrix $\boldsymbol{K}$, which is still missing, to be determined. Using (5.47), we obtain the following:

$$\begin{aligned} E\{([\boldsymbol{I} - \boldsymbol{K}(k+1)\boldsymbol{C}(k+1)][\boldsymbol{x}(k+1) - \boldsymbol{x}^*(k+1)] \\ - \boldsymbol{K}(k+1)\boldsymbol{w}(k+1))[\boldsymbol{x}'(k+1)\boldsymbol{C}'(k+1) + \boldsymbol{w}'(k+1)]\} = \boldsymbol{0} \end{aligned} \tag{5.48}$$

The variables $\boldsymbol{x}(k + 1)$ and $\boldsymbol{x}^*(k + 1)$ are not correlated with $\boldsymbol{w}(k + 1)$. The other expectations can be transformed in a simple manner. In order to do this, equation (5.44b) is used in terms of $\boldsymbol{x}^*(k + 1)$ so that the cross-covariance between $\boldsymbol{x}(k + 1) - \boldsymbol{x}^*(k + 1)$ and $\boldsymbol{x}(k + 1)$ can be replaced by $\boldsymbol{P}^*(k + 1)$. The result is

$$[\boldsymbol{I} - \boldsymbol{K}(k + 1)\boldsymbol{C}(k + 1)]\boldsymbol{P}^*(k + 1)\boldsymbol{C}'(k + 1) - \boldsymbol{K}(k + 1)\boldsymbol{R}(k + 1) = \boldsymbol{0} \quad (5.49)$$

It follows from this, on rearranging and solving for $\boldsymbol{K}$, that

$$\begin{aligned}&\boldsymbol{K}(k + 1)\\ &= \boldsymbol{P}^*(k + 1)\boldsymbol{C}'(k + 1)\{\boldsymbol{C}(k + 1)\boldsymbol{P}^*(k + 1)\boldsymbol{C}'(k + 1) + \boldsymbol{R}(k + 1)\}^{-1}\end{aligned} \quad (5.\text{IV})$$

Finally, we determine $\tilde{\boldsymbol{P}}(k + 1)$ by postmultiplying equation (5.47) by $\boldsymbol{x}'(k + 1)$ and forming the expectations on both sides. Again, equation (5.44b) is used in terms of $\tilde{\boldsymbol{x}}(k + 1)$ and $\boldsymbol{x}^*(k + 1)$ as appropriate. We immediately obtain

$$\tilde{\boldsymbol{P}}(k + 1) = \{\boldsymbol{I} - \boldsymbol{K}(k + 1)\boldsymbol{C}(k + 1)\}\boldsymbol{P}^*(k + 1) \quad (5.\text{V})$$

This completes a second derivation of the Kalman filter algorithm which is more general and rigorous. We summarize results (5.I)–(5.V) in the following theorem.

*Theorem 5.1 (Kalman filter).* The linear unbiased minimum variance estimate for the state $\boldsymbol{x}(k)$ of the discrete-time stochastically disturbed system

$$\boldsymbol{x}(k + 1) = \boldsymbol{A}(k)\boldsymbol{x}(k) + \boldsymbol{v}(k) \quad E\{\boldsymbol{x}(k_0)\} = \boldsymbol{0}$$

$$\boldsymbol{y}(k) = \boldsymbol{C}(k)\boldsymbol{x}(k) + \boldsymbol{w}(k)$$

for which the sample sequence $\boldsymbol{y}(k_0), \ldots, \boldsymbol{y}(k)$ and the *a priori* knowledge according to equations (5.36a)–(5.36f) are available, is given by the following recursive algorithm:

(i) *Filter (Figure 5.1)*

$$\boldsymbol{x}^*(k + 1) = \boldsymbol{A}(k)\hat{\boldsymbol{x}}(k) \quad \boldsymbol{x}^*(k_0) = \boldsymbol{0} \quad (5.50\text{a})$$

$$\hat{\boldsymbol{x}}(k) = \boldsymbol{x}^*(k) + \boldsymbol{K}(k)\{\boldsymbol{y}(k) - \boldsymbol{C}(k)\boldsymbol{x}^*(k)\} \quad (5.50\text{b})$$

(ii) *Error covariance matrices and gain matrix*

$$\boldsymbol{P}^*(k + 1) = \boldsymbol{A}(k)\tilde{\boldsymbol{P}}(k)\boldsymbol{A}'(k) + \boldsymbol{Q}(k) \quad (5.51\text{a})$$

$$\boldsymbol{K}(k) = \boldsymbol{P}^*(k)\boldsymbol{C}'(k)\{\boldsymbol{C}(k)\boldsymbol{P}^*(k)\boldsymbol{C}'(k) + \boldsymbol{R}(k)\}^{-1} \quad (5.51\text{b})$$

$$\tilde{\boldsymbol{P}}(k) = \boldsymbol{P}^*(k) - \boldsymbol{K}(k)\boldsymbol{C}(k)\boldsymbol{P}^*(k) \tag{5.51c}$$

(iii) *Initial condition*

$$\boldsymbol{P}^*(k_0) = \boldsymbol{P}(k_0) \tag{5.52}$$

where $\boldsymbol{P}(k_0)$ is the covariance matrix of the initial state $\boldsymbol{x}(k_0)$ and expresses the relevant *a priori* knowledge.

Comparison with the Gauss–Markov estimation (5.41) and (5.42) shows that the linear unbiased minimum variance estimation is the generalization of the case

$$\boldsymbol{P}(k_0) = \infty$$

to any finite values of this symmetric positive semidefinite matrix.

Designs for first- and second-order filters can readily be produced using a simple pocket calculator. A programmable computer is needed for higher-order filters. Most of the effort involves calculating $\boldsymbol{K}$ from equation (5.51b). To do so, an equation of the form

$$\{\boldsymbol{C}\boldsymbol{P}^*\boldsymbol{C}' + \boldsymbol{R}\}\boldsymbol{K}' = \boldsymbol{C}\boldsymbol{P}^*$$

is to be solved for $\boldsymbol{K}'$ in every interval (Cholesky algorithm (Zurmuehl and Falk, 1984)). Fortunately the matrix in braces is only of the $m \times m$ type, i.e., sometimes only a scalar. $\boldsymbol{K}'$ and $\boldsymbol{C}\boldsymbol{P}^*$ each have $n$ columns with $m$ elements.

The calculated $\boldsymbol{K}$ is then postmultiplied by the $\boldsymbol{C}\boldsymbol{P}^*$ already obtained. This gives the symmetric product $\boldsymbol{K}\boldsymbol{C}\boldsymbol{P}^*$ which is subtracted from $\boldsymbol{P}^*$ in order to obtain $\tilde{\boldsymbol{P}}$. It can be seen that $\boldsymbol{K}\boldsymbol{C}\boldsymbol{P}^*$ must be symmetric by postmultiplying (5.51b) by $\boldsymbol{C}(k)\boldsymbol{P}^*(k)$. Factoring out the matrix $\boldsymbol{P}^*$ on the right in (5.51c) is not recommended because finding the product of $\boldsymbol{I} - \boldsymbol{K}\boldsymbol{C}$ with $\boldsymbol{P}^*$ is much more cumbersome than multiplying $\boldsymbol{K}$ by $\boldsymbol{C}\boldsymbol{P}^*$.

Checks are essential for extensive numerical calculations. Apart from the usual checks, for example, with column and row sums (Zurmuehl and Falk, 1984), two simple additional or alternative checks can be made here:

(i) checking the matrices $\boldsymbol{P}^*$ and $\tilde{\boldsymbol{P}}$ for symmetry;
(ii) checking with the help of the equation

$$\tilde{\boldsymbol{P}}(k)\boldsymbol{C}'(k) = \boldsymbol{K}(k)\boldsymbol{R}(k) \tag{5.53}$$

which follows directly from the intermediate result (5.49) if $k + 1$ is replaced by $k$ and the term $[\ldots]\boldsymbol{P}^*$ is simplified by using (5.51c).

*Example 5.1.* Suppose that the error of a sensor consists of a systematic component constant over time (bias) and a stochastic component uncorrelated in time (white noise). For the purposes of calibration, the output variable of the sensor is observed at times 1,2,3, . . . in the absence of the input signal. We define the systematic error as the state variable $x$ and the stochastic error as the measurement noise $w$. Thus the model of the "observed system" is (Figure 5.2):

$$x(k+1) = x(k) \qquad k = 1,2,3,\ldots$$
$$y(k) = x(k) + w(k)$$

The parameters $A$ and $C$ of this model are obviously equal to unity. The variable $v$ is absent so that $\boldsymbol{Q} \equiv 0$. Let the standard deviation of the white noise $w(k)$ be constant and equal to 2 so that $R \equiv 4$. The systematic error is known to have a standard deviation of 3. We therefore set $P(k_0) = 9$. A Kalman filter which can be used to estimate the systematic error is to be designed.

In this example, equation (5.50a) reduces to $x^*(k+1) = \hat{x}(k)$. The filter therefore has the form (Figure 5.2)

$$\hat{x}(k) = \hat{x}(k-1) + K(k)\{y(k) - \hat{x}(k-1)\} \quad x(0) = 0$$

The sequence of gain values $\boldsymbol{K}(k)$ and the variance of the estimate error $\tilde{x}(k)$ is obtained from the set (5.51). Here we have

$$P^*(k+1) = \tilde{P}(k) \quad P^*(1) = 9$$
$$K(k) = \frac{P^*(k)}{P^*(k) + 4}$$
$$\tilde{P}(k) = \{1 - K(k)\}P^*(k)$$

In this simple case a difference equation can be given directly in terms of $K$. It follows from the above equations that

$$K(k+1) = P^*(k+1)\{P^*(k+1) + 4\}^{-1} = \tilde{P}(k)\{\tilde{P}(k) + 4\}^{-1}$$

Because of equation (5.53), we have $\tilde{P}(k) = 4K(k)$ so that

$$K(k+1) = K(k)\{K(k) + 1\}^{-1}$$

The numerical results for $K(k)$ and $\tilde{P}(k)$ are shown on the left-hand side of Table 5.1 and in Figures 5.2 and 5.3.

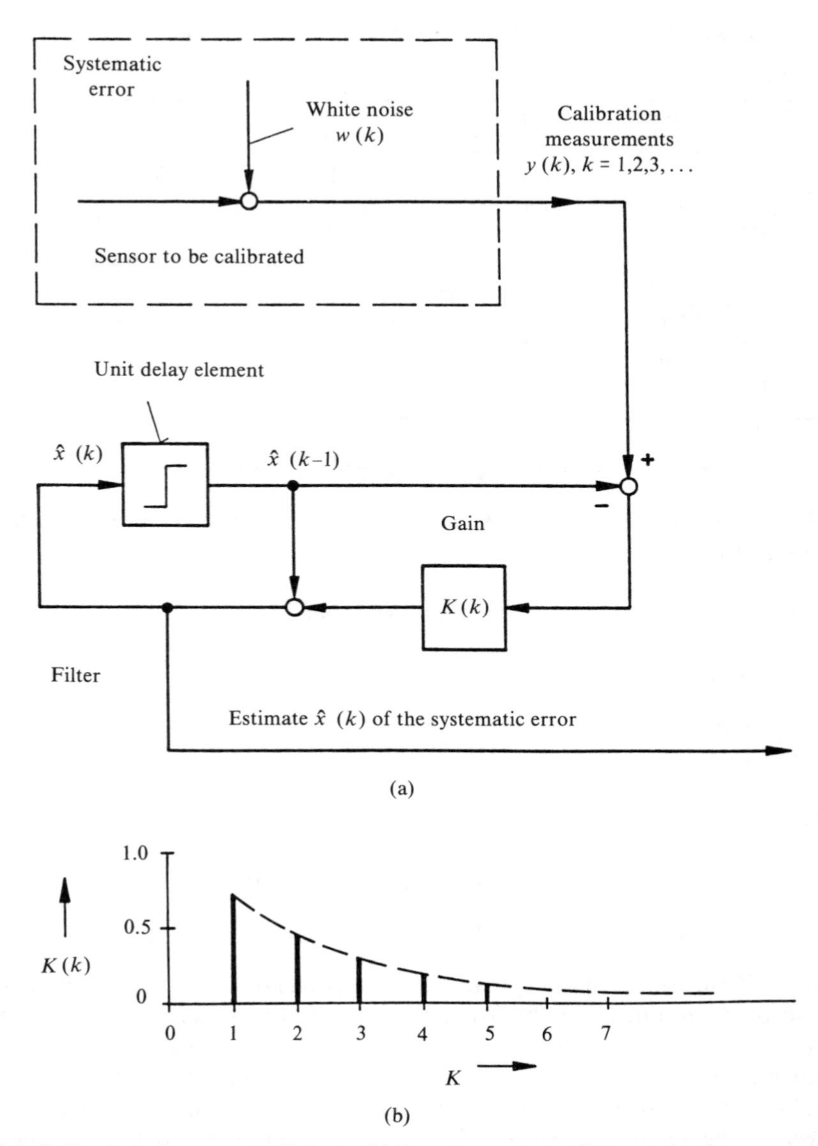


**Figure 5.2** Calibration of a sensor by Kalman filtering: (a) sensor and filter; (b) sequence of filter gain values.

**Table 5.1** Results of the Filter Algorithm (5.50) and (5.51) in Example 5.1

| $k$ | $P^*$ | $K$ | $\tilde{P}$ | $w$ | $y$ | $\hat{x}$ |
|---|---|---|---|---|---|---|
| 1 | 9.00 | 0.69 | 2.78 | −2 | 1 | 0.69 |
| 2 | 2.78 | 0.41 | 1.64 | −2 | 1 | 0.82 |
| 3 | 1.64 | 0.29 | 1.16 | +2 | 5 | 2.03 |
| 4 | 1.16 | 0.225 | 0.90 | +2 | 5 | 2.71 |
| 5 | 0.90 | 0.18 | 0.73 | −2 | 1 | 2.40 |
| | | | | | | |
| $\infty$ | 0 | 0 | 0 | | | |

The difference equation for $K$ indicates that the solution can only settle to an equilibrium for $k \to \infty$. With $K(k + 1) = K(k) = K$, we obtain

$$K = K(K + 1)^{-1}$$

It follows from this that

$$K(\infty) = 0$$

and also

$$P^*(\infty) = (P^* + 4)K = 0$$
$$\tilde{P}(\infty) = 0$$

Therefore, if measurements are made for a sufficiently long time, the systematic error can be determined exactly. After five sample times, its standard deviation is already reduced from 3 to $(0.73)^{1/2} = 0.85$. The actual evolution of the sequence $\hat{x}(k)$ depends on the received sample sequence $y(k)$. In order to simulate this, a random value for $x$ must be generated and combined with a white sequence $w(k)$ from a corresponding random generator, or white noise can be produced manually by tossing a coin. We obtain the required variance of 4 by saying that heads $\to w = 2$, tails $\to w = -2$. Let the random variable $x$ take the value 3. We then obtain the set of sample sequences for $w$, $y$ and $\hat{x}$ given on the right-hand side of Table 5.1 (see also Figure 5.3).

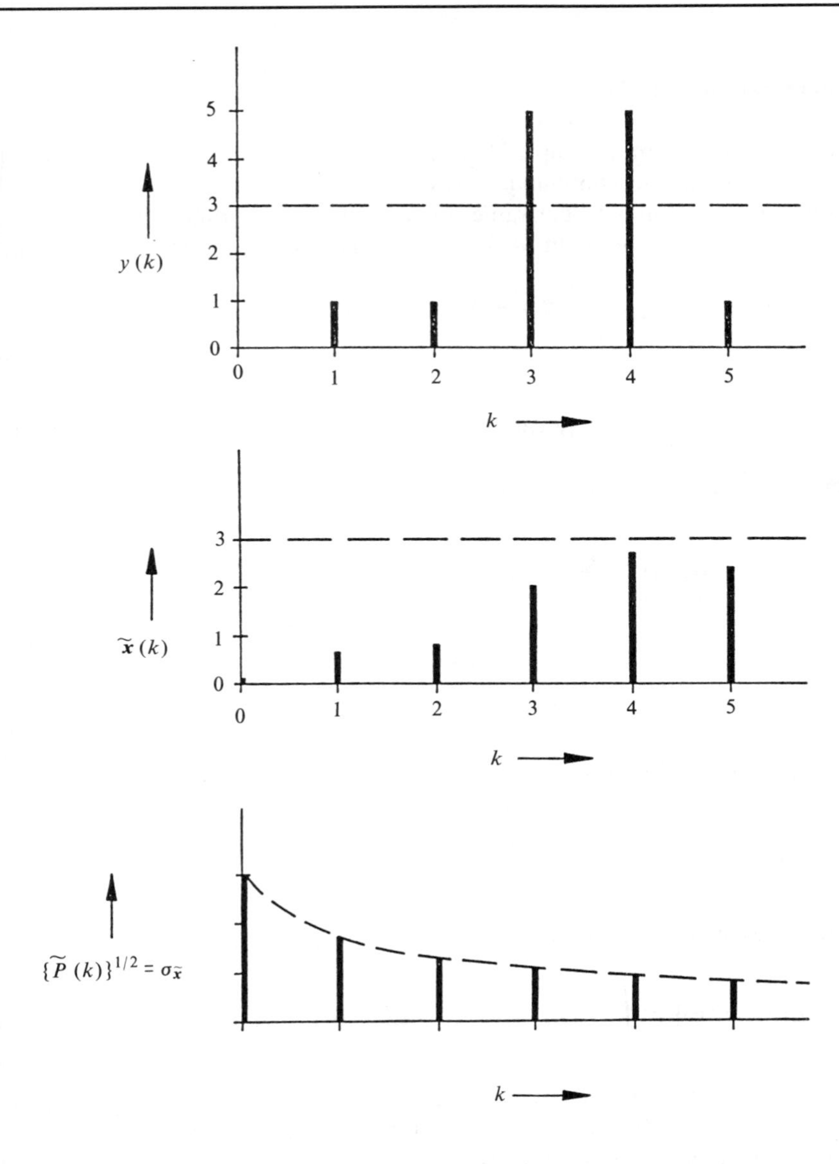


**Figure 5.3** Sample sequences for the measurements and the estimate for the sensor calibration (Figure 5.2) and evolution of the standard deviation of the estimation error.

### 5.3.4 Determination of $Q(k)$

If a continuous-time system is sampled, the dynamic matrix and the control matrix have different values in the continuous and discrete models (see Section 4.6). As will be shown in what follows, the same applies for the covariance matrix of the white disturbance noise at the input to the system. The starting point is the system

$$\dot{\boldsymbol{x}}(t) = \bar{\boldsymbol{A}}(t)\boldsymbol{x}(t) + \bar{\boldsymbol{v}}(t)$$

where

$$E\{\bar{\boldsymbol{v}}(t)\bar{\boldsymbol{v}}'(\tau)\} = \bar{\boldsymbol{Q}}(t)\delta(t - \tau) \tag{5.54}$$

The movement of the state from one sampling point to the next is described by

$$\boldsymbol{x}(t_{k+1}) = \boldsymbol{\phi}(t_{k+1}, t_k)\boldsymbol{x}(t_k) + \int_{t_k}^{t_{k+1}} \boldsymbol{\phi}(t_{k+1}, \tau)\bar{\boldsymbol{v}}(\tau)\mathrm{d}\tau$$

or

$$\boldsymbol{x}(k + 1) = \boldsymbol{A}(k)\boldsymbol{x}(k) + \boldsymbol{v}(k)$$

As a continuation of Section 4.6, equations (4.51)–(4.54), we therefore define

$$\boldsymbol{v}(k) := \int_{t_k}^{t_{k+1}} \boldsymbol{\phi}(t_{k+1}, \tau)\bar{\boldsymbol{v}}(\tau)\mathrm{d}\tau \tag{5.55}$$

The required covariance matrix of $\boldsymbol{v}$ is

$$E\{\boldsymbol{v}(k)\boldsymbol{v}'(\kappa)\} = \int_{t_k}^{t_{k+1}} \boldsymbol{\phi}(t_{k+1}, \tau) \int_{t_\kappa}^{t_{\kappa+1}} E\{\bar{\boldsymbol{v}}(\tau)\bar{\boldsymbol{v}}'(\sigma)\}\boldsymbol{\phi}'(t_{\kappa+1}, \sigma)\mathrm{d}\sigma\mathrm{d}\tau$$

Here we have written the product of two single integrals of the form (5.55) as a double integral and taken the expectation into the inner integral. If $k \neq \kappa$, then $\tau$ and $\sigma$ lie in two different intervals and the expectation vanishes because of (5.54). For $k = \kappa$, the masking property of the $\delta$ function is used with the result

$$\boldsymbol{Q}(k) = \int_{t_k}^{t_{k+1}} \boldsymbol{\phi}(t_{k+1}, \tau)\bar{\boldsymbol{Q}}(\tau)\boldsymbol{\phi}'(t_{k+1}, \tau)\mathrm{d}\tau \tag{5.56}$$

*Example 5.2*. An aircraft starts from a point which is known exactly and is subject to unknown disturbance accelerations during its flight. It takes position measurements every second but these contain errors. We require optimum estimates for the current position and speed.

We choose as state variables the position error $x_1$ and the speed error $x_2$. The disturbance acceleration is taken as normalized white noise in continuous time. The differential equation of the observed system is therefore

$$\frac{d}{dt}\begin{bmatrix} x_1 \\ x_2 \end{bmatrix} = \begin{bmatrix} 0 & 1 \\ 0 & 0 \end{bmatrix}\begin{bmatrix} x_1 \\ x_2 \end{bmatrix} + \begin{bmatrix} 0 \\ 1 \end{bmatrix}\bar{v} \quad E\{\bar{v}(t)\bar{v}(\tau)\} = \delta(t - \tau)$$

The transition matrix of this system is determined as in Example 4.3. Because $\boldsymbol{A}^2 = \boldsymbol{0}$, we have

$$\boldsymbol{\phi}(t) = \boldsymbol{I} + \bar{\boldsymbol{A}}t = \begin{bmatrix} 1 & t \\ 0 & 1 \end{bmatrix}$$

and

$$\boldsymbol{A} = \boldsymbol{\phi}(1) = \begin{bmatrix} 1 & 1 \\ 0 & 1 \end{bmatrix}$$

The covariance matrix $\boldsymbol{Q}$ is obtained from equation (5.56) by setting $t_k = 0$, $t_{k+1} = 1$ and

$$\bar{\boldsymbol{Q}}(t) = \begin{bmatrix} 0 \\ 1 \end{bmatrix} \times 1 \times [0 \quad 1]$$

Then,

$$\boldsymbol{Q} = \int_0^1 \boldsymbol{\phi}(1 - \tau)\begin{bmatrix} 0 \\ 1 \end{bmatrix}[0 \quad 1]\boldsymbol{\phi}'(1 - \tau)d\tau = \begin{bmatrix} 1/3 & 1/2 \\ 1/2 & 1 \end{bmatrix}$$

On sampling the position, the measurement error $w(k)$ occurs. Let it be normalized white noise in discrete time. The discrete-time model of the observed process is now complete and can be written as follows:

$$\boldsymbol{x}(k + 1) = \begin{bmatrix} 1 & 1 \\ 0 & 1 \end{bmatrix}\boldsymbol{x}(k) + \boldsymbol{v}(k) \quad \boldsymbol{Q}(k) = \begin{bmatrix} 1/3 & 1/2 \\ 1/2 & 1 \end{bmatrix}$$

$$y(k) = [1 \quad 0]\boldsymbol{x}(k) + w(k) \quad R(k) = 1$$

The position and speed errors at the starting point are zero, i.e., $\boldsymbol{P}(0) = \boldsymbol{0}$.

The Kalman filter for this problem has the form (Figure 5.4)

$$\boldsymbol{x}^*(k+1) = \begin{bmatrix} 1 & 1 \\ 0 & 1 \end{bmatrix} \hat{\boldsymbol{x}}(k) \quad \boldsymbol{x}^*(0) = \boldsymbol{0}$$

$$\hat{\boldsymbol{x}}(k) = \boldsymbol{x}^*(k) + \boldsymbol{k}(k)\{y(k) - x_1^*(k)\}$$

The two gain factors $k_1$ and $k_2$ are obtained from the algorithm (5.51) with the initial condition $\boldsymbol{P}^*(0) = \boldsymbol{0}$. Therefore $\boldsymbol{k}(0)$ and $\tilde{\boldsymbol{P}}(0)$ are equal to zero. The numerical results for $k = 1, \ldots, 7$ are given in Table 5.2, and the variances of the estimation errors for position and speed are plotted in Figure 5.6 in connection with Example 5.3. The checking equation (5.53) states that in this case we must have

$$\tilde{P}(k)\begin{bmatrix} 1 \\ 0 \end{bmatrix} = \boldsymbol{k}(k) \times 1$$

This means that the first column of $\tilde{\boldsymbol{P}}$ is always identical with $\boldsymbol{k}$. In the seventh interval, the covariance matrix $\tilde{\boldsymbol{P}}$ is essentially unchanged with respect to its previous value. Since $\boldsymbol{A}$ and $\boldsymbol{Q}$ are constant over time, we obtain $\boldsymbol{P}^*(8) = \boldsymbol{P}^*(7)$ in the next step. Then, because of the time invariance of $\boldsymbol{C}$ and $\boldsymbol{R}$, $\boldsymbol{k}(8) = \boldsymbol{k}(7)$. Thus, we finally obtain the old value for $\tilde{\boldsymbol{P}}$, i.e., $\tilde{\boldsymbol{P}}(8) = \tilde{\boldsymbol{P}}(7)$.

This state of the equation system (5.51) will be maintained in the future; it is the steady state. It is worth noting that in this example the steady state was already established at the seventh sampling point. The standard deviations of the steady state position and speed errors are

$$\sigma_{x_1} = \{\tilde{P}_{11}(7)\}^{1/2} = 0.87$$

$$\sigma_{x_2} = \{\tilde{P}_{22}(7)\}^{1/2} = 1.015$$

### 5.3.5 Biased Initial Values and Known Input Variables for the Observed System

Until now, it has been assumed that the random processes $\{\boldsymbol{x}(k)\}$ and $\{y(k)\}$ have zero expectation for all $k$ (see the formulation of the problem and the subsequent remark in Section 5.3.1). As soon as $E\{\boldsymbol{x}(k_0)\}$ takes known nonzero values these processes are no longer unbiased. The same applies if, in addition to the stochastic disturbance variables $\boldsymbol{v}(k)$, known input variables $\boldsymbol{u}(k)$ enter the observed system. Now the state

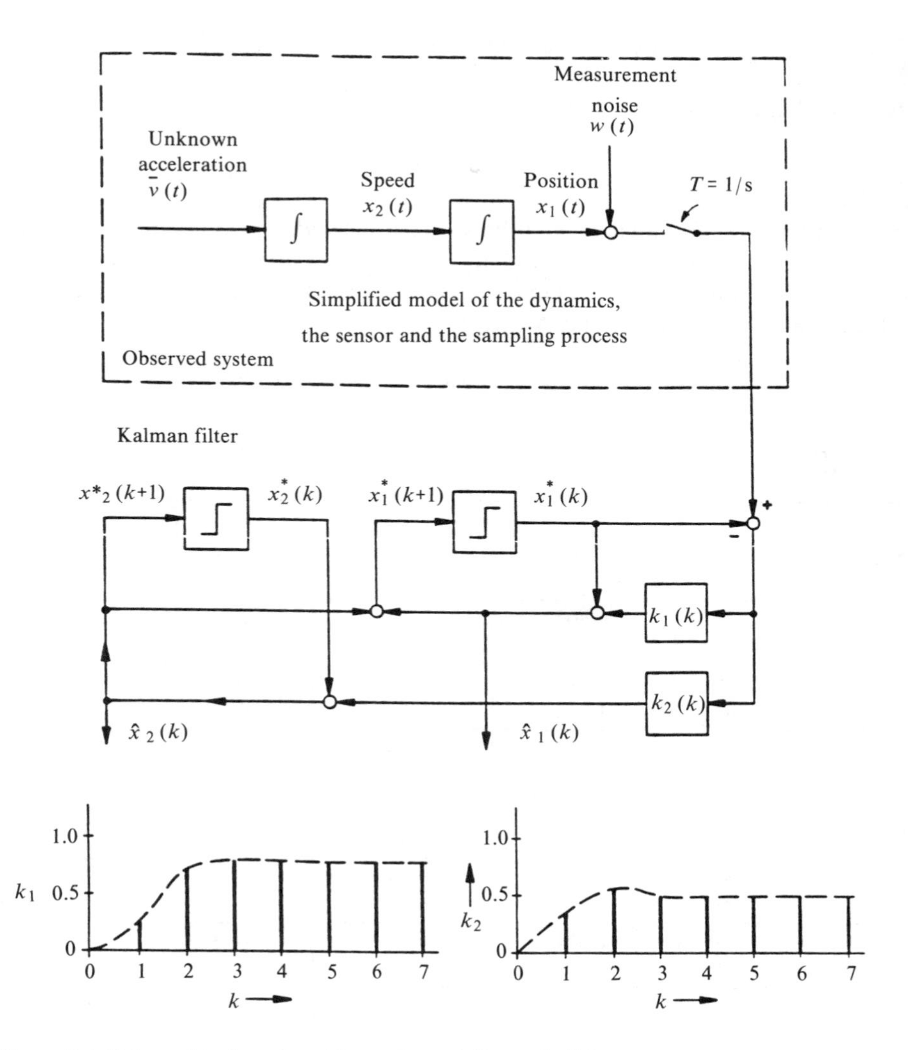


**Figure 5.4** Kalman filtering of position and speed (Example 5.2).

$\boldsymbol{x}(k)$ and the measurement variable $\boldsymbol{y}(k)$ contain deterministic known components and the estimate $\hat{\boldsymbol{x}}(k)$ must be modified accordingly.

The observation situation considered here is described by the following extended model:

$$\boldsymbol{x}(k+1) = \boldsymbol{A}(k)\boldsymbol{x}(k) + \boldsymbol{B}(k)\boldsymbol{u}(k) + \boldsymbol{v}(k) \quad k \geqslant k_0$$

$$y(k) = C(k)x(k) + w(k)$$

Let the initial condition itself or its expectation be given:

$$E\{x(k_0)\} = \xi$$

where $\xi$ can take any finite value. Let the sequence of $p$-dimensional input variables $u(k_0)$, $u(k_0 + 1), \ldots, u(k)$ be known exactly. Otherwise, the same assumptions apply as in the formulation of the problem in Section 5.3.1. We again seek a linear unbiased minimum variance estimate for $x(k)$.

**Table 5.2** Results of Algorithm (5.51) for Example 5.2

| $k$ | $P^*$ | | $k$ | $\tilde{P}$ | |
|---|---|---|---|---|---|
| 0 | 0<br>0 | 0<br>0 | 0<br>0 | 0<br>0 | 0<br>0 |
| 1 | 0.333<br>0.500 | 0.500<br>1.000 | 0.250<br>0.375 | 0.250<br>0.375 | 0.375<br>0.812 |
| 2 | 2.145<br>1.687 | 1.687<br>1.812 | 0.682<br>0.536 | 0.682<br>0.536 | 0.536<br>0.908 |
| 3 | 2.995<br>1.944 | 1.944<br>1.908 | 0.750<br>0.485 | 0.750<br>0.485 | 0.485<br>0.964 |
| 4 | 3.017<br>1.949 | 1.949<br>1.964 | 0.751<br>0.485 | 0.751<br>0.485 | 0.485<br>1.019 |
| 5 | 3.073<br>2.004 | 2.004<br>2.019 | 0.755<br>0.493 | 0.755<br>0.493 | 0.493<br>1.031 |
| 6 | 3.105<br>2.024 | 2.024<br>2.031 | 0.756<br>0.493 | 0.756<br>0.493 | 0.493<br>1.031 |
| 7 | 3.106<br>2.024 | 2.024<br>2.031 | 0.756<br>0.493 | 0.756<br>0.493 | 0.493<br>1.031 |

It was established in Chapter 4 that the effect of known variables on the state $x(k)$ and the measurement variable $y(k)$ can be calculated separately because of the linearity of the observed system. This "calculation" could of course be done by using a deterministic model of the system. The state and output of this model will be denoted by the subscript 0:

$$x_0(k + 1) = A(k)x_0(k) + B(k)u(k) \quad x_0(k_0) = \xi \tag{5.57a}$$

$$\boldsymbol{y}_0(k) = \boldsymbol{C}(k)\boldsymbol{x}_0(k) \tag{5.57b}$$

On filtering, we would use the unbiased value $\boldsymbol{y}(k) - \boldsymbol{y}_0(k)$ instead of $\boldsymbol{y}(k)$ to estimate the unknown and unbiased random variable

$$\boldsymbol{x}_1(k) := \boldsymbol{x}(k) - \boldsymbol{x}_0(k) \tag{5.58}$$

Then the known value $\boldsymbol{x}_0(k)$ would have to be added to the unbiased estimate $\hat{\boldsymbol{x}}_1(k)$. As we shall immediately see, these seemingly awkward manipulations can be achieved by a simple modification of the filter equations.

For this, (5.50b) is formulated for the unbiased variables $\hat{\boldsymbol{x}}_1$ and $\boldsymbol{y} - \boldsymbol{y}_0$. At the first sampling point, with $\boldsymbol{x}^*(k_0) = \boldsymbol{0}$, we have

$$\hat{\boldsymbol{x}}_1(k_0) = \boldsymbol{K}(k_0)\{\boldsymbol{y}(k_0) - \boldsymbol{y}_0(k_0)\}$$

Addition of $\boldsymbol{x}_0(k_0)$ to both sides gives the estimate for $\boldsymbol{x}(k_0)$:

$$\hat{\boldsymbol{x}}(k_0) = \boldsymbol{x}_0(k_0) + \hat{\boldsymbol{x}}_1(k_0) = \boldsymbol{x}_0(k_0) + \boldsymbol{K}(k_0)\{\boldsymbol{y}(k_0) - \boldsymbol{C}(k_0)\boldsymbol{x}_0(k_0)\}$$

Comparison with (5.50b) shows that only $\boldsymbol{x}^*(k_0)$ has changed:

$$\boldsymbol{x}^*(k_0) = \boldsymbol{x}_0(k_0) = \boldsymbol{\xi}$$

For additional sampling points, equation (5.50b) is taken at time $k + 1$ and the variable $\boldsymbol{x}^*(k + 1)$ is eliminated by using (5.50a):

$$\begin{aligned} \hat{\boldsymbol{x}}_1(k + 1) = {} & \boldsymbol{A}(k)\hat{\boldsymbol{x}}_1(k) \\ & + \boldsymbol{K}(k + 1)\{\boldsymbol{y}(k + 1) - \boldsymbol{y}_0(k + 1) - \boldsymbol{C}(k + 1)\boldsymbol{A}(k)\hat{\boldsymbol{x}}_1(k)\} \end{aligned} \tag{5.59}$$

Addition of (5.57a) and (5.59) using $\boldsymbol{x}_0 + \hat{\boldsymbol{x}}_1 = \hat{\boldsymbol{x}}$ gives

$$\begin{aligned} \hat{\boldsymbol{x}}(k + 1) = {} & \boldsymbol{A}(k)\hat{\boldsymbol{x}}(k) + \boldsymbol{B}(k)\boldsymbol{u}(k) \\ & + \boldsymbol{K}(k + 1)\{\boldsymbol{y}(k + 1) - \boldsymbol{C}(k + 1)\, \boldsymbol{x}_0(k + 1) - \boldsymbol{C}(k + 1)\boldsymbol{A}(k)\hat{\boldsymbol{x}}_1(k)\} \end{aligned}$$

The variable $\boldsymbol{x}_0(k + 1)$ in braces is replaced by the right-hand side of (5.57a). Then terms with $\boldsymbol{x}_0$ and $\hat{\boldsymbol{x}}_1$ are combined:

$$\begin{aligned} \hat{\boldsymbol{x}}(k + 1) = {} & \{\boldsymbol{A}(k)\hat{\boldsymbol{x}}(k) + \boldsymbol{B}(k)\boldsymbol{u}(k)\} \\ & + \boldsymbol{K}(k + 1)[\boldsymbol{y}(k + 1) - \boldsymbol{C}(k + 1)\{\boldsymbol{A}(k)\hat{\boldsymbol{x}}(k) + \boldsymbol{B}(k)\boldsymbol{u}(k)\}] \end{aligned}$$

Another comparison with (5.50b) shows that the form of the equation, and in particular the gain matrix $\boldsymbol{K}$, has remained the same but that the extrapolated estimate has changed to

$$\boldsymbol{x}^*(k+1) = \boldsymbol{A}(k)\hat{\boldsymbol{x}}(k) + \boldsymbol{B}(k)\boldsymbol{u}(k)$$

Intuitively, this result is also immediately obvious. We are now in the position to formulate the following theorem (see also Figure 5.5).

*Theorem 5.2.*

(i) The algorithm for the linear unbiased minimum variance estimate for the state $\boldsymbol{x}(k)$ in the system

$$\boldsymbol{x}(k+1) = \boldsymbol{A}(k)\boldsymbol{x}(k) + \boldsymbol{B}(k)\boldsymbol{u}(k) + \boldsymbol{v}(k) \quad E\{\boldsymbol{x}(k_0)\} = \boldsymbol{\xi} \tag{5.60a}$$

$$\boldsymbol{y}(k) = \boldsymbol{C}(k)\boldsymbol{x}(k) + \boldsymbol{w}(k) \tag{5.60b}$$

for which the sample sequence $\boldsymbol{y}(k_0), \ldots, \boldsymbol{y}(k)$, the sequence of the input variables $\boldsymbol{u}(k_0), \ldots, \boldsymbol{u}(k-1)$ and *a priori* knowledge according to (5.36a)–(5.36f) are given, is

$$\boldsymbol{x}^*(k+1) = \boldsymbol{A}(k)\hat{\boldsymbol{x}}(k) + \boldsymbol{B}(k)\boldsymbol{u}(k) \quad \boldsymbol{x}^*(k_0) = \boldsymbol{\xi} \tag{5.61a}$$

$$\hat{\boldsymbol{x}}(k) = \boldsymbol{x}^*(k) + \boldsymbol{K}(k)\{\boldsymbol{y}(k) - \boldsymbol{C}(k)\boldsymbol{x}^*(k)\} \tag{5.61b}$$

(ii) The gain matrix $\boldsymbol{K}(k)$ is determined by (5.51) with the initial condition (5.52).

*Proof.* The linearity of the estimate $\hat{\boldsymbol{x}}(k)$ is obvious. In order to prove the freedom from bias, the errors of the extrapolated and instantaneous estimates are formed. Subtraction of equation (5.61a) from (5.60a) gives

$$\boldsymbol{x}(k+1) - \boldsymbol{x}^*(k+1) = \boldsymbol{A}(k)\tilde{\boldsymbol{x}}(k) + \boldsymbol{v}(k) \tag{5.62}$$

with the initial value

$$\boldsymbol{x}(k_0) - \boldsymbol{x}^*(k_0) = \boldsymbol{x}(k_0) - \boldsymbol{\xi} \tag{5.63}$$

With equations (5.61b) and (5.60b), we obtain

$$\begin{aligned} \tilde{\boldsymbol{x}}(k) = \boldsymbol{x}(k) - \hat{\boldsymbol{x}}(k) &= \boldsymbol{x}(k) - \boldsymbol{x}^*(k) - \boldsymbol{K}(k)\{\boldsymbol{C}(k)\boldsymbol{x}(k) \\ &\quad + \boldsymbol{w}(k) - \boldsymbol{C}(k)\boldsymbol{x}^*(k)\} \\ &= \{\boldsymbol{I} - \boldsymbol{K}(k)\boldsymbol{C}(k)\}\{\boldsymbol{x}(k) - \boldsymbol{x}^*(k)\} - \boldsymbol{K}(k)\boldsymbol{w}(k) \end{aligned} \tag{5.64}$$

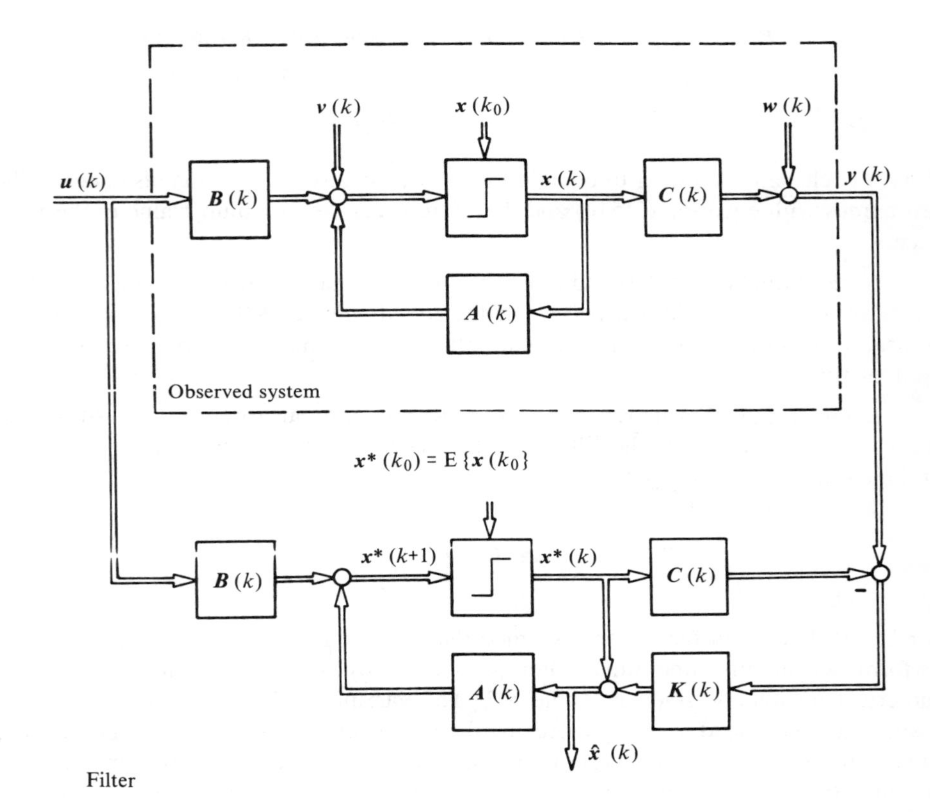


**Figure 5.5** Observed system with measurable input variables, e.g., control variables, and the Kalman filter (the area enclosed by the broken line is inaccessible).

We form the expectations of equations (5.63), (5.64) and (5.62) and go from $k$ to $k + 1$ by induction. In this way, it can be shown that the estimation errors $\boldsymbol{x}(k) - \boldsymbol{x}^*(k)$ and $\tilde{\boldsymbol{x}}(k)$ have zero expectation for all $k$ regardless of how large $\boldsymbol{\xi}$ and $\boldsymbol{u}$ are. The estimation is therefore unbiased.

According to (5.62), the covariance matrix of $\boldsymbol{x}(k + 1) - \boldsymbol{x}^*(k + 1)$ has the value

$$\boldsymbol{P}^*(k + 1) = \boldsymbol{A}(k)\tilde{\boldsymbol{P}}(k)\boldsymbol{A}'(k) + \boldsymbol{Q}(k) \tag{5.65}$$

Its initial value formed from equation (5.63) is equal to $\boldsymbol{P}(k_0)$ and its subsequent course is governed by (5.51a). According to equation (5.64), the covariance matrix of $\tilde{\boldsymbol{x}}(k)$ is given by

$$\tilde{\boldsymbol{P}}(k) = \{\boldsymbol{I} - \boldsymbol{K}(k)\boldsymbol{C}(k)\}\boldsymbol{P}^*(k)\{\boldsymbol{I} - \boldsymbol{C}'(k)\boldsymbol{K}'(k)\} + \boldsymbol{K}(k)\boldsymbol{R}(k)\boldsymbol{K}'(k)$$
$$= \boldsymbol{P}^*(k) - \boldsymbol{K}(k)\boldsymbol{C}(k)\boldsymbol{P}^*(k) - \boldsymbol{P}^*(k)\boldsymbol{C}'(k)\boldsymbol{K}'(k)$$
$$+ \boldsymbol{K}(k)\{\boldsymbol{C}(k)\boldsymbol{P}^*(k)\boldsymbol{C}'(k) + \boldsymbol{R}(k)\}\boldsymbol{K}'(k) \tag{5.66}$$

If $\boldsymbol{K}(k)$ is chosen according to equation (5.51b), the last two summands cancel. The rest agrees with equation (2.51c) and therefore gives the minimum value of the variance.

*Note*. Equations (5.65) and (5.66) for the covariance matrices of the estimation errors are generally valid irrespective of the actual value of $\boldsymbol{K}(k)$. They can therefore be used to determine the quality of a simplified suboptimal filter compared with the optimal filter.

*Example 5.3*. Let us return to Example 5.2 (Kalman filtering of position and speed). Now after $k = 1$, the filter gain factors $k_1$ and $k_2$ are set constant and equal to their steady state in order to simplify the design:

$$\boldsymbol{k} = \begin{bmatrix} 3/4 \\ 1/2 \end{bmatrix}$$

For $k = 0$, both gain factors are assigned the optimal value zero (open switch). The performance of this suboptimal filter is assessed by using (5.65) and (5.66). The numerical results are given in Table 5.3. The variances of the estimation errors of position $x_1$ and speed $x_2$ are plotted in Figure 5.6 and compared with the optimal values. It can be seen that the optimal variances are naturally always smaller than the suboptimal values. However, the difference is only important in the first interval and almost completely disappears after a few sampling points.

**Table 5.3** Variances of the Suboptimal Filter (Example 5.3)

| $k$ | $P^*$ | | $\tilde{P}$ | |
|---|---|---|---|---|
| 0 | 0<br>0 | 0<br>0 | 0<br>0 | 0<br>0 |
| 1 | 0.333<br>0.500 | 0.500<br>1.000 | 0.583<br>0.458 | 0.458<br>0.833 |
| 2 | 2.665<br>1.791 | 1.791<br>1.833 | 0.727<br>0.491 | 0.491<br>0.960 |
| 3 | 3.002<br>1.951 | 1.951<br>1.960 | 0.751<br>0.485 | 0.485<br>1.008 |

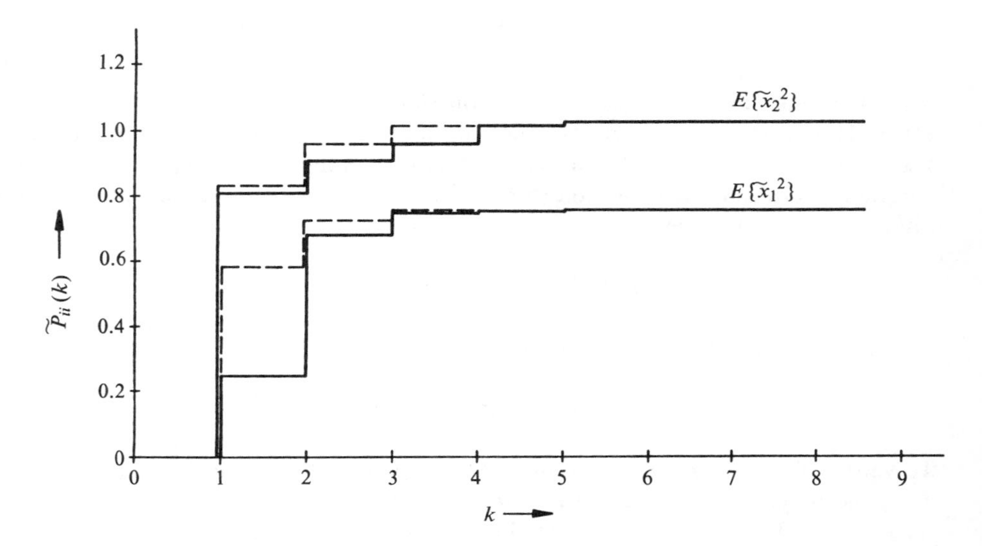


**Figure 5.6** Variances of the estimation errors of position $x_1$ and speed $x_2$: ——, optimal values (Example 5.2); ---, suboptimal values (Example 5.3).

### 5.3.6 Prediction (extrapolation)

We return to the original formulation of the problem in Section 5.3.1. In particular, let the initial condition $\boldsymbol{x}(k_0)$ again be unbiased and the input sequence $\boldsymbol{u}(k)$ be absent so that the processes $\{\boldsymbol{x}(k)\}$ and $\{\boldsymbol{y}(k)\}$ have zero expectation. We now require the linear unbiased minimum variance estimate of the *future* state vector $\boldsymbol{x}(K)$ with $K > k$. It is to be computed on the basis of the data $\boldsymbol{y}(k_0), \ldots, \boldsymbol{y}(k)$ ($k$ present). The required estimate is denoted by $\hat{\boldsymbol{x}}(K|k)$. This estimate must also satisfy the discrete Wiener–Hopf equation (5.14). As in Section 5.3.3 it is used in the version (5.16). We adapt this condition to the prediction problem which we have here and obtain

$$E\{[\boldsymbol{x}(K) - \hat{\boldsymbol{x}}(K|k)][\boldsymbol{y}'(k_0), \boldsymbol{y}'(k_0 + 1) \ldots \boldsymbol{y}'(k)]\} = \boldsymbol{0} \tag{5.67}$$

Note that for the estimate extrapolated by one sampling point we have the identity

$$\hat{\boldsymbol{x}}(k + 1|k) \equiv \boldsymbol{x}^*(k + 1)$$

This estimate has already been determined from (5.45b). The approach used there suggests reducing all subsequent predictions to $\boldsymbol{x}^*(k + 1)$ which meanwhile has been determined. For this, we express $\boldsymbol{x}(K)$ in terms of $\boldsymbol{x}(k + 1)$ and the input sequence $\boldsymbol{v}(k + 1), \ldots, \boldsymbol{v}(K - 1)$ with the help of the general solution equation (4.55d):

$$\boldsymbol{x}(K) = \boldsymbol{\phi}(K, k+1)\boldsymbol{x}(k+1) + \sum_{\kappa=k+1}^{K-1} \boldsymbol{\phi}(K, \kappa+1)\, \boldsymbol{v}(\kappa) \tag{5.68}$$

The given data $\boldsymbol{y}(k_0), \ldots, \boldsymbol{y}(k)$ in equation (5.67) depend only on $\boldsymbol{x}(k_0)$, $\boldsymbol{v}(k_0), \ldots, \boldsymbol{v}(k-1)$ and $\boldsymbol{w}(k_0), \ldots, \boldsymbol{w}(k)$. The data are therefore not correlated with the values $\boldsymbol{v}(k+1), \ldots, \boldsymbol{v}(K-1)$ under the summation in equation (5.68). Therefore, by substituting (5.68) into the condition (5.67), the entire sum can be cancelled. The condition therefore becomes

$$E\{[\boldsymbol{\phi}(K, k+1)\boldsymbol{x}(k+1) - \hat{\boldsymbol{x}}(K|k)][\boldsymbol{y}'(k_0), \boldsymbol{y}'(k_0+1) \ldots \boldsymbol{y}'(k)]\} = \boldsymbol{0}$$

This rule can be satisfied by choosing

$$\hat{\boldsymbol{x}}(K|k) = \boldsymbol{\phi}(K, k+1)\boldsymbol{x}^*(k+1)$$

To verify this the common factor $\boldsymbol{\phi}(K, k+1)$ is brought in front of the expectation. Then the remaining expectation is identical with the statement (5.45b) which is already valid. The solution of the prediction problem is therefore given by a simple extension to the filter equations. This is summarized as follows.

*Theorem 5.3*. The linear unbiased minimum variance estimate for the future state $\boldsymbol{x}(K)$, $K > k$, of the discrete-time stochastically disturbed system

$$\boldsymbol{x}(k+1) = \boldsymbol{A}(k)\boldsymbol{x}(k) + \boldsymbol{v}(k) \quad E\{\boldsymbol{x}(k_0)\} = \boldsymbol{0}$$
$$\boldsymbol{y}(k) = \boldsymbol{C}(k)\boldsymbol{x}(k) + \boldsymbol{w}(k)$$

for which the sample sequence $\boldsymbol{y}(k_0), \ldots, \boldsymbol{y}(k)$ and *a priori* knowledge according to (5.36a)–(5.36f) are available, is given by

$$\hat{\boldsymbol{x}}(K|k) = \boldsymbol{\phi}(K, k+1)\boldsymbol{x}^*(k+1) \tag{5.69}$$

where $\boldsymbol{\phi}$ represents the transition matrix to $\boldsymbol{A}(k)$ and $\boldsymbol{x}^*(k+1)$ is to be calculated by using the recursion algorithm of Theorem 5.1.

Of course, nonzero control variables and biased initial conditions can be accounted for as in Theorem 5.2. However, for optimal prediction, the present and future control variables $\boldsymbol{u}(k)$, $\boldsymbol{u}(k+1), \ldots, \boldsymbol{u}(K-1)$ must be known (or vanish). In this connection, prediction by a single interval is of special interest and is very popular for didactic purposes. Here, apart from the past values of the control variable, only its present value $\boldsymbol{u}(k)$ is needed and this can always be regarded as given. This result is already implicitly contained in Theorem 5.2. However, it can be formulated more elegantly by substitution of equation (5.61b) in (5.61a):

$$\boldsymbol{x}^*(k+1) = \boldsymbol{A}(k)[\boldsymbol{x}^*(k) + \boldsymbol{K}(k)\{\boldsymbol{y}(k) - \boldsymbol{C}(k)\boldsymbol{x}^*(k)\}] + \boldsymbol{B}(k)\boldsymbol{u}(k)$$

With the conventions

$$\begin{aligned} \boldsymbol{x}^*(k) &:= \hat{\boldsymbol{x}}(k|k-1) \\ \boldsymbol{K}^*(k) &:= \boldsymbol{A}(k)\boldsymbol{K}(k) \end{aligned} \tag{5.70}$$

this equation yields the solution

$$\begin{aligned} \hat{\boldsymbol{x}}(k+1|k) = {} & \boldsymbol{A}(k)\hat{\boldsymbol{x}}(k|k-1) \\ & + \boldsymbol{B}(k)\boldsymbol{u}(k) + \boldsymbol{K}^*(k)\{\boldsymbol{y}(k) - \boldsymbol{C}(k)\hat{\boldsymbol{x}}(k|k-1)\} \end{aligned} \tag{5.71}$$

This result is frequently quoted. It provides a complete analogy to the model of the observed system on the one hand (Figure 5.7) and to the filter for continuous time on the other hand (see Section 5.4 and Figure 5.8). As before, the matrix $\boldsymbol{K}(k)$ and thus the new gain factor $\boldsymbol{K}^*(k)$ are determined using (5.51b). Let us substitute the $\boldsymbol{K}(k)$ from (5.51b) in (5.51c). Further, let us substitute the result in (5.51a). We then obtain a difference equation for the covariance matrix $\boldsymbol{P}^*(k)$ of the error $\tilde{\boldsymbol{x}}(k|k-1)$:

$$\begin{aligned} \boldsymbol{P}^*(k+1) = {} & \boldsymbol{A}(k)\boldsymbol{P}^*(k)\boldsymbol{A}'(k) \\ & - \boldsymbol{A}(k)\boldsymbol{P}^*(k)\boldsymbol{C}'(k)\{\boldsymbol{C}(k)\boldsymbol{P}^*(k)\boldsymbol{C}'(k) + \boldsymbol{R}(k)\}^{-1} \\ & \times \boldsymbol{C}(k)\boldsymbol{P}^*(k)\boldsymbol{A}'(k) + \boldsymbol{Q}(k) \\ \boldsymbol{P}^*(k_0) = {} & \boldsymbol{P}(k_0) \end{aligned} \tag{5.72}$$

This is the discrete-time counterpart of the famous matrix Riccati differential equation in continuous-time filtering.

### 5.3.7 Summary and Final Remarks

The Kalman filter and prediction algorithm for discrete time has been constructed in stages for increasing amounts of statistical *a priori* knowledge.

(i) The starting point was Section 4.6 where the deterministic observation problem was considered. Its solution was represented there as a special case of the *Gaussian least-squares method*. It was subsequently shown that this formulation of the problem corresponds to a situation in which there is no *a priori* knowledge. First, the initial values of the state variables have zero expectation and infinite standard deviation. Second, the disturbance variables at the input of the observed system are neglected. Third, all the measurement errors at the output are considered as unbiased, normalized and mutually uncorrelated. In mathematical terms

$$E\{\boldsymbol{x}(k_0)\} = \boldsymbol{0} \quad \boldsymbol{P}^{-1}(k_0) = \boldsymbol{0}$$

$$\boldsymbol{Q} \equiv \boldsymbol{0} \quad \boldsymbol{R}(k) \equiv \boldsymbol{I}$$

(ii) The next stage was the *Gauss–Markov method*. As before, no assumptions were made with respect to the initial state. However, *a priori* statistical knowledge of the disturbance and measurement noise can be taken into account. This is possible provided that they are mutually independent white random processes with given covariance matrices. In mathematical terms

$$E\{\boldsymbol{x}(k_0)\} = \boldsymbol{0} \quad \boldsymbol{P}^{-1}(k_0) = \boldsymbol{0}$$

with $\boldsymbol{Q}(k)$ and $\boldsymbol{R}(k)$ suitably defined.

(iii) The third stage was the *minimum variance method*. As a generalization of the previous methods it allows *a priori* knowledge of the initial state $\boldsymbol{x}(k_0)$ to be taken into account. This is done by specifying its covariance matrix. In mathematical terms

$$E\{\boldsymbol{x}(k_0)\} = \boldsymbol{0}$$

with $\boldsymbol{P}(k_0)$, $\boldsymbol{Q}(k)$ and $\boldsymbol{R}(k)$ suitably defined.

(iv) Finally, the minimum variance method was extended to biased initial values and known input variables:

$$E\{\boldsymbol{x}(k_0)\} \quad \text{and} \quad \boldsymbol{u}(k_0), \boldsymbol{u}(k_0 + 1), \ldots, \boldsymbol{u}(k - 1) \quad \text{arbitrary}$$

(v) In addition, the recursive minimum variance filter was augmented by a formula for prediction.

In this way, the Kalman filter problem was solved in its basic form. The remaining questions are as follows.

(a) *Correlation between disturbance and measurement process:*

$$E\{\boldsymbol{v}(k)\boldsymbol{w}'(k)\} \neq \boldsymbol{0}$$

The filter algorithm can be generalized to this case without any fundamental difficulty. However, a few extra terms appear in the evaluation of the Wiener–Hopf equation (Brammer, 1967; Sage and Melsa, 1971).

(b) *Colored disturbance and measurement noises:* These are dealt with by setting up a suitable shaping filter. This has the form of a vector difference equation which is excited by white noise. The model of this shaping filter is added to the original model of the observed system. In doing so, the original state vector is augmented by the state variables of the shaping filter, and the disturbance vector is

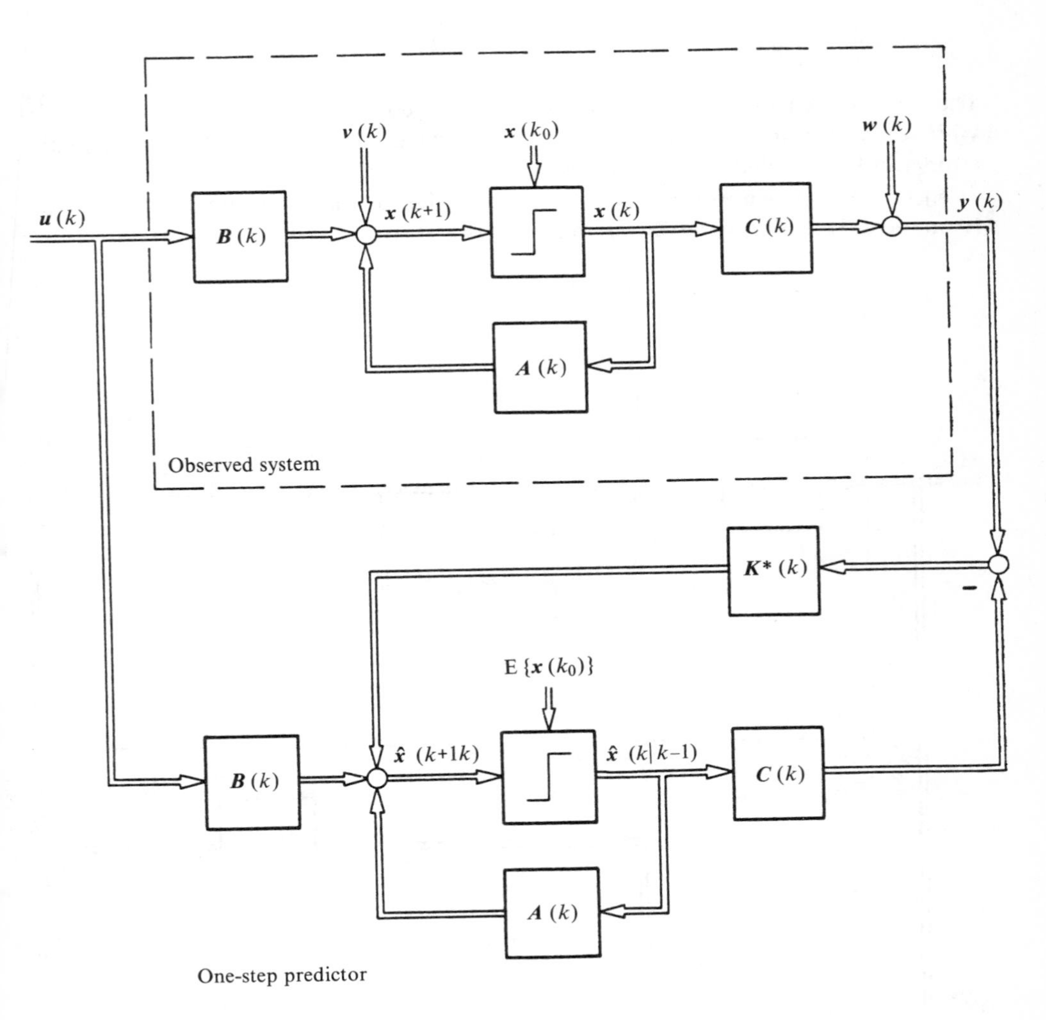


**Figure 5.7** Prediction by one interval: $\hat{\boldsymbol{x}}(k/k-1) = \boldsymbol{x}^*(k)$, $\boldsymbol{K}^*(k) = \boldsymbol{A}(k)\boldsymbol{K}(k)$ (compare Figure 5.5).

augmented by the exciting variables. The matrices $\boldsymbol{A}$, $\boldsymbol{C}$ and $\boldsymbol{Q}$ are enlarged accordingly. The synthesis of the shaping filter from a given matrix covariance function is described in Section 6.6.

(c) *Systematic disturbances and measurement errors:* In a similar manner to the case of colored noise, this situation is handled by defining extra state variables. However, these $x_i$ are now time invariant; i.e., their exciting variables $v_i$ are identically zero. Thus the *shaping filter* here is given by

$$x_i(k+1) = x_i(k) \quad i > n$$

The matrices $\boldsymbol{A}$ and $\boldsymbol{C}$ are correspondingly enlarged and the matrix $\boldsymbol{Q}$ is completed with zeros. *A priori* knowledge of the mean value and variance of the systematic errors can be specified and included in the normal manner: $E\{x_i(k_0)\} = \xi_i$ and $E\{[x_i(k_0) - \xi_i]^2\} = p_{ii}(k_0)$. An example for this problem has already been treated above (see Example 5.1 dealing with sensor calibration).

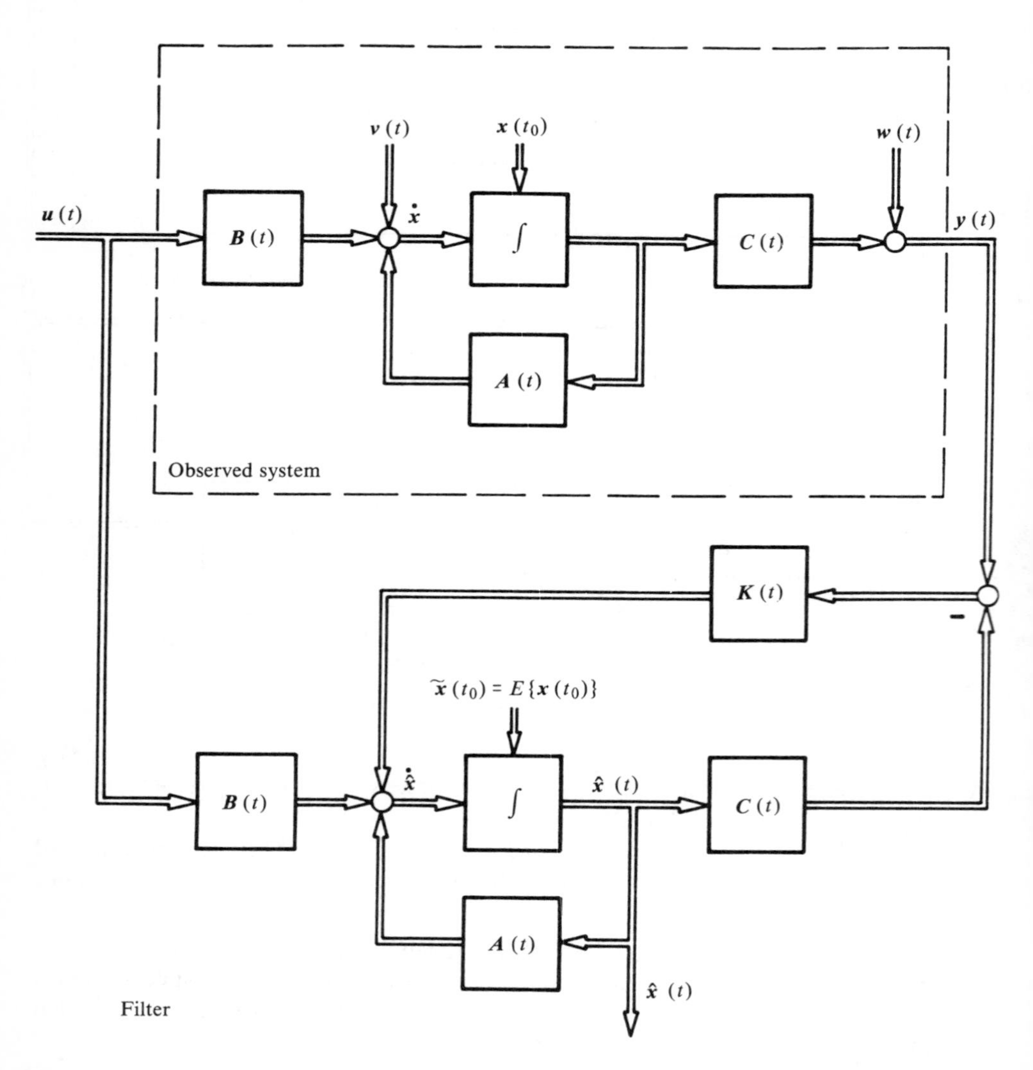


**Figure 5.8** Observed system in continuous time and Kalman–Bucy filter (the area enclosed by the broken line is inaccessible).

At this point it must be said that the augmentation of the state vector for colored noise and systematic errors often makes the main contribution to the computational complexity of the filter problem. For example, let us assume that the position and speed of a vehicle are being measured. Both measurements have systematic and colored errors. In addition, the vehicle is subject to unknown colored disturbance acceleration. The augmented model of this observation situation then provides the following state variables:

position $x_1$ } original state variables per dimension
speed $x_2$ }

colored acceleration $x_3$

colored position measurement error $x_4$

colored speed measurement error $x_5$

systematic position measurement error (bias) $x_6$

systematic speed measurement error $x_7$

Note that the order of the problem has increased from two to seven in each dimension! Nevertheless, the simplest possible model was assumed for all colored noise; in each case the shaping filter was only of order unity.

(d) *Singular **R** matrix:* The covariance matrix of the measurement errors is singular if one or more of its rows, and the corresponding columns, vanish or are linearly dependent on each other. In the first case we have exact measurements, i.e., $w_i = 0$. In the second case some elements of $z$ are perfectly correlated. For example, let us consider the three-component measurement equation

$$\begin{bmatrix} y_1 \\ y_2 \\ y_3 \end{bmatrix} = \begin{bmatrix} \boldsymbol{c}^1 \\ \boldsymbol{c}^2 \\ \boldsymbol{c}^3 \end{bmatrix} \boldsymbol{x} + \begin{bmatrix} w_1 \\ \alpha w_1 \\ 0 \end{bmatrix}$$

Here, the $\boldsymbol{c}^i$ are the rows of $\boldsymbol{C}$ and $\alpha$ is a given coefficient. Now we have

$$\boldsymbol{R} = E\left\{ \begin{bmatrix} w_1 \\ \alpha w_1 \\ 0 \end{bmatrix} [w_1, \alpha w_1, 0] \right\} = \begin{bmatrix} r_{11} & \alpha r_{11} & 0 \\ \alpha r_{11} & \alpha^2 r_{11} & 0 \\ 0 & 0 & 0 \end{bmatrix}$$

The matrix $\boldsymbol{R}$ in this example therefore has rank 1.

In the case of perfectly correlated measurements, exact measurement values can be obtained by suitable linear combinations. In the above example, $y_1$ and $y_2$ are perfectly correlated and a suitable linear combination consists of the difference between $\alpha y_1$ and $y_2$. We have

$$\alpha y_1 - y_2 = (\alpha \boldsymbol{c}^1 - \boldsymbol{c}^2)\boldsymbol{x} + 0 := y_2^0$$

Suppose that in this example we use the set of measurements $y_1$, $y_2^0$ and $y_3$, and that the second row of the measurement matrix $\boldsymbol{C}$ is correspondingly altered. Then we only have a single measurement with errors and two exact measurements.

The performance of the Kalman filter algorithm is not affected by the singularity of $\boldsymbol{R}$ as long as the sum $\boldsymbol{R} + \boldsymbol{C}\boldsymbol{P}^*\boldsymbol{C}'$ remains nonsingular (see equation (5.51b)). However, the order of the Kalman filter is now unnecessarily high. It can be shown that the order can be reduced by the number of exact measurements as in the case of the Luenberger observer (Section 4.2 and Luenberger, 1966). Corresponding studies have been published for continuous time (Bryson and Johansen, 1965; Bucy, 1967) and discrete time (Bryson and Henrikson, 1967; Brammer, 1968a,b; Sage and Melsa, 1971). We will return to this in Section 6.7.

(e) *Smoothing (interpolation):* This is understood as the estimation of past values of the state on the basis of measurements made up to the present time; thus we require the estimate $\hat{\boldsymbol{x}}(K|k)$ with $K < k$. The case of smoothing is considerably more complicated than pure filtering although the approach is similar in principle. There are now considerably more terms which do not vanish in the Wiener–Hopf equation. Results are available in the literature (Mayne, 1966; Meditch, 1967; Sage and Melsa, 1971).

## 5.4 THE KALMAN–BUCY FILTER (CONTINUOUS TIME)

Mathematically speaking, filtering and prediction in continuous time is a more advanced problem than filtering in discrete time. This is why we dealt first with the discrete-time filter problem. It could be formulated and solved with elementary concepts taken from probability theory and pure algebraic methods. Conceptually, a vector random process in discrete time can be simply interpreted as a compound multidimensional random variable. Then its compound dimension is equal to the number of vector elements multiplied by the number of points in time considered. Consequently, the treatment of stochastic difference equations and the time evolution of the covariances is also elementary. White noise does not cause any problems in discrete time.

On transferring to continuous time, difference equations and sums are replaced by differential equations and integrals. The conceptual specification of a random process is more problematic for continuous time because the number of points in time considered becomes a continuum (noncountable infinity). The stochastic differential equations are particularly tricky; strictly speaking, they must be handled mathematically using Ito calculus (Doob, 1953; Bucy and Joseph, 1968; Jazwinski, 1970; Sage and Melsa, 1971). Ideal white noise in continuous time is a fiction; it has infinite power. Only its integral, Brownian motion or the Wiener process exists in the mathematical and physical sense.

Therefore if the filter problem for continuous time is to be formulated and solved with complete mathematical rigor, this has to be done in terms of Ito differentials and Brownian motion. However, few readers of this kind of book are acquainted with this representation. Fortunately, in the linear case, which we are mainly considering here, this approach can easily be avoided if certain precautions are taken. Therefore, in what follows, we will retain the conventional notation with ordinary differential equations and white noise.

Just as with the observation problem in Chapter 4, we will see in the filtering problem that the continuous-time case has strong parallels with the discrete-time case. We will emphasize these parallels once more by using suitable notation. In particular, the same symbols will be chosen for corresponding discrete-time and continuous-time matrices in the model of the observed system and in the filter algorithm. If discrete-time and continuous-time relationships occur anywhere in close proximity, the risk of confusion will be eliminated as before by placing bars over one of the two sets of symbols.

### 5.4.1 Formulation of the Problem

By analogy with the discrete version of the problem, we first consider a linear system without known input variables or in which their effect has already been eliminated (this precondition will be lifted later).

We will again consider stochastic disturbances at the input and stochastic measurement errors at the output. Thus the model of the observed system is (see upper part of Figure 5.8 with $\boldsymbol{Bu} \equiv \boldsymbol{0}$)

$$\dot{\boldsymbol{x}}(t) = \boldsymbol{A}(t)\boldsymbol{x}(t) + \boldsymbol{v}(t) \quad t \geqslant t_0 \tag{5.73a}$$

$$\boldsymbol{y}(t) = \boldsymbol{C}(t)\boldsymbol{x}(t) + \boldsymbol{w}(t) \tag{5.73b}$$

The state vector $\boldsymbol{x}(t)$ to be estimated is $n$-dimensional. Let its initial value $\boldsymbol{x}(t_0)$ be a random vector with zero expectation (this precondition will be lifted later) and given covariance matrix $\boldsymbol{P}(t_0)$:

$$E\{[\boldsymbol{x}(t_0) - E\{\boldsymbol{x}(t_0)\}][\boldsymbol{x}(t_0) - E\{\boldsymbol{x}(t_0)\}]'\} = \boldsymbol{P}(t_0) \tag{5.74a}$$

The measurement variable $\boldsymbol{y}(t)$ has $m$ components. The elements $a_{ik}(t)$ and $c_{ik}(t)$ of the matrices $\boldsymbol{A}$ and $\boldsymbol{C}$ are known continuous functions of time $t$.

The $n$-dimensional disturbance $\boldsymbol{v}(t)$ and the $m$-component measurement error $\boldsymbol{w}(t)$ are vector-valued white random processes with given covariance matrices (the precondition that the processes are white can be lifted (Bryson and Johansen, 1965; Bucy, 1967; Sage and Melsa, 1971)):

$$E\{\boldsymbol{v}(t)\} \equiv \boldsymbol{0} \quad E\{\boldsymbol{v}(t)\boldsymbol{v}'(\tau)\} = \boldsymbol{Q}(t)\delta(t-\tau) \tag{5.74b}$$

$$E\{\boldsymbol{w}(t)\} \equiv \boldsymbol{0} \quad E\{\boldsymbol{w}(t)\boldsymbol{w}'(\tau)\} = \boldsymbol{R}(t)\delta(t-\tau) \tag{5.74c}$$

where $\delta(t - \tau)$ is the Dirac $\delta$ function. $\boldsymbol{Q}$ and $\boldsymbol{R}$ are symmetric and their elements are continuous functions of $t$. $\boldsymbol{Q}$ is positive semidefinite, but $\boldsymbol{R}$ must be positive definite (nonsingular) (this precondition can be lifted (Bryson and Johansen, 1965; Bucy, 1967; Sage and Melsa, 1971)). Let the initial state, the disturbance process and the measurement process be mutually uncorrelated:

$$E\{\boldsymbol{x}(t_0)\boldsymbol{v}'(t)\} \equiv \boldsymbol{0} \tag{5.74d}$$

$$E\{\boldsymbol{x}(t_0)\boldsymbol{w}'(t)\} \equiv \boldsymbol{0} \tag{5.74e}$$

$$E\{\boldsymbol{v}(t)\boldsymbol{w}'(\tau)\} \equiv \boldsymbol{0} \tag{5.74f}$$

We seek a *linear unbiased* estimate of $\boldsymbol{x}(t)$ based on the sample function $\boldsymbol{y}(\tau)$, $t_0 \leqslant \tau \leqslant t$. This estimate is denoted by $\hat{\boldsymbol{x}}(t)$. The estimation error is defined as usual:

$$\tilde{\boldsymbol{x}}(t) := \boldsymbol{x}(t) - \hat{\boldsymbol{x}}(t) \tag{5.75}$$

For unbiased estimation, its covariance matrix is defined by

$$\tilde{\boldsymbol{P}}(t) := E\{\tilde{\boldsymbol{x}}(t)\tilde{\boldsymbol{x}}'(t)\} \tag{5.76}$$

The estimate must be optimal in the sense that the components of the estimation error have *minimum variances*. This is equivalent to the requirement

$$\mathrm{tr}\tilde{\boldsymbol{P}}(t) \rightarrow \text{minimum}$$

### 5.4.2 The Matrix Wiener–Hopf Equation

In what follows, a necessary and sufficient condition will be derived for the required linear unbiased *minimum variance* estimate. In Section 5.2 for the discrete-time case, this estimate had the form $\boldsymbol{Gz}$ (see equation (5.11)). By a suitable partitioning of the weighting matrix $\boldsymbol{G}$ and the accumulated observations $\boldsymbol{z}$, it can be brought into an equivalent summation form

$$\hat{\boldsymbol{x}}(K|k) = \boldsymbol{G}(K, k)\boldsymbol{z} := \sum_{\kappa=k_0}^{k} \bar{\boldsymbol{G}}(K, \kappa)\boldsymbol{y}(\kappa)$$

In continuous time an equivalent approach is adopted for the estimate of the state $\boldsymbol{x}(T)$ based on the measurements up to time $t$:

$$\hat{\boldsymbol{x}}(T|t) = \int_{t_0}^{t} \boldsymbol{G}(T, \tau)\boldsymbol{y}(\tau)\mathrm{d}\tau \quad T, t \geqslant t_0 \tag{5.77}$$

where $[t_0, t]$ is the observation interval, $T < t$ means smoothing, $T = t$ means pure filtering and $T > t$ means prediction. The matrix weighting function $\boldsymbol{G}(t, \tau)$ must be $n \times m$, real and continuous. Otherwise it is arbitrary.

The chosen form of (5.77) is obviously linear. We next verify that the estimate is unbiased. For this, we note that the process $\{\boldsymbol{x}(t)\}$ which is governed by the differential equation (5.73a) has zero expectation for all $t$ since $E\{\boldsymbol{x}(t_0)\} = \boldsymbol{0}$ and $E\{\boldsymbol{v}(t)\} \equiv \boldsymbol{0}$. Furthermore, because we have $E\{\boldsymbol{w}(t)\} \equiv \boldsymbol{0}$, the process $\{\boldsymbol{y}(t)\}$ according to (5.73b) also has zero expectation, i.e., $E\{\boldsymbol{y}(\tau)\} = \boldsymbol{0}$ for all $\tau \in [t_0, t]$. Thus the expectation of the integral in equation (5.77) vanishes and we have $E\{\hat{\boldsymbol{x}}(T|t)\} = \boldsymbol{0} = E\{\boldsymbol{x}(T)\}$. The estimate is therefore indeed unbiased.

Note that in principle, equation (5.77) has the same form as the observation law (4.17). There the weighting function is given by $\boldsymbol{G}(t_1, \tau) = \boldsymbol{M}^{-1}(t_1, t_0)\boldsymbol{\phi}'(\tau, t_1)\boldsymbol{C}'(\tau)$. In both cases, the problem is only meaningfully formulated if the observed system is completely observable in the considered interval.

In order to find a condition for the optimal value of the weight function $\boldsymbol{G}(T, \tau)$, we adopt a procedure which some readers will already know from Wiener filter theory (Wiener, 1949, appendix by Levinson). The weighting function is divided into two summands:

$$\boldsymbol{G}(T, \tau) = \boldsymbol{G}_0(T, \tau) + \lambda\boldsymbol{G}_1(T, \tau) \tag{5.78}$$

Here $\boldsymbol{G}_0$ is the required optimal function while $\lambda$ and the elements of $\boldsymbol{G}_1$ can take any real continuous values. The estimate now also consists of two parts:

$$\hat{\boldsymbol{x}}(T|t) = \hat{\boldsymbol{x}}_0(T|t) + \lambda\hat{\boldsymbol{x}}_1(T|t)$$

where $\hat{\boldsymbol{x}}_0$ and $\hat{\boldsymbol{x}}_1$ are to be formed according to (5.77) with $\boldsymbol{G}_0$ and $\boldsymbol{G}_1$ respectively. The estimation error therefore becomes

$$\begin{aligned}\tilde{\boldsymbol{x}}(T|t) &:= \boldsymbol{x}(T) - \hat{\boldsymbol{x}}(T|t) = \boldsymbol{x}(T) - \hat{\boldsymbol{x}}_0(T|t) - \lambda\hat{\boldsymbol{x}}_1(T|t) \\ &= \tilde{\boldsymbol{x}}_0(T|t) - \lambda\hat{\boldsymbol{x}}_1(T|t)\end{aligned}$$

where

$$\tilde{\boldsymbol{x}}_0(T|t) := \mathbf{x}(T) - \hat{\boldsymbol{x}}_0(T|t)$$

The quality index is now (with the arguments $T$ and $t$ suppressed)

$$\begin{aligned} E\{\tilde{\boldsymbol{x}}'\tilde{\boldsymbol{x}}\} &= E\{(\tilde{\boldsymbol{x}}_0 - \lambda\hat{\boldsymbol{x}}_1)'(\tilde{\boldsymbol{x}}_0 - \lambda\hat{\boldsymbol{x}}_1)\} \\ &= E\{\tilde{\boldsymbol{x}}_0'\tilde{\boldsymbol{x}}_0\} - 2\lambda E\{\tilde{\boldsymbol{x}}_0'\hat{\boldsymbol{x}}_1\} + \lambda^2 E\{\hat{\boldsymbol{x}}_1'\hat{\boldsymbol{x}}_1\} \end{aligned} \tag{5.79}$$

Because $\boldsymbol{G}_0$ in (5.78) should be the optimal value, the index (5.79) must have its extreme value for all $\boldsymbol{G}_1$ at the point $\lambda = 0$:

$$\frac{\partial}{\partial\lambda} E\{\tilde{\boldsymbol{x}}'\tilde{\boldsymbol{x}}\}\bigg|_{\lambda=0} = -2E\{\tilde{\boldsymbol{x}}_0'\hat{\boldsymbol{x}}_1\} = 0 \tag{5.80a}$$

or

$$\begin{aligned} \mathrm{tr}E\{\tilde{\boldsymbol{x}}_0\hat{\boldsymbol{x}}_1'\} &= \mathrm{tr}E\left\{\tilde{\boldsymbol{x}}_0 \int_{t_0}^{t} \boldsymbol{y}'(\tau)\boldsymbol{G}_1'(T, \tau)\mathrm{d}\tau\right\} \\ &= \mathrm{tr}\int_{t_0}^{t} E\{\tilde{\boldsymbol{x}}_0\boldsymbol{y}'(\tau)\}\boldsymbol{G}_1'(T, \tau)\mathrm{d}\tau = 0 \end{aligned} \tag{5.80b}$$

The trace is a summation and as such can be interchanged with the integral. We now only need to consider the main diagonal of the integrand $E\{\ldots\}\boldsymbol{G}_1'$. The $i$th main element is obtained by multiplying the $i$th row of the expectation by the $i$th column of $\boldsymbol{G}_1'$ (i.e., by the $i$th row of $\boldsymbol{G}_1$ itself). Therefore we must have

$$\int_{t_0}^{t} \sum_{i=1}^{n} \sum_{j=1}^{m} E\{\tilde{\boldsymbol{x}}_0\boldsymbol{y}'(\tau)\}_{ij}[\boldsymbol{G}_1(T, \tau)]_{ij}\mathrm{d}\tau = 0 \tag{5.80c}$$

The condition must be satisfied for all $\boldsymbol{G}_1$ where $\boldsymbol{G}_1$ can be completely arbitrarily chosen. For example, let us specify the elements of $\boldsymbol{G}_1$ so that

$$E\{\ldots\}_{ij} \equiv [\boldsymbol{G}_1]_{ij} \quad \text{for all } i, j$$

Then the integral in (5.80c) extends over a sum of squares of continuous functions. For $t - t_0 > 0$, the integral is therefore always positive unless all functions over the entire integration interval are zero. The condition (5.80c) can therefore only be satisfied for all $\boldsymbol{G}_1$ if and only if all elements of the expectation vanish identically in the integration interval:

$$E\{\tilde{\boldsymbol{x}}_0(T|t)\boldsymbol{y}'(\tau)\} = \boldsymbol{0} \quad t_0 < \tau < t \tag{5.81a}$$

This result is the exact equivalent to equation (5.16). Here too, it is of central importance. It states that the optimal estimation error must not be correlated with the measurements (orthogonality condition). If we set $\tilde{\boldsymbol{x}}_0 = \boldsymbol{x} - \hat{\boldsymbol{x}}_0$ and express $\hat{\boldsymbol{x}}_0$ by means of (5.77), we obtain

$$E\left\{\left[\boldsymbol{x}(T) - \int_{t_0}^{t} \boldsymbol{G}_0(T, \sigma)\boldsymbol{y}(\sigma)\mathrm{d}\sigma\right]\boldsymbol{y}'(\tau)\right\} = \boldsymbol{0}$$

or*

$$E\{\boldsymbol{x}(T)\boldsymbol{y}'(\tau)\} - \int_{t_0}^{t} \boldsymbol{G}_0(T, \sigma)E\{\boldsymbol{y}(\sigma)\boldsymbol{y}'(\tau)\}\mathrm{d}\sigma = \boldsymbol{0} \quad t_0 < \tau < t \qquad (5.81\text{b})$$

Equation (5.81b) is the generalized form of the well-known *Wiener–Hopf integral equation* (Wiener, 1949). We are no longer restricted to scalar stationary filtering with $t_0 = -\infty$. We are now dealing with vector-valued filtering of nonstationary processes for a finite observation interval.

As has just been shown, equations (5.81a) and (5.81b) represent two equivalent forms of a necessary condition for the linear unbiased minimum variance estimate. That they are also sufficient is shown as follows. If equation (5.81a) is valid, (5.80b) is also satisfied. Proceeding further backward to (5.80a) we can see that the penultimate summand in equation (5.79) vanishes. The last summand is never negative so that the quality index takes its minimum value for $\lambda = 0$. It follows that $\boldsymbol{G}_0$ must be optimal.

Remember that condition (5.81a) corresponds to equation (5.16) for discrete observations. Following this analogy, (5.81a) is postmultiplied by $\boldsymbol{G}_0'(T, \tau)$ and integrated over $\tau$ from $t_0$ to $t$:

$$\int_{t_0}^{t} E\{\tilde{\boldsymbol{x}}_0(T|t)\boldsymbol{y}'(\tau)\}\boldsymbol{G}_0'(T, \tau)\mathrm{d}\tau = \boldsymbol{0}$$

After simple rearrangement, it follows that

$$E\left\{\tilde{\boldsymbol{x}}_0(T|t) \int_{t_0}^{t} \boldsymbol{y}'(\tau)\boldsymbol{G}_0'(T, \tau)\mathrm{d}\tau\right\} = \boldsymbol{0}$$

*If $\boldsymbol{y}$ involves white noise as, for example, in (5.73b), then $E\{\boldsymbol{y}(\sigma)\boldsymbol{y}'(\tau)\}$ contains a $\delta$ function, i.e., $\boldsymbol{R}(\sigma)\delta(\sigma - \tau)$, see (5.74c). According to this, the integral $\int_{t_0}^{t}\delta(\sigma - \tau)\mathrm{d}\sigma$ occurs in the expectation $E\{\tilde{\boldsymbol{x}}_0\boldsymbol{y}'(\tau)\}$ in (5.81a) and (5.80c). This integral is equal to unity, provided that $t_0 < \tau < t$, but jumps to the value 1/2 for $\tau = t_0$ and $\tau = t$. Because of this discontinuity of the covariance $E\{\tilde{\boldsymbol{x}}_0\boldsymbol{y}'(\tau)\}$ at the limits of the observation interval, we have restricted the validity of the conditions (5.81a) and (5.81b) to the open interval.

or, according to (5.77)

$$E\{\tilde{x}_0(T|t)\hat{x}_0'(T|t)\} = 0 \tag{5.82}$$

This means that the optimal estimation error is not correlated with the optimal estimate either. This relationship can be used to transform the covariance matrix of the optimal estimation error:

$$\begin{aligned} \mathrm{E}\{\tilde{x}_0(T|t)\tilde{x}_0'(T|t)\} &= \mathrm{E}\{\tilde{x}_0(T|t)[x(T) - \hat{x}_0(T|t)]'\} \\ &= \mathrm{E}\{\tilde{x}_0(T|t)x'(T)\} \end{aligned} \tag{5.83}$$

This equation will subsequently be needed in various places.

### 5.4.3 The Solution for Pure Filtering ($T = t$)

In the classical case of filtering for stationary processes, the Wiener–Hopf integral equation is solved by transformation to the frequency domain. In doing so, a number of spectral densities occur. These must be factored in a certain manner and recombined. The result is the frequency response of the filter, i.e., the Fourier transform of the required impulse response (Wiener, 1949; Davenport and Root, 1956; Laning and Battin, 1956; Schlitt, 1960). It is impossible to use this method for nonstationary processes and finite observation duration. Rather, we use the analogy of the recursion solution for discrete time and derive a system of differential equations for the filter, its gain factor and the error variances.

To simplify the notation, we will omit the subscript 0 for the optimum values and use the following conventions

$$\hat{x}(t|t) := \hat{x}(t) \quad \tilde{x}(t|t) := \tilde{x}(t)$$

At the starting time $t_0$, equation (5.77) for the optimal estimate reduces to

$$\hat{x}(t_0) = 0$$

The associated estimation error is therefore equal to $x(t_0)$ itself. The initial value of the covariance matrix of the estimation error is therefore given by

$$\tilde{P}(t_0) = E\{x(t_0)x'(t_0)\} = P(t_0)$$

Proceeding in a recursive manner, we now assume that the minimum variance estimate has already been determined at time $t$. We now look for the new estimate at time $t + \mathrm{d}t$.

Because the old estimate $\hat{\boldsymbol{x}}(t)$ was required to be optimal, the Wiener–Hopf equation already should be satisfied for it. Hence, according to (5.81a) we have

$$E\{\tilde{\boldsymbol{x}}(t)\boldsymbol{y}'(\tau)\} = \boldsymbol{0} \quad \text{for} \quad t_0 < \tau < t \tag{5.84a}$$

In a similar manner to the discrete-time case, we use prediction by the increment d$t$ as an interim solution. For this, also according to (5.81a), we have

$$E\{\tilde{\boldsymbol{x}}(t + \mathrm{d}t|t)\boldsymbol{y}'(\tau)\} = \boldsymbol{0} \quad t_0 < \tau < t \tag{5.84b}$$

The new estimate must satisfy the following version of the Wiener–Hopf condition:

$$E\{\tilde{\boldsymbol{x}}(t + \mathrm{d}t)\boldsymbol{y}'(\tau)\} = \boldsymbol{0} \quad t_0 < \tau < t + \mathrm{d}t \tag{5.84c}$$

In both conditions (5.84a) and (5.84b), $\tau$ only runs as far as $t$. To satisfy the last condition on the basis of the former, we construct a suitable expression for $\hat{\boldsymbol{x}}(t + \mathrm{d}t|t)$. For this, we first consider the state $\boldsymbol{x}(t + \mathrm{d}t)$ itself. According to the differential equation (5.73a), it is given by

$$\boldsymbol{x}(t + \mathrm{d}t) = \{\boldsymbol{I} + \boldsymbol{A}(t)\mathrm{d}t\}\boldsymbol{x}(t) + \int_t^{t+\mathrm{d}t} \boldsymbol{v}(\sigma)\mathrm{d}\sigma \tag{5.85a}$$

For $t < \sigma < t + \mathrm{d}t$, the disturbance $\boldsymbol{v}(\sigma)$ is not correlated with $\boldsymbol{y}'(\tau)$ for $t_0 < \tau < t$. The contribution of the integral over $\boldsymbol{v}(\sigma)$ to the expectation in (5.84b) will therefore vanish. This suggests choosing the extrapolated estimate as follows:

$$\hat{\boldsymbol{x}}(t + \mathrm{d}t|t) = \{\boldsymbol{I} + \boldsymbol{A}(t)\mathrm{d}t\}\hat{\boldsymbol{x}}(t) \tag{5.85b}$$

When the estimate is extrapolated in this way its error is obtained by subtracting equation (5.85b) from (5.85a):

$$\tilde{\boldsymbol{x}}(t + \mathrm{d}t|t) = \{\boldsymbol{I} + \boldsymbol{A}(t)\mathrm{d}t\}\tilde{\boldsymbol{x}}(t) + \int_t^{t+\mathrm{d}t} \boldsymbol{v}(\sigma)\mathrm{d}\sigma \tag{5.85c}$$

Substituting this in the left-hand side of (5.84b), canceling the integral over $\boldsymbol{v}(\sigma)$ and placing $\boldsymbol{I} + \boldsymbol{A}\mathrm{d}t$ in front of the expectation produces the left-hand side of (5.84a). As this was already equal to zero, the above choice for $\hat{\boldsymbol{x}}(t + \mathrm{d}t|t)$ was in fact correct.

Following the conventions for discrete time, the covariance matrix of the extrapolated estimate is denoted by $\boldsymbol{P}^*$:

$$P^*(t + dt|t) = E\{\tilde{x}(t + dt|t)\tilde{x}'(t + dt|t)\}$$

With (5.85c), we obtain

$$P^*(t + dt|t) = E\left\{\left([I + A(t)dt]\tilde{x}(t) + \int_t^{t+dt} v(\sigma)d\sigma\right)\left(\tilde{x}'(t)[I + A'(t)dt] + \int_t^{t+dt} v'(\lambda)d\lambda\right)\right\}$$

The disturbance $v$ in the interval $[t, t + dt]$ is not correlated with $x(t)$ or $y(t)$ or $\hat{x}(t)$. Therefore the cross-covariance between $v(\sigma)$ or $v(\lambda)$ on the one hand and $\tilde{x}(t)$ on the other hand completely vanishes. We are left with

$$P^*(t + dt|t) = [I + A(t)dt]E\{\tilde{x}(t)\tilde{x}'(t)\}[I + A'(t)dt] + \int_t^{t+dt}\int_t^{t+dt} E\{v(\sigma)v'(\lambda)\}d\sigma d\lambda$$

The expectation in the first summand is the covariance matrix of the old estimation error, i.e., $\tilde{P}(t)$. The covariance under the double integral is equal to $Q(\sigma)\delta(\sigma - \lambda)$ (see (5.74b)). On integrating over $\sigma$, using the masking property of the $\delta$ function, we obtain

$$\int_t^{t+dt}\int_t^{t+dt} Q(\sigma)\delta(\sigma - \lambda)d\sigma d\lambda = \int_t^{t+dt} Q(\lambda)d\lambda = Q(t)dt + 0(dt)^2$$

Therefore the above equation for $P^*$ becomes

$$P^*(t + dt|t) = \tilde{P}(t) + A(t)\tilde{P}(t)dt + \tilde{P}(t)A'(t)dt + Q(t)dt + 0(dt)^2 \tag{5.86}$$

This result corresponds to (5.51a) for discrete time.

Now that we have the optimal estimates $\hat{x}(t)$ and $\hat{x}(t + dt|t)$, the next step is to determine the new estimate $\hat{x}(t + dt)$. Following the discrete-time case, we attempt to satisfy condition (5.84c) with the following expression:

$$\hat{x}(t + dt) = \hat{x}(t + dt|t) + K(t)\{y(t) - C(t)\hat{x}(t)\}dt \tag{5.87}$$

In order to derive the associated estimation error from this equation we multiply it by $-1$ and add $x(t + dt)$ to both sides:

$$\tilde{\boldsymbol{x}}(t + \mathrm{d}t) = \tilde{\boldsymbol{x}}(t + \mathrm{d}t|t) - \boldsymbol{K}(t)\{\boldsymbol{y}(t) - \boldsymbol{C}(t)\hat{\boldsymbol{x}}(t)\}\mathrm{d}t$$

It follows from this, with (5.73b), that

$$\tilde{\boldsymbol{x}}(t + \mathrm{d}t) = \tilde{\boldsymbol{x}}(t + \mathrm{d}t|t) - \boldsymbol{K}(t)\boldsymbol{C}(t)\tilde{\boldsymbol{x}}(t)\mathrm{d}t - \int_t^{t+\mathrm{d}t} \boldsymbol{K}(\lambda)\boldsymbol{w}(\lambda)\mathrm{d}\lambda \qquad (5.88)$$

This expression must satisfy equation (5.84c). We first only consider this condition in the old interval. According to (5.84b) and (5.84a), $\tilde{\boldsymbol{x}}(t + \mathrm{d}t|t)$ and $\tilde{\boldsymbol{x}}(t)$ are not correlated with the old observations. Furthermore, the cross-covariance $E\{\boldsymbol{w}(\lambda)\boldsymbol{y}'(\tau)\}$ vanishes for the values of $\lambda$ and $\tau$ considered. Therefore condition (5.84c) is already satisfied in the old interval $(t_0, t)$. In the next interval $(t, t + \mathrm{d}t)$, we write it in the form

$$\mathrm{E}\{\tilde{\boldsymbol{x}}(t + \mathrm{d}t)\boldsymbol{y}'(t + \mathrm{d}\tau)\} = \boldsymbol{0} \quad 0 < \mathrm{d}\tau < \mathrm{d}t$$

or, with (5.73b),

$$E\{\tilde{\boldsymbol{x}}(t + \mathrm{d}t)[\boldsymbol{x}'(t + \mathrm{d}\tau)\boldsymbol{C}'(t + \mathrm{d}\tau) + \boldsymbol{w}'(t + \mathrm{d}\tau)]\} = \boldsymbol{0}$$

On multiplying out, $\tilde{\boldsymbol{x}}(t + \mathrm{d}t)$ is left unchanged in the first product but in the second product it is expressed by using equation (5.88):

$$E\{\tilde{\boldsymbol{x}}(t + \mathrm{d}t)\boldsymbol{x}'(t + \mathrm{d}\tau)\}\boldsymbol{C}'(t + \mathrm{d}\tau)$$
$$+ E\left\{\left[\tilde{\boldsymbol{x}}(t + \mathrm{d}t|t) - \boldsymbol{K}(t)\boldsymbol{C}(t)\tilde{\boldsymbol{x}}(t)\mathrm{d}t - \int_t^{t+\mathrm{d}t} \boldsymbol{K}(\lambda)\boldsymbol{w}(\lambda)\mathrm{d}\lambda\right]\boldsymbol{w}'(t + \mathrm{d}\tau)\right\} = \boldsymbol{0}$$

In the second expectation, the correlation of $\boldsymbol{w}'(t + \mathrm{d}\tau)$ with $\tilde{\boldsymbol{x}}(t + \mathrm{d}t|t)$ and $\tilde{\boldsymbol{x}}(t)$ vanishes. The remainder can be written as follows (see (5.74c)):

$$- \int_t^{t+\mathrm{d}t} \boldsymbol{K}(\lambda)E\{\boldsymbol{w}(\lambda)\boldsymbol{w}'(t + \mathrm{d}\tau)\}\mathrm{d}\lambda$$
$$= - \int_t^{t+\mathrm{d}t} \boldsymbol{K}(\lambda)\boldsymbol{R}(\lambda)\delta(\lambda - t - \mathrm{d}\tau)\mathrm{d}\lambda$$
$$= - \boldsymbol{K}(t + \mathrm{d}\tau)\boldsymbol{R}(t + \mathrm{d}\tau)$$

In the last rearrangement, the masking property of the $\delta$ function has been used. The above condition therefore becomes

$$E\{\tilde{\boldsymbol{x}}(t + \mathrm{d}t)\boldsymbol{x}'(t + \mathrm{d}\tau)\}\boldsymbol{C}'(t + \mathrm{d}\tau) - \boldsymbol{K}(t + \mathrm{d}\tau)\boldsymbol{R}(t + \mathrm{d}\tau) = \boldsymbol{0}$$

Here the increment $d\tau$ lies between zero and $dt$. Remember that the matrices $\boldsymbol{C}$, $\boldsymbol{K}$ and $\boldsymbol{R}$ are, like $E\{\tilde{\boldsymbol{x}}\boldsymbol{x}'\}$, continuous functions of time. Therefore this condition must also be valid for $dt$, $d\tau \to \boldsymbol{0}$:

$$E\{\tilde{\boldsymbol{x}}(t)\boldsymbol{x}'(t)\}\boldsymbol{C}'(t) - \boldsymbol{K}(t)\boldsymbol{R}(t) = \boldsymbol{0}$$

Here, according to (5.83), the expectation is equal to $E\{\tilde{\boldsymbol{x}}(t)\tilde{\boldsymbol{x}}'(t)\}$. We therefore obtain

$$\tilde{\boldsymbol{P}}(t)\boldsymbol{C}'(t) = \boldsymbol{K}(t)\boldsymbol{R}(t) \tag{5.89}$$

This is a condition for the optimal value of the gain matrix $\boldsymbol{K}(t)$. Because it has been assumed that $\boldsymbol{R}(t)$ is nonsingular, equation (5.89) can be solved for $\boldsymbol{K}(t)$ if $\tilde{\boldsymbol{P}}(t)$ is known. With this value of $\boldsymbol{K}(t)$, rule (5.87) for the new estimate satisfies the Wiener–Hopf equation (5.84c) and is thus optimal.

$\tilde{\boldsymbol{P}}(t)$ is still missing. We will establish a differential equation for it. With equations (5.83) and (5.88), we obtain

$$\begin{aligned}\tilde{P}(t + dt) &= E\{\tilde{\boldsymbol{x}}(t + dt)\boldsymbol{x}'(t + dt)\}\\ &= E\left\{\left[\tilde{\boldsymbol{x}}(t + dt|t) - \boldsymbol{K}(t)\boldsymbol{C}(t)\tilde{\boldsymbol{x}}(t)dt - \int_t^{t+dt} \mathrm{K}(\lambda)\boldsymbol{w}(\lambda)d\lambda\right]\boldsymbol{x}'(t + dt)\right\}\end{aligned}$$

The correlation of the integral with $\boldsymbol{x}'(t + dt)$ vanishes; in the middle term, $\boldsymbol{x}'(t + dt)$ is expressed using (5.85a). Then (5.83) is used again. The result is

$$\tilde{\boldsymbol{P}}(t + dt) = \boldsymbol{P}^*(t + dt|t) - \boldsymbol{K}(t)\boldsymbol{C}(t)\tilde{P}(t)dt + \boldsymbol{0}(dt)^2$$

We now substitute $\boldsymbol{P}^*$ according to (5.86) and $\boldsymbol{K}(t)$ according to (5.89). In this way we obtain the well-known *matrix Riccati differential equation* of filter theory:

$$\begin{aligned}\tilde{\boldsymbol{P}}(t + dt) = \tilde{\boldsymbol{P}}(t) + \{&\boldsymbol{A}(t)\tilde{\boldsymbol{P}}(t) + \tilde{\boldsymbol{P}}(t)\boldsymbol{A}'(t) + \boldsymbol{Q}(t)\\ &- \tilde{\boldsymbol{P}}(t)\boldsymbol{C}'(t)\boldsymbol{R}^{-1}(t)\boldsymbol{C}(t)\tilde{\boldsymbol{P}}(t)\}dt + \boldsymbol{0}(dt)^2\end{aligned} \tag{5.90}$$

Finally, by substituting (5.85b) in equation (5.87), we obtain the following differential equation for the estimate $\hat{\boldsymbol{x}}(t)$:

$$\hat{\boldsymbol{x}}(t + dt) = \hat{\boldsymbol{x}}(t) + [\boldsymbol{A}(t)\hat{\boldsymbol{x}}(t) + \boldsymbol{K}(t)\{\boldsymbol{y}(t) - \boldsymbol{C}(t)\hat{\boldsymbol{x}}(t)\}]dt \tag{5.91}$$

We have now reached our objective. The solution of the filter problem consists of equations (5.91), (5.90) and (5.89) with the initial conditions given at the beginning. We summarize as follows.

*Theorem 5.4 (Kalman–Bucy filter):*

(i) The linear unbiased minimum variance estimate for the state $\boldsymbol{x}(t)$ of the continuous-time stochastically disturbed system

$$\dot{\boldsymbol{x}}(t) = \boldsymbol{A}(t)\boldsymbol{x}(t) + \boldsymbol{v}(t) \quad E\{\boldsymbol{x}(t_0)\} = \boldsymbol{0}$$
$$\boldsymbol{y}(t) = \boldsymbol{C}(t)\boldsymbol{x}(t) + \boldsymbol{w}(t)$$

for which the sample function $\boldsymbol{y}(\tau)$, $t_0 \leqslant \tau \leqslant t$, and *a priori* knowledge according to (5.74a)–(5.74f) are available, is given by the differential equation

$$\frac{\mathrm{d}}{\mathrm{d}t}\hat{\boldsymbol{x}}(t) = \boldsymbol{A}(t)\hat{\boldsymbol{x}}(t) + \boldsymbol{K}(t)\{\boldsymbol{y}(t) - \boldsymbol{C}(t)\hat{\boldsymbol{x}}(t)\} \quad \hat{\boldsymbol{x}}(t_0) = \boldsymbol{0} \tag{5.92}$$

(ii) The gain matrix is given by

$$\boldsymbol{K}(t) = \tilde{\boldsymbol{P}}(t)\boldsymbol{C}'(t)\boldsymbol{R}^{-1}(t) \tag{5.93}$$

(iii) The covariance matrix of the estimation error obeys the matrix Riccati differential equation

$$\frac{\mathrm{d}}{\mathrm{d}t}\tilde{\boldsymbol{P}}(t) = \boldsymbol{A}(t)\tilde{\boldsymbol{P}}(t) + \tilde{\boldsymbol{P}}(t)\boldsymbol{A}'(t) - \tilde{\boldsymbol{P}}(t)\boldsymbol{C}'(t)\boldsymbol{R}^{-1}(t)\boldsymbol{C}(t)\tilde{\boldsymbol{P}}(t) + \boldsymbol{Q}(t)$$
$$\tilde{\boldsymbol{P}}(t_0) = \boldsymbol{P}(t_0) \tag{5.94}$$

A block diagram of the filter with $\boldsymbol{Bu} = \boldsymbol{0}$ is shown in the lower part of Figure 5.8. It has the now familiar form of a model of the system with weighted feedback of the differences between the output variables of the system and the model.

A comparison with the deterministic observer (Section 4.3, equations (4.29)–(4.31)) shows that the observer is a special case of the filter with

$$\boldsymbol{M}^{-1}(t, t_0) = \tilde{\boldsymbol{P}}(t)$$
$$\boldsymbol{Q}(t) = \boldsymbol{0}$$
$$\boldsymbol{R}(t) = \boldsymbol{I}$$

For time-invariant matrices $\boldsymbol{A}$, $\boldsymbol{C}$, $\boldsymbol{Q}$ and $\boldsymbol{R}$ and $t_0 \to -\infty$, the observed process is stationary and we are dealing with a Wiener filter problem. The gain matrix of the Wiener filter is determined by the corresponding equilibrium steady state solution of the Riccati differential equation (for more details, see Section 6.5).

*Example 5.4 (first-order Wiener filter).* The measurement $y(t)$ of a desired signal $x(t)$ is disturbed by additive noise $w(t)$ (Figure 5.9). The power spectral densities of $x$ and $w$ are

$$S_{xx}(\omega) = \frac{2}{1 + \omega^2}, \quad S_{ww}(\omega) = 1$$

We seek the equations of the corresponding Kalman–Bucy filter and the numerical value of its steady state gain (Wiener filter).

First, the model of the observed process must be formulated in state space representation. For this, the power spectral density of $x$ is decomposed into two complex conjugate factors:

$$S_{xx}(\omega) = \frac{\sqrt{2}}{1 + j\omega} \times \frac{\sqrt{2}}{1 - j\omega}$$

The left-hand factor is a stable frequency response which represents the appropriate shaping filter. The corresponding differential equation generating $x(t)$ from hypothetical white noise is therefore

$$\dot{x} + x = v$$

where

$$S_{vv}(\omega) = (\sqrt{2})^2 = 2$$

Therefore the required model of the observed process has the form

$$\dot{x}(t) = -x(t) + v(t)$$

where

$$E\{v(t)v(\tau)\} = 2\delta(t - \tau)$$

and

$$y(t) = x(t) + w(t)$$

where

$$E\{w(t)w(\tau)\} = 1\delta(t - \tau)$$

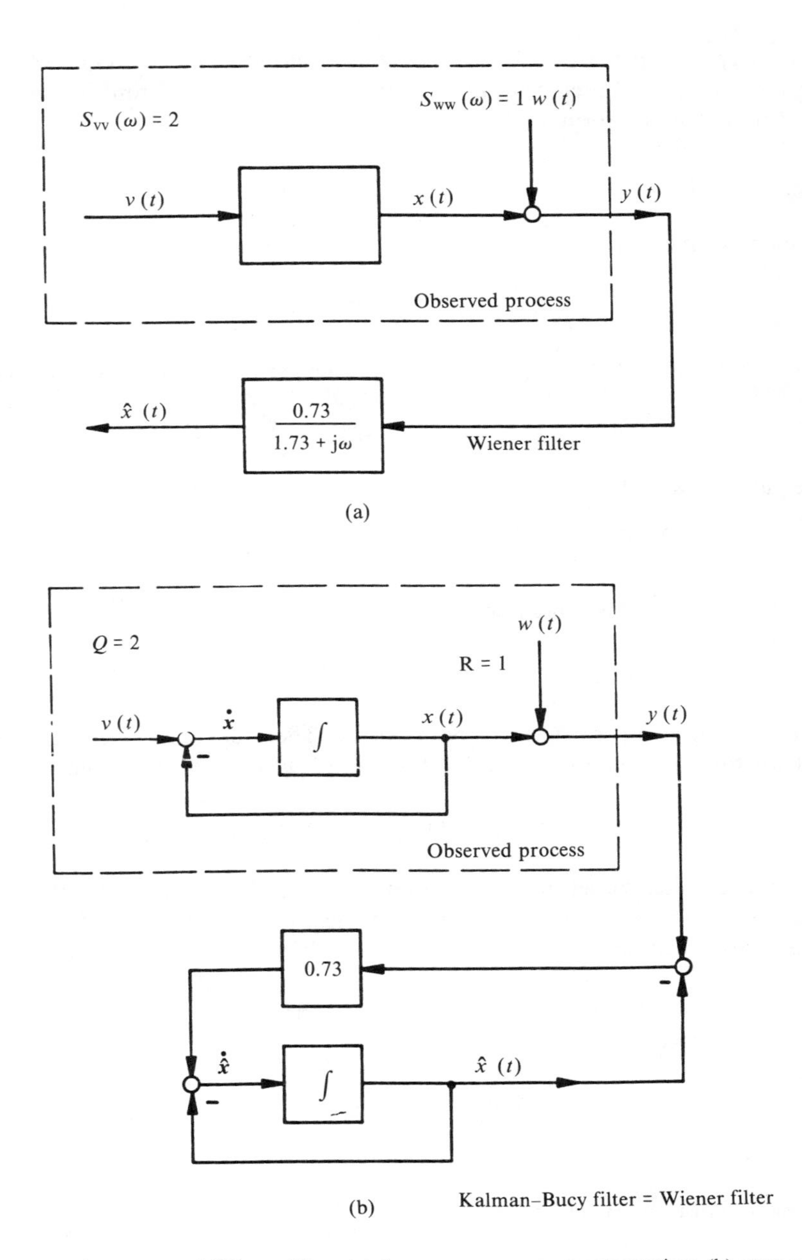


**Figure 5.9** Observed process and Wiener filter: (a) frequency response representation; (b) state space representation.

Obviously we have $A = -1$, $C = 1$, $Q = 2$ and $R = 1$. The required equations of the Kalman–Bucy filter can now immediately be inferred from Theorem 5.4. The filter itself has the form

$$\dot{\hat{x}}(t) = -\hat{x}(t) + k(t)\{y(t) - \hat{x}(t)\}$$

Its gain is given by

$$k(t) = \tilde{p}(t)$$

where $\tilde{p}$ is the solution of the following scalar Riccati differential equation:

$$\dot{\tilde{p}}(t) = -2\tilde{p}(t) - \tilde{p}^2(t) + 2$$

The steady state of the Ricatti differential equation is defined by $\dot{\tilde{p}} = 0$, i.e.,

$$\tilde{p}^2 + 2\tilde{p} - 2 = 0$$

The two roots of this quadratic equation are

$$\tilde{p}_{12} = -1 \pm \sqrt{3}$$

As $\tilde{p}$ is a variance, it can never be negative. Therefore we must select the positive root for the equilibrium solution of the Ricatti differential equation:

$$\tilde{p}(\infty) = -1 + \sqrt{3} = 0.73$$

The Wiener filter therefore has the gain $k(\infty) = 0.73$ (see Figure 5.9). The Wiener filter can be specified as a frequency response by rearranging and transforming its differential equation:

$$\begin{aligned}\dot{\hat{x}}(t) &= -\hat{x}(t) + 0.73\{y(t) - \hat{x}(t)\} \\ &= -1.73\hat{x}(t) + 0.73y(t)\end{aligned}$$

$$\frac{\hat{X}(\mathrm{j}\omega)}{Y(\mathrm{j}\omega)} = \frac{0.73}{1.73 + \mathrm{j}\omega}$$

The bandwidth (3 dB decrease) is $\omega_0 = 1.73$.

### 5.4.4 Biased Initial Values and Measurable Input Variables

This case has been comprehensively discussed for discrete time in Section 5.3.5. The arguments provided there can be simply taken over. We can immediately formulate the following theorem.

*Theorem 5.5*

(i) The linear unbiased minimum variance estimate for the state $\boldsymbol{x}(t)$ of the system

$$\dot{\boldsymbol{x}}(t) = \boldsymbol{A}(t)\boldsymbol{x}(t) + \boldsymbol{B}(t)\boldsymbol{u}(t) + \boldsymbol{v}(t) \quad E\{\boldsymbol{x}(t_0)\} = \boldsymbol{\xi} \tag{5.95a}$$

$$\boldsymbol{y}(t) = \boldsymbol{C}(t)\boldsymbol{x}(t) + \boldsymbol{w}(t) \tag{5.95b}$$

for which the sample function $\boldsymbol{y}(\tau)$ and the input variable $\boldsymbol{u}(\tau)$, $t_0 \leqslant \tau \leqslant t$, and *a priori* knowledge according to (5.74a)–(5.74f) are available, is determined by the differential equation

$$\begin{aligned} \frac{\mathrm{d}}{\mathrm{d}t}\hat{\boldsymbol{x}}(t) &= \boldsymbol{A}(t)\hat{\boldsymbol{x}}(t) + \boldsymbol{B}(t)\boldsymbol{u}(t) + \boldsymbol{K}(t)\{\boldsymbol{y}(t) - \boldsymbol{C}(t)\hat{\boldsymbol{x}}(t)\} \\ \hat{\boldsymbol{x}}(t_0) &= \boldsymbol{\xi} \end{aligned} \tag{5.96}$$

(ii) The gain matrix $\boldsymbol{K}(t)$ is given by equation (5.93) and the covariance matrix of the estimation error obeys the matrix Riccati differential equation (5.94).

*Proof*. The estimate according to (5.96) is obviously linear in $\boldsymbol{y}$. To prove the unbiased property, we take the difference between the differential equations (5.95a) and (5.96). Using (5.95b), we then obtain the differential equation for the estimation error:

$$\begin{aligned} \frac{\mathrm{d}}{\mathrm{d}t}\tilde{\boldsymbol{x}}(t) &= \{\boldsymbol{A}(t) - \boldsymbol{K}(t)\boldsymbol{C}(t)\}\tilde{\boldsymbol{x}}(t) + \boldsymbol{v}(t) - \boldsymbol{K}(t)\boldsymbol{w}(t) \\ \tilde{\boldsymbol{x}}(t_0) &= \boldsymbol{x}(t_0) - \boldsymbol{\xi} \end{aligned} \tag{5.97}$$

The initial condition $\tilde{\boldsymbol{x}}(t_0)$ and the excitations $\boldsymbol{v}$ and $\boldsymbol{w}$ have zero expectation. Therefore $E\{\tilde{\boldsymbol{x}}(t)\}$ vanishes for all $t \geqslant t_0$, irrespective of the values of $\boldsymbol{\xi}$ and $\boldsymbol{u}(t)$. This means that $\hat{\boldsymbol{x}}(t)$ is unbiased.

The initial value of the estimation error in equation (5.97) has the covariance matrix:

$$\tilde{\boldsymbol{P}}(t_0) = E\{[\boldsymbol{x}(t_0) - \boldsymbol{\xi}][\boldsymbol{x}(t_0) - \boldsymbol{\xi}]'\} = \boldsymbol{P}(t_0)$$

Remember that the evolution of $\tilde{\boldsymbol{P}}(t)$ must be subject to the same matrix Riccati equation as in the case $\boldsymbol{u}(t) \equiv \boldsymbol{0}$ and $\boldsymbol{\xi} = \boldsymbol{0}$. If $\boldsymbol{P}(t_0)$ is the same, then $\tilde{\boldsymbol{P}}(t)$ also has the same value for all $t$ in both cases. The variances of $\tilde{\boldsymbol{x}}(t)$, which are situated in the main diagonal of $\tilde{\boldsymbol{P}}(t)$, are therefore also minimal in the case of Theorem 5.5.

The known input variables $\boldsymbol{u}(t)$ and the given expectation of $\boldsymbol{x}(t_0)$ have already been included in the filter shown in Figure 5.8.

The matrix Riccati differential equation has a closed solution in only a few trivial cases. Usually it has to be solved by numerical integration methods. Therefore it is barely possible to make complete calculations of examples without computers. This is in contrast with the discrete-time case where examples up to third order can be solved manually or with a pocket calculator. We will return to the numerical treatment of the Riccati differential equation in the final chapter.

### 5.4.5 Prediction ($T > t$)

We now seek the estimate $\hat{\boldsymbol{x}}(T|t)$ of the future state $\boldsymbol{x}(T)$ on the basis of the sample function $\boldsymbol{y}(\tau)$ observed up to the present time $t$. This estimate must also satisfy a Wiener–Hopf equation. Since the latter is only valid for unbiased random variables, we will again first assume here that the expectation of the initial condition $\boldsymbol{x}(t_0)$ and the input variable $\boldsymbol{u}(t)$ of the observed system vanish. We start from the Wiener–Hopf equation in the version of equation (5.81a). In a slightly modified form it can be written as

$$E\{[\boldsymbol{x}(T) - \hat{\boldsymbol{x}}(T|t)]\boldsymbol{y}'(\tau)\} = \boldsymbol{0} \quad t_0 < \tau < t \tag{5.98}$$

The future state $\boldsymbol{x}(T)$ can be expressed in terms of $\boldsymbol{x}(t)$ by using the general solution formula of the differential equation $\dot{\boldsymbol{x}} = \boldsymbol{A}\boldsymbol{x} + \boldsymbol{v}$:

$$\boldsymbol{x}(T) = \boldsymbol{\phi}(T, t)\boldsymbol{x}(t) + \int_t^T \boldsymbol{\phi}(T, \sigma)\boldsymbol{v}(\sigma)\mathrm{d}\sigma \tag{5.99}$$

This expression is substituted into condition (5.98). Here, the cross-covariance between the integral and $\boldsymbol{y}'(\tau)$ vanishes because the future disturbances $\boldsymbol{v}(\sigma)$ are not correlated with the past measurements. We are left with

$$E\{[\boldsymbol{\phi}(T, t)\boldsymbol{x}(t) - \hat{\boldsymbol{x}}(T|t)]\boldsymbol{y}'(\tau)\} = \boldsymbol{0} \quad t_0 < \tau < t$$

This condition can be satisfied by choosing

$$\hat{\boldsymbol{x}}(T|t) = \boldsymbol{\phi}(T, t)\hat{\boldsymbol{x}}(t)$$

where $\hat{\boldsymbol{x}}(t)$ is the optimal filtered value obeying Theorem 5.4. To prove this, we bring $\boldsymbol{\phi}(T, t)$ in front of the expectation and remember that the remaining part of the expectation, i.e.,

$$E\{[\boldsymbol{x}(t) - \hat{\boldsymbol{x}}(t)]\boldsymbol{y}'(\tau)\} = E\{\tilde{\boldsymbol{x}}(t)\boldsymbol{y}'(\tau)\}$$

is already zero for all $\tau$ in the interval $(t_0, t)$. We can add the following lemma to Theorem 5.4.

*Theorem 5.6* The linear unbiased minimum variance estimate for the future state $\boldsymbol{x}(T)$, $T > t$, in the system

$$\dot{\boldsymbol{x}}(t) = \boldsymbol{A}(t)\boldsymbol{x}(t) + \boldsymbol{v}(t) \quad E\{\boldsymbol{x}(t_0)\} = \boldsymbol{0}$$

$$\boldsymbol{y}(t) = \boldsymbol{C}(t)\boldsymbol{x}(t) + \boldsymbol{w}(t)$$

for which the sample function $\boldsymbol{y}(\tau)$, $t_0 \leq \tau \leq t$, and *a priori* knowledge according to (5.74a)–(5.74f) are available, is given by

$$\hat{\boldsymbol{x}}(T|t) = \boldsymbol{\phi}(T, t)\hat{\boldsymbol{x}}(t) \tag{5.100}$$

where $\boldsymbol{\phi}(T, t)$ is the transition matrix of $\boldsymbol{A}(t)$ and $\hat{\boldsymbol{x}}(t)$ represents the estimate of the present state $\boldsymbol{x}(t)$ according to Theorem 5.4. When including biased initial values $\boldsymbol{x}(t_0)$ and known input variables $\boldsymbol{u}(\tau)$, $t_0 \leq \tau \leq t$, $\hat{\boldsymbol{x}}(t)$ must be computed according to Theorem 5.5.

The evaluation of the Wiener–Hopf equation for interpolation ($T < t$, smoothing) is considerably more difficult because the correlation of the integral in equation (5.99) with the measurement variable $\boldsymbol{y}(\tau)$ does not vanish. The result is correspondingly complicated (Mayne, 1966; Meditch, 1967; Sage and Melsa, 1971).

### 5.4.6 Final Remarks

Starting from two previous cases (i.e., the observer for continuous time and the filter for discrete time), we have derived the Kalman–Bucy filter by a limiting procedure in time. In this way complete analogy with the Kalman filter could be preserved.

It was necessary to make the usually restrictive assumption here that $\boldsymbol{R}(t)$ is nonsingular. This means that all components of the measurement vector $\boldsymbol{y}$ contain white noise. This prerequisite is not necessary for discrete time. It can also be removed for continuous time whereby the order of the filter is reduced by the number of measurement components without white noise (Bryson and Johansen, 1965; Bucy, 1967; Sage and Melsa, 1971).

A certain difficulty exists for continuous time in the correct treatment of white noise. One expression of this problem is that $\delta$ functions occur in the corresponding covariances. By taking suitable precautions, in particular by only allowing $\delta$ functions under integrals and using the masking property, tricky situations could be circumvented. We will briefly look at the rigorous treatment by using the Ito calculus in Section 6.8.

For questions relating to correlated disturbance and measurement noises, colored noise and systematic errors, reference is made to Section 5.3.7. The remarks made there for discrete time apply to continuous time in an analogous manner.

## REFERENCES


Bode, H.W. and Shannon, C.E. (1950). A simplified derivation of linear least square smoothing and prediction theory, *Proc. IRE,* **38**, 417–424.

Booton, R.C. (1952). An optimization theory for time-varying linear systems with nonstationary statistical inputs, *Proc. IRE,* **40**, 977–981.

Brammer, K. (1967). Optimale Filterung und Vorhersage instationarer stochastischer Folgen, *Nachrichtentech. Fachber.,* **33**, 103–110.

Brammer, K. (1968a). Lower order optimal linear filtering of nonstationary random sequences, *IEEE Trans. Autom. Control,* **13**, 198–199.

Brammer, K. (1968b). Zur optimalen linearen Filterung und Vorhersage instationarer Zufallsprozesse in diskreter Zeit, *Regelungstechnik,* **16**, 105–110.

Brammer, K. (1971). Gausssche Ausgleichsrechnung und Kalman-filterung, *Regelungstech. Prozess-Datenverarb.,* **19**, 215–217.

Bryson, A.E. and Henrikson, L.J. (1967). Estimation using sampled-data containing sequentially correlated noise, *Tech. Rep. 533,* Division of Engineering and Applied Physics, Harvard University, Cambridge, MA, June.

Bryson, A.E. and Johansen, D.E. (1965). Linear filtering for time-varying systems using measurements containing colored noise, *IEEE Trans. Autom. Control,* **10**, 4–10.

Bucy, R.S. (1959). Optimum finite time filters for a special non-stationary class of inputs, *Internal Rep. BBD-600,* Applied Physics Laboratory, Johns Hopkins University.

Bucy, R.S. (1967). Optimal filtering for correlated noise, *J. Math. Anal. Appl.,* **20** (1).

Bucy, R.S. and Joseph, P.D. (1968). *Filtering for Stochastic Processes, with Applications to Guidance,* Interscience, New York.

Davenport, W.B. and Root, W.L. (1956). *An Introduction to the Theory of Random Signals and Noise,* McGraw-Hill, New York.

Doob, J.L. (1953). *Stochastic Processes,* Wiley, New York.

Follin, J.W. and Carlton, A.G. (1956). Recent developments in fixed and adaptive filtering, *AGARDograph 21.*

Hanson, J.E. (1957). Some notes on the application of the calculus of variations to smoothing for finite time, *Intern Memo. BBD-346,* Applied Physics Laboratory, Johns Hopkins University.

Jazwinski, A.H. (1970). *Stochastic Processes and Filtering Theory,* Academic Press, New York.

Kalman, R.E. (1960). A new approach to linear filtering and prediction problems, *Trans. ASME, Ser. D, J. Basic Eng.* **82**, 35–45.

Kalman, R.E. and Bucy, R.S. (1961). New results in linear filtering and prediction theory, *Trans. ASME, Ser. D, J. Basic Eng.,* **83**, 95–108.

Kolmogorov, A.N. (1941). Interpolation and extrapolation of stationary random sequences, *Bull. Acad. Sci. USSR, Math. Ser.*, **5**, 3–14.

Laning, J.H. and Battin, R.H. (1956). *Random Processes in Automatic Control,* McGraw-Hill, New York.

Luenberger, D.G. (1966). Observers for multivariable systems, *IEEE Trans. Autom. Control,* ***11***, 190–197.

Mayne, D.Q. (1966). A solution of the smoothing problem for linear dynamic systems, *Automatica,* **4**, 73–92.

Meditch, J.S. (1967). Orthogonal projection and discrete optimal linear smoothing, *SIAM J. Control,* **5**, 74–80.

Plackett, R.L. (1950). Some theorems in least squares, *Biometrika,* **37**, 149–157.

Sage, A.P. and Melsa, J.L. (1971). *Estimation Theory, with Applications to Communications and Control,* McGraw-Hill, New York.

Schlitt, H. (1960). *Systemstheorie fuer regellose Vorgaenge,* Springer, Berlin.

Swerling, P. (1959). First-order error propagation in a stagewise smoothing procedure for satellite observations, *J. Astronaut. Sci.,* **6**, 46–52.

Wiener, N. (1949). *Extrapolation, Interpolation, and Smoothing of Stationary Time Series,* Wiley, New York.

Zurmuehl, R. and Falk, S. (1984). *Matrices and their Applications,* Part 1 (5th edn), Springer, Berlin.

## BIBLIOGRAPHY

Anderson, B.D.O. and Moore, J.B. (1979). *Optimal Filtering,* Prentice-Hall, Englewood Cliffs, NJ.

Gelb, A. (ed.) (1978). *Applied Optimal Estimation* (4th edn), MIT Press, Cambridge, MA.

Kalman, R.E. (1963). New methods and results in linear prediction and filtering theory. In Bogdanov, J.L. and Kozin, F. (eds), *Proc. 1st Symp. on Engineering Applications of Random Function Theory and Probability, 1960,* Wiley, New York.

Leondes, C.T. (ed.) (1970). Theory and applications of Kalman filtering, *AGARDograph 139,* February.

Maybeck, P.S. (1979, 1982, 1983). *Stochastic Models, Estimation and Control,* Academic Press, New York, Vol. 1, 1979, Vol. 2, 1982, Vol. 3, 1983.

Ruymgaart, P.A. and Soong, T.T. (1985). *Mathematics of Kalman–Bucy Filtering,* Springer, Berlin.

Schrick, K.-W. (ed.) (1977). *Anwendungen der Kalman-Filter Technik; Anleitung und Beispiele,* Oldenbourg, Munich.

# *Chapter 6*
# *Practical Problems in Filter Synthesis*

For completeness we will make a few remarks on the practical synthesis and application of observers and filters. In the first three sections, we will examine the application of observers and filters in a closed control loop. Then, the matrix Riccati equation and its equilibrium solution will be discussed. A brief introduction to the synthesis of shaping-filter models follows. Finally, we take a brief look at the synthesis of filters for partly noise-free measurement variables and nonlinear processes.

## 6.1 DETERMINISTIC CONTROL WITH FEEDBACK OF THE STATE VECTOR

Referring to the concept of controllability (Section 4.5), we consider here an $n$th-order controlled process in continuous time with a completely measurable state vector and no disturbance variables (Figure 6.1):

$$\dot{\boldsymbol{x}}(t) = \boldsymbol{A}(t)\boldsymbol{x}(t) + \boldsymbol{B}(t)\boldsymbol{u}(t) \quad t_0 \leqslant t \leqslant t_1 \tag{6.1}$$

Let the $n \times n$ matrix $\boldsymbol{A}$ and the $n \times p$ matrix $\boldsymbol{B}$ be known. The $p$-dimensional control variable $\boldsymbol{u}(t)$ is to be determined so that the state $\boldsymbol{x}(t)$ goes to zero in a manner which will be specified in more detail later. In this connection, the variable $\boldsymbol{x}(t)$ should be regarded as a control deviation or as the deviation from an operating point or a reference trajectory.

In many control problems, the relevant control law is linear and requires the feedback of the state vector. It therefore has the form (Figure 6.1)

$$\boldsymbol{u}(t) = -\boldsymbol{L}(t)\boldsymbol{x}(t) \tag{6.2}$$

An example of this is the control law (4.48) in Section 4.5 where the feedback gain $\boldsymbol{L}$ is given by

$$\boldsymbol{L}(t) = \boldsymbol{B}'(t)\boldsymbol{W}^{-1}(t_1, t)$$

The matrix $\boldsymbol{W}^{-1}$ is the solution of the backward integrated matrix Riccati differential equation (4.49). However, this form of the control law is only valid in the first subinterval $t_0 \leqslant t < t_r$ in which $\boldsymbol{W}(t_1, t)$ is nonsingular. For $t_r \leqslant t \leqslant t_1$, equation (4.43) has to be chosen as the control law with $t_0 = t_r$, while $\boldsymbol{W}(t_1, t_r)$ has to be calculated by backward integration of the bilinear matrix equation (4.47). This type of control has the advantage that $\boldsymbol{x}$ becomes exactly equal to zero at time $t_1$. However, it has the disadvantage that in the last subinterval $[t_r, t_1]$ only open-loop control is carried out. It is therefore often customary to formulate the linear optimal control problem as follows. Choose the control variable $\boldsymbol{u}(\tau)$ of the linear system (6.1) so that the quadratic performance index

$$J(t_1, t_0) := \boldsymbol{x}'(t_1)\bar{\boldsymbol{P}}(t_1)\boldsymbol{x}(t_1) + \int_{t_0}^{t_1} \{\boldsymbol{x}'(\tau)\bar{\boldsymbol{Q}}(\tau)\boldsymbol{x}(\tau) + \boldsymbol{u}'(\tau)\bar{\boldsymbol{R}}(\tau)\boldsymbol{u}(\tau)\}\mathrm{d}\tau \qquad (6.3)$$

becomes a minimum. Here, $\bar{\boldsymbol{P}}$, $\bar{\boldsymbol{Q}}$ and $\bar{\boldsymbol{R}}$ are symmetric matrices; $\bar{\boldsymbol{P}}$ and $\bar{\boldsymbol{Q}}$ are positive semidefinite, but $\bar{\boldsymbol{R}}$ must be positive *definite*. Otherwise these three matrices can be chosen arbitrarily. By using the calculus of variations (Carathéodory, 1935), dynamic programming (Bellman and Dreyfus, 1962) or the maximum principle (Pontryagin *et al.*, 1964) it can be shown that the resulting optimal control law again has the general linear form (6.2). However, the gain matrix can now be expressed in an alternative manner as

$$\boldsymbol{L}(t) = \bar{\boldsymbol{R}}^{-1}(t)\boldsymbol{B}'(t)\boldsymbol{P}(t) \qquad (6.4)$$

where $\boldsymbol{P}$ is the solution of the matrix Riccati equation

$$\begin{aligned} -\frac{\mathrm{d}}{\mathrm{d}t}\boldsymbol{P}(t) &= \boldsymbol{A}'(t)\boldsymbol{P}(t) + \boldsymbol{P}(t)\boldsymbol{A}(t) - \boldsymbol{P}(t)\boldsymbol{B}(t)\bar{\boldsymbol{R}}^{-1}(t)\boldsymbol{B}'(t)\boldsymbol{P}(t) + \bar{\boldsymbol{Q}}(t) \\ \boldsymbol{P}(t_1) &= \bar{\boldsymbol{P}}(t_1) \end{aligned} \qquad (6.5)$$

which is to be integrated backward. The minimum value of the index (6.3) which results from the application of the optimal control law (6.4) and (6.5) is (Athans and Falb, 1966):

$$J_{\min} = \boldsymbol{x}'(t_0)\boldsymbol{P}(t_0)\boldsymbol{x}(t_0) \qquad (6.6)$$

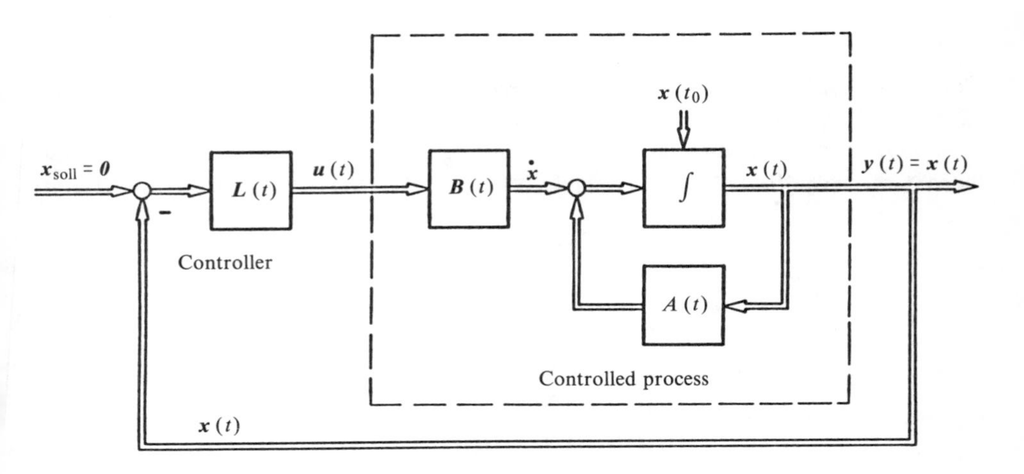


**Figure 6.1** Linear control with feedback of the state of the process.

By comparing the solution (6.4) and (6.5) with the control law given by (4.47), (4.48) and (4.49), the following can be established.

(i) The requirement that $\boldsymbol{x}$ must be exactly zero at the end of the control interval corresponds to the choice

$$\bar{\boldsymbol{P}}(t_1) = \infty$$

in the index (6.3).

(ii) The evolution of the state $\boldsymbol{x}(t)$ during the control process was not taken into account at all in Section 4.5. In order to express this in the index (6.3) we must set

$$\bar{\boldsymbol{Q}}(t) \equiv \boldsymbol{0} \quad t_0 \leqslant t \leqslant t_1$$

(iii) In Section 4.5 the cost of the control energy was set equal for all components $u_i(t)$ and all considered $t$ (compare equation (4.44)). In the index (6.3), this is expressed as follows:

$$\bar{\boldsymbol{R}}(t) \equiv \boldsymbol{I} \quad t_0 \leqslant t \leqslant t_1$$

(iv) If we now examine equations (4.49) and (6.5) placed side by side and take these relationships into account, we can conclude that

$$W^{-1}(t_1, t) = P(t)$$

Continuing with this analogy, we again arrive at the question of duality. An obvious idea is that the linear optimal control law could be dual to the Kalman–Bucy filter law. This is in fact confirmed by comparing equations (5.93) and (5.94) with (6.4) and (6.5). The following duality relationships apply (Kalman and Bucy, 1961):

| filtering | | control |
|---|---|---|
| $\boldsymbol{K}$ | $\triangleq$ | $\boldsymbol{L}'$ |
| $\tilde{\boldsymbol{P}}$ | $\triangleq$ | $\boldsymbol{P}$ |
| $\boldsymbol{C}^{\mathrm{T}}$ | $\triangleq$ | $\boldsymbol{B}$ |
| $\boldsymbol{R}$ | $\triangleq$ | $\bar{\boldsymbol{R}}$ |
| $\boldsymbol{A}$ | $\triangleq$ | $\boldsymbol{A}'$ |
| $\boldsymbol{Q}$ | $\triangleq$ | $\bar{\boldsymbol{Q}}$ |
| $t_0$ | $\triangleq$ | $t_1$ |
| $t$ | $\rightarrow$ | $-t$ |

These duality relationships have far-reaching practical consequences since optimal filters and optimal controllers can be designed by using the same computer program!

However, an important timing difference should be noted. The Riccati differential equation of the filter problem is integrated from the beginning of the observation interval (i.e., forward in time from $t_0$). Therefore the filter gain $\boldsymbol{K}$ at the current time $t$ is based on the past parameter values ($\boldsymbol{A}$, $\boldsymbol{C}$, $\boldsymbol{Q}$ and $\boldsymbol{R}$) of the observed process. In contrast, in the control problem the Riccati differential equation must be integrated from the end of the control interval (i.e., backward in time from $t_1$). This means that the control gain $\boldsymbol{L}$ at the current time $t$ depends on the future values of the system parameters ($\boldsymbol{A}$ and $\boldsymbol{B}$) and the weight matrices in the performance index ($\bar{\boldsymbol{Q}}$ and $\bar{\boldsymbol{R}}$). This theoretical difference has fundamental practical consequences with respect to adaptive systems. As is well known, adaptive systems automatically adapt to changes in the parameters of the observed or controlled process. For this purpose, we usually start by estimating the parameters which vary in an unknown manner. This estimation is in principle a filter problem even though in most cases it is a nonlinear problem (see, for example, Section 6.8). In real time, this estimation or filtering can only be done on the basis of past measurement values.

For an adaptive filter, the Riccati differential equation (5.94) is integrated by progressing along in real time using the estimates of the parameters. Therefore, apart from questions related to the computer requirements, an adaptive optimal filter is basically feasible. However, this is not the case for the adaptive optimal controller. In order to integrate the Riccati differential equation here, we need the estimates of the future process parameters and these values must be taken as unknown. Thus the only recourse is to predict (extrapolate) the unknown process parameters while assuming a certain structure for the future parameter variations.

We now return to general linear control with state vector feedback, i.e., the control law of the form (6.2). We now seek the differential equation of the closed control loop. This is given by substituting equation (6.2) in the differential equation (6.1) for the control system. After combining terms, we obtain

$$\dot{\boldsymbol{x}}(t) = \{\boldsymbol{A}(t) - \boldsymbol{B}(t)\boldsymbol{L}(t)\}\boldsymbol{x}(t) \tag{6.7}$$

The dynamics of the closed control loop is now determined by the matrix $\boldsymbol{A} - \boldsymbol{BL}$. More specifically, in the case of constant parameters the eigenvalues of $\boldsymbol{A} - \boldsymbol{BL}$ are relevant.

If the matrix $\boldsymbol{A}$ is in the control canonical form or is similar to this form, the feedback of all state variables by the linear controller $\boldsymbol{Lx}$ causes closed-loop behavior with derivative feedback up to $(n - 1)$th order. This is because the $n - 1$ derivatives of $x_1$ are now either directly equal to $x_2, \ldots, x_n$ or can be reconstructed from the state variables by a similarity transformation.

## 6.2 OBSERVER IN THE CONTROL LOOP AND ALGEBRAIC SEPARATION

In the previous section it was assumed that all state variables can be directly measured for the purpose of control. However, this ideal case seldom occurs. Usually, only $m$ output variables $\boldsymbol{y}(t)$ can be measured. Then the differential equation (6.1) for the process must have the following measurement equation added to it:

$$\boldsymbol{y}(t) = \boldsymbol{C}(t)\boldsymbol{x}(t) \tag{6.8}$$

Here the rank $m$ of the matrix $\boldsymbol{C}$ is less than $n$. The inaccessible state vector must now be estimated. We already know from Chapters 4 and 5 that this can be done with the help of an observer of the form

$$\frac{\mathrm{d}}{\mathrm{d}t}\hat{\boldsymbol{x}}(t) = \boldsymbol{A}(t)\hat{\boldsymbol{x}}(t) + \boldsymbol{B}(t)\boldsymbol{u}(t) + \boldsymbol{K}(t)\{\boldsymbol{y}(t) - \boldsymbol{C}(t)\hat{\boldsymbol{x}}(t)\} \tag{6.9}$$

In order to obtain the differential equation of the associated observation error $\tilde{\boldsymbol{x}}(t) = \boldsymbol{x}(t) - \hat{\boldsymbol{x}}(t)$, we subtract the observation equation (6.9) from the state equation (6.1) and obtain

$$\frac{\mathrm{d}}{\mathrm{d}t}\tilde{\boldsymbol{x}}(t) = \boldsymbol{A}(t)\tilde{\boldsymbol{x}}(t) - \boldsymbol{K}(t)\{\boldsymbol{y}(t) - \boldsymbol{C}(t)\hat{\boldsymbol{x}}(t)\}$$

After substituting (6.8) and collecting terms with $\tilde{\boldsymbol{x}}(t)$, we obtain

$$\frac{\mathrm{d}}{\mathrm{d}t}\tilde{\boldsymbol{x}}(t) = \{\boldsymbol{A}(t) - \boldsymbol{K}(t)\boldsymbol{C}(t)\}\tilde{\boldsymbol{x}}(t), \quad \tilde{\boldsymbol{x}}(t_0) \neq \boldsymbol{0} \tag{6.10}$$

This result corresponds to equation (5.97) of the filter error with $\boldsymbol{v}(t)$ and $\boldsymbol{w}(t)$ set equal to zero. In transposed form, the differential equation (6.10) is dual to the equation of the closed control loop (equation (6.7)). Seen in this way, the observer is a control loop which tries to drive the observation error to zero. If the differential equation (6.10) is asymptotically stable, the observation error goes to zero from any initial value and in general does so more rapidly for larger $\boldsymbol{K}$.

In the observation law (4.30), the gain matrix was given by

$$\boldsymbol{K}(t) = \boldsymbol{M}^{-1}(t, t_0)\boldsymbol{C}'(t)$$

However, in this section $\boldsymbol{K}(t)$ may also take other values.

The observer (6.9) is now connected between the measurement variable $\boldsymbol{y}$ and the input to the controller (Figure 6.2). Thus, instead of being fed with the state itself, the controller is fed with its estimate

$$\boldsymbol{u}(t) = -\boldsymbol{L}(t)\hat{\boldsymbol{x}}(t) \tag{6.11}$$

We now examine the dynamics of the closed loop. This obviously has order $2n$. We choose the state $\boldsymbol{x}(t)$ of the process and the observation error $\tilde{\boldsymbol{x}}(t)$ as the overall state variable. Just as for $\boldsymbol{K}(t)$, we do not need preconditions for the gain matrix $\boldsymbol{L}(t)$ of the controller here. In the closed loop, equations (6.10), (6.11) and (6.1) apply. Combination of the latter two gives

$$\dot{\boldsymbol{x}}(t) = \boldsymbol{A}(t)\boldsymbol{x}(t) - \boldsymbol{B}(t)\boldsymbol{L}(t)\{\boldsymbol{x}(t) - \tilde{\boldsymbol{x}}(t)\} \tag{6.12}$$

The loop consisting of the process, observer and controller therefore has the combined differential equation:

$$\frac{\mathrm{d}}{\mathrm{d}t}\begin{bmatrix}\boldsymbol{x}(t)\\ \tilde{\boldsymbol{x}}(t)\end{bmatrix} = \begin{bmatrix}\boldsymbol{A}(t) - \boldsymbol{B}(t)\boldsymbol{L}(t) & \boldsymbol{B}(t)\boldsymbol{L}(t)\\ \boldsymbol{0} & \boldsymbol{A}(t) - \boldsymbol{K}(t)\boldsymbol{C}(t)\end{bmatrix}\begin{bmatrix}\boldsymbol{x}(t)\\ \tilde{\boldsymbol{x}}(t)\end{bmatrix} \tag{6.13}$$

It is important to note the zero matrix in the bottom left-hand corner. It expresses the decoupling of the dynamic properties of the observation and control processes.

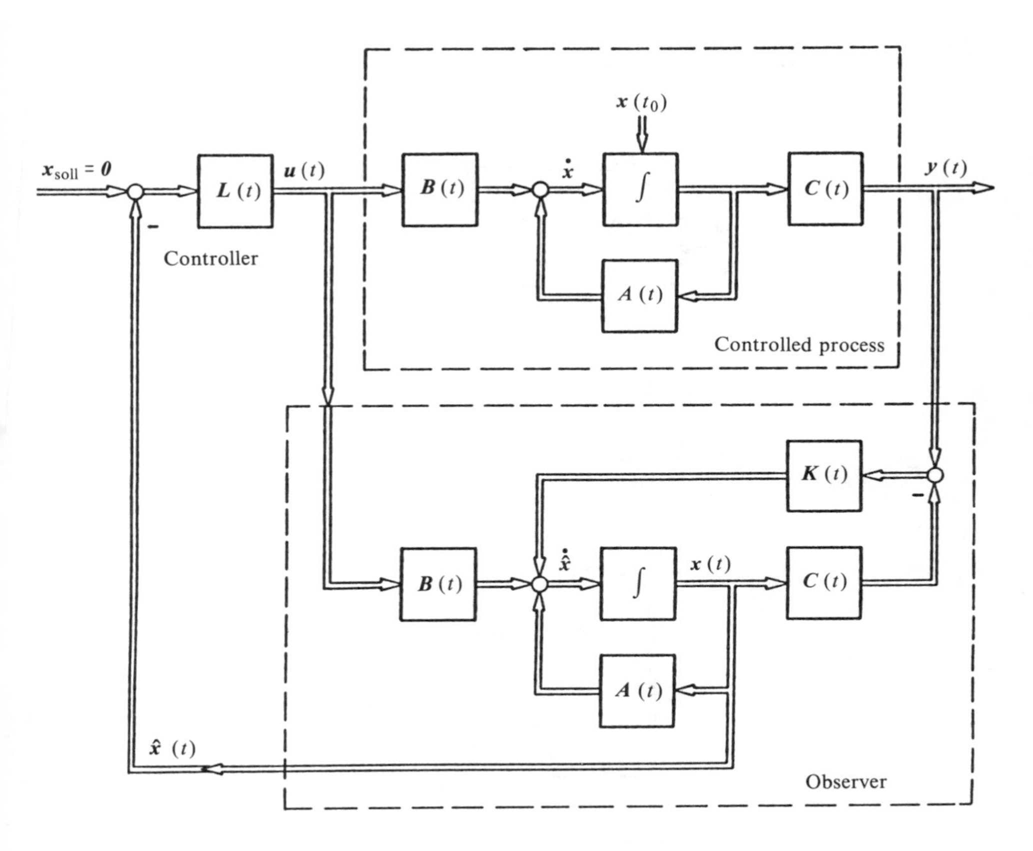


**Figure 6.2** Observer in the control loop.

This decoupling is called *algebraic separation*. It is always valid irrespective of the values of $\boldsymbol{K}$ and $\boldsymbol{L}$ whenever the process has the form (6.1), the observer has the form (6.9) and the controller has the form (6.11). This fact is particularly striking if we consider Figure 6.3 which corresponds to the differential equation (6.13) and is equivalent to Figure 6.2. Starting from the initial value $\tilde{\boldsymbol{x}}(t_0)$, the observer error performs movements that are not influenced at all by the control process and the controller gain. In its turn, the dynamics of the control process is completely independent of the observer and its gain factor. In terms of the control process, the estimation error $\tilde{\boldsymbol{x}}$ only acts as a disturbance. If the observer is designed so that $\tilde{\boldsymbol{x}}(t) \rightarrow \boldsymbol{0}$ for some $t_1$, the control process is also completely autonomous from $t_1$ onward. Then the characteristic motions of the observer no longer appear at all in the control loop. In Figure 6.2, the observer including the matrix $\boldsymbol{C}(t)$ of the process is dynamically shunted by the relationship $\hat{\boldsymbol{x}}(t) \equiv \boldsymbol{x}(t)$ from $t_1$ on. However, this total *vanishing*

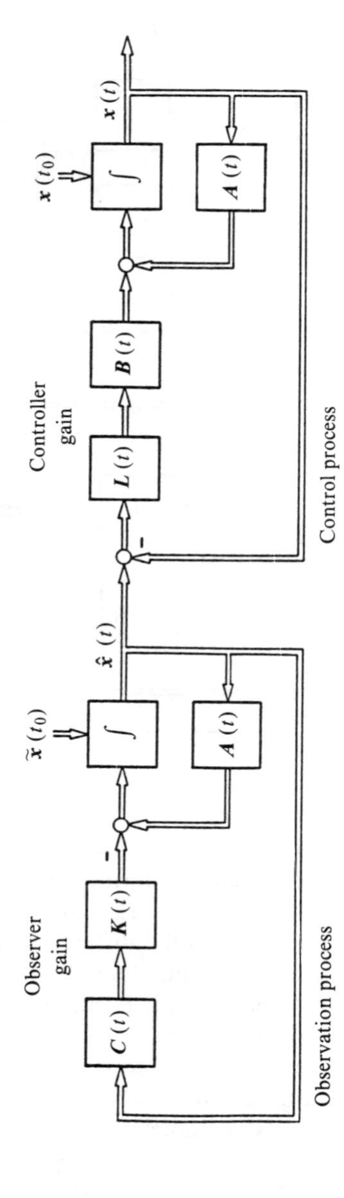


**Figure 6.3** Algebraic separation of observation and control.

of the observer only occurs when the system matrices $\boldsymbol{A}$, $\boldsymbol{B}$ and $\boldsymbol{C}$ are exactly known, and when disturbance and measurement noises are absent. Stochastic variables are considered in the next section.

## 6.3 FILTER IN THE CONTROL LOOP AND STOCHASTIC SEPARATION

We are dealing with a stochastic control problem if at least one of the following three cases is given:

(i) the initial state of the process is a random variable;

(ii) there is a stochastic disturbance variable $\boldsymbol{v}$ at the input to the process;

(iii) there is measurement noise $\boldsymbol{w}$ associated with the output variable of the process.

In the case of discrete time, the stochastic control problem for linear processes is usually formulated as follows. We are given the system

$$\boldsymbol{x}(k+1) = \boldsymbol{A}(k)\boldsymbol{x}(k) + \boldsymbol{B}(k)\boldsymbol{u}(k) + \boldsymbol{v}(k) \tag{6.14a}$$

$$\boldsymbol{y}(k) = \boldsymbol{C}(k)\boldsymbol{x}(k) + \boldsymbol{w}(k), \quad k_0 \leqslant k \leqslant k_1 \tag{6.14b}$$

The expectation and the covariance of the initial state $\boldsymbol{x}(k_0)$ are known. The parameters $\boldsymbol{A}$, $\boldsymbol{B}$ and $\boldsymbol{C}$ and the covariance matrices $\boldsymbol{Q}$ and $\boldsymbol{R}$ of the white noise $\boldsymbol{v}$ and $\boldsymbol{w}$ are given for all $\kappa$ in the interval $[k_0, k_1]$. We seek a control law of the form

$$\boldsymbol{u}(\kappa) = f\{\kappa, \boldsymbol{y}(k_0), \boldsymbol{y}(k_0+1), \ldots, \boldsymbol{y}(k)\} \tag{6.14c}$$

for $k \leqslant \kappa \leqslant k_1 - 1$ so that the conditional expectation

$$\hat{J}(k_1, k) = E\left\{ \sum_{\kappa=k+1}^{k_1} [\boldsymbol{x}'(\kappa)\bar{\boldsymbol{Q}}(\kappa)\boldsymbol{x}(\kappa) + \boldsymbol{u}'(\kappa-1)\bar{\boldsymbol{R}}(\kappa-1)\boldsymbol{u}(\kappa-1)] \,\middle|\, \boldsymbol{y}(k_0), \ldots, \boldsymbol{y}(k) \right\} \tag{6.14d}$$

becomes a minimum.

*Remark 6.1.* The control law (6.14c) is to be evaluated at time $k$. Therefore it must be formulated so that the required control sequence in the present ($k$) and the future ($\kappa > k$) depends only on the present and past measurement values (Figure 6.4). What is most interesting here is the current value $\boldsymbol{u}(k)$. The last value of the

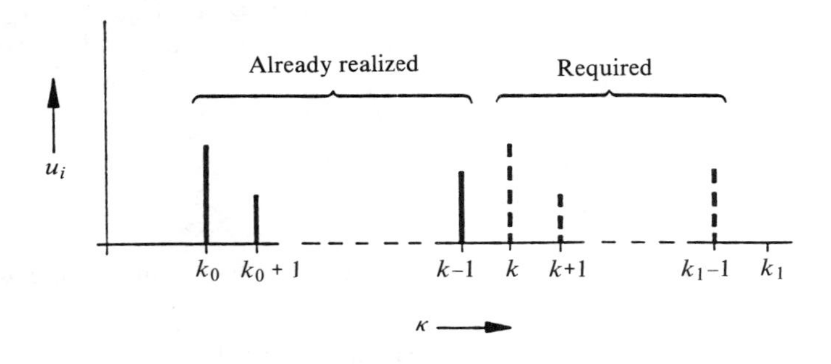


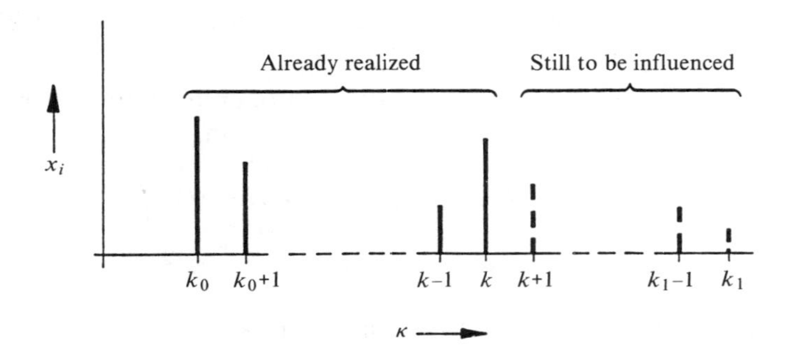


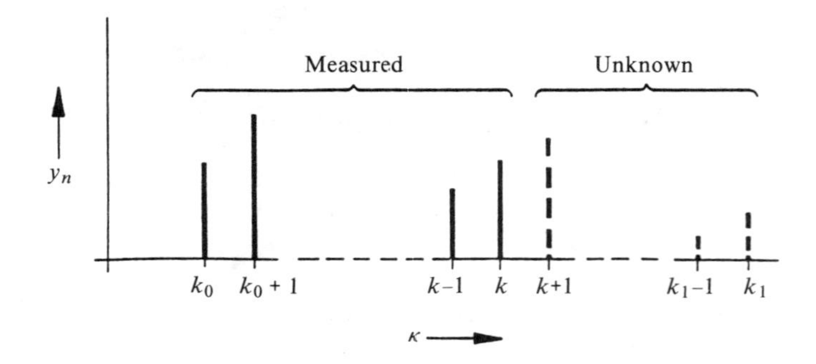


**Figure 6.4** On the stochastic control problem in discrete time.

measurement variable on which it is based is $\boldsymbol{y}(k)$. According to (6.14a) and (6.14b) this is determined by

$$\boldsymbol{y}(k) = \boldsymbol{C}(k)\{\boldsymbol{A}(k-1)\boldsymbol{x}(k-1) + \boldsymbol{B}(k-1)\boldsymbol{u}(k-1) + \boldsymbol{v}(k-1)\} + \boldsymbol{w}(k)$$

where $\boldsymbol{u}(k-1)$ has already been realized and is therefore known.

*Remark 6.2*. The sum in the performance index (6.14d) consists of quadratic forms. Only the sequence $\boldsymbol{x}(k+1), \ldots, \boldsymbol{x}(k_1)$ is weighted without the current value $\boldsymbol{x}(k)$ because this can no longer be influenced by the choice of $\boldsymbol{u}(k)$. The control sequence is only of importance in the index for the times $k, \ldots, k_1 - 1$ since $\boldsymbol{u}(k_1)$ no longer affects the end value $\boldsymbol{x}(k_1)$.

*Remark 6.3*. Apart from $\boldsymbol{u}(\kappa)$, the sequence $\boldsymbol{x}(k+1), \ldots, \boldsymbol{x}(k_1)$ also depends on $\boldsymbol{v}(\kappa)$. The sum in the index (6.14d) is therefore a stochastic variable even for given $\boldsymbol{y}(k_0), \ldots, \boldsymbol{y}(k)$ and its value is determined by the random sequence $\boldsymbol{v}(k), \ldots, \boldsymbol{v}(k_1 - 1)$. Therefore it would make little sense to use this sum as a performance index because it would scatter stochastically from experiment to experiment. Instead, a mean value is formed. The averaging can be conditional with respect to the known sequence $\boldsymbol{y}(k_0), \ldots, \boldsymbol{y}(k)$. The remaining expectation should be taken with respect to the distribution of the initial state $\boldsymbol{x}(k_0)$ and the processes $\{\boldsymbol{v}(\kappa)\}$ and $\{\boldsymbol{w}(\kappa)\}$.

*Remark 6.4*. Let us assume in (6.14a) that $\boldsymbol{u}(k) \equiv \boldsymbol{0}$ and that the initial state $\boldsymbol{x}(k_0)$ and the disturbance variables $\boldsymbol{v}$ of the uncontrolled process are Gaussian random variables. Then the sequence $\boldsymbol{x}(\kappa)$, $k_0 \leqslant \kappa \leqslant k_1$, is a sample of a vector Gaussian process because of the linearity of the system. If, in addition, the noise $\boldsymbol{w}$ in the linear measurement equation (6.14b) is also normally distributed, the measurement sequence $\boldsymbol{y}(\kappa)$, $k_0 \leqslant \kappa \leqslant k_1$, also forms a Gaussian process. Now, if $u(\kappa) \neq \boldsymbol{0}$, the Gaussian property is retained for the controlled process provided that the control law (6.14c) is linear. It can be shown (Doob, 1953) that in the case of Gaussian variables the linear minimum variance estimate ("optimum in the wide sense") is equal to the conditional expectation of the required variable with respect to the measurements ("optimum in the strict sense"). In the Gaussian case, we therefore have for the estimate from Section 5.3.6:

$$\hat{\boldsymbol{x}}(\kappa|k) = E\{\boldsymbol{x}(\kappa)|\, \boldsymbol{y}(k_0), \boldsymbol{y}(k_0+1), \ldots, \boldsymbol{y}(k)\} \tag{6.15}$$

After these remarks, we are in the position to give a heuristic explanation of the separation principle. We look more closely at the first quadratic form in the index (6.14d). We decompose $\boldsymbol{x}(\kappa)$ into the prediction value $\hat{\boldsymbol{x}}(\kappa|k)$ on the basis of the data $\boldsymbol{y}(k_0), \ldots, \boldsymbol{y}(k)$ and into the associated prediction error. In the notation of Section 5.3.6, we therefore have

$$\boldsymbol{x}(\kappa) = \hat{\boldsymbol{x}}(\kappa|k) + \tilde{\boldsymbol{x}}(\kappa|k)$$

so that

$$\begin{aligned}
&\boldsymbol{x}'(\kappa)\bar{\boldsymbol{Q}}(\kappa)\boldsymbol{x}(\kappa)\\
&= \{\hat{\boldsymbol{x}}'(\kappa|k) + \tilde{\boldsymbol{x}}'(\kappa|k)\}\bar{\boldsymbol{Q}}(\kappa)\{\hat{\boldsymbol{x}}(\kappa|k) + \tilde{\boldsymbol{x}}(\kappa|k)\}\\
&= \hat{\boldsymbol{x}}'(\kappa|k)\bar{\boldsymbol{Q}}(\kappa)\hat{\boldsymbol{x}}(\kappa|k) + \tilde{\boldsymbol{x}}'(\kappa|k)\bar{\boldsymbol{Q}}(\kappa)\tilde{\boldsymbol{x}}(\kappa|k) + 2\hat{\boldsymbol{x}}'(\kappa|k)\bar{\boldsymbol{Q}}(\kappa)\tilde{\boldsymbol{x}}(\kappa|k)
\end{aligned}$$

Because $\boldsymbol{a}'\boldsymbol{b} = \text{tr}\{\boldsymbol{b}\boldsymbol{a}'\}$, the last summand can also be written as

$$2\ \text{tr}\{\bar{\boldsymbol{Q}}(\kappa)\tilde{\boldsymbol{x}}(\kappa|k)\hat{\boldsymbol{x}}'(\kappa|k)\}$$

The conditional expectation of this expression, which should be taken according to (6.14d), vanishes for all $\kappa$ because

$$E\{\tilde{\boldsymbol{x}}(\kappa|k)\hat{\boldsymbol{x}}'(\kappa|k)\} = \boldsymbol{0}$$

The index (6.14d) can now be split into two parts:

$$\begin{aligned}\hat{J}_1(k_1, k) = {} & E\left\{\sum_{\kappa=k+1}^{k_1} \hat{\boldsymbol{x}}'(\kappa|k)\bar{\boldsymbol{Q}}(\kappa)\hat{\boldsymbol{x}}(\kappa|k) + \boldsymbol{u}'(\kappa-1)\bar{\boldsymbol{R}}(\kappa-1)\boldsymbol{u}(\kappa-1)\right\} \\ & + E\left\{\sum_{\kappa=k+1}^{k_1} \tilde{\boldsymbol{x}}'(\kappa|k)\bar{\boldsymbol{Q}}(\kappa)\tilde{\boldsymbol{x}}(\kappa|k)\right\}\end{aligned}$$

The first part reflects the contribution which is to be minimized by optimal choice of the sequence $\boldsymbol{u}(k), \ldots, \boldsymbol{u}(k_1 - 1)$. According to Sections 5.3.5 and 5.3.6, $\hat{\boldsymbol{x}}(\kappa|k)$ depends on $\boldsymbol{u}(k), \ldots, \boldsymbol{u}(\kappa - 1)$ (compare Theorems 5.2 and 5.3). However, the second part is minimized if $\boldsymbol{x}(\kappa)$ is estimated in an optimal manner.

The stochastic control problem for a linear process, quadratic performance index, linear control law and Gaussian disturbance and measurement noise can therefore be separated into (a) a problem of deterministic optimal control and (b) a problem of stochastic optimal filtering. In such a case, controller and filter can be designed completely separately from each other. The resulting closed loop consisting of process, filter and controller, in which instead of the process state $\boldsymbol{x}$ its optimal estimation value $\hat{\boldsymbol{x}}$ is fed back to the controller, is the required optimal system in the sense of the index (6.14d).

The separation theorem was discovered by Joseph and Tou in 1961 for the discrete-time case (Joseph, 1961; Joseph and Tou, 1961). The separation principle was immediately applied to systems in continuous time (Kalman *et al.*, 1962), but it was only in 1968 that Bucy provided a rigorous proof for this (Bucy and Joseph, 1968).

In continuous time, the linear stochastic control problem is formulated as follows. We are given the process

$$\dot{\boldsymbol{x}}(t) = \boldsymbol{A}(t)\boldsymbol{x}(t) + \boldsymbol{B}(t)\boldsymbol{u}(t) + \boldsymbol{v}(t) \tag{6.16a}$$

$$\boldsymbol{y}(t) = \boldsymbol{C}(t)\boldsymbol{x}(t) + \boldsymbol{w}(t), \quad t_0 \leq t \leq t_1 \tag{6.16b}$$

where $\boldsymbol{x}(t_0)$ is a normally distributed random vector with expectation $\boldsymbol{\xi}$ while $\{\boldsymbol{v}(t)\}$

and $\{w(t)\}$ are vector-valued Gaussian white random processes. We require a linear control law of the form

$$\boldsymbol{u}(\tau) = F\{\tau; \boldsymbol{y}(\sigma), \quad t_0 \leq \sigma \leq t\}, \quad t \leq \tau \leq t_1 \tag{6.16c}$$

so that the quadratic performance index

$$\hat{J}(t_1, t) = E\left\{\boldsymbol{x}'(t_1)\bar{\boldsymbol{P}}(t_1)\boldsymbol{x}(t_1) + \int_t^{t_1} [\boldsymbol{x}'(\tau)\bar{\boldsymbol{Q}}(\tau)\boldsymbol{x}(\tau) + \boldsymbol{u}'(\tau)\bar{\boldsymbol{R}}(\tau)\boldsymbol{u}(\tau)]\mathrm{d}\tau \,\middle|\, \boldsymbol{y}(\sigma), \quad t_0 \leq \sigma \leq t\right\} \tag{6.16d}$$

takes a minimum.

*Note 6.1*. The formulation of the problem is very similar to the discrete-time case. The only difference worth noting is the term with $\bar{\boldsymbol{P}}(t_1)$ in equation (6.16d). It is specially introduced to weigh the control result at the end of the interval with finite cost. For discrete time, this weighting is performed by virtue of $\bar{\boldsymbol{Q}}(k_1)$ in the index (6.14d). For continuous time, the matrix $\bar{\boldsymbol{Q}}(\tau)$ would have to be provided with a $\delta$ function at $\tau = t_1$ if we wish to neglect the term with $\bar{\boldsymbol{P}}(t_1)$.

The solution of the continuous-time stochastic control problem as determined by the separation principle is given by (a) the deterministic optimal control law (6.11), (6.4) and (6.5), and (b) the stochastic optimal filter law (5.96), (5.93) and (5.94). The resulting control system is shown in Figure 6.5.

Stochastic separation is of central importance for the optimal synthesis of linear stochastically disturbed control loops. This class of problems is largely solved by decomposition into deterministic linear control and stochastic linear filtering which both have a fully developed theory. So far, the separation theorem has only been proved for linear systems. It cannot be assumed that it is generally valid in the strict sense for nonlinear systems as well. Nevertheless in the synthesis of nonlinear stochastic control systems, there is seldom any other choice available than to proceed as if separation were valid. Nonlinear filters and nonlinear controllers are also designed separately and subsequently connected together. Although this is not a completely satisfactory method mathematically, it is at least feasible. However, there is no guarantee for absolute optimization.

We now consider a linear control loop which has been optimally designed according to the stochastic separation principle and examine it in terms of algebraic separation. In the closed loop, the following equations apply: equation (5.97) for

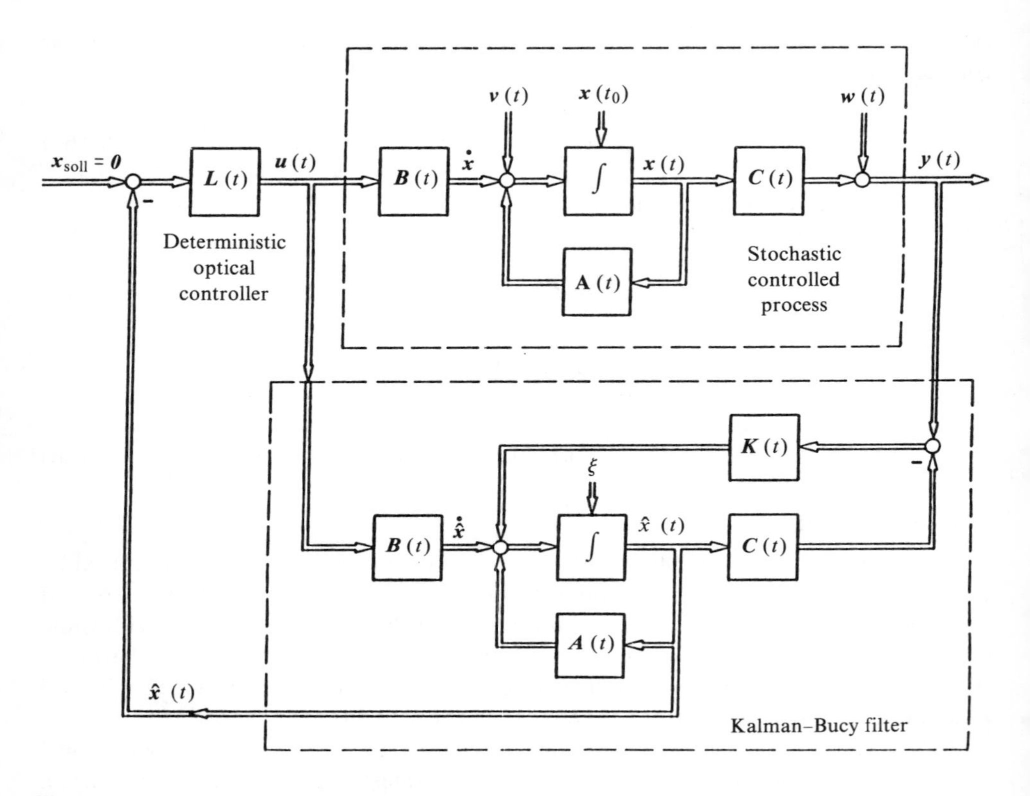


**Figure 6.5** The optimal linear stochastic control system.

$\tilde{x}(t)$, equation (6.11) with $\hat{x}(t) = x(t) - \tilde{x}(t)$ and equation (6.16a). After substitution of (6.11) into (6.16a), we obtain

$$\dot{x}(t) = A(t)x(t) - B(t)L(t)\{x(t) - \tilde{x}(t)\} + v(t) \tag{6.17}$$

We repeat equation (5.97), i.e.,

$$\dot{\tilde{x}}(t) = \{A(t) - K(t)C(t)\}\tilde{x}(t) + v(t) - K(t)w(t)$$

Writing these equations as a combined differential equation yields

$$\frac{\mathrm{d}}{\mathrm{d}t}\begin{bmatrix} x \\ \tilde{x} \end{bmatrix} = \begin{bmatrix} A - BL & BL \\ 0 & A - KC \end{bmatrix}\begin{bmatrix} x \\ \tilde{x} \end{bmatrix} + \begin{bmatrix} I \\ I \end{bmatrix} v + \begin{bmatrix} 0 \\ -K \end{bmatrix} w \tag{6.18}$$

Here as well the dynamics of the error $\tilde{\boldsymbol{x}}(t)$ is completely decoupled from the dynamics of the state $\boldsymbol{x}(t)$: $\tilde{\boldsymbol{x}}$ is influenced neither by $\boldsymbol{x}$ nor by $\boldsymbol{L}$. The characteristic motions of $\boldsymbol{x}$, conversely, are only determined by $\boldsymbol{A}$, $\boldsymbol{B}$ and $\boldsymbol{L}$, while $\boldsymbol{BL}\tilde{\boldsymbol{x}}$ and $\boldsymbol{v}$ merely act as external excitations. The algebraic separation principle is therefore also valid here.

## 6.4 COMMENTS ON THE MATRIX RICCATI DIFFERENTIAL EQUATION

The properties of the Kalman–Bucy filter and the process $\{\hat{\boldsymbol{x}}(t)\}$ are largely determined by the solution $\tilde{\boldsymbol{P}}(t)$ of the matrix Riccati differential equation (5.94). First of all, $\tilde{\boldsymbol{P}}(t)$ is the only unknown factor in the filter gain factor $\boldsymbol{K}(t)$ and so the only free parameter in the filter equation (5.92). Furthermore, the main diagonal of $\tilde{\boldsymbol{P}}(t)$ contains the variances of the estimation errors which are the key measures for the quality of filtering.

Remember that in the case of a Gaussian process the output variable $\hat{\boldsymbol{x}}(t)$ of the Kalman–Bucy filter represents the conditional expectation of $\boldsymbol{x}(t)$ while $\tilde{\boldsymbol{P}}(t)$ is the conditional covariance of $\boldsymbol{x}(t)$. For both, the conditioning is made with respect to the measured sample function $\boldsymbol{y}(\tau)$, $t_0 \leqslant \tau \leqslant t$. Also note that a Gaussian distribution is completely defined by its expectation and covariance. Thus, for Gaussian processes, the specification of $\hat{\boldsymbol{x}}(t)$ and $\tilde{\boldsymbol{P}}(t)$ is equivalent to the knowledge of the complete $n$-dimensional conditional joint distribution of $\boldsymbol{x}(t)$! On this basis, all other conditional expectations of $\boldsymbol{x}(t)$ of any interest can be calculated by ordinary integration.

The matrix Riccati differential equation of the filter problem had the following form:

$$\frac{\mathrm{d}}{\mathrm{d}t}\tilde{\boldsymbol{P}}(t) = \boldsymbol{A}(t)\tilde{\boldsymbol{P}}(t) + \tilde{\boldsymbol{P}}(t)\boldsymbol{A}'(t) - \tilde{\boldsymbol{P}}(t)\boldsymbol{C}'(t)\boldsymbol{R}^{-1}(t)\boldsymbol{C}(t)\tilde{\boldsymbol{P}}(t) + \boldsymbol{Q}(t)$$

$$\tilde{\boldsymbol{P}}(t_0) = \boldsymbol{P}(t_0) \quad t_0 \leqslant t \tag{6.19}$$

The matrices $\boldsymbol{A}$, $\boldsymbol{C}$, $\boldsymbol{P}$, $\boldsymbol{Q}$ and $\boldsymbol{R}$ were defined in Section 5.4.1. The matrix Riccati differential equation constitutes a system of $n^2$ coupled differential equations of the first order for the $n \times n$ elements $\tilde{p}_{ij}(t)$. The term $\tilde{\boldsymbol{P}}\boldsymbol{C}'\boldsymbol{R}^{-1}\boldsymbol{C}\tilde{\boldsymbol{P}}$ leads to terms of the form $\tilde{p}_{ij}\tilde{p}_{kl}$ or $\tilde{p}_{ij}^2$ so that the system of differential equations is nonlinear. Since $\tilde{\boldsymbol{P}}$ is symmetric, the number of distinct elements reduces to $(n^2 + n)/2$, but in many practical cases where $n$ is of the order of magnitude of 10 this is still a considerable number. This makes it clear that, apart from establishing the mathematical model of the observed process, the main effort in filter synthesis is the solution of the matrix Riccati differential equation.

With the specifications for $\boldsymbol{P}(t_0)$, $\boldsymbol{Q}$ and $\boldsymbol{R}$ from Section 5.4.1 and the assumption that $\boldsymbol{A}(t)$, $\boldsymbol{C}(t)$, $\boldsymbol{Q}(t)$ and $\boldsymbol{R}(t)$ are continuous in $t$, there always exists a unique solution to the Riccati differential equation such that $\tilde{\boldsymbol{P}}(t)$ is positive semidefinite throughout. Stability criteria are also available in the relevant literature (e.g., Bucy and Joseph, 1968).

As a rule, the Riccati differential equation must be solved numerically. Two problems are likely to occur as a result of truncation errors: (i) loss of symmetry for $\tilde{\boldsymbol{P}}$ and (ii) violation of the positive semidefinite structure of $\tilde{\boldsymbol{P}}$.

In order to avoid (i), it is recommended that all $n^2$ elements of $\tilde{\boldsymbol{P}}$ are retained and $\tilde{\boldsymbol{P}}$ is made symmetric after every integration or computational interval by arithmetic averaging of $\tilde{p}_{ij}$ and $\tilde{p}_{ji}$. In order to eliminate problem (ii) which occurs in particular for weakly controllable processes ($\boldsymbol{Q}$ small and/or positive semidefinite), it is suggested that the elements of $\boldsymbol{Q}$ are artificially enlarged (Bucy and Joseph, 1968).

The most obvious and commonly used method for solving the Riccati differential equation is direct integration with the help of one of the relevant numerical integration procedures. However, sometimes a method for analytical or numerical solution of the Riccati differential equation is used which is based on the following considerations. Generally, the *adjoint system* of a differential equation $\dot{\boldsymbol{x}} = \boldsymbol{F}(t)\boldsymbol{x}$ is defined by the differential equation $\dot{\boldsymbol{\xi}} = -\boldsymbol{F}'(t)\boldsymbol{\xi}$ (see Section 4.4, equation (4.33a)). We now take the homogeneous part of the differential equation of the filter error (5.97) and replace $\boldsymbol{K}(t)$ by its optimal value according to equation (5.93) (all time arguments are equal to $t$ and are omitted):

$$\dot{\tilde{\boldsymbol{x}}} = (\boldsymbol{A} - \tilde{\boldsymbol{P}}\boldsymbol{C}'\boldsymbol{R}^{-1}\boldsymbol{C})\tilde{\boldsymbol{x}}$$

The adjoint system to this is equation (6.20). We next postmultiply the Riccati differential equation by $\boldsymbol{\xi}$ and place it in the row beneath:

$$\dot{\boldsymbol{\xi}} = (-\qquad \boldsymbol{A}' + \quad \boldsymbol{C}'\boldsymbol{R}^{-1}\boldsymbol{C}\tilde{\boldsymbol{P}})\qquad \boldsymbol{\xi} \tag{6.20}$$

$$\dot{\tilde{\boldsymbol{P}}}\boldsymbol{\xi} = (\boldsymbol{A}\tilde{\boldsymbol{P}} + \tilde{\boldsymbol{P}}\boldsymbol{A}' - \tilde{\boldsymbol{P}}\boldsymbol{C}'\boldsymbol{R}^{-1}\boldsymbol{C}\tilde{\boldsymbol{P}} + \boldsymbol{Q})\boldsymbol{\xi} \tag{6.21}$$

Premultiplication of (6.20) by $\tilde{\boldsymbol{P}}$ and addition of both equations gives

$$\tilde{\boldsymbol{P}}\dot{\boldsymbol{\xi}} + \dot{\tilde{\boldsymbol{P}}}\boldsymbol{\xi} = (\boldsymbol{A}\tilde{\boldsymbol{P}} + \boldsymbol{Q})\boldsymbol{\xi} \tag{6.22}$$

Fortunately, a particular effect of this procedure has been to remove the nonlinear term of the Riccati differential equation! With the definition

$$\tilde{\boldsymbol{P}}(t)\boldsymbol{\xi}(t) := \boldsymbol{\eta}(t) \tag{6.23}$$

equations (6.20) and (6.22) become the 2nd-order linear Hamiltonian system

$$\frac{\mathrm{d}}{\mathrm{d}t}\begin{bmatrix}\boldsymbol{\xi}\\ \boldsymbol{\eta}\end{bmatrix} = \begin{bmatrix}-\boldsymbol{A}'(t) & \boldsymbol{C}'(t)\boldsymbol{R}^{-1}(t)\boldsymbol{C}(t)\\ \boldsymbol{Q}(t) & \boldsymbol{A}(t)\end{bmatrix}\begin{bmatrix}\boldsymbol{\xi}\\ \boldsymbol{\eta}\end{bmatrix} \tag{6.24}$$

With this interim result, the integration of the inhomogeneous ($\boldsymbol{Q} \neq 0$) nonlinear Riccati differential equation can be reduced to the solution of this homogeneous linear differential equation. Obviously, we can integrate (6.24) successively $n$ times: as the initial condition for the $k$th integration we choose the $k$th column of the unit matrix for $\boldsymbol{\xi}(t_0)$ and the $k$th column of $\tilde{\boldsymbol{P}}(t_0) = \boldsymbol{P}(t_0)$ for $\boldsymbol{\eta}(t_0)$ according to (6.23). The pairs of solutions $\boldsymbol{\xi}_k(t)$, $\boldsymbol{\eta}_k(t)$ obtained for $k = 1, 2, \ldots, n$ are combined to form the system

$$\begin{bmatrix}\boldsymbol{X}(t)\\ \boldsymbol{Y}(t)\end{bmatrix} := \begin{bmatrix}\boldsymbol{\xi}_1(t), \ldots, \boldsymbol{\xi}_n(t)\\ \boldsymbol{\eta}_1(t), \ldots, \boldsymbol{\eta}_n(t)\end{bmatrix}$$

with

$$\begin{bmatrix}\boldsymbol{X}(t_0)\\ \boldsymbol{Y}(t_0)\end{bmatrix} = \begin{bmatrix}\boldsymbol{I}\\ \tilde{\boldsymbol{P}}(t_0)\end{bmatrix}$$

Because of the definition (6.23) we always have the relationship $\tilde{\boldsymbol{P}}(t)\boldsymbol{\xi}_k(t) = \boldsymbol{\eta}_k(t)$, or in combined form

$$\tilde{\boldsymbol{P}}(t)\boldsymbol{X}(t) = \boldsymbol{Y}(t)$$

Solving for $\tilde{\boldsymbol{P}}$ gives

$$\tilde{\boldsymbol{P}}(t) = \boldsymbol{Y}(t)\boldsymbol{X}^{-1}(t) \tag{6.25}$$

The system of solutions $\boldsymbol{X}(t)$, $\boldsymbol{Y}(t)$ can also be expressed in terms of the transition matrix of the Hamiltonian system. Let $\boldsymbol{H}(t)$ be the $2n \times 2n$ matrix in equation (6.24), i.e.,

$$\boldsymbol{H}(t) := \begin{bmatrix}-\boldsymbol{A}'(t) & \boldsymbol{C}'(t)\boldsymbol{R}^{-1}(t)\boldsymbol{C}(t)\\ \boldsymbol{Q}(t) & \boldsymbol{A}(t)\end{bmatrix} \tag{6.26}$$

The $2n \times 2n$ transition matrix $\boldsymbol{\theta}(t, t_0)$ associated with $\boldsymbol{H}(t)$ is defined by

$$\frac{\mathrm{d}}{\mathrm{d}t}\boldsymbol{\theta}(t, t_0) = \boldsymbol{H}(t)\boldsymbol{\theta}(t, t_0) \quad \boldsymbol{\theta}(t_0, t_0) = \boldsymbol{I}_{2n} \tag{6.27}$$

$\boldsymbol{\theta}(t, t_0)$ is now subdivided into four $n \times n$ submatrices $\theta_{ij}(t, t_0)$. In this way the general solution of the Hamiltonian system (6.24) is given by

$$\begin{bmatrix} \boldsymbol{\xi}(t) \\ \boldsymbol{\eta}(t) \end{bmatrix} = \begin{bmatrix} \boldsymbol{\theta}_{11}(t, t_0) & \boldsymbol{\theta}_{12}(t, t_0) \\ \boldsymbol{\theta}_{21}(t, t_0) & \boldsymbol{\theta}_{22}(t, t_0) \end{bmatrix} \begin{bmatrix} \boldsymbol{\xi}(t_0) \\ \boldsymbol{\eta}(t_0) \end{bmatrix}$$

With the above initial conditions, the combined system of solutions $\boldsymbol{X}(t)$, $\boldsymbol{Y}(t)$ can be written as

$$\begin{bmatrix} \boldsymbol{X}(t) \\ \boldsymbol{Y}(t) \end{bmatrix} = \begin{bmatrix} \boldsymbol{\theta}_{11}(t, t_0) & \boldsymbol{\theta}_{12}(t, t_0) \\ \boldsymbol{\theta}_{21}(t, t_0) & \boldsymbol{\theta}_{22}(t, t_0) \end{bmatrix} \begin{bmatrix} \boldsymbol{I} \\ \tilde{\boldsymbol{P}}(t_0) \end{bmatrix}$$

Substitution in (6.25) gives an elegant solution formula for the matrix Riccati differential equation (6.19):

$$\tilde{\boldsymbol{P}}(t) = \{\boldsymbol{\theta}_{21}(t, t_0) + \boldsymbol{\theta}_{22}(t, t_0)\tilde{\boldsymbol{P}}(t_0)\}\{\boldsymbol{\theta}_{11}(t, t_0) + \boldsymbol{\theta}_{12}(t, t_0)\tilde{\boldsymbol{P}}(t_0)\}^{-1} \tag{6.28}$$

It can be shown (Bucy and Joseph, 1968) that the matrix in the second pair of braces is nonsingular for all $t \geqslant t_0$ and all positive semidefinite matrices $\tilde{\boldsymbol{P}}(t_0)$ so that this solution exists for every filter problem.

The result (6.28) is of great practical importance, in particular for time-invariant systems or stationary processes ($\boldsymbol{A}$, $\boldsymbol{C}$, $\boldsymbol{Q}$ and $\boldsymbol{R}$ constant over time). In these cases $\boldsymbol{H}$ is also constant and the transition matrix $\boldsymbol{\theta}$ is only a function of the difference $\Delta t$ of its arguments. It can be calculated numerically as exactly as required by the well-known exponential formula

$$\begin{aligned} \boldsymbol{\theta}(\Delta t) &= \exp(\boldsymbol{H}\Delta t) \\ &= \boldsymbol{I} + \boldsymbol{H}\Delta t + \ldots + \boldsymbol{H}^k \frac{(\Delta t)^k}{k!} + \ldots \end{aligned} \tag{6.29}$$

for any given value of $\Delta t$. With the help of equation (6.28), we finally obtain the corresponding value of $\tilde{\boldsymbol{P}}$ (without any integration). In this manner, $\tilde{\boldsymbol{P}}$ can be calculated in steps proceeding from one point in time to the next. (Do not forget symmetrization!) In the automatic synthesis program of Kalman and Englar (1965) this method is used for Riccati differential equations up to order 15, with equation (6.29) being of the type $30 \times 30$.

In order to complete the treatment of the Hamiltonian system, we introduce the $2n \times 2n$ matrix

$$\boldsymbol{J} := \begin{bmatrix} \boldsymbol{0} & \boldsymbol{I} \\ -\boldsymbol{I} & \boldsymbol{0} \end{bmatrix} \tag{6.30}$$

where $\boldsymbol{I}$ is the unit matrix of $n$th order. This matrix $\boldsymbol{J}$ constitutes the generalization of the imaginary number $j$ to $2n$ dimensions because we obviously have

$$\boldsymbol{J}^2 = -\boldsymbol{I}_{2n} \tag{6.31}$$

It can easily be shown by multiplying that

$$\boldsymbol{J}\boldsymbol{H}(t) + \boldsymbol{H}'(t)\boldsymbol{J} = \boldsymbol{0} \tag{6.32}$$

We premultiply this equation by $\boldsymbol{\theta}'(t, t_0)$ and postmultiply it by $\boldsymbol{\theta}(t, t_0)$, and then we substitute (6.27):

$$\boldsymbol{\theta}'(t, t_0)\boldsymbol{J}\frac{\mathrm{d}}{\mathrm{d}t}\boldsymbol{\theta}(t, t_0) + \frac{\mathrm{d}}{\mathrm{d}t}\boldsymbol{\theta}'(t, t_0)\boldsymbol{J}\boldsymbol{\theta}(t, t_0) = \boldsymbol{0}$$

It immediately follows from this that

$$\frac{\mathrm{d}}{\mathrm{d}t}\{\boldsymbol{\theta}'(t, t_0)\boldsymbol{J}\boldsymbol{\theta}(t, t_0)\} = \boldsymbol{0}$$

The product in braces is therefore constant over time. At time $t = t_0$, it is equal to $\boldsymbol{J}$ because $\boldsymbol{\theta}(t_0, t_0) = \boldsymbol{I}$. Therefore we have

$$\boldsymbol{\theta}'(t, t_0)\boldsymbol{J}\boldsymbol{\theta}(t, t_0) = \boldsymbol{J} \quad \text{for all } t \tag{6.33}$$

A $2n \times 2n$ matrix $\boldsymbol{D}$ with the property $\boldsymbol{D}'\boldsymbol{J}\boldsymbol{D} = \boldsymbol{J}$ is called *symplectic*. The transition matrix $\boldsymbol{\theta}(t, t_0)$ of the Hamiltonian system $\boldsymbol{H}(t)$ is therefore symplectic for all $t$. This property is also of practical use. If we premultiply (6.33) by $\boldsymbol{J}$ on both sides, we obtain $\boldsymbol{J}\boldsymbol{\theta}'\boldsymbol{J}\boldsymbol{\theta} = -\boldsymbol{I}_{2n}$ because of (6.31). Postmultiplication by $\boldsymbol{\theta}^{-1}$ gives

$$\boldsymbol{\theta}^{-1}(t, t_0) = -\boldsymbol{J}\boldsymbol{\theta}'(t, t_0)\boldsymbol{J} \tag{6.34}$$

With this result, it is very simple to invert a given transition matrix of a Hamiltonian system numerically.

Finally, we make a comment on the poles of Hamiltonian systems with constant parameters. It is well known from the relevant literature that the characteristic equation of $\boldsymbol{H}$ can be decomposed as follows:

$$\det(s\boldsymbol{I} - \boldsymbol{H}) = (-1)^n h(s)h(-s) \tag{6.35}$$

Here, $h(s)$ is a polynomial of $n$th degree in $s$, all of whose roots have a nonpositive real part (Bucy and Joseph, 1968, p. 105). The coefficients of $h(s)$ are real so that its complex roots are conjugated. The roots of $h(-s)$ are $-1$ times the roots of $h(s)$. Thus the eigenvalues (poles) of the Hamiltonian system $\boldsymbol{H}$ always lie on the $s$ plane in pairs and are reflected about the imaginary axis. Apart from the special case when all the poles lie directly on the imaginary axis, a constant Hamiltonian system always has both decreasing and increasing characteristic oscillations.

## 6.5 STATIONARY CONDITIONS AND WIENER FILTERS

In this section we consider the filter problem from the point of view of stationary processes. In the model of the observed process

$$\dot{\boldsymbol{x}}(t) = \boldsymbol{A}\boldsymbol{x}(t) + \boldsymbol{v}(t) \quad E\{\boldsymbol{x}(t_0)\} = \boldsymbol{0} \tag{6.36a}$$

$$\mathbf{y}(t) = \boldsymbol{C}\boldsymbol{x}(t) + \boldsymbol{w}(t) \tag{6.36b}$$

all parameters (i.e., the matrices $\boldsymbol{A}$ and $\boldsymbol{C}$ as well as the covariances $\boldsymbol{Q}$ and $\boldsymbol{R}$), are constant over time. According to Theorem 5.4, the corresponding Kalman–Bucy filter has the form

$$\dot{\hat{\boldsymbol{x}}}(t) = \boldsymbol{A}\hat{\boldsymbol{x}}(t) + \boldsymbol{K}(t)\{\boldsymbol{y}(t) - \boldsymbol{C}\hat{\boldsymbol{x}}(t)\} \quad \hat{\boldsymbol{x}}(t_0) = \boldsymbol{0} \tag{6.37a}$$

where

$$\boldsymbol{K}(t) = \tilde{\boldsymbol{P}}(t)\boldsymbol{C}'\boldsymbol{R}^{-1} \tag{6.37b}$$

and $\tilde{\boldsymbol{P}}$ is given by the Riccati differential equation with constant coefficients:

$$\begin{aligned} \dot{\tilde{\boldsymbol{P}}}(t) &= \boldsymbol{A}\tilde{\boldsymbol{P}}(t) + \tilde{\boldsymbol{P}}(t)\boldsymbol{A}' - \tilde{\boldsymbol{P}}(t)\boldsymbol{C}'\boldsymbol{R}^{-1}\boldsymbol{C}\tilde{\boldsymbol{P}}(t) + \boldsymbol{Q} \\ \tilde{\boldsymbol{P}}(t_0) &= \boldsymbol{P}(t_0) \end{aligned} \tag{6.37c}$$

Equations (6.37a–c) show that the filter has a time-varying gain factor $\boldsymbol{K}(t)$ as long as $\tilde{\boldsymbol{P}}(t)$ is still varying. A filter with constant parameters can only emerge when the Riccati differential equation has settled into an equilibrium state. The only exception is the unimportant special case when the initial condition $\tilde{\boldsymbol{P}}(t_0)$ corresponds to the stationary solution of the Riccati differential equation right from the beginning.

The transient motion of $\tilde{\boldsymbol{P}}(t)$ is governed by the asymptotic behavior of the matrix Riccati differential equation (6.37c). It can be shown that, if the model of the observed process is completely controllable with respect to $\boldsymbol{Q}$ and completely observable with respect to $\boldsymbol{R}$, all trajectories of the Riccati differential equation (6.37c) finally converge to the same end value $\bar{\boldsymbol{P}}$ (Bucy and Joseph, 1968). Hence the initial condition $\tilde{\boldsymbol{P}}(t_0)$ has no effect at all on the stationary solution. Therefore, the latter can be found in this case by starting at $\tilde{\boldsymbol{P}}(t_0) = \boldsymbol{0}$ and integrating the Riccati differential equation until $\tilde{\boldsymbol{P}}(t)$ reaches the steady state.

*Remark 6.5.* In the model (6.36a) the "control matrix" associated with $\boldsymbol{v}$ is the unit matrix. For example, complete controllability with respect to $\boldsymbol{Q}$ can be established by decomposing $\boldsymbol{Q}$ into its eigenvector matrix $\boldsymbol{V}$ and its Jordan canonical form $\boldsymbol{N}$ (see equation (5.33)) and substituting the transformed control matrix $\boldsymbol{V}\boldsymbol{N}^{1/2}$ in the controllability criterion (4.45). Because $\boldsymbol{V}$ is always nonsingular, complete controllability of the process $\boldsymbol{x}(t)$ with respect to $\boldsymbol{v}(t)$ is guaranteed according to Theorem 4.5 if $\boldsymbol{N}$ is nonsingular (i.e., if $\boldsymbol{Q}$ is nonsingular).

*Remark 6.6.* The equilibrium state $\bar{\boldsymbol{P}}$ is characterized by $\mathrm{d}\tilde{\boldsymbol{P}}/\mathrm{d}t = \boldsymbol{0}$ in (6.37c) and therefore satisfies the purely algebraic equation

$$\boldsymbol{0} = \boldsymbol{A}\bar{\boldsymbol{P}} + \bar{\boldsymbol{P}}\boldsymbol{A}' - \bar{\boldsymbol{P}}\boldsymbol{C}'\boldsymbol{R}^{-1}\boldsymbol{C}\bar{\boldsymbol{P}} + \boldsymbol{Q} \tag{6.38}$$

Because this is a system of $n^2$ coupled nonlinear equations, it is certainly not suitable for calculating $\bar{\boldsymbol{P}}$. This is all the more true because it is very difficult to extract from the multiple roots those which satisfy the positive semidefinite structure of $\bar{\boldsymbol{P}}$. However, equation (6.38) can be used to verify a solution for $\bar{\boldsymbol{P}}$ which has been found by using some other method.

In what follows, a better algebraic method will be given in which the unique positive semidefinite solution for $\bar{\boldsymbol{P}}$ is determined by using a linear equation system. It was discovered by Roth and Bass independently (Bass, 1967). We start from the characteristic equation (6.35) of the Hamiltonian system (6.26) which in this case is time invariant. In the polynomial $h(s)$ defined there, the Laplace variable $s$ is replaced by the matrix $\boldsymbol{H}$. We have

$$[-\bar{\boldsymbol{P}}, \boldsymbol{I}]h(\boldsymbol{H}) = \boldsymbol{0} \tag{6.39a}$$

or

$$h(-\boldsymbol{H})\begin{bmatrix} \boldsymbol{I} \\ \bar{\boldsymbol{P}} \end{bmatrix} = \boldsymbol{0} \tag{6.39b}$$

The proof of these interesting relationships can be found in Bass (1967) or Bucy and Joseph (1968, p. 105). The characteristic equation for $\boldsymbol{H}$, for example, can be conveniently determined on a digital computer using the Souriau–Fadeeva algorithm (Section A.5.3). Computer-aided methods are also available for the factorization in $h(s)$ and $h(-s)$. The solution of (6.39a) or (6.39b) can finally be obtained using the Gaussian or Cholesky algorithm. We should note in passing that each of the two systems (6.39a) and (6.39b) provides $2n^2$ scalar equations for the $n^2$ elements of $\bar{\boldsymbol{P}}$, half of which are redundant and can be omitted (see (6.45) in the following example).

*Example 6.1.* We consider the stochastic version of the observation problem in Example 4.1. An aircraft continuously takes the position measurement $y(t)$, which now has white noise $w(t)$ superimposed on it (here also, we only consider one geometric dimension). Speed measurements are not implemented. The acceleration $u(t)$ can be crudely measured; the errors of the accelerometer are modeled as white noise $v_1(t)$. The problem consists in designing a stationary Kalman–Bucy filter which provides the optimal estimates for position and speed.

As in Example 4.1, we define the state variables of position and speed as $x_2$ and $x_1$ respectively. In this way, the model of the observed process is as follows:

$$\begin{bmatrix} \dot{x}_1 \\ \dot{x}_2 \end{bmatrix} = \begin{bmatrix} 0 & 0 \\ 1 & 0 \end{bmatrix}\begin{bmatrix} x_1 \\ x_2 \end{bmatrix} + \begin{bmatrix} 1 \\ 0 \end{bmatrix} u + \begin{bmatrix} 1 \\ 0 \end{bmatrix} v_1 \tag{6.40a}$$

$$y = [0 \quad 1]\begin{bmatrix} x_1 \\ x_2 \end{bmatrix} + w \tag{6.40b}$$

We assume that the variance of $v_1$ is $q_{11} = 2$, $v_2 = 0$ and the variance of $w$ is $r = 0.5$.

We first examine the controllability and observability of the process. The controllability matrix (4.45) with respect to $v$ is

$$\bar{\boldsymbol{W}} = [\boldsymbol{B}, \boldsymbol{AB}] = \begin{bmatrix} 1 & 0 \\ 0 & 1 \end{bmatrix}$$

and is clearly nonsingular. The observability matrix (4.22) is

$$\bar{\boldsymbol{M}} = \begin{bmatrix} \boldsymbol{C} \\ \boldsymbol{CA} \end{bmatrix} = \begin{bmatrix} 0 & 1 \\ 1 & 0 \end{bmatrix}$$

and is also nonsingular. As $q$ and $r$ are positive numbers, we also have complete controllability and observability with respect to $q$ and $r$. The Riccati differential equation of the problem therefore has a unique stationary solution.

To calculate $\bar{\boldsymbol{P}}$, the Hamiltonian matrix (6.26) is formed:

$$\boldsymbol{H} = \begin{bmatrix} 0 & -1 & 0 & 0 \\ 0 & 0 & 0 & 2 \\ 2 & 0 & 0 & 0 \\ 0 & 0 & 1 & 0 \end{bmatrix} \tag{6.41}$$

The characteristic equation of $\boldsymbol{H}$ is

$$\det(s\boldsymbol{I} - \boldsymbol{H}) = \begin{vmatrix} s & 1 & 0 & 0 \\ 0 & s & 0 & -2 \\ -2 & 0 & s & 0 \\ 0 & 0 & -1 & s \end{vmatrix} = s^4 + 4 = h(s)h(-s) \tag{6.42}$$

Factorization gives

$$\begin{aligned} s^4 + 4 &= (s^2 + \mathrm{j}2) \times (s^2 - \mathrm{j}2) \\ &= (s + 1 - \mathrm{j})(s - 1 + \mathrm{j}) \times (s + 1 + \mathrm{j})(s - 1 - \mathrm{j}) \end{aligned}$$

Combining the first and the third factors yields $h(s)$:

$$h(s) = 2 + 2s + s^2 \tag{6.43}$$

The matrix polynomial $h(\boldsymbol{H})$ is accordingly

$$h(\boldsymbol{H}) = 2\boldsymbol{I} + 2\boldsymbol{H} + \boldsymbol{H}^2$$

where

$$\boldsymbol{H}^2 = \begin{bmatrix} 0 & & & -2 \\ & & 2 & \\ & -2 & & \\ 2 & & & 0 \end{bmatrix}$$

Hence we obtain

$$h(\boldsymbol{H}) = \begin{bmatrix} 2 & -2 & 0 & -2 \\ 0 & 2 & 2 & 4 \\ 4 & -2 & 2 & 0 \\ 2 & 0 & 2 & 2 \end{bmatrix} \tag{6.44}$$

The calculation procedure used so far can be verified using the Cayley–Hamilton theorem (Section A.5.2) since, according to (6.42), we must have $h(\boldsymbol{H})h(-\boldsymbol{H}) = \boldsymbol{0}$ (this is left as an exercise for the reader).

Equation (6.39a) means that $\bar{\boldsymbol{P}}$ times the top half of $h(\boldsymbol{H})$ is equal to $\boldsymbol{I}$ times the bottom half of $h(\boldsymbol{H})$:

$$\bar{\boldsymbol{P}}\left[\begin{array}{cc|cc} 2 & -2 & 0 & -2 \\ 0 & 2 & 2 & 4 \end{array}\right] = \left[\begin{array}{cc|cc} 4 & -2 & 2 & 0 \\ 2 & 0 & 2 & 2 \end{array}\right] \tag{6.45}$$

These are eight equations for the four elements of $\bar{\boldsymbol{P}}$. We only use the left-hand side of the broken line and transpose this so that we obtain the standard form of a linear system of equations:

$$\begin{bmatrix} 2 & 0 \\ -2 & 2 \end{bmatrix} \bar{\boldsymbol{P}} = \begin{bmatrix} 4 & 2 \\ -2 & 0 \end{bmatrix}$$

The solution is

$$\bar{\boldsymbol{P}} = \begin{bmatrix} 2 & 1 \\ 1 & 1 \end{bmatrix} \tag{6.46}$$

Verification with equation (6.38) confirms this result (exercise!). The quadratic form

$$\boldsymbol{x}'\bar{\boldsymbol{P}}\boldsymbol{x} = 2x_1^2 + 2x_1x_2 + x_2^2 = x_1^2 + (x_1 + x_2)^2$$

is positive for all $\boldsymbol{x} \neq \boldsymbol{0}$ and zero for $\boldsymbol{x} = \boldsymbol{0}$; $\bar{\boldsymbol{P}}$ is therefore positive definite. The variance of the estimation error of the speed is $\bar{p}_{11} = E\{\tilde{x}_1^2\} = 2$, and the error variance of the position is $\bar{p}_{22} = E\{\tilde{x}_2^2\} = 1$. It is now only a minor matter to specify the Kalman–Bucy filter. The gain is

$$\bar{\boldsymbol{K}} = \bar{\boldsymbol{P}}\boldsymbol{C}'\boldsymbol{R}^{-1} = \begin{bmatrix} 2 \\ 2 \end{bmatrix}$$

and so, according to (6.37b), we have

$$\dot{\hat{x}}(t) = \begin{bmatrix} 0 & 0 \\ 1 & 0 \end{bmatrix} \hat{x}(t) + \begin{bmatrix} 2 \\ 2 \end{bmatrix} \{y(t) - \hat{x}_2(t)\} \tag{6.47a}$$

This form has been chosen for didactic reasons. For technical implementation, it is written in the simpler form $\dot{\hat{x}} = (A - \bar{K}C)\hat{x} + \bar{K}y$:

$$\dot{\hat{x}}(t) = \begin{bmatrix} 0 & -2 \\ 1 & -2 \end{bmatrix} \hat{x}(t) + \begin{bmatrix} 2 \\ 2 \end{bmatrix} y(t) \tag{6.47b}$$

This solves the synthesis problem for the example given. The filter has the characteristic polynomial $s^2 + 2s + 2$ and the poles $s_{12} = -1 \pm j$; it is therefore asymptotically stable.

The procedure for filter synthesis using equations (6.39), which has just been demonstrated, provides the optimal filter for the stationary state of the observed process. It is therefore a new purely algebraic method for the design of Wiener filters.

In some situations, we would wish to be optimally observing the initial transients of the process. Then, the time-varying gain matrix $K(t)$ which is based on the transient motion of $\tilde{P}(t)$, must be realized. Of course, the trajectory of $\tilde{P}(t)$ can be determined by numerical solution of the Riccati differential equation (6.37c) as mentioned above. However, for lower-order systems, it can also be determined by using the Laplace transformation. According to equation (6.28), with $t_0 = 0$, we have

$$\tilde{P}(t) = \{\theta_{21}(t) + \theta_{22}(t)\tilde{P}(0)\}\{\theta_{11}(t) + \theta_{12}(t)\tilde{P}(0)\}^{-1} \tag{6.48}$$

It is known that the Laplace transform of the transition matrix $\theta(t)$ for the system $H$ is given by

$$\mathcal{L}\{\theta(t)\} = (sI - H)^{-1} \tag{6.49}$$

$$= \frac{1}{\det(sI - H)} \operatorname{adj}(sI - H) \tag{6.50}$$

The characteristic polynomial of the $2n \times 2n$ matrix $H$ (compare Section A.5.1) is given by

$$\det(sI - H) = s^{2n} + \alpha_{2n-1}s^{2n-1} + \ldots + \alpha_1 s + \alpha_0 \tag{6.51}$$

According to equation (A.42b) the adjoint matrix can be written

$$\operatorname{adj}(sI - H) = C_{2n-1}s^{2n-1} + C_{2n-2}s^{2n-2} + \ldots + C_1 s + C_0 \tag{6.52}$$

The coefficients $\alpha_i$ and $C_i$ are calculated by using the Souriau–Fadeeva algorithm (replace $\boldsymbol{A}$ by $\boldsymbol{H}$ and $n$ by $2n$ in (A.43a) and (A.43b)). The main work now consists in performing the inverse transformation, element by element, of equation (6.50) to the time domain and the evaluation of the solution formula (6.48).

*Example 6.2.* A first-order Wiener filter was designed in Example 5.4. Here, we examine the same process but we are now interested in the optimal filtering of the transient process starting from $t = 0$. The model for the observed scalar process was given by

$$\dot{x}(t) = -x(t) + v(t)$$
$$y(t) = x(t) + w(t)$$

where $Q = 2$, $R = 1$ and $t \geqslant 0$ (Figure 6.6). Let the initial state be exactly zero, i.e., $P(0) = 0$. The associated Hamiltonian matrix (6.26) has the form

$$\boldsymbol{H} = \begin{bmatrix} -A' & C^2R^{-1} \\ Q & A \end{bmatrix} = \begin{bmatrix} 1 & 1 \\ 2 & -1 \end{bmatrix}$$

The $2 \times 2$ transition matrix $\boldsymbol{\theta}(t)$ of $\boldsymbol{H}$ is defined by

$$\dot{\boldsymbol{\theta}}(t) = \boldsymbol{H}\boldsymbol{\theta}(t) \quad \boldsymbol{\theta}(0) = \boldsymbol{I} \tag{6.53}$$

Its Laplace transform is given by

$$\mathscr{L}\{\boldsymbol{\theta}(t)\} = \frac{1}{s^2 + \alpha_1 s + \alpha_0}(\boldsymbol{C}_1 s + \boldsymbol{C}_0) \tag{6.54}$$

The coefficients $\alpha_1$ and $\alpha_0$ as well as $\boldsymbol{C}_1$ and $\boldsymbol{C}_0$ follow from the algorithm (A.43) with $n = 2$.

*Initial value:*

$$\boldsymbol{C}_1 = \boldsymbol{I}$$

$k = 1$:

$$\alpha_1 = -\mathrm{tr}(\boldsymbol{C}_1\boldsymbol{H}) = -\mathrm{tr}\boldsymbol{H} = 0$$
$$\boldsymbol{C}_0 = \boldsymbol{C}_1\boldsymbol{H} + \alpha_1\boldsymbol{I} = \boldsymbol{H}$$

$k = 2$:

$$\begin{aligned} \alpha_0 &= -\tfrac{1}{2}\,\mathrm{tr}(C_0H) \\ &= -\tfrac{1}{2}\,\mathrm{tr}\boldsymbol{H}^2 \\ &= -\frac{1}{2}\mathrm{tr}\begin{bmatrix} 3 & 0 \\ 0 & 3 \end{bmatrix} = -3 \end{aligned}$$

$$\boldsymbol{C}-1 = \boldsymbol{C}_0\boldsymbol{H} + \alpha_0\boldsymbol{I}$$
$$= \boldsymbol{H}^2 - 3\boldsymbol{I} = \boldsymbol{0}$$

The last equation is the checking equation. With these results, (6.54) becomes

$$\mathcal{L}\{\boldsymbol{\theta}(t)\} = \frac{1}{s^2 - 3}\begin{bmatrix} s+1 & 1 \\ 2 & s-1 \end{bmatrix} \tag{6.55}$$

The poles of $\boldsymbol{\theta}(t)$ are at $s_1 = -\sqrt{3}$ and $s_2 = +\sqrt{3}$. Inverse transformation gives

$$\boldsymbol{\theta}(t) = \frac{\sqrt{3}}{6}\cdot [(\sqrt{3} - 1)\exp(-\sqrt{3}t) + (\sqrt{3} + 1)\exp(\sqrt{3}t) - \exp(-\sqrt{3}t) + \exp(\sqrt{3}t)$$
$$-2\exp(-\sqrt{3}t) + 2\exp(\sqrt{3}b), (\sqrt{3} + 1)\exp(-\sqrt{3}t) + (\sqrt{3} - 1)\exp(\sqrt{3}t)] \tag{6.56}$$

Of course, $\boldsymbol{\theta}(0) = \boldsymbol{I}$. With $\tilde{P}(0) = P(0) = 0$, it follows from (6.48) that

$$\tilde{P}(t) = \frac{-2\exp(-\sqrt{3}t) + 2\exp(\sqrt{3}t)}{(\sqrt{3} - 1)\exp(-\sqrt{3}t) + (\sqrt{3} + 1)\exp(\sqrt{3}t)} \tag{6.57}$$

The time constant of the decaying components is $T = 1/\sqrt{3} = 0.578$. For $t \gg T$, the increasing components predominate in the numerator and denominator so that

$$\tilde{P}(\infty) = \bar{P} = \frac{2}{\sqrt{3} + 1} = \sqrt{3} - 1 = 0.73$$

This result agrees with the solution of Example 5.4. The required Kalman–Bucy filter is

$$\dot{\hat{x}}(t) = -\hat{x}(t) + K(t)\{y(t) - \hat{x}(t)\}$$

The gain factor is equal to $\tilde{P}(t)C'R^{-1}$ so that

$$K(t) = \tilde{P}(t)$$

The filter is shown in Figure 6.6 together with the time evolution of $K(t)$. The transient process of the gain factor is approximately complete after three time constants. Then we are confronted with the Wiener filter. More details on the stationary state are given in Example 5.4.

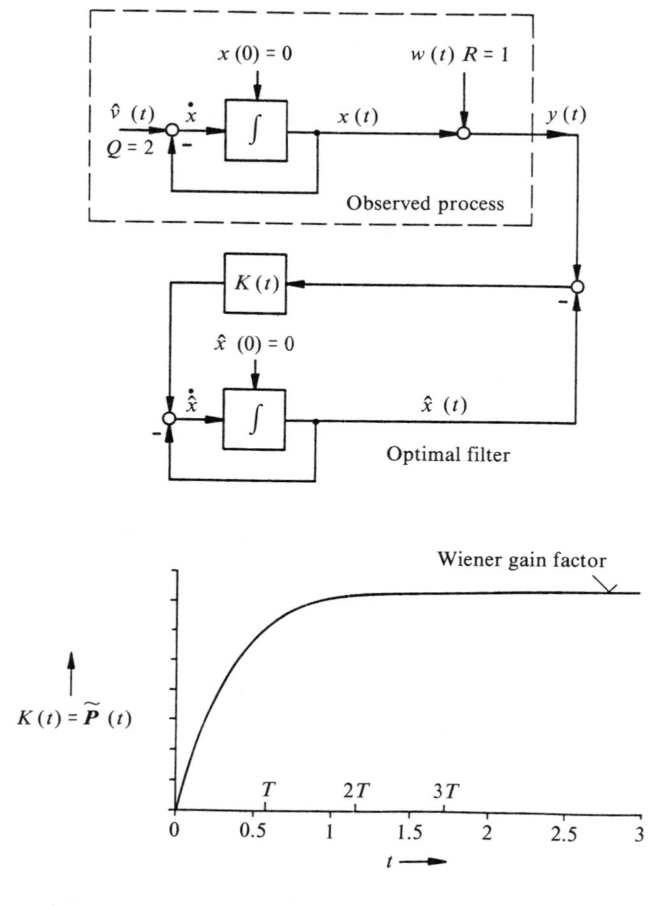


**Figure 6.6** Kalman–Bucy filter for a scalar process (Example 6.2).

We have seen that the Wiener filter is a special case of the Kalman–Bucy filter for stationary processes after the transient processes have decayed. In the Wiener theory further restrictions are that only one scalar measurement variable $y(t)$ is present and that $\boldsymbol{v}(t)$ has at most two elements, one for the useful signal and possibly a second for colored noise. If colored noise is present, we can also dispense with white noise $w(t)$ in the Wiener method. This is not allowed in the Kalman–Bucy method because $E\{w^2(t)\}$ must be positive. However, this apparent greater generality of the Wiener method has the effect that differentiation elements, which are undesirable and not strictly feasible, are required in the filter. Bryson and Johansen (1965) have solved this case, for which $\boldsymbol{R}$ is positive semidefinite, within the framework of the

Kalman–Bucy theory and also obtain differentiation elements (see also Bucy (1967) and Section 6.7).

Furthermore, it must be assumed for the Wiener method that the observed process itself is stable; i.e., the matrix $\boldsymbol{A}$ may only have eigenvalues with a negative real part. This limitation is unnecessary in the design of time-invariant Kalman–Bucy filters. Under the assumptions made at the beginning of this section, the stationary Kalman–Bucy filter can always be realized and is even asymptotically stable (all poles of the filter itself lie in the left-hand $s$ half-plane (see Example 6.1)).

The considerations of this section also apply correspondingly to discrete time. Synthesis problems in discrete time (sampled-data systems) are basically simpler to handle than the corresponding continuous-time cases. Instead of differential equations and integrals, only difference equations and sums have to be evaluated, which means that right from the beginning only purely algebraic operations are involved. A stationary Kalman filter can always be calculated in a straightforward manner from the algorithm (5.51) in Theorem 5.1. For systems of lower order ($n \leqslant 3$), it can even be solved using a pocket calculator (see Example 5.2 where the equilibrium solution of the discrete-time Riccati equation was already established for $k = 6$ (cf. Table 5.2 and Figure 5.4)).

## 6.6 SHAPING FILTERS FOR VECTOR-VALUED MARKOV PROCESSES

So far it has been assumed that the mathematical model of the observed process is already given in the standard form $\dot{\boldsymbol{x}} = \boldsymbol{A}\boldsymbol{x} + \boldsymbol{v}$ with the covariance matrix $\boldsymbol{Q}$. This is indeed realistic for many practical design problems. In particular, if the observed system is a stochastically disturbed control process with lumped parameters, it is likely that its state space model is known or can be obtained relatively easily. For example, we almost always find this situation for aircraft and spacecraft.

However, it can happen that only the statistical characteristics of the process $\{\boldsymbol{x}\}$ are known. Even more important for practical filter synthesis is the situation when the disturbance and measurement noises contain colored components. In both cases a shaping filter must be set up for $\{\boldsymbol{x}\}$ or for that part of the augmented vector $\boldsymbol{x}$ which represents the colored noise. This must be done before the actual filter design is started (cf. Section 5.3.7, point (b)).

In view of the limited scope of this treatment, we limit ourselves to Markov processes in discrete time with expectation zero and given covariance matrices. If, in addition, the process is Gaussian, it is completely specified by this information.

Let $\{\boldsymbol{x}(k),\ k_0 \leq k\}$ be the process under consideration where $\boldsymbol{x}$ either represents the overall state of the required model or only the part of the state belonging to the colored noise. Let the following sequence of covariance matrices of $\{\boldsymbol{x}(k)\}$ be given:

$$\boldsymbol{P}(k_2,k_1) := E\{\boldsymbol{x}(k_2)\boldsymbol{x}'(k_1)\} \quad k_0 \leq k_1 \leq k_2 \tag{6.58}$$

We need only consider the pairs of arguments $k_1 \leq k_2$ because obviously $\boldsymbol{P}(k_1,k_2) = \boldsymbol{P}'(k_2,k_1)$. We make the following assumptions:

(i) $\boldsymbol{P}(k,k)$ is finite and symmetric for all $k$;
(ii) $\boldsymbol{P}(k,k)$ is always positive definite (nonsingular);
(iii) The relationship

$$\boldsymbol{P}(k_3,k_2)\boldsymbol{P}^{-1}(k_2,k_2)\boldsymbol{P}(k_2,k_1) = \boldsymbol{P}(k_3,k_1) \tag{6.59}$$

holds for all $k_1 \leq k_2 \leq k_3$.

These specifications are necessary and sufficient for $\{\boldsymbol{x}(k)\}$ to be a nonsingular Markov process in the wide sense, i.e., that for all $k_1 < k_2 < \ldots < k_n$ it has the property

$$E\{\boldsymbol{x}(k_n)|\boldsymbol{x}(k_1), \quad \boldsymbol{x}(k_2), \ldots, \boldsymbol{x}(k_{n-1})\} = E\{\boldsymbol{x}(k_n)|\boldsymbol{x}(k_{n-1})\} \tag{6.60a}$$

This theorem has been formulated and proved for scalar processes by Doob (1964, Chapter V, Theorem 8.1). We can make it plausible for vector-valued processes by means of the following argument. The property (6.60a) is certainly valid if the model of the process has the form

$$\boldsymbol{x}(k + 1) = \boldsymbol{A}(k)\boldsymbol{x}(k) + \boldsymbol{v}(k) \quad k_0 \leq k \tag{6.60b}$$

where $\{\boldsymbol{v}(k)\}$ represents white noise that is not correlated with $\boldsymbol{x}(k_0)$. The covariance of $\{\boldsymbol{v}(k)\}$ is

$$E\{\boldsymbol{v}(k)\boldsymbol{v}'(\kappa)\} = \boldsymbol{Q}(k)\delta_{k\kappa} \tag{6.60c}$$

Let the transition matrix of $\boldsymbol{A}(k)$ be $\boldsymbol{\phi}(k,\kappa)$. We have already established in Section 4.6 that the following relationships apply (see (4.55)):

$$\boldsymbol{\phi}(k_2 + 1,k_1) = \boldsymbol{A}(k_2)\boldsymbol{\phi}(k_2,k_1) \quad \boldsymbol{\phi}(k_1,k_1) = \boldsymbol{I} \quad k_1 \leq k_2 \tag{6.61}$$

$$\boldsymbol{\phi}(k_3,k_2)\boldsymbol{\phi}(k_2,k_1) = \boldsymbol{\phi}(k_3,k_1) \quad k_1 \leq k_2 \leq k_3 \tag{6.62}$$

$$\boldsymbol{x}(k_2) = \boldsymbol{\phi}(k_2,k_1)\boldsymbol{x}(k_1) + \sum_{\kappa=k_1}^{k_2-1} \boldsymbol{\phi}(k_2,\kappa + 1)\boldsymbol{v}(\kappa) \quad k_1 < k_2 \tag{6.63}$$

The last of these equations is substituted in definition (6.58). Because of the lack of correlation, all terms $E\{\boldsymbol{v}(\kappa)\boldsymbol{x}'(k_1)\}$ originating from the sum vanish and we are left with

$$\boldsymbol{P}(k_2,k_1) = E\{\boldsymbol{\phi}(k_2,k_1)\boldsymbol{x}(k_1)\boldsymbol{x}'(k_1)\}$$

so that

$$\boldsymbol{P}(k_2,k_1) = \boldsymbol{\phi}(k_2,k_1)\boldsymbol{P}(k_1,k_1) \quad k_0 \leq k_1 \leq k_2 \tag{6.64}$$

Postmultiplication by the inverse of $\boldsymbol{P}(k_1,k_1)$ gives

$$\boldsymbol{\phi}(k_2,k_1) = \boldsymbol{P}(k_2,k_1)\boldsymbol{P}^{-1}(k_1,k_1) \tag{6.65}$$

Substitution in equation (6.62) gives

$$\boldsymbol{P}(k_3,k_2)\boldsymbol{P}^{-1}(k_2,k_2)\boldsymbol{P}(k_2,k_1)\boldsymbol{P}^{-1}(k_1,k_1) = \boldsymbol{P}(k_3,k_1)\boldsymbol{P}^{-1}(k_1,k_1) \tag{6.66}$$

Postmultiplication by $\boldsymbol{P}(k_1,k_1)$ produces the above condition (6.59).

We now take condition (6.59), as given, as a starting point, and determine the required matrices $\boldsymbol{A}(k)$ and $\boldsymbol{Q}(k)$ of the model of $\{\boldsymbol{x}(k)\}$. Because $\boldsymbol{P}(k,k)$ is assumed to be always positive definite, equation (6.66) and therefore (6.65) follow. With $k_2 = k + 1$ and $k_1 = k$, equation (6.65) takes the form

$$\boldsymbol{\phi}(k + 1,k) = \boldsymbol{P}(k + 1,k)\boldsymbol{P}^{-1}(k,k)$$

The left-hand side is equal to $\boldsymbol{A}(k)$ (see (6.61) with $k_2 = k_1 = k$). We therefore obtain

$$\boldsymbol{A}(k) = \boldsymbol{P}(k + 1,k)\boldsymbol{P}^{-1}(k,k) \tag{6.67}$$

To find $\boldsymbol{Q}(k)$, we consider $\boldsymbol{P}(k + 1,k + 1)$ and substitute equation (6.60b) in (6.58).

$$\boldsymbol{P}(k + 1,k + 1) = E\{[\boldsymbol{A}(k)\boldsymbol{x}(k) + \boldsymbol{v}(k)][\boldsymbol{x}'(k)\boldsymbol{A}'(k) + \boldsymbol{v}'(k)]\}$$

The mixed expectations vanish because $\boldsymbol{x}(k)$ is not correlated with $\boldsymbol{v}(k)$, and we are left with

$$\boldsymbol{P}(k + 1,k + 1) = E\{\boldsymbol{A}(k)\boldsymbol{x}(k)\boldsymbol{x}'(k)\boldsymbol{A}'(k) + \boldsymbol{v}(k)\boldsymbol{v}'(k)\}$$

or

$$\boldsymbol{P}(k + 1,k + 1) = \boldsymbol{A}(k)\boldsymbol{P}(k,k)\boldsymbol{A}'(k) + \boldsymbol{Q}(k) \tag{6.68}$$

As $\boldsymbol{P}$ is given and $\boldsymbol{A}(k)$ is already determined by (6.67), the matrix $\boldsymbol{Q}(k)$ in (6.68) forms the only unknown for which a solution can easily be obtained.

In summary, we may conclude that, under the assumptions made at the beginning of this section, we can always determine a sequence $\boldsymbol{A}(k)$ according to (6.67) and subsequently the associated sequence $\boldsymbol{Q}(k)$ according to (6.68). In this way, the required shaping filter (6.60) is completely defined. This shaping filter is an abstract dynamic system with the property that it synthesizes the Markov process $\{\boldsymbol{x}(k)\}$ from white noise $\{\boldsymbol{v}(k)\}$ such that it displays the specified statistical properties. The solution of the shaping-filter problem is completely analogous for continuous time (details can be found in the literature). However, from the practical point of view the problem has less to do with the evaluation of equations (6.67) and (6.68) than with the main task of determining the covariance sequence $\boldsymbol{P}(k_2,k_1)$ in terms of measurements.

The synthesis of the shaping filter in state space considered here is equivalent to the factorization of the spectral density in classical theory.

## 6.7 REDUCTION OF THE ORDER OF THE FILTER

A large number of references have already been made to the fact that a special situation exists when one or more elements of the measurement vector $\boldsymbol{y}$ are free of white noise. These elements $y_i$ therefore either contain no noise at all or only colored noise generated by a shaping filter. The case when the elements $w_i$ are linearly dependent on each other is essentially equivalent to this case. Both cases have the effect that the covariance matrix $\boldsymbol{R}$ of the measurement noise is no longer strictly positive definite but is singular (compare Section 5.3.7, point (d)).

In continuous time, the inverse matrix $\boldsymbol{R}^{-1}$ occurs directly in the Riccati differential equation and as a coefficient in the gain factor $\boldsymbol{K}(t)$ (see (6.37)). Therefore, if measurements without white noise are present, the Kalman–Bucy filter can no longer be implemented in the original form. As mentioned previously, Bryson and Johansen (1965) and Bucy (1967) were able to solve this problem by modifying the procedure. The basic idea of the solution procedure is to differentiate measurements which contain no noise or only colored noise sufficiently often for white noise to be produced. This white noise can now be traced back to some of the components of $\boldsymbol{v}$ so that $\boldsymbol{Q}$ and the new matrix $\boldsymbol{R}^*$ are correlated with each other. This produces extra terms in the filter algorithm which, however, do not cause any serious difficulty. The problems which make the procedure questionable for many practical purposes are much more involved with the technical implementation. Certainly, the last differentiation, i.e., the one which would result in white noise, can be bypassed in the practical design. This is done by feeding the signal to be differentiated not to the input but to the output of the integrator of the filter. Thus, for example, measurements of the form

$$y_i(t) = c_{i1}x_1 + \ldots + c_{ij}x_j + \ldots c_{in}x_n$$

with the first-order colored noise

$$\dot{x}_j(t) = \alpha_j x_j(t) + v_j(t)$$

need not be differentiated at all in practice. However, if the measurements lie higher than one order above white noise, the remaining differentiation terms must be technically implemented in the actual filter. We will therefore not examine the continuous-time case any further. Instead, we turn to the discrete-time case which is simpler to handle (Leondes and Novak, 1974).

We consider an $n$th-order system which has $p$ measurement variables without white noise:

$$\boldsymbol{x}(k + 1) = \boldsymbol{A}(k)\boldsymbol{x}(k) + \boldsymbol{v}(k) \quad k_0 \leq k \tag{6.69a}$$

$$\bar{\boldsymbol{y}}(k) = \bar{\boldsymbol{C}}(k)\boldsymbol{x}(k) \tag{6.69b}$$

$$\boldsymbol{y}^*(k) = \boldsymbol{C}^*(k)\boldsymbol{x}(k) + \boldsymbol{w}^*(k) \tag{6.69c}$$

Here, $\bar{\boldsymbol{y}}$ is a $p$ vector and $\boldsymbol{y}^*$ and $\boldsymbol{w}^*$ are $(m - p)$-vectors. Let the $p \times n$ matrix $\bar{\boldsymbol{C}}(k)$ have rank $p$ for all $k$ (i.e., its $p$ rows are linearly independent; otherwise, first eliminate the redundant rows). All elements vanish in the covariance matrix $\boldsymbol{R}(k)$ with the exception of those in the right-hand bottom corner of order $(m - p)(m - p)$. The other quantities are the same as specified in Section 5.3.1.

The measurements $\bar{\boldsymbol{y}}(k)$ are exact observations of certain linear combinations of the state variables $x_i(k)$. The latter either represent the required variables (useful signals) or colored noise. For colored noise, let the associated shaping filter already be included in (6.69a). If we had $p = n$, we could immediately solve (6.69b) for the $x_i(k)$. However, because $p$ is almost always smaller than $n$, the first $n - p$ components of $\boldsymbol{x}$ are combined to form a vector $\bar{\boldsymbol{x}}$ which is obtained by estimation. We organize the relationships as follows:

$$\begin{bmatrix} \bar{\boldsymbol{x}}(k) \\ \bar{\boldsymbol{y}}(k) \end{bmatrix} := \begin{bmatrix} \boldsymbol{I}_{n-p} & \boldsymbol{o} \\ \bar{\boldsymbol{C}}_1 & \bar{\boldsymbol{C}}_2 \end{bmatrix} \boldsymbol{x}(k) := \begin{bmatrix} \boldsymbol{B} \\ \bar{\boldsymbol{C}}(k) \end{bmatrix} \boldsymbol{x}(k) \tag{6.70}$$

The matrix $\boldsymbol{B}$ is defined as the top part consisting of $n - p$ rows of the total matrix in (6.70). The matrices $\bar{\boldsymbol{C}}_1(k)$ and $\bar{\boldsymbol{C}}_2(k)$ are defined by (6.70) as submatrices of the $(n - p) \times n$ matrix $\bar{\boldsymbol{C}}(k)$ with $n - p$ and $p$ columns. As $\bar{\boldsymbol{C}}$ has rank $p$, $\bar{\boldsymbol{C}}_2$ can be assumed to be nonsingular. If necessary, this must be arranged by renumbering the variables (we assume that this structure is then maintained for all $k$). The inverse of the combined matrix in (6.70) can easily be formed and we obtain

$$\boldsymbol{x}(k) = \begin{bmatrix} \boldsymbol{I}_{n-p} & \boldsymbol{o} \\ -\bar{\boldsymbol{C}}_2^{-1}\bar{\boldsymbol{C}}_1 & \bar{\boldsymbol{C}}_2^{-1} \end{bmatrix} \begin{bmatrix} \bar{\boldsymbol{x}}(k) \\ \bar{\boldsymbol{y}}(k) \end{bmatrix} \tag{6.71a}$$

or

$$\boldsymbol{x}(k) = \boldsymbol{M}(k)\bar{\boldsymbol{x}}(k) + \boldsymbol{N}(k)\bar{\boldsymbol{y}}(k) \tag{6.71b}$$

Here $\boldsymbol{M}$ and $\boldsymbol{N}$ are defined by direct comparison of (6.71a) and (6.71b). Since $\boldsymbol{M}$, $\boldsymbol{N}$ and $\bar{\boldsymbol{y}}$ are known, only $\bar{\boldsymbol{x}}$ remains to be estimated. We have

$$\hat{\boldsymbol{x}}(k) = \boldsymbol{M}(k)\hat{\bar{\boldsymbol{x}}}(k) + \boldsymbol{N}(k)\bar{\boldsymbol{y}}(k) \tag{6.72}$$

and so

$$\tilde{\boldsymbol{x}}(k) = \boldsymbol{M}(k)\tilde{\bar{\boldsymbol{x}}}(k) \tag{6.73}$$

The estimate (6.72) is optimal because the estimation error (6.73) satisfies the Wiener–Hopf equation in version (5.16). The following formulas are visually simpler if we combine the two equations (6.69b) and (6.69c):

$$\boldsymbol{y}(k) := \begin{bmatrix} \bar{\boldsymbol{y}}(k) \\ \boldsymbol{y}^*(k) \end{bmatrix} \quad \boldsymbol{w}(k) := \begin{bmatrix} \boldsymbol{o} \\ \boldsymbol{w}^*(k) \end{bmatrix} \tag{6.74}$$

$$\boldsymbol{C}(k) := \begin{bmatrix} \bar{\boldsymbol{C}}(k) \\ \boldsymbol{C}^*(k) \end{bmatrix} \tag{6.75}$$

We can now write

$$\bar{\boldsymbol{x}}(k) = \boldsymbol{B}\boldsymbol{x}(k) \tag{6.76}$$

$$\boldsymbol{y}(k) = \boldsymbol{C}(k)\boldsymbol{x}(k) + \boldsymbol{w}(k) \tag{6.77}$$

From Theorem 5.1, equations (5.50a) and (5.50b), the Kalman filter has the form

$$\boldsymbol{x}^*(k+1) = \boldsymbol{A}(k)\hat{\boldsymbol{x}}(k) \quad \boldsymbol{x}^*(k_0) = \boldsymbol{0} \tag{6.78}$$

$$\hat{\boldsymbol{x}}(k) = \boldsymbol{x}^*(k) + \boldsymbol{K}(k)\{\boldsymbol{y}(k) - \boldsymbol{C}(k)\boldsymbol{x}^*(k)\} \tag{6.79}$$

This filter still has order $n$. However, the following transformation makes it possible to reduce the order to $n - p$. To eliminate the $n$-dimensional filter state $\boldsymbol{x}^*$, the terms with $\boldsymbol{x}^*$ in equation (6.79) are combined and both sides are premultiplied by $\boldsymbol{B}$:

$$\boldsymbol{B}\hat{\boldsymbol{x}}(k) = \boldsymbol{B}\{\boldsymbol{I} - \boldsymbol{K}(k)\boldsymbol{C}(k)\}\boldsymbol{x}^*(k) + \boldsymbol{B}\boldsymbol{K}(k)\boldsymbol{y}(k) \tag{6.80}$$

On the left-hand side, we have the estimate $\hat{\bar{\boldsymbol{x}}}(k)$ (see equation (6.76)). The right-hand side can be simplified by introducing the following definitions:

$$\boldsymbol{z}^*(k) := \boldsymbol{B}\{\boldsymbol{I} - \boldsymbol{K}(k)\boldsymbol{C}(k)\}\boldsymbol{x}^*(k) \tag{6.81}$$

$$\bar{\boldsymbol{K}}(k) := \boldsymbol{B}\boldsymbol{K}(k) \tag{6.82}$$

In this way, (6.80) is transformed into the new filter equation (6.83b) given below. The other filter equation is obtained by replacing $k$ by $k + 1$ in (6.81), using (6.82) and substituting (6.78):

$$\boldsymbol{z}^*(k + 1) = \{\boldsymbol{B} - \bar{\boldsymbol{K}}(k + 1)\boldsymbol{C}(k + 1)\}\boldsymbol{A}(k)\hat{\boldsymbol{x}}(k)$$

Then, using (6.72), we finally obtain the filter equation (6.83a). Summarizing the result, we have

$$\boldsymbol{z}^*(k + 1) = \{\boldsymbol{B} - \bar{\boldsymbol{K}}(k + 1)\boldsymbol{C}(k + 1)\}\boldsymbol{A}(k)\{\boldsymbol{M}(k)\hat{\bar{\boldsymbol{x}}}(k) + \boldsymbol{N}(k)\bar{\boldsymbol{y}}(k)\} \tag{6.83a}$$

$$\hat{\bar{\boldsymbol{x}}}(k) = \boldsymbol{z}^*(k) + \bar{\boldsymbol{K}}(k)\boldsymbol{y}(k) \quad \boldsymbol{z}^*(k_0) = \boldsymbol{0} \tag{6.83b}$$

Equations (6.83) form the new filter of order $n - p$. It should be noted that the matrices in equation (6.83a) can be multiplied:

$$\boldsymbol{F}(k) := \{\boldsymbol{B} - \bar{\boldsymbol{K}}(k + 1)\boldsymbol{C}(k + 1)\}\boldsymbol{A}(k)\boldsymbol{M}(k) \tag{6.84a}$$

$$\boldsymbol{G}(k) := \{\boldsymbol{B} - \bar{\boldsymbol{K}}(k + 1)\boldsymbol{C}(k + 1)\}\boldsymbol{A}(k)[\boldsymbol{N}(k),\boldsymbol{0}\} \tag{6.84b}$$

They are reduced to the type $(n - p) \times (n - p)$ and $(n - p) \times m$ respectively. In $\boldsymbol{G}(k)$, only the first $p$ columns are nonzero. Equation (6.83a) can now be written in the form

$$\boldsymbol{z}^*(k + 1) = \boldsymbol{F}(k)\hat{\bar{\boldsymbol{x}}}(k) + \boldsymbol{G}(k)\boldsymbol{y}(k) \tag{6.84c}$$

The $(n - p)$-vector $\boldsymbol{z}^*(k)$ is the new filter state (Figure 6.7). This becomes particularly clear when we substitute (6.83b) in (6.84c):

$$z^*(k+1) = \boldsymbol{F}(k)z^*(k) + \{\boldsymbol{F}(k)\bar{\boldsymbol{K}}(k) + \boldsymbol{G}(k)\}\boldsymbol{y}(k) \tag{6.84d}$$

The matrix $\boldsymbol{F}(k)$ determines the dynamics of the filter.

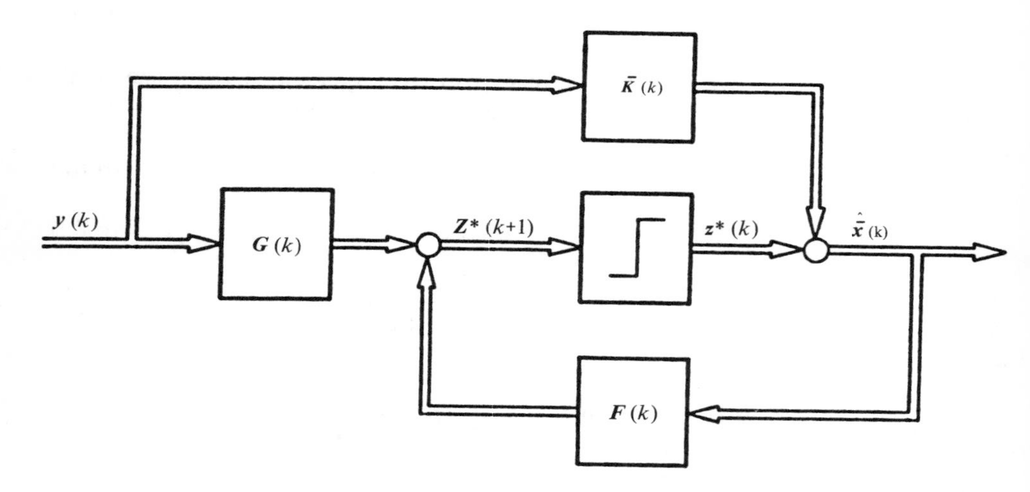


**Figure 6.7** Filter of order $n - p$ for $p$ measurement elements without white noise.

The state variables $x_{n-p+1}, \ldots, x_n$ of the observed process no longer occur in the filter. These often include state variables which belong to colored noise and are of no interest at all. In any case, the estimate of the entire state vector can always be constructed using equation (6.72).

The above filter algorithm was first derived by Brammer (1967) in a somewhat different manner, and was also published in Brammer (1968a,b). The only matrix in the filter which is still unknown is the gain matrix $\bar{\boldsymbol{K}}$ which, according to (6.70) and (6.82), is the top part of the Kalman gain matrix $\boldsymbol{K}(k)$. This follows from the discrete Riccati algorithm (5.51). Provided that this is already available as a digital computer program, the easiest way is to take this over without any modifications. However, for manual calculations it is worth simplifying the Riccati algorithm a little. This is done by first introducing the covariance matrix of the estimation error $\tilde{\bar{\boldsymbol{x}}}$:

$$\boldsymbol{S}(k) := E\{\tilde{\bar{\boldsymbol{x}}}(k)\tilde{\bar{\boldsymbol{x}}}'(k)\} \tag{6.85}$$

According to (6.76), $\tilde{\bar{\boldsymbol{x}}} = \boldsymbol{B}\tilde{\boldsymbol{x}}$, and so we obviously have

$$\boldsymbol{S}(k) = \boldsymbol{B}E\{\tilde{\boldsymbol{x}}(k)\tilde{\boldsymbol{x}}'(k)\}\boldsymbol{B}' = \boldsymbol{B}\tilde{\boldsymbol{P}}(k)\boldsymbol{B}' \tag{6.86}$$

This means that $\boldsymbol{S}(k)$ represents the top left-hand corner of $\tilde{\boldsymbol{P}}(k)$. Conversely, it follows from (6.73) that

$$\tilde{\boldsymbol{P}}(k) = \boldsymbol{M}(k)\boldsymbol{S}(k)\boldsymbol{M}'(k) \tag{6.87}$$

Equations (6.86) and (6.87) are substituted in the Kalman algorithm (5.51). We also take into consideration (6.82) and immediately obtain

$$\boldsymbol{P}^*(k+1) = \boldsymbol{A}(k)\boldsymbol{M}(k)\boldsymbol{S}(k)\boldsymbol{M}'(k)\boldsymbol{A}'(k) + \boldsymbol{Q}(k) \quad \boldsymbol{P}^*(k_0) = \boldsymbol{P}(k_0) \tag{6.88a}$$

$$\bar{\boldsymbol{K}}(k) = \boldsymbol{B}\boldsymbol{P}^*(k)\boldsymbol{C}'(k)\{\boldsymbol{C}(k)\boldsymbol{P}^*(k)\boldsymbol{C}'(k) + \boldsymbol{R}(k)\}^{-1} \tag{6.88b}$$

$$\boldsymbol{S}(k) = \{\boldsymbol{B} - \bar{\boldsymbol{K}}(k)\boldsymbol{C}(k)\}\boldsymbol{P}^*(k)\boldsymbol{B}' \tag{6.88c}$$

This modified algorithm has also been derived by Brammer (1967) (see also Brammer, 1968a,b).* The elements of $\boldsymbol{K}(k)$ and $\tilde{\boldsymbol{P}}(k)$ from the Kalman algorithm which are not needed have been eliminated here so that the computing requirements are correspondingly less. This applies in particular for time-invariant $\boldsymbol{C}$ because then the matrices $\boldsymbol{M}$ and $\boldsymbol{N}$ are also time invariant and only need to be calculated once.

The difference equation (6.88a) is still of order $n$. As already indicated by Brammer (1967), it can be completely eliminated by the following procedure, just as $\boldsymbol{x}^*(k+1)$ was eliminated from the filter. For the sake of simplicity, we assume that

$$\boldsymbol{A}, \boldsymbol{C}, \boldsymbol{Q} \text{ and } \boldsymbol{R} \text{ are constant} \tag{6.89}$$

In this case, the elimination is particularly worthwhile. We now replace $k$ by $k + 1$ in equations (6.88b) and (6.88c) and substitute (6.88a):

$$\bar{\boldsymbol{K}}(k+1) = \{\boldsymbol{BAMS}(k)(\boldsymbol{CAM})' + \boldsymbol{BQC}'\} \times \{\boldsymbol{CAMS}(k)(\boldsymbol{CAM})' + \boldsymbol{CQC}' + \boldsymbol{R}\}^{-1} \tag{6.90a}$$

$$\boldsymbol{S}(k+1) = \boldsymbol{BAMS}(k)(\boldsymbol{BAM})' + \boldsymbol{BQB}' - \bar{\boldsymbol{K}}(k+1)\{\boldsymbol{CAMS}(k)(\boldsymbol{BAM})' + \boldsymbol{CQB}'\} \tag{6.90b}$$

The initial condition $\boldsymbol{S}(k_0)$ follows from (6.88b) and (6.88c). However complicated this algorithm may appear at first sight, it is still considerably more economical in computing terms than either the original Kalman algorithm or the system (6.88). This is because the order of the algorithm (6.90) is only $n - p$ and the necessary

*If the inverse in (6.88b) does not exist, we use the pseudo-inverse (see, for example, Penrose, 1955; Zadeh and Desoer, 1963).

coefficients $\boldsymbol{BAM}$, $\boldsymbol{CAM}$, $\boldsymbol{BQB'}$, $\boldsymbol{BQC'}$ and $\boldsymbol{CQC'}$ can be relatively simply determined. Furthermore, under the assumption (6.89), they remain constant once and for all. The premultiplication of a matrix by $\boldsymbol{B}$ is particularly trivial because this merely means selecting the first $n - p$ rows from the matrix.

*Example 6.3* (Brammer, 1967). A second-order signal process is observed at sampling times $k = 0, 1, 2, \ldots$ . The scalar measurement variable consists of the state variable $x_2$ and first-order colored noise $x_3$ which is not correlated with the signal. The model of the overall system including the shaping filter is

$$\boldsymbol{x}(k+1) = \begin{bmatrix} 0.7 & 0 & 0 \\ 0.4 & 0.3 & 0 \\ 0 & 0 & 0.1 \end{bmatrix} \boldsymbol{x}(k) + \boldsymbol{v}(k)$$

$$\boldsymbol{Q}(k) = \begin{bmatrix} 1 & 2 & 0 \\ 2 & 4 & 0 \\ 0 & 0 & 0.8 \end{bmatrix}$$

$$y(k) = [0 \quad 1 \quad 1]\boldsymbol{x}(k) \qquad R(k) = 0$$

$$\boldsymbol{P}(0) = \boldsymbol{0} \quad \boldsymbol{x}(0) = \boldsymbol{0}$$

We require the corresponding optimal filter. Its order is $3 - 1 = 2$. First we calculate $\boldsymbol{M}$ and $\boldsymbol{N}$ according to the specifications (6.70) and (6.71a). We have

$$\begin{bmatrix} \boldsymbol{B} \\ \\ \bar{\boldsymbol{C}} \end{bmatrix} = \left[\begin{array}{ccc} 1 & 0 & 0 \\ 0 & 1 & 0 \\ \hdashline 0 & 1 & 1 \end{array}\right]$$

The associated inverse matrix is

$$[\boldsymbol{M}, \boldsymbol{N}] = \left[\begin{array}{cc:c} 1 & 0 & 0 \\ 0 & 1 & 0 \\ 0 & -1 & 1 \end{array}\right]$$

The filter is given by (6.83) or (6.84) with the above values of $\boldsymbol{B}$, $\boldsymbol{C}$, $\boldsymbol{A}$, $\boldsymbol{M}$ and $\boldsymbol{N}$. We calculate the missing matrix $\bar{\boldsymbol{K}}(k)$ by using algorithm (6.90). The initial values $\bar{\boldsymbol{K}}(0)$ and $\boldsymbol{S}(0)$ vanish according to (6.88b) and (6.88c) because $\boldsymbol{P}^*(0) = \boldsymbol{P}(0) = \boldsymbol{0}$. Furthermore, we have

$$\boldsymbol{BAM} = \begin{bmatrix} 0.7 & 0 \\ 0.4 & 0.3 \end{bmatrix} \qquad \boldsymbol{CAM} = [0.4 \quad 0.2]$$

$$\boldsymbol{BQB}' = \begin{bmatrix} 1 & 2 \\ 2 & 4 \end{bmatrix} \qquad \boldsymbol{BQB}' = \begin{bmatrix} 2 \\ 4 \end{bmatrix}$$

$$\boldsymbol{CQC}' = 4.8$$

Thus algorithm (6.90) can be implemented. The resulting recursive sequence is

$$\bar{\boldsymbol{K}}(0) = \boldsymbol{0} \qquad \boldsymbol{S}(0) = \boldsymbol{0}$$

$$\bar{\boldsymbol{K}}(1) = \begin{bmatrix} 0.417 \\ 0.834 \end{bmatrix} \qquad \boldsymbol{S}(1) = \begin{bmatrix} 0.166 & 0.332 \\ 0.332 & 0.664 \end{bmatrix}$$

$$\bar{\boldsymbol{K}}(2) = \begin{bmatrix} 0.427 \\ 0.843 \end{bmatrix} \qquad \boldsymbol{S}(2) = \begin{bmatrix} 0.187 & 0.351 \\ 0.351 & 0.684 \end{bmatrix}$$

$$\bar{\boldsymbol{K}}(3) = \begin{bmatrix} 0.427 \\ 0.843 \end{bmatrix}$$

The algorithm has already settled into its equilibrium state at $k = 3$. The short *time constant* of the algorithm is caused by the high value of $\boldsymbol{Q}$. In the system (6.90) terms with $\boldsymbol{Q}$ predominate over all others. In the equilibrium state, the filter (6.83) takes the form

$$z^*(k+1) = \begin{bmatrix} 0.529 & -0.086 \\ 0.063 & 0.131 \end{bmatrix} \hat{\hat{\boldsymbol{x}}}(k) + \begin{bmatrix} -0.043 \\ -0.084 \end{bmatrix} y(k) \tag{6.91a}$$

$$\hat{\hat{\boldsymbol{x}}}(k) = z^*(k) + \begin{bmatrix} 0.427 \\ 0.843 \end{bmatrix} y(k) \tag{6.91b}$$

In the version (6.84d) and (6.83b) we obtain an alternative to this result:

$$z^*(k+1) = \begin{bmatrix} 0.529 & -0.086 \\ 0.063 & 0.131 \end{bmatrix} z^*(k) + \begin{bmatrix} 0.111 \\ 0.052 \end{bmatrix} y(k)$$

$$\hat{\hat{\boldsymbol{x}}}(k) = z^*(k) + \begin{bmatrix} 0.427 \\ 0.843 \end{bmatrix} y(k)$$

The eigenvalues of the filter are $\lambda_1 = 0.511$ and $\lambda_2 = 0.150$. Their magnitude is less than unity so that the filter is asymptotically stable.

## 6.8 BRIEF LOOK AT ITO CALCULUS AND NONLINEAR FILTERING

In the treatment of nonlinear differential equations with statistical excitation, it is essential to replace white noise (that can be tolerated in the linear theory) by Brownian motion and to apply Ito calculus. Otherwise, incorrect results are likely to occur owing to the special properties of white noise. Therefore, before we formulate the nonlinear filter problem, we will first look at Brownian motion as well as stochastic integration and differentiation. We first consider the scalar case and then generalize to $n$ dimensions. More details on the material in this chapter can be found in the literature (e.g., Arnold, 1973).

### 6.8.1 The Brownian Process

The Brownian random process is a mathematical idealization of actual Brownian motion. It is thoroughly discussed in standard texts on random processes (e.g., Doob, 1964; Krickeberg, 1963). Brown made the first experimental observations in 1826. The theoretical treatment was begun by Bachelier in 1900, and resumed by Einstein and Smoluchowski in 1906. The first rigorous examination of the process was that of Wiener (1923), which is why it is also commonly called the Wiener process. The theory was advanced further by Lévy (1948). A new trend is to turn from the collective properties of the entire ensemble to the individual properties of single sample functions (Ito and McKean, 1965).

Within the framework of this book, we have a twofold interest in Brownian motion: on the one hand as a means of modeling stochastic disturbance variables at the input of the observed system, and on the other hand as a means of modeling the measurement errors at its output.

*Definition*. The Brownian motion process is a random process $\{\beta(t), t_0 \leq t < \infty\}$ in continuous time which satisfies the following three conditions:

(i) $\beta(t)$ is always real;

(ii) for $t_0 \leq t_1 < t_2 < \ldots < t_n$, the increments $\beta(t_2) - \beta(t_1)$, $\beta(t_3) - \beta(t_2)$, $\ldots$, $\beta(t_n) - \beta(t_{n-1})$ are mutually independent ("process with independent increments");

(iii) the increments between any two fixed points in time $t_1$ and $t_2$ are normally distributed with zero expectation so that

$$E\{\beta(t_2) - \beta(t_1)\} = 0 \tag{6.92}$$

and with variance

$$E\{\beta(t_2) - \beta(t_1)\}^2 = \sigma^2|t_2 - t_1| \tag{6.93}$$

where $\sigma$ is a fixed positive parameter (Doob, 1964).

*Remark 6.7.* Equations (6.92) and (6.93) are often written with infinitesimal increments as $E\{d\beta\} = 0$ and $E\{d\beta\}^2 = \sigma^2 dt$. The distribution of the increments $\beta(t_2) - \beta(t_1)$ obviously does not depend on the times $t_1$ and $t_2$ themselves but only on their difference; the increments are therefore stationary (in the strict sense). If $\sigma = 1$, we refer to a normalized Brownian process. It is usually agreed that $\beta(t_0) = 0$. This does not mean any loss in generality since every Brownian sample function $\beta^*(t)$ with $\beta^*(t_0) \neq 0$ can be decomposed into $\beta(t) + \beta^*(t_0)$ where $\beta(t) := \beta^*(t) - \beta^*(t_0)$. In this way $\beta(t_0)$ vanishes and the increments of the process $\{\beta^*(t)\}$ are equal to those of $\{\beta(t)\}$.

*Properties* (a) The independence of the increments means (Bayesian rule) that the conditional probability density of future increments for given previous increments is equal to the unconditional density.

(b) The Brownian process has the Markov property.

(c) The sample functions of the Brownian process are continuous (Wiener).

(d) They are almost nowhere differentiable (Wiener).

(e) They are of unlimited variation in every finite time interval (Lévy).

The last three properties hold with a probability of unity. For their proof, further properties and detailed arguments, see, for example, Krickeberg (1963) or Doob (1964).

*Note 6.2.* The time "derivative" of Brownian motion is the Gaussian "white noise" common in communications and control technology. In the language of probability theory, the Wiener theorem (d) expresses the well-known fact that white noise can only be a conceptual virtual signal.

### 6.8.2 Stochastic Integration

In the next section we will concern ourselves with differential equations that are excited by Brownian motion and have the form

$$dx(t) = f(x,t)dt + v(x,t)d\beta(t) \tag{6.94}$$

where $f(x,t)$ and $v(x,t)$ are ordinary scalar functions of $x$ and $t$. We will interpret these differential equations by integration of both sides from $t = a$ to $t = b$:

$$x(b) - x(a) = \int_a^b f(x,t)dt + \int_a^b v(x,t)d\beta(t) \tag{6.95}$$

The first integral is an ordinary Riemann integral. Its calculation and properties are well known. However, the second integral does not make any sense at first sight: it is not an ordinary integral since $d\beta(t)/dt$ does not exist, and it is not a Stieltjes integral because $\beta(t)$ does not have limited variation. Instead, we are dealing with a *stochastic integral*. An elementary event $\omega$ produces a pair of sample functions $\beta(t;\omega)$ and $x(t;\omega)$ in the interval $[a,b]$. The integral itself is therefore also a random variable which depends on $\omega$:

$$V(\omega) := \int_a^b v(x(t;\omega),t)d\beta(t;\omega) \quad a \leq b \tag{6.96}$$

This type of stochastic integral was first defined and treated by Ito (1944) (see also Doob, 1964). In what follows, let $V(\omega)$ and $W(\omega)$ be the Ito integrals over $v(x,t)$ and $w(x,t)$ respectively, both performed for the same Brownian process $\{\beta(t)\}$. No limitation is involved if we further assume that $\{\beta(t)\}$ is *normalized* (i.e., $E\{d\beta\}^2 = dt$).

*Properties*

(a) $V(\omega)$ is uniquely defined for every $\omega$.

(b) $cV(\omega) + dW(\omega) = \int_a^b (cv + dw)d\beta(t)$

where $c$ and $d$ are constants.

(c) $E\{V\} = 0$

$$\text{(d)}E\{VW\} = \int_a^b E\{vw\}dt \tag{6.97}$$

The proofs are given by Ito (1944) and Doob (1964). Properties (c) and (d) are particularly important. We obtain the following interesting relationship from (d) with $v = w = 1$.

$$E\left\{\left[\int_a^b d\beta(t)\right]^2\right\} = b - a, \quad a \leq b \tag{6.98}$$

This means that the variance of the increment $\beta(b) - \beta(a)$ of a normalized Brownian process is equal to the elapsed time span $b - a$. Of course, the dispersion is then equal to $(b - a)^{1/2}$.

### 6.8.3 Stochastic Differential Equations

It is customary in the theory of nonlinear filtering to describe the observed random signals by a mathematical model in the form of a differential equation whose right-hand side consists of an ordinary part and an additional part excited by Brownian motion. Since all the basic problems are present in the scalar case, we will limit ourselves in this section to first-order differential equations:

$$\mathrm{d}x(t) = f(x,t)\mathrm{d}t + v(x,t)\mathrm{d}\beta(t) \quad t_0 \leq t \tag{6.99}$$

This equation is called a *stochastic differential equation of the diffusion type*. Here, $\{\beta(t),\ t_0 \leq t\}$ is a normalized Brownian process, i.e., it is real, and it has independent and normally distributed increments with $E\{\mathrm{d}\beta(t)\} = 0$ and $E\{\mathrm{d}\beta(t)\}^2 = \mathrm{d}t$. The scaling of the noise can be included in the function $v$. The initial condition $x(t_0)$ is either deterministic or a random variable independent of the increments $\beta(t_2) - \beta(t_1)$. Equation (6.99) is the strict notational form for the commonly used loose formulation

$$\dot{x}(t) = f(x,t) + v(x,t)\dot{\beta}(t)$$

Here, we have formally divided by $\mathrm{d}t$ and $\dot{\beta}(t)$ is Gaussian white noise.

Let $t_0 \leq t \leq T < \infty$. Following the theory of ordinary differential equations, we interpret the stochastic differential equation (6.99) in the following manner. A solution passing through the initial point $x(t_0)$, $t_0$ is a sample function $x(t)$ which satisfies the equation

$$x(t) - x(t_0) = \int_{t_0}^{t} f(x(\tau),\tau)\mathrm{d}\tau + \int_{t_0}^{t} v(x(\tau),\tau)\mathrm{d}\beta(\tau) \tag{6.100}$$

where the second integral is of the stochastic type (6.96). Under certain conditions (see Doob, 1964) it can be shown that a solution of (6.100) exists and that it is unique with a probability of unity. The random process consisting of the solutions $\{x(t),\ t_0 \leq t \leq T\}$ has the following properties.

*Properties*

(a) The sample functions $x(t)$ are continuous with a probability of unity.

(b) $\displaystyle\int_{t_0}^{T} E\{x^2(t)\}\mathrm{d}t < \infty$

(c) For $t_0 < t \leq \tau < T$, $x(t) - x(t_0)$ is independent of the ensemble of increments $\{\beta(T) - \beta(\tau)\}$.

(d) The ensemble of all sample functions $\{x(t),\ t_0 \leq t \leq T\}$ constitutes a Markov process.

(e) The microscopic structure of the process is determined by

$$E\{x(t+h) - x(t)|x(t) = \xi\} = \int_t^{t+h} f(\xi,\tau)\mathrm{d}\tau + 0(h^{3/2}) \qquad (6.101\mathrm{a})$$

$$E\{[x(t+h) - x(t)]^2|x(t) = \xi\} = \int_t^{t+h} v^2(\xi,\tau)\mathrm{d}\tau + 0(h^2) \qquad (6.101\mathrm{b})$$

and the Gaussian nature of the increments $\mathrm{d}\beta(t)$.

It should be noted that the correction term in (6.101a) has order $h^{3/2}$, which is surprising and not trivial. For $h = \mathrm{d}t \rightarrow 0$, the conditional density of $\mathrm{d}x(t)$ for given $x(t) = \xi$ is still Gaussian and determined by the conditional expectation and variance according to equations (6.101a) and (6.101b). It is

$$p(\mathrm{d}x(t)|\xi) = \frac{1}{v(\xi,t)(2\pi\mathrm{d}t)^{1/2}} \exp\left[\frac{-\{\mathrm{d}x - f(\xi,t)\mathrm{d}t\}^2}{2v^2(\xi,t)\mathrm{d}t}\right] \qquad (6.101\mathrm{c})$$

### 6.8.4 Stochastic Differentials Along a Solution Curve

Sometimes we need the total differential $\mathrm{d}\Phi$ of a function $\Phi(x,t)$ where $x(t)$ is the solution of a stochastic differential equation of type (6.99). For example, this is the case in the calculation of the mean square value of $x(t)$ as a function of time: as a first stage, we calculate the differential of $x^2(t)$, i.e., $\mathrm{d}(x^2)$.

The situation for deterministic differential equations is well known. Let

$$\Phi = \Phi(x,t)$$

with

$$\mathrm{d}x(t) = f(x,t)\mathrm{d}t$$

Then the total differential of $\Phi$ along the solution $x(t)$ is given by

$$\mathrm{d}\Phi(x,t) = \frac{\partial\Phi}{\partial t}\mathrm{d}t + \frac{\partial\Phi}{\partial x}\mathrm{d}x(t) \qquad (6.102)$$

However, for stochastic differential equations, (6.102) leads to wrong results! An extra term in $\mathrm{d}t$ must be added to the right-hand side. This can be made plausible by the following considerations. Let

$$\Phi = \Phi(x,t)$$

with

$$\mathrm{d}x(t) = f(x,t)\mathrm{d}t + v(x,t)\mathrm{d}\beta(t)$$

be given where the Brownian increments $\mathrm{d}\beta(t)$ are normally distributed with mean value zero and variance $\mathrm{d}t$. In order to obtain the required $\mathrm{d}\Phi$, we expand $\Phi$ in a Taylor series with respect to the point $x(t),t$:

$$\Delta\Phi = \frac{\partial\Phi}{\partial t} h + \frac{\partial\Phi}{\partial x} \Delta x + \frac{1}{2}\frac{\partial^2\Phi}{\partial x^2} (\Delta x)^2 + O(h^2) + O(\Delta x)^3 + \ldots$$

where $\Delta\Phi = \Phi\{x(t + h),t + h\} - \Phi\{x(t),t\}$ and $\Delta x = x(t + h) - x(t)$. Every meaningful expression for the differential must now be such that at least the conditional expectation with respect to $x(t)$ is correct in terms of the first power of $h$. With this in mind, we consult (6.101a) and (6.101b).

$$E\{\Delta x | x(t) = \xi\} = f(\xi,t)\, h + O(h^{3/2})$$
$$E\{(\Delta x)^2 | x(t) = \xi\} = v^2(\xi,t)\, h + O(h^2)$$

Higher powers of $\Delta x$ do not produce any more terms in $h$. Combining and passing to the limit $h = \mathrm{d}t \rightarrow 0$, we obtain

$$\mathrm{d}\Phi(x,t) = \frac{\partial\Phi}{\partial t} \mathrm{d}t + \frac{\partial\Phi}{\partial x} \mathrm{d}x(t) + \frac{1}{2}\frac{\partial^2\Phi}{\partial x^2} v^2(x,t)\mathrm{d}t \qquad (6.103)$$

The term with $\mathrm{d}x(t)$ contains $\mathrm{d}\beta(t)$; therefore (6.103) is again a stochastic differential equation. The new final term in the differential equation (6.103) stems from the fact that the $(\mathrm{d}x)^2$ term in the Taylor expansion makes a significant contribution to the conditional mean value of $\mathrm{d}\Phi$. The result (6.103), which has only been derived here in a heuristic manner, has been rigorously proved by Ito (1944).

The above considerations can easily be generalized to the differentials of scalar functions of vector variables $\boldsymbol{x}$, where $\boldsymbol{x}$ is the solution to the following vector stochastic differential equation:

$$\mathrm{d}\boldsymbol{x}(t) = \boldsymbol{f}(\boldsymbol{x},t)\,\mathrm{d}t + \boldsymbol{V}(\boldsymbol{x},t)\,\mathrm{d}\boldsymbol{\beta}(t) \tag{6.104}$$

Here, let $\boldsymbol{x}$ be an $n$-vector and $\boldsymbol{\beta}$ an $m$-vector. Its components $\beta_i$ are mutually independent normalized Brownian motions so that $E\{\mathrm{d}\boldsymbol{\beta}\} = \boldsymbol{0}$ and $E\{\mathrm{d}\boldsymbol{\beta}\mathrm{d}\boldsymbol{\beta}'\} = I_m\mathrm{d}t$; $\boldsymbol{f}$ is an $n$-dimensional vector function and $\boldsymbol{V}$ is an $n \times m$ matrix function.

The multidimensional equivalents of the first and second partial derivatives of $\Phi$ with respect to $x$ are the *gradient* and the *Hessian matrix* respectively (see Sections A.9.1 and A.9.2).

The colon product of an $n \times m$ matrix $\boldsymbol{A}$ and an $m \times n$ matrix $\boldsymbol{B}$ is defined as

$$\boldsymbol{A} : \boldsymbol{B} := \mathrm{tr}(\boldsymbol{A}\,\boldsymbol{B}) = \sum_{i=1}^{n} (\boldsymbol{A}\,\boldsymbol{B})_{ii} \tag{6.105}$$

If we again use the Taylor expansion in a formal manner, we obtain

$$\mathrm{d}\Phi = \frac{\partial \Phi}{\partial t}\,\mathrm{d}t + \frac{\partial \Phi}{\partial \boldsymbol{x}}\,\mathrm{d}\boldsymbol{x} + \frac{1}{2}\frac{\partial^2 \Phi}{\partial \boldsymbol{x}^2} : \mathrm{d}\boldsymbol{x}\,\mathrm{d}\boldsymbol{x}' + \ldots$$

In order to set the conditional mean value of $\mathrm{d}\Phi$ correctly, we examine the conditional expectation of the last term for a given $\boldsymbol{x}(t)$. The only significant contribution to this is provided by the component $1/2(\partial^2\Phi/\partial\boldsymbol{x}^2) : (\boldsymbol{V}\mathrm{d}\boldsymbol{\beta}\mathrm{d}\boldsymbol{\beta}'\boldsymbol{V}')$. For the conditional expectation of this component, $\partial^2\Phi/\partial\boldsymbol{x}^2$ and $\boldsymbol{V}$ are constants whereas $\mathrm{d}\boldsymbol{\beta}\mathrm{d}\boldsymbol{\beta}'$ leads to $\boldsymbol{I}_m\mathrm{d}t$. The exact formulation provides the following theorem due to Ito (1961): let $\Phi(\boldsymbol{x},t)$ be continuously differentiable at least twice in every $x_i$ and at least once in $t$. Let $\boldsymbol{x}(t)$ be a unique solution in the Ito sense of the stochastic differential equation (6.104) with $\boldsymbol{V}\boldsymbol{V}' = \boldsymbol{Q}$. Then we have

$$\mathrm{d}\Phi(\boldsymbol{x},t) = \frac{\partial \Phi}{\partial t}\,\mathrm{d}t + \frac{\partial \Phi}{\partial \boldsymbol{x}}\,\mathrm{d}\boldsymbol{x}(t) + \frac{1}{2}\frac{\partial^2 \Phi}{\partial \boldsymbol{x}^2} : \boldsymbol{Q}(\boldsymbol{x},t)\mathrm{d}t \tag{6.106}$$

### 6.8.5 The Fokker–Planck Equation

We are given the vector stochastic differential equation (6.104). Let the $n$-dimensional vector function $\boldsymbol{f}$ be continuously differentiable at least once in all components $x_i$ and the $n \times m$ matrix function $\boldsymbol{V}$ be continuously differentiable at least twice in all components $x_i$. We consider the conditional probability density of the state $\boldsymbol{x}(t)$ for a given state $\boldsymbol{x}(\tau)$ for $\tau \leqslant t$. We have

$$\frac{\partial}{\partial t} p(\boldsymbol{\xi},t \,|\, \boldsymbol{x}(\tau) = \boldsymbol{\eta}) = -\sum_{i=1}^{n} \frac{\partial(f_i p)}{\partial \xi_i} + \frac{1}{2}\sum_{i=1}^{n}\sum_{j=1}^{n} \frac{\partial^2(q_{ij} p)}{\partial \xi_i\,\partial \xi_j} \tag{6.107}$$

where $f_i$ is the $i$th component of the vector $\boldsymbol{f}(\boldsymbol{\xi},t)$ and $q_{ij}$ is an element of the matrix $\boldsymbol{Q}(\boldsymbol{\xi},t) = \boldsymbol{V}(\boldsymbol{\xi},t)\boldsymbol{V}'(\boldsymbol{\xi},t)$. Equation (6.107) is the multidimensional *Fokker–Planck equation* or *Kolmogorov forward diffusion equation*. Its scalar version was first established by Fokker and Planck in studies of diffusion theory and was later rigorously proved by Kolmogorov (1931) (together with his backward diffusion equation). Feller (1936) proved the existence and uniqueness theorems for its solutions. The Kolmogorov proof was later generalized to $n$ dimensions (e.g., Gnedenko, 1957). The Fokker–Planck or Kolmogorov differential equation is needed as a preliminary step in the solution of the nonlinear filter problem.

### 6.8.6 The Nonlinear Filter Problem and the Kushner–Stratonovitch Equation

We are given a vector stochastic differential equation of the diffusion type:

$$\mathrm{d}\boldsymbol{x}(t) = \boldsymbol{f}(\boldsymbol{x},t)\mathrm{d}t + \boldsymbol{V}(\boldsymbol{x},t)\mathrm{d}\boldsymbol{\beta}(t) \quad 0 \leqslant t \tag{6.108}$$

The state vector $\boldsymbol{x}(t)$ cannot be directly measured. Instead, we observe the vector-valued variable $z(t)$ which is formed as follows from $\boldsymbol{x}(t)$ and the measurement noise $\boldsymbol{W}\mathrm{d}\boldsymbol{\gamma}$:

$$\mathrm{d}z(t) = \boldsymbol{h}(\boldsymbol{x},t)\mathrm{d}t + \boldsymbol{W}(t)\mathrm{d}\boldsymbol{\gamma}(t) \quad 0 \leqslant t \tag{6.109}$$

Let $\boldsymbol{x}$ and $\boldsymbol{f}$ be $n$-vectors and $z$ and $\boldsymbol{h}$ be $m$-vectors. The noise vectors $\boldsymbol{\beta}$ and $\boldsymbol{\gamma}$ consist of mutually independent normalized Brownian motions:

$$E\{\mathrm{d}\boldsymbol{\beta}\mathrm{d}\boldsymbol{\beta}'\} = \boldsymbol{I}\mathrm{d}t \quad E\{\mathrm{d}\boldsymbol{\gamma}\mathrm{d}\boldsymbol{\gamma}'\} = \boldsymbol{I}\mathrm{d}t$$

Any scaling and mutual correlation among the components of $\boldsymbol{\beta}$ and $\boldsymbol{\gamma}$ is assumed to be incorporated in the matrices $\boldsymbol{V}$ and $\boldsymbol{W}$. For simplicity, we further make the nonessential assumption that $\boldsymbol{\beta}$ and $\boldsymbol{\gamma}$ are independent:

$$E\{\mathrm{d}\boldsymbol{\beta}\mathrm{d}\boldsymbol{\gamma}'\} \equiv \boldsymbol{0}$$

Two essential conditions now follow:

(a) the matrix $\boldsymbol{W}$ is not a function of the state vector $\boldsymbol{x}$;

(b) it has rank $m$ so that there are at least as many disturbance variables $\gamma_i$ as measurement variables $z_i$.

Let the covariance matrix of the measurement errors $\boldsymbol{W}\mathrm{d}\boldsymbol{\gamma}$ be the symmetric $m \times m$ matrix $\boldsymbol{R}$:

$$\boldsymbol{W}(t)\boldsymbol{W}'(t) := \boldsymbol{R}(t) \quad \boldsymbol{R} \text{ positive definite}$$

Condition (a) enables a significant simplification of the theory to be made; violation of condition (b) lets the filter problem degenerate to a singular case.

We require $\hat{p}(\boldsymbol{\xi},t|z_{[0,t]})$ which is the conditional probability density of the state $\boldsymbol{x}(t)$ when $z_{[0,t]}$; the measurement sample $z(\tau)$ is given in the complete interval $0 \leqslant \tau \leqslant t$. This problem is solved using the following theorem.

*Theorem 6.1* (Kushner 1964). If $\boldsymbol{x}(t)$ and $z(t)$ are solutions of the stochastic differential equations (6.108) and (6.109), with $\boldsymbol{VV}' = \boldsymbol{Q}$ and $\boldsymbol{WW}' = \boldsymbol{R}$, we have

$$\begin{aligned} &\hat{p}(\boldsymbol{\xi},t + \mathrm{d}t|z_{[0,t+\mathrm{d}t]}) - \hat{p}(\boldsymbol{\xi},t|z_{[0,t]}) \\ &\quad = \mathrm{d}\hat{p} = -\sum_{i=1}^{n} \frac{\partial(f_i\hat{p})}{\partial\xi_i}\,\mathrm{d}t + \frac{1}{2}\sum_{i=1}^{n}\sum_{j=1}^{n} \frac{\partial^2(q_{ij}\hat{p})}{\partial\xi_i\,\partial\xi_j}\,\mathrm{d}t \\ &\quad + \hat{p} \times \{\boldsymbol{h}(\boldsymbol{\xi},t) - \hat{\boldsymbol{h}}(t)\}'\boldsymbol{R}^{-1}(t)\{\mathrm{d}z(t) - \hat{\boldsymbol{h}}(t)\,\mathrm{d}t\} \end{aligned} \tag{6.110}$$

where $\hat{\boldsymbol{h}}$ is the conditional expectation of $\boldsymbol{h}$ for given $z_{[0,t]}$.

*Note 6.3*. The middle row of (6.110) is a Fokker–Planck equation for $\hat{p}$ (compare equation (6.107)). This part describes the evolution of the conditional density $\hat{p}$ in the absence of usable measurements, i.e., $\|\boldsymbol{R}\| = \infty$. The last row of (6.110) represents the improvement to the density based on the observations $z(t)$. In a meaningful filter situation, this influence will act towards concentrating the density around the actual value $\boldsymbol{\xi} = \boldsymbol{x}(t)$. The ideal filtering result would be that $\hat{p}$ is completely concentrated on $\boldsymbol{x}(t)$ so that it would have the mean value $\boldsymbol{x}(t)$ and zero variance (singular density). This is counteracted by the fact that all the measurements are continuously subject to noise in the form of $\boldsymbol{\gamma}$.

*Note 6.4*. The term $\mathrm{d}z(t)$ in the last row of (6.110) contains the term $\mathrm{d}\boldsymbol{\gamma}(t)$ of the Brownian motion. Because of this, (6.110) is a stochastic differential equation in contrast with the Fokker–Planck equation. Hence, it is not possible to divide by $\mathrm{d}t$ in order to obtain $\partial\hat{p}/\partial t$.

*Note 6.5*. Stratonovitch (1960) was the first to apply this approach to solving the nonlinear filter problem. However, his version of (6.110) did not take into account the peculiarities of Brownian motion. Kushner (1964) provided the correct result by adequate handling of the stochastic differentials, but without reference to Ito's theorem. Bucy (1965) proved Kushner's result in another more rigorous manner: he first established a closed expression for $\hat{p}$ and then obtained formula (6.110) from this by differentiation.

*Note 6.6*. The stochastic partial differential equation (6.110) associated with the names of Stratonovitch, Kushner and Bucy describes the evolution of the conditional probability density of the state $\boldsymbol{x}(t)$ with respect to a given measurement sample $z_{[0,t]}$. It solves the nonlinear filter problem with complete generality. However, this differential equation is very cumbersome for practical purposes, so that

Kushner (1964) has proposed replacing it with a system of ordinary differential equations for the conditional expectation and the conditional central moments of $\boldsymbol{x}(t)$. Strictly speaking, this system has infinite order but, if the filter error $\boldsymbol{x}(t) - E\{\boldsymbol{x}(t)|z_{[0,t]}\}$ is small, it can be truncated after the second conditional central moment (covariance matrix) as an approximation. Bucy (1965) has done this for scalar $x(t)$.

Following this, Bass *et al.* (1966) treated the general $n$-dimensional case. Their result consists of a coupled system of stochastic differential equations for the conditional expectation $\hat{\boldsymbol{x}}$ and the conditional covariance matrix $\tilde{\boldsymbol{P}}$ of $\boldsymbol{x}$. It is written as follows:

$$\frac{\mathrm{d}}{\mathrm{d}t}\hat{\boldsymbol{x}}(t) = \boldsymbol{f}(\hat{\boldsymbol{x}},t) + \frac{1}{2}\boldsymbol{f}_{xx}(\hat{\boldsymbol{x}},t) : \tilde{P}$$
$$+ \tilde{\boldsymbol{P}}\,\boldsymbol{h}_x'(\hat{\boldsymbol{x}},t)\,\boldsymbol{R}^{-1}(t)\left\{\boldsymbol{y}(t) - \boldsymbol{h}(\hat{\boldsymbol{x}},t) - \frac{1}{2}\boldsymbol{h}_{xx}(\hat{\boldsymbol{x}},t) : \tilde{\boldsymbol{P}}\right\} \tag{6.111}$$

$$\frac{\mathrm{d}}{\mathrm{d}t}\tilde{\boldsymbol{P}}(t) = \tilde{\boldsymbol{P}}\boldsymbol{f}_x'(\hat{\boldsymbol{x}},t) + \boldsymbol{f}_x(\hat{\boldsymbol{x}},t)\tilde{\boldsymbol{P}}$$
$$- \tilde{\boldsymbol{P}}\{\boldsymbol{h}_x'(\hat{\boldsymbol{x}},t)\boldsymbol{R}^{-1}(t)\boldsymbol{h}_x(\hat{\boldsymbol{x}},t)\}\tilde{\boldsymbol{P}} + \boldsymbol{Q}(\hat{\boldsymbol{x}},t) + \frac{1}{2}\boldsymbol{Q}_{xx}(\hat{\boldsymbol{x}},t) : \tilde{\boldsymbol{P}}$$
$$- \frac{1}{2}\left[\tilde{\boldsymbol{P}} : \boldsymbol{h}_{xx}'(\hat{\boldsymbol{x}},t)\boldsymbol{R}^{-1}(t)\left\{\boldsymbol{y}(t) - \boldsymbol{h}(\hat{\boldsymbol{x}},t) - \frac{1}{2}\boldsymbol{h}_{xx}(\hat{\boldsymbol{x}},t) : \tilde{\boldsymbol{P}}\right\}\right]\tilde{\boldsymbol{P}} \tag{6.112}$$

Here, $\boldsymbol{f}_x$ is the Jacobian matrix and $\boldsymbol{f}_{xx}$ consists of the stacked Hessian matrices of the elements $f_i$ (see Sections A.9.2 and A.9.3). The expressions $\boldsymbol{f}_{xx}$: $\tilde{\boldsymbol{P}}$ and $\boldsymbol{Q}_{xx} : \tilde{\boldsymbol{P}}$ consist of the colon products of $f_{i,xx}$ and $q_{ij,xx}$ respectively with $\tilde{\boldsymbol{P}}$. The initial conditions are

$$\hat{\boldsymbol{x}}(t_0) = E\{\boldsymbol{x}(t_0)\} \quad \text{and} \quad \tilde{\boldsymbol{P}}(t_0) = \boldsymbol{P}(t_0) \tag{6.113}$$

where $\boldsymbol{P}(t_0)$ is the covariance matrix of the initial state (see (5.74a)). The terms with the colon products originate from second-order terms in the Taylor expansion and show the importance of Ito's theorem (6.106) for nonlinear filtering. It should be noted that (6.112) is no longer independent of $\boldsymbol{y}$!

When equations (6.111) and (6.112) are applied, it is recommended that checks for convergence and stability are done in each case because they only form an approximate solution to a problem for which there are no general test criteria. For linear systems (i.e., $\boldsymbol{f} = \boldsymbol{A}\boldsymbol{x}$, $\boldsymbol{h} = \boldsymbol{C}\boldsymbol{x}$ and $\boldsymbol{Q}$ independent of $\boldsymbol{x}$), we of course obtain the Kalman–Bucy filter and the Riccati differential equation.

Brammer (1969, 1970a) has applied the nonlinear filtering theory to real-time identification of control systems. Adaptive filters can also be designed in this way (Brammer, 1970b). A practical example of a nonlinear filter problem in radar systems is described by Brammer (1983).

## REFERENCES


Arnold, L. (1973). *Stochastische Differentialgleichungen*, Oldenbourg, Munich.

Athans, M. and Falb, P.L. (1966). *Optimal Control,* McGraw-Hill, New York.

Bass, R.W. (1967). Machine solution of high order matrix Riccati equations, *Tech. Rep.,* Missiles and Space Systems Division, McDonnell Douglas Aircraft, Santa Monica, CA.

Bass, R.W., Norum, V.D. and Schwartz, L. (1966). Optimal multichannel nonlinear filtering, *J. Math. Anal. Appl.,* **16,** 152–164.

Bellman, R. and Dreyfus, S.E. (1962). *Applied Dynamic Programming,* Princeton University Press, Princeton, NJ.

Brammer, K. (1967). Lower order linear filtering and prediction of nonstationary random sequences, *Tech. Rep. SRL 67-0003,* OAR, Frank J. Seiler Research Laboratory, Colorado Springs, CO, February.

Brammer, K. (1968a). Lower order optimal linear filtering of nonstationary random sequences, *IEEE Trans. Autom. Control,* **13,** 198–199.

Brammer, K. (1968b). Zur optimalen linearen Filterung und Vorhersage instationer Zufallsprozesse in diskreter zeit, *Regelungstechnik,* **16,** 105–110.

Brammer, K. (1969). Parametererkennung geregelter Strecken durch nichtlineare Filterung, *Dissertation,* Technische Hochschule Darmstadt, April.

Brammer, K. (1970a). Schätzung von Parametern und Zustandsvariablen linearer Regelstrecken durch nichtlineare Filterung, *Regelungstech. Prozess-Datenverarbeit.,* **18,** 255–261.

Brammer, K. (1970b). Input-adaptive Kalman–Bucy filtering, *IEEE Trans. Autom. Control,* **15,** 157–158.

Brammer. K. (1983). Stochastic filtering problems in multiradar tracking. In Bucy, R.S. and Moura, J.M.F. (eds.), *Nonlinear Stochastic Problems,* Reidel, Dordrecht.

Bryson, A.E. and Johansen, D.E. (1965). Linear filtering for time-varying systems using measurements containing colored noise. *IEEE Trans. Autom. Control,* **10,** 4–10.

Bucy, R.S. (1965). Nonlinear filtering theory, *IEEE Trans. Autom. Control,* **10,** 198.

Bucy, R.S. (1967). Optimal filtering for correlated noise, *J. Math. Anal. Appl.,* **20,** (1).

Bucy, R.S. and Joseph, P.D. (1968). *Filtering for Stochastic Processes, with Applications to Guidance,* Interscience, New York.

Carathéodory, C.C. (1935). *Variationsrechnung und Partielle Differentialgleichungen erster Ordnung,* Teubner, Leipzig.

Doob, J.L. (1964). *Stochastic Processes* (5th edn), Wiley, New York.

Feller, W. (1936). Zur Theorie der stochastischen Prozesse (Existenz- und Eindeutigkeitssätze), *Math. Ann.,* **113,** 113–160.

Gnedenko, B.W. (1957). Lehrbuch der Wahrscheinlichkeitsrechnung, *Textbook on Probability Calculus* (in German), Akademie-Verlag, Berlin.

Ito, K. (1944). Stochastic integral, *Proc. Imp. Acad. Tokyo,* **20,** 519–524.

Ito, K. (1961). On stochastic processes, *Lecture Notes,* Tata Institute for Fundamental Research, Bombay.

Ito, K. and McKean, H.P. (1965). *Diffusion Processes and their Sample Paths,* Springer, Berlin.

Joseph, P.D. (1961). Optimum design of linear multivariate digital control systems, *Dissertation,* Purdue University, August.

Joseph, P.D. and Tou, J.T. (1961). On linear control theory. *AIEE Trans. Appl. Ind., Pt II,* **80,** 193–196.

Kalman, R.E. and Bucy, R.S. (1961). New results in linear filtering and prediction theory, *Trans. ASME, Ser. D, J. Basic Eng.,* **83,** 95–108.

Kalman, R.E. and Englar, T.S. (1965). A user's manual for the automatic synthesis program (ASP-C), *Tech. Rep. NASA-CR-475,* Ames Research Center, Moffet Field, CA.

Kalman, R.E., Englar, T.S., and Bucy, R.S. (1962). Fundamental study of adaptive control systems, *Tech. Rep. ASD-TR-61-27,* Vol. I, Flight Control Laboratory, Aeronautical Systems Division, AFSC, Wright-Patterson Air Force Base, Dayton, OH, April.

Kolmogorov, A. (1931). Über die analytischen Methoden in der Wahrscheinlichkeitsrechnung, *Math. Ann.,* **104,** 415–458.

Krickeberg, K. (1963). Wahrscheinlichkeitstheorie, *Probability Theory* (in German), Teubner, Stuttgart.

Kushner, H.J. (1964). On the differential equations satisfied by conditional probability densities of Markov processes, with applications, *J. SIAM Control, Ser. A,* **2,** 106–119.

Leondes, C.T. and Novak, L.M. (1974). Reduced-order observers for linear discrete-time systems, *IEEE Trans. Autom. Control,* **19,** 42–46.

Lévy, P. (1948). Processus stochastiques et mouvement Brownien, Gauthiers-Villars, Paris.

Penrose, R. (1955). A generalized inverse for matrices, *Proc. Cambridge Phil. Soc.,* **51,** 406–413.

Pontryagin, L.S., Boltyanskii, V.G., Gamkrelidze, R.V., and Misčenko, E.F. (1964). *Mathematische Theorie Optimaler Prozesse,* Oldenbourg, Munich.

Schmidt, G.T. (ed.) (1976). Practical aspects of Kalman filtering implementation, *AGARD Lect. Ser.,* **82,** March.

Stratonovitch, R.L. (1960). Conditional Markov processes, *Theory Probab. Its Appl. (USSR),* **5,** 156–178.

Wiener, N. (1923). Differential space, *J. Math. Phys. Inst. Tech.,* **2,** 131–174.

Zadeh, L.A. and Desoer, C.A. (1963). *Linear System Theory,* McGraw-Hill, New York.

## BIBLIOGRAPHY

Bucy, R.S. and Moura, J.M.F. (eds) (1983). *Nonlinear Stochastic Problems,* Reidel, Dordrecht.

Krebs, V. (1980). *Nichtlineare Filterung,* Oldenbourg, Munich.

Leondes, C.T. (ed.) (1982). Advances in the techniques and technology of the application of nonlinear filters and Kalman filters, *AGARDograph 256,* March.

# *Appendix*
# *Some Basic Principles of Matrix Theory*

Like many texts on control systems, this book makes extensive use of matrix theory. Therefore the most important principles of this calculus are summarized below.

## A.1 THE CONCEPTS OF VECTOR AND MATRIX

The concept of a vector is basically very general. In the broadest sense, it is defined as follows: the elements $\boldsymbol{v}$ of a certain space are *vectors* in a vector space if and only if they all satisfy the following eight conditions.

(i) *Addition*
The commutative law holds:

$$\boldsymbol{v}_1 + \boldsymbol{v}_2 = \boldsymbol{v}_2 + \boldsymbol{v}_1 \tag{A.1}$$

The associative law holds:

$$\boldsymbol{v}_1 + (\boldsymbol{v}_2 + \boldsymbol{v}_3) = (\boldsymbol{v}_1 + \boldsymbol{v}_2) + \boldsymbol{v}_3 \tag{A.2}$$

There is a uniquely defined *zero vector* $\boldsymbol{0}$ such that

$$\boldsymbol{v} + \boldsymbol{0} = \boldsymbol{v} \tag{A.3}$$

For every vector $\boldsymbol{v}$, there exists a unique *additive inverse* $-\boldsymbol{v}$ such that

$$\boldsymbol{v} + (-\boldsymbol{v}) = \boldsymbol{0} \tag{A.4}$$

(ii) *Multiplication by scalars c*
The associative law holds:

$$c_1(c_2\boldsymbol{v}) = (c_1c_2)\boldsymbol{v} \tag{A.5}$$

The distributive law holds for scalars:

$$(c_1 + c_2)\boldsymbol{v} = c_1\boldsymbol{v} + c_2\boldsymbol{v} \tag{A.6}$$

The distributive law holds for vectors:

$$c(\boldsymbol{v}_1 + \boldsymbol{v}_2) = c\boldsymbol{v}_1 + c\boldsymbol{v}_2 \tag{A.7}$$

Multiplication by unity leaves $\boldsymbol{v}$ unchanged:

$$1 \times \boldsymbol{v} = \boldsymbol{v} \tag{A.8}$$

Examples of vectors in this general sense are physical quantities with direction (e.g., forces, moments, velocities), the set of all continuous functions in a certain interval and matrices.

An $m \times n$ *matrix* or a matrix of the $m \times n$ type is a "tuple" of $mn$ numbers arranged in the form of a rectangle with $m$ rows and $n$ columns. In this book, matrices are denoted by bold italic letters. Matrices with more than one row or column are written in capital letters. We therefore have

$$\boldsymbol{A} := \begin{bmatrix} a_{11} & a_{12} & \dots & a_{1n} \\ a_{21} & a_{22} & \dots & a_{2n} \\ \vdots & & & \\ a_{m1} & a_{m2} & \dots & a_{mn} \end{bmatrix} := (a_{ik}) \tag{A.9}$$

The real or complex numbers $a_{ik}$ are called *elements* of the matrix $\boldsymbol{A}$. The first subscript of an element indicates the row occupied by the element and the second indicates the column.

In this book the term vector, apart from the above definition, is only used in a quite restricted sense. For us, a vector is a single-row or single-column matrix. Vectors in this sense are denoted by lower case bold italic letters. Where necessary, a distinction is made between *column vectors* ($m \times 1$ matrix or $m$ vector, designated by $\boldsymbol{x}$, for example) and *row vectors* ($1 \times n$ matrix, designated by a prime, e.g., $\boldsymbol{x}'$). A pure and simple "vector" is always understood here as a column vector. Every column of the above matrix $\boldsymbol{A}$ is a (column) vector. The $k$th column, consisting of the elements $a_{1k}, a_{2k}, \dots, a_{mk}$, is sometimes denoted by $\boldsymbol{a}_k$. The $i$th row of the matrix $\boldsymbol{A}$ is a row vector so that $\boldsymbol{a}^i = [a_{i1}, a_{i2}, \dots, a_{in}]$. (The use of the superscript $i$ is intended to remind us of the prime for row vectors.)

A matrix which has as many rows as columns is called a *square matrix*. Its elements $a_{11}, a_{22}, \dots, a_{nn}$ are called *main elements* and form the *main diagonal*. The *trace* of a square matrix is the sum of its main elements:

$$\operatorname{tr} \boldsymbol{A} := \sum_{i=1}^{n} a_{ii} \tag{A.10}$$

A *diagonal matrix* is a square matrix in which all elements outside the main diagonal are zero. If all the main elements of a diagonal matrix are equal, we have a *scalar matrix*. A *unit matrix* $\boldsymbol{I}$ is a diagonal matrix whose main elements are all equal to unity.

All types of matrices in which all the elements vanish without exception are called *zero matrices* (symbol $\mathbf{0}$).

Two $m \times n$ matrices are *equal* if and only if every element of one matrix is equal to its corresponding counterpart in the other matrix.

The *transpose* of an $m \times n$ matrix $\boldsymbol{A}$ is the $n \times m$ matrix $\boldsymbol{A}'$ which is obtained by interchanging the rows and columns. A square matrix which is equal to its transpose is called *symmetric*. If $\boldsymbol{A} = -A'$, we have a *skew symmetric* matrix; its main elements are all equal to zero.

A *complex matrix* is a matrix with complex elements. Any complex matrix $\boldsymbol{C}$ is converted to its *conjugate transpose* $\boldsymbol{C}^*$ if it is transposed and all elements are substituted by their complex conjugates. The conjugate transpose is usually more useful than the normal transpose for complex matrices and vectors. In the real case, they are both the same.

The *determinant* of a square matrix is formed from the elements using the normal rules (Bodewig 1959; Hildebrand 1952):

$$\det \boldsymbol{A} := \begin{vmatrix} a_{11} & a_{12} & \dots & a_{1n} \\ a_{21} & a_{22} & \dots & a_{2n} \\ \vdots & & & \\ a_{n1} & a_{n2} & \dots & a_{nn} \end{vmatrix} \tag{A.11}$$

A square matrix is *nonsingular* if its determinant is nonzero. If the determinant vanishes, the matrix is *singular*.

The vectors $\boldsymbol{a}_1, \boldsymbol{a}_2, \dots, \boldsymbol{a}_n$ are *linearly dependent* if there is at least one $c_k \neq 0$, so that

$$c_1\boldsymbol{a}_1 + c_2\boldsymbol{a}_2 + \dots + c_n\boldsymbol{a}_n = \boldsymbol{0}$$

Otherwise they are *linearly independent* (see Section A.2 for the execution of the arithmetic operations). According to this, a tuple of vectors containing a zero vector is always linearly dependent. The number of linearly independent rows or columns of a matrix is known as its *rank*.

Advantages of the concise notation for matrices are the saving of space and the greater clarity. In particular, many analogies with the scalar case become directly apparent. However, the main advantage is that meaningful arithmetic operations can be defined for matrices so that compound matrix expressions can be transformed in an elegant manner. This considerably facilitates the treatment of multidimensional systems.

## A.2 THE BASIC OPERATIONS

The basic operations of matrix theory are the addition of two matrices and the multiplication of a matrix by a scalar. Both are defined in such a way that the eight conditions (A.1)–(A.8) are satisfied for vectors in the broad sense. Vectors in the strict sense, i.e., single-row or single-column matrices, are naturally included.

*Matrix addition* is only defined for matrices of the same type. The *sum* of two $m \times n$ matrices is another $m \times n$ matrix which is formed by pairwise addition of the corresponding elements of the summands. In formal terms, $\boldsymbol{A} + \boldsymbol{B} = \boldsymbol{C}$ means $a_{ik} + b_{ik} = c_{ik}$ for all $i$, $k$. Matrix addition is obviously commutative and associative (cf. (A.1) and (A.2)). When any matrix is added to a suitable zero matrix, $\boldsymbol{A} + \boldsymbol{0} = \boldsymbol{A}$ is always true (compare (A.3)).

*Multiplication of a matrix* by a scalar is defined for any type of matrix and should be carried out so that every element of the matrix is multiplied by the number in question, i.e.,

$$c\boldsymbol{A} = \boldsymbol{B} \quad \text{means} \quad ca_{ik} = b_{ik} \quad \text{for all} \quad i,k$$

Obviously, the associative law (A.5) and the distributive laws ((A.6) and (A.7)) apply here for scalars and matrices. Multiplication by unity leaves matrices unchanged (cf. (A.8)). Of course, the commutative law also applies:

$$c\boldsymbol{A} = \boldsymbol{A}c$$

The additive inverse matrix $-\boldsymbol{A}$ is defined as the product $(-1)\boldsymbol{A}$. This means that

$$\boldsymbol{A} + (-\boldsymbol{A}) = \boldsymbol{0}$$

is always true (compare (A.4)). Finally, the *subtraction of two matrices* is defined as follows:

$$\boldsymbol{A} - \boldsymbol{B} := \boldsymbol{A} + (-\boldsymbol{B})$$

*Note*. The set of all $m \times n$ matrices forms a vector space in the broad sense.

## A.3 MATRIX MULTIPLICATION

Matrix multiplication forms the central core of matrix theory. This operation is only defined for *concatenable* matrices, i.e., the number of columns of the left-hand factor must be equal to the number of rows of the right-hand factor. The *product* of an $m \times n$ matrix $\boldsymbol{A}$ and an $n \times p$ matrix $\boldsymbol{B}$ is the $m \times p$ matrix $\boldsymbol{C}$ whose elements $c_{ik}$ are produced by passing along the $i$th row of $\boldsymbol{A}$ and the $k$th column of $\boldsymbol{B}$, multiplying each pair of elements and then adding. In formal terms

$$\boldsymbol{A}\,\boldsymbol{B} = \boldsymbol{C} \quad \text{means} \quad \sum_{j=1}^{n} a_{ij} b_{jk} = c_{ik} \quad \text{for all} \quad i,k \qquad \text{(A.12)}$$

For manual calculation it is best to use Falk's scheme which is shown in Figure A.1 (Zurmuehl, 1965; Zurmuehl and Falk, 1984, 1986). For postmultiplication by additional matrices, the diagram is simply extended on the top right and for further premultiplication it is extended on the bottom left. In this way every element only needs to be written down once.

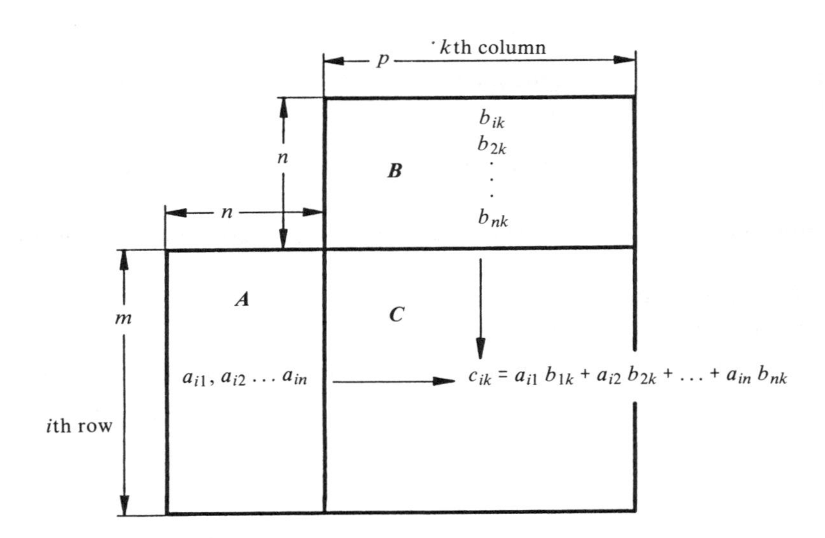


**Figure A.1** Falk's scheme for the multiplication $\boldsymbol{AB} = \boldsymbol{C}$.

In a matrix product, care must be taken to distinguish between multiplication on the left (premultiplication) and on the right (postmultiplication) because the commutative law is not generally valid here. This is obvious when the factors cannot be concatenated on both sides. In other cases the lack of commutativeness can easily be illustrated by a suitable example. If, exceptionally, the product is indeed commutative, the two factors are called *interchangeable*. For example, diagonal matrices of the same order are always interchangeable, just like the factors $\boldsymbol{A}$ and $\boldsymbol{A}^k$.

Matrix multiplication is *associative:*

$$(\boldsymbol{A}\,\boldsymbol{B})\,\boldsymbol{C} = \boldsymbol{A}(\boldsymbol{B}\,\boldsymbol{C}) \tag{A.13}$$

Right-hand and left-hand multiplication are *distributive:*

$$(\boldsymbol{A} + \boldsymbol{B})\boldsymbol{C} = \boldsymbol{A}\,\boldsymbol{C} + \boldsymbol{B}\,\boldsymbol{C} \tag{A.14a}$$

$$\boldsymbol{C}(\boldsymbol{A} + \boldsymbol{B}) = \boldsymbol{C}\,\boldsymbol{A} + \boldsymbol{C}\,\boldsymbol{B} \tag{A.14b}$$

Each of these properties can be proved by forming the elements in the same position on both sides according to the rules and comparing them.

When any matrix is multiplied by a suitable type of unit matrix, it remains unchanged:

$$\boldsymbol{I}\,\boldsymbol{A} = \boldsymbol{A}, \quad \boldsymbol{A}\,\boldsymbol{I} = \boldsymbol{A} \tag{A.15}$$

As for an ordinary product, the matrix product vanishes when a factor is zero. However, it is also possible that

$$\boldsymbol{A}\,\boldsymbol{B} = \boldsymbol{0} \quad \text{although} \quad \boldsymbol{A} \neq \boldsymbol{0} \quad \text{and} \quad \boldsymbol{B} \neq \boldsymbol{0} \tag{A.16}$$

This remarkable case occurs when every row of $\boldsymbol{A}$ is orthogonal* to all columns of $\boldsymbol{B}$†.

Examples of the application of the above computational rules are given below (the proofs are left as an exercise for the reader).

*Example A.1.* The following notations are equivalent for a linear system of equations:

$$\begin{gathered}\sum_{k=1}^{n} a_{ik}x_k = y_i, \quad i = 1 \ldots m \\ \begin{bmatrix} a_{11} & \ldots & a_{1n} \\ \vdots & & \vdots \\ a_{m1} & \ldots & a_{mn} \end{bmatrix} \begin{bmatrix} x_1 \\ \vdots \\ x_n \end{bmatrix} = \begin{bmatrix} y_1 \\ \vdots \\ y_m \end{bmatrix} \\ \boldsymbol{A}\boldsymbol{x} = \boldsymbol{y}\end{gathered} \tag{A.17}$$

*See scalar product.

†If $\boldsymbol{A}$ is nonsingular and square, however, it follows from $\boldsymbol{AB} = \boldsymbol{0}$ that $\boldsymbol{B} = \boldsymbol{0}$ (see equations (A.34) and (A.35) with $\boldsymbol{Y} = \boldsymbol{0}$). If $\boldsymbol{B}$ is nonsingular and square, it follows that $\boldsymbol{A} = \boldsymbol{0}$.

*Example A.2*. An often useful decomposition of the left-hand side of the above system is

$$\boldsymbol{A}\boldsymbol{x} = \boldsymbol{a}_1 x_1 + \boldsymbol{a}_2 x_2 + \ldots + \boldsymbol{a}_n x_n \qquad \text{(A.18)}$$

where $\boldsymbol{a}_i$ is the $i$th column of $\boldsymbol{A}$. In a corresponding manner, a product $\boldsymbol{z}'\boldsymbol{A}$ can be written as the sum of the rows $\boldsymbol{a}^k$.

*Example A.3*. The *scalar product* of two $n$ vectors $\boldsymbol{x}$ and $\boldsymbol{y}$ is defined as $\boldsymbol{x}'\boldsymbol{y}$. Obviously, the two vectors can be interchanged in the scalar product. Two vectors are *orthogonal* if their scalar product is equal to zero.

*Example A.4*. The *dyadic product* of two vectors $\boldsymbol{x}$ and $\boldsymbol{y}$ is defined as $\boldsymbol{x}\,\boldsymbol{y}'$. A dyad always has a rank of unity. The trace of the dyad of two vectors of the same dimension is equal to their scalar product:

$$\operatorname{tr}(\boldsymbol{x}\,\boldsymbol{y}') = \boldsymbol{x}'\boldsymbol{y} \qquad \text{(A.19)}$$

*Example A.5*. The *cross product* of three-dimensional vectors can be written as the product of a skew symmetric matrix and a column vector:

$$\boldsymbol{x} \times \boldsymbol{y} = \begin{bmatrix} 0 & -x_3 & x_2 \\ x_3 & 0 & -x_1 \\ -x_2 & x_1 & 0 \end{bmatrix} \begin{bmatrix} y_1 \\ y_2 \\ y_3 \end{bmatrix} \qquad \text{(A.20)}$$

*Example A.6*. The *transpose of a product* is equal to the product of the transposes in reverse order:

$$(\boldsymbol{A}\,\boldsymbol{B})' = \boldsymbol{B}'\boldsymbol{A}' \qquad \text{(A.21)}$$

This rule, which can easily be derived from the definitions, is frequently applied. For complex matrices, we also have

$$(\boldsymbol{A}\boldsymbol{B})^* = \boldsymbol{B}^*\boldsymbol{A}^*$$

*Example A.7*. The *determinant of a product* of two $n \times n$ matrices is equal to the product of the individual determinants:

$$\det(\boldsymbol{A}\,\boldsymbol{B}) = \det \boldsymbol{A} \det \boldsymbol{B} \qquad \text{(A.22)}$$

The general proof is not simple. It is given, for example, by Gantmacher (1958, 1959). Therefore the product of square matrices is nonsingular if and only if all factors are nonsingular.

## A.4 SYSTEMS OF LINEAR EQUATIONS AND THE INVERSE MATRIX

### A.4.1 Solution of a Single System of Equations

We have a system of $n$ linear equations of the form

$$\boldsymbol{A}\,\boldsymbol{x} = \boldsymbol{y} \tag{A.23}$$

Let the $n \times n$ matrix $\boldsymbol{A}$ have known elements and be nonsingular. The $n$-vector $\boldsymbol{x}$ consists of the unknowns $x_1, \ldots, x_n$. Let the components $y_1, \ldots, y_n$ of the $n$-vector $\boldsymbol{y}$ be the given values. In what follows, some methods of solving this system for the unknowns are discussed.

The *Gaussian algorithm* is a numerical method in which the equation system (A.23) is first transformed into a triangular system by stepwise elimination of the unknowns $x_1, \ldots, x_{n-1}$. Working backwards, this can easily be solved for the unknowns $x_n, \ldots, x_1$ (Dietrich and Stahl, 1978; Zurmuehl and Falk, 1984, 1986). If the *Gauss–Jordan algorithm,* which is a modified version of the Gaussian algorithm, is used, elimination and solution of the unknowns is carried out in one sequence (Hildebrand, 1952). The *Cholesky algorithm* is recommended for symmetric matrices $\boldsymbol{A}$ (Zurmuehl and Falk, 1984, 1986).

*Cramer's rule* is less suitable for numerical calculations but is of key importance in general theoretical statements (Hildebrand, 1952). Its vector matrix form is as follows:

$$\boldsymbol{x} = \frac{1}{\det \boldsymbol{A}} \begin{bmatrix} D_1 \\ D_2 \\ \vdots \\ D_n \end{bmatrix} \quad \text{where} \quad D_k = \begin{vmatrix} a_{11} & \cdots & y_1 & \cdots & a_{1n} \\ a_{21} & \cdots & y_2 & \cdots & a_{2n} \\ \vdots & & \vdots & & \vdots \\ a_{n1} & \cdots & y_n & \cdots & a_{nn} \end{vmatrix} \tag{A.24}$$

where $D_k$, $k = 1, \ldots, n$, is a determinant of order $n$ which is formed after replacing the $k$th column of $\boldsymbol{A}$ by $\boldsymbol{y}$. It can easily be seen from equation (A.24) that the nonsingularity of $\boldsymbol{A}$ is necessary and sufficient for the equation system (A.23) to be solvable.

Cramer's rule takes another form when $D_k$ is expanded with respect to the $k$th column. To do this we need the cofactors. The *cofactor* $A_{ik}$ of the element $a_{ik}$ of an $n \times n$ matrix $\boldsymbol{A}$ is formed as follows: the $i$th row and the $k$th column of $\boldsymbol{A}$ are deleted, and the remaining minor of order $(n - 1)$ is calculated and multiplied by $(-1)^{i+k}$. According to the *Laplace expansion theorem* for determinants (Gantmacher, 1958, 1959; Zurmuehl, 1965; Hildebrand, 1952), the $D_k$ in (A.24) can be expressed as

$$D_k = A_{1k}y_1 + A_{2k}y_2 + \ldots + A_{nk}y_n \tag{A.25}$$

Note that the $A_{1k}, \ldots, A_{nk}$ are not just the cofactors of the $k$th column of $\boldsymbol{A}$ but are also the cofactors of the $k$th column of $D_k$. Expression (A.25) is now substituted into Cramer's rule (A.24) for $k = 1, \ldots, n$. We have

$$\boldsymbol{x} = \frac{1}{\det \boldsymbol{A}} \begin{bmatrix} A_{11}y_1 & + & A_{21}y_2 & + & \ldots & + & A_{n1}y_n \\ A_{12}y_1 & + & A_{22}y_2 & + & \ldots & + & A_{n2}y_n \\ \vdots & & & & & & \\ A_{1n}y_1 & + & A_{2n}y_2 & + & \ldots & + & A_{nn}y_n \end{bmatrix}$$

or

$$\boldsymbol{x} = \frac{1}{\det \boldsymbol{A}} \begin{bmatrix} A_{11} & A_{21} & \ldots & A_{n1} \\ A_{12} & A_{22} & \ldots & A_{n2} \\ \vdots & & & \\ A_{1n} & A_{2n} & \ldots & A_{nn} \end{bmatrix} \boldsymbol{y} \tag{A.26}$$

This version of Cramer's rule differs from (A.24) in that the determinants to be evaluated no longer depend on the $y_i$. It is therefore superior for fixed $\boldsymbol{A}$ and variable $\boldsymbol{y}$.

### A.4.2 The (Multiplicative) Inverse Matrix

If we start from rule (A.26), the introduction of the inverse matrix is only a matter of definition. To ensure that the numbering of the elements corrresponds to the usual convention, we write the matrix of the cofactors from (A.26) in transposed form and define:

$$\boldsymbol{A}^{-1} := \frac{1}{\det \boldsymbol{A}} \begin{bmatrix} A_{11} & A_{12} & \ldots & A_{1n} \\ A_{21} & A_{22} & \ldots & A_{2n} \\ \vdots & & & \\ A_{n1} & A_{n2} & \ldots & A_{nn} \end{bmatrix}' \tag{A.27}$$

where

$$A_{ik} := (-1)^{i+k} \operatorname{minor}(\boldsymbol{A})_{ik}$$

The inverse $\boldsymbol{A}^{-1}$ exists if and only if $\boldsymbol{A}$ is square and nonsingular. When the above definition is used, Cramer's solution of the equation system (A.23) takes the compact form

$$x = A^{-1}y \tag{A.28}$$

The inverse matrix has the following properties.

(i) The product of a nonsingular $n \times n$ matrix $A$ with its inverse is the $n \times n$ unit matrix:

$$A\,A^{-1} = I \tag{A.29}$$

$$A^{-1}A = I \tag{A.30}$$

This can be proved without difficulty by applying definitions (A.12) and (A.27) and the expansion theorem for determinants (this is left as an exercise for the reader). Some authors have reversed this approach, considering the above property as the definition equation for $A^{-1}$ and then proving (A.27).

(ii) The *inverse matrix of the product* of nonsingular square matrices is equal to the product of the inverse factors in reverse order:

$$(A\,B)^{-1} = B^{-1}\,A^{-1} \tag{A.31}$$

*Proof.* We have $B^{-1}(A^{-1}A)B = I$. Postmultiplication of both sides by $(AB)^{-1}$ gives (A.31). The proof for more than two factors is trivial.

The above property, like rule (A.21) for transposing a product, is extremely useful.

(iii) The *inverse matrix of a transpose* is equal to the transpose of the inverse:

$$(A')^{-1} = (A^{-1})' \tag{A.32}$$

*Proof.* We have $I' = (A^{-1}A)'$ so that $I = A'(A^{-1})'$. Premultiplication by $(A')^{-1}$ gives (A.32).

(iv) The *determinant of an inverse matrix* is equal to the inverse determinant of the original matrix:

$$\det(A^{-1}) = (\det A)^{-1} \tag{A.33}$$

*Proof.* $1 = \det I = \det(A^{-1}A) = \det(A^{-1})\det A$. For nonsingular finite $A$ there is a finite value different from zero on the right-hand side of (A.33). Examination of the left-hand side shows that $A^{-1}$ is also nonsingular.

### A.4.3 Multiple Equation Systems, Matrix Division and Computational Requirements

Given a system of $n$ equations in the form $\boldsymbol{A}\,\boldsymbol{x}_k = \boldsymbol{y}_k$ with nonsingular $\boldsymbol{A}$. This system is to be solved $m$ times for $\boldsymbol{x}_k$ with different values for $\boldsymbol{y}_k$ each time but always the same $\boldsymbol{A}$. The $m$ individual systems can be combined using the rules of matrix multiplication (Figure A.2). The right-hand sides $\boldsymbol{y}_1, \ldots, \boldsymbol{y}_m$ are written as columns of an $n \times m$ matrix $\boldsymbol{Y}$ and the corresponding unknown vectors $\boldsymbol{x}_1, \ldots, \boldsymbol{x}_m$ are combined to form an $n \times m$ matrix $\boldsymbol{X}$. In this way the following multiple system is produced:

$$\boldsymbol{A}\,\boldsymbol{X} = \boldsymbol{Y} \tag{A.34}$$

The solution for $\boldsymbol{X}$ can be obtained formally by premultiplying both sides by $\boldsymbol{A}^{-1}$ and applying property (A.30) to the left-hand side. The result is

$$\boldsymbol{X} = \boldsymbol{A}^{-1}\boldsymbol{Y} \tag{A.35}$$

In this way the process of going from (A.34) to (A.35) represents *division* by the matrix $\boldsymbol{A}$. Matrix division is only possible with a nonsingular square matrix. Of course, we must also distinguish here between left and right division. To avoid errors in matrix division, we therefore recommend considering it as premultiplication or postmultiplication with the corresponding inverse matrix.

However, this method is only suitable for numerical calculations when $\boldsymbol{A}^{-1}$ is known anyway or needed elsewhere. This is because to calculate $\boldsymbol{A}^{-1}$ for a fully occupied asymmetric matrix $n^3$ multiplications must be carried out (Zurmuehl and Falk, 1984, 1986) (without checks). The product of $\boldsymbol{A}^{-1}$ and $\boldsymbol{Y}$ requires another $n^2m$ multiplications as can be seen from Falk's scheme. Altogether, therefore, the formal solution using (A.35) requires $n^3 + n^2m$ multiplications. If, however, the Gauss algorithm is applied directly to the multiple system (A.34), a total of $n^3/3 + n^2m - n/3$ multiplications is sufficient for the solution (Zurmuehl and Falk, 1984, 1986). This number is smaller than that given above for all $n$ and $m$. For example, for $n = 10$ and $m = 5$, the solution using (A.35) requires 1500 multiplications whereas only 830, a little more than half, are needed with the Gaussian algorithm.

To determine an $n$th-order inverse using the definition equation (A.27), $n^2$ minors of order $n - 1$ must be formed. The further $n$ multiplications required for the expansion of det $\boldsymbol{A}$ and the $n^2$ divisions by det $\boldsymbol{A}$ are almost negligible for $n \gg 1$. The evaluation of one determinant of order $n - 1$ requires approximately $(n - 1)^3/3$ multiplications using triangular decomposition with the concatenated Gauss algorithm (Zurmuehl and Falk, 1984, 1986). Therefore to form the inverse matrix

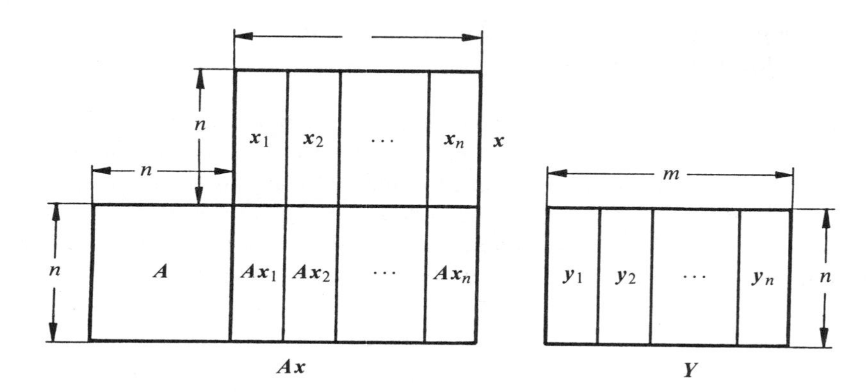


**Figure A.2** $AX = Y$, an $m$-fold system of $n$th-order equations.

using (A.27), a total of $n^2(n-1)^3/3 \approx n^5/3$ multiplications is required for large $n$. However, if the inverse is determined according to (A.29) by solving the equation system $\boldsymbol{AX} = \boldsymbol{I}$ with the Gauss algorithm, only $n^3$ multiplications are involved. For $n = 10$, for example, the first method requires 24 300 multiplications but the second method requires only 1000!

Therefore, before carrying out the numerical evaluation of higher-order matrix expressions, it is worth minimizing the amount of calculation by comparing alternative methods. As a final example for this we consider the chain product $\boldsymbol{A}\ \boldsymbol{B}\ \boldsymbol{c}$ (all factors of $n$th order and all matrices fully occupied). Execution in the sequence $(\boldsymbol{AB})\boldsymbol{c}$ requires $n^3 + n^2$ multiplications, whereas only $2n^2$ multiplications are needed in the sequence $\boldsymbol{A}(\boldsymbol{Bc})$.

## A.5 EIGENVALUE PROBLEMS

In control theory eigenvalue problems occur in the calculation of the poles and the characteristic frequencies of a dynamic system. In matrix theory they are involved in the diagonalization of matrices.

The basic form of the eigenvalue problem is as follows. Given a system of linear equations

$$\boldsymbol{A}\ \boldsymbol{v} = s\ \boldsymbol{v} \tag{A.36}$$

with a known $n \times n$ matrix $\boldsymbol{A}$. First the free, possibly complex scalar $s$ is determined so that nontrivial solutions are possible for $\boldsymbol{v}$; the equation system is then solved for $\boldsymbol{v}$. The numbers $s_i$ defined in this way are called *eigenvalues* or *characteristic values* of the matrix $\boldsymbol{A}$, and the associated solutions $\boldsymbol{v}_i$ are called *eigenvectors* or *charac-*

*teristic vectors*. They have a special property according to equation (A.36): on transformation with $\boldsymbol{A}$, an eigenvector of the matrix $\boldsymbol{A}$ is only stretched or shortened in $n$-dimensional space ($s$ real) and possibly inverted ($s < 0$) but not rotated in any other way.

### A.5.1 The Characteristic Equation

To solve the eigenvalue problem, both sides of equation (A.36) are combined:

$$(s\,\boldsymbol{I} - \boldsymbol{A})\,\boldsymbol{v} = \boldsymbol{0} \tag{A.37}$$

We are therefore dealing with a homogenous system of equations. This only has nontrivial solutions ($\boldsymbol{v} \neq \boldsymbol{0}$) when the matrix of the coefficients is singular. Therefore $s$ must be chosen so that

$$\det(s\,\boldsymbol{I} - \boldsymbol{A}) = 0 \tag{A.38a}$$

i.e., more explicitly,

$$\begin{vmatrix} s - a_{11} & -a_{12} & \ldots & -a_{1n} \\ -a_{21} & s - a_{22} & \ldots & -a_{2n} \\ \vdots & & & \\ -a_{n1} & -a_{n2} & \ldots & s - a_{nn} \end{vmatrix} = 0 \tag{A.38b}$$

In the case of a diagonal matrix $\boldsymbol{A}$, this determinant can easily be formed:

$$\det(s\boldsymbol{I} - \boldsymbol{A}) = (s - a_{11})(s - a_{22}), \ldots, (s - a_{nn}) = 0$$

In this case, the eigenvalues are equal to the diagonal elements $a_{11}, \ldots, a_{nn}$. In the general case, evaluation of the above determinant produces an $n$th-order equation in $s$:

$$s^n + \alpha_{n-1}s^{n-1} + \ldots + \alpha_1 s + \alpha_0 = 0 \tag{A.38c}$$

The coefficients $\alpha_i$ are functions of the elements $a_{ik}$. For example,

$$\alpha_0 = \det(-\boldsymbol{A}) \tag{A.39}$$

and

$$\alpha_{n-1} = \text{tr}(-\boldsymbol{A}) \tag{A.40}$$

The first of these relationships is obtained by setting $s$ equal to zero in (A.38a) and (A.38c), and the second by successively expanding (A.38b) with respect to the main elements. The remaining coefficients $\alpha_i$ have a less simple form.

Condition (A.38c) is called the *characteristic equation* of the matrix $\boldsymbol{A}$. Sometimes this name is also applied to the equivalent versions (A.38b) and (A.38a). The $s_i$ must satisfy this condition, i.e., the eigenvalues of the matrix $\boldsymbol{A}$ are identical with the $n$ roots of the characteristic equation. They can be single or multiple, real or complex. If the matrices are real, the coefficients $\alpha_i$ are also real; complex roots are then pairwise conjugate. Real symmetric matrices only have real eigenvalues (Hildebrand, 1952).

The coefficient $\alpha_0$ is equal to the product of all the roots of (A.38c), i.e., with (A.39) $s_1 s_2 \ldots s_n = \det(-\boldsymbol{A})$. It follows from this that eigenvalues vanish if and only if $\boldsymbol{A}$ is singular.

### A.5.2 The Cayley–Hamilton Theorem

This very well known and useful theorem can be stated as follows. Every $n \times n$ matrix $\boldsymbol{A}$ satisfies its characteristic equation, i.e.,

$$\boldsymbol{A}^n + \alpha_{n-1}\boldsymbol{A}^{n-1} + \ldots + \alpha_1\boldsymbol{A} + \alpha_0\boldsymbol{I} = \boldsymbol{0} \tag{A.41}$$

*Proof.* According to (A.27)

$$(s\boldsymbol{I} - \boldsymbol{A})^{-1} = \frac{1}{\det(s\boldsymbol{I} - \boldsymbol{A})}\,\text{adj}(s\boldsymbol{I} - \boldsymbol{A}) \tag{A.42a}$$

where $\text{adj}(s\boldsymbol{I} - \boldsymbol{A})$ is the transposed matrix of the cofactors of $(s\boldsymbol{I} - \boldsymbol{A})$. The latter, apart from a factor of $(-1)^{i+k}$, are minors of order $n - 1$ of $s\boldsymbol{I} - \boldsymbol{A}$, i.e., they are polynomials in $s$ of maximum order $n - 1$. We can therefore make the statement

$$\text{adj}(s\boldsymbol{I} - \boldsymbol{A}) = \boldsymbol{C}_{n-1}s^{n-1} + \boldsymbol{C}_{n-2}s^{n-2} + \ldots + \boldsymbol{C}_1 s + \boldsymbol{C}_0 \tag{A.42b}$$

The $\boldsymbol{C}_i$ are square coefficient matrices of order $n$; their values are of no interest here. Equation (A.42a) is now postmultiplied by $(s\boldsymbol{I} - \boldsymbol{A})\det(s\boldsymbol{I} - \boldsymbol{A})$:

$$\boldsymbol{I}\det(s\boldsymbol{I} - \boldsymbol{A}) = \text{adj}(s\boldsymbol{I} - \boldsymbol{A})(s\boldsymbol{I} - \boldsymbol{A})$$

On substituting the characteristic polynomial (left-hand side of (A.38c)) and using statement (A.42b), it follows that

$$\begin{aligned} & I(s^n + \alpha_{n-1}s^{n-1} + \ldots + \alpha_1 s + \alpha_0) \\ & \quad = (C_{n-1}s^{n-1} + C_{n-2}s^{n-2} + \ldots + C_1 s + C_0)(sI - A) \end{aligned}$$

Comparison of the coefficients of powers in $s$ gives

$$\begin{array}{lcrcl} s^n & : & I & = & C_{n-1} \\ s^{n-1} & : & \alpha_{n-1} I & = & -C_{n-1}A + C_{n-2} \\ & & & \vdots & \\ s & : & \alpha_1 I & = & \qquad - C_1 A \quad + C_0 \\ s^0 & : & \alpha_0 I & = & \qquad\qquad - C_0 A \end{array} \tag{A.42c}$$

The penultimate row of the equation system is postmultiplied by $A, \ldots,$ the second by $A^{n-1}$ and the first by $A^n$. On subsequent addition, all of the terms on the right-hand side cancel each other in pairs and the matrix polynomial of equation (A.41) appears on the left-hand side. This proves (A.41).

One of the consequences of the Cayley–Hamilton theorem is that any power of $A$, even a negative power, can be written as a linear combination of the powers $A^0, \ldots, A^{n-1}$! For example, $A^{n+1}$ is obtained by multiplying equation (A.41) by $A$, solving for $A^{n+1}$ and expressing $A^n$ according to (A.41). If $\alpha_0 = \det(-A) \neq 0$, $A^{-1}$ is obtained by multiplying equation (A.41) by $A$ and dividing by $\alpha_0$.

### A.5.3 The Souriau–Fadeeva Algorithm

For every $n \times n$ matrix $A$, the coefficients $\alpha_{n-k}$ of the characteristic polynomial (A.38c) and the coefficients $C_{n-k}$ of the matrix $\mathrm{adj}(sI - A)$ in equation (A.42b) can be determined as follows for $k = 1, 2, \ldots, n$:

$$C_{n-k-1} = C_{n-k}A + \alpha_{n-k}I, \quad C_{n-1} = I \quad (C_{-1} = 0) \tag{A.43a}$$

$$\alpha_{n-k} = -\frac{1}{k}\mathrm{tr}\,(C_{n-k}A) \tag{A.43b}$$

*Proof.* The recursion formula (A.43a) follows directly from equation system (A.42c); the version for $k = n$ can be used for a numerical check. The starting point for the proof of (A.43b) is the following relationship (its derivation is rather laborious (see Zadeh and Desoer, 1963)):

$$\frac{\mathrm{d}}{\mathrm{d}s}\det(s\boldsymbol{I} - \boldsymbol{A}) = \mathrm{tr}\{\mathrm{adj}(s\boldsymbol{I} - \boldsymbol{A})\} \tag{A.44}$$

The characteristic polynomimal of (A.38c) is substituted on the left-hand side and differentiated with respect to $s$; equation (A.42b) is substituted on the right-hand side. Comparision of coefficients gives for $s^{n-k-1}$

$$(n - k)\alpha_{n-k} = \mathrm{tr}\, \boldsymbol{C}_{n-k-1} \quad k = 0, \ldots, n - 1$$

Equation (A.43b) follows from this by substitution of (A.43a).

Algorithm (A.43a) and (A.43b) allows a comparatively simple calculation of $\det(s\boldsymbol{I} - \boldsymbol{A})$ and $\mathrm{adj}(s\boldsymbol{I} - \boldsymbol{A})$ (Souriau, 1948; Fadeeva, 1959). According to (A.42a), this also gives the inverse $(s\boldsymbol{I} - \boldsymbol{A})^{-1}$. This is the Laplace transform of the transition matrix of the differential equation:

$$\dot{\boldsymbol{x}}(t) = \boldsymbol{A}\boldsymbol{x}(t).$$

### A.5.4 The Modal Matrix

After the eigenvalues $s_i$ have been calculated as roots of the characteristic equation, they can be substituted successively in equation (A.37):

$$(s_i\boldsymbol{I} - \boldsymbol{A})\boldsymbol{v}_i = \boldsymbol{0} \quad i = 1, 2, 3, \ldots \tag{A.45}$$

The $n$-dimensional eigenvector $\boldsymbol{v}_i$ is not uniquely defined by this $n$th-order system of equations because it is singular. If the rank of $(s_i\boldsymbol{I} - \boldsymbol{A})$ is equal to $n - 1$ for a given $s_i$, equation (A.45) contains $n - 1$ linearly independent scalar equations and a unique solution $\boldsymbol{v}_i$ can be constructed by making an additional stipulation, e.g., by a normalization of the type $\boldsymbol{v}_i' \boldsymbol{v}_i = 1$. If $\boldsymbol{A}$ and $s_i$ are real, $\boldsymbol{v}_i$ is also real; otherwise it is complex. In addition, the following theorems are useful (for proofs see Gantmacher, 1958, 1959; Zurmuehl and Falk, 1984, 1986).

(i) For a single eigenvalue, rank $(s_i\boldsymbol{I} - \boldsymbol{A}) = n - 1$ so that this $s_i$ has exactly *one* linearly independent eigenvector.

(ii) A $p$-fold eigenvalue has at least one and at most $p$ linearly independent eigenvectors.

(iii) Eigenvectors of different eigenvalues are linearly independent.

It follows from (i) and (iii) that $n \times n$ matrices with single eigenvalues have exactly $n$ linearly independent eigenvectors. If there is a $p$-fold eigenvalue which does not produce $p$ distinct eigenvectors the situation becomes quite cumbersome (see Zurmuehl and Falk, 1984, 1986). Multiple or complex eigenvalues (poles) are not infrequent in control engineering. Thus it often happens that there are less than

$n$ linearly independent eigenvectors and that they may be complex. Because of these drawbacks, opinions are divided as to the usefulness of eigenvalue methods.

Assume that an $n \times n$ matrix $\boldsymbol{A}$ has $n$ linear independent eigenvectors. This is always true, for example, if all its eigenvalues are distinct (single) or if it is symmetric. Then $\boldsymbol{A}$ can be transformed to a diagonal matrix. To do this, the eigenvectors $\boldsymbol{v}_1, \ldots, \boldsymbol{v}_n$ are combined to form the *eigenvector matrix* or *modal matrix* $\boldsymbol{V}$, which in our case is always regular:

$$\boldsymbol{V} := [\boldsymbol{v}_1, \boldsymbol{v}_2, \ldots, \boldsymbol{v}_n]$$

The $n$ versions of (A.36) with the $n$ eigenvalues $s_i$ are written in combined form:

$$[\boldsymbol{A}\boldsymbol{v}_1, \boldsymbol{A}\boldsymbol{v}_2, \ldots, \boldsymbol{A}\boldsymbol{v}_n] = [s_1\boldsymbol{v}_1, s_2\boldsymbol{v}_2, \ldots, s_n\boldsymbol{v}_n]$$

or

$$\boldsymbol{A}\boldsymbol{V} = \boldsymbol{V}\boldsymbol{J} \tag{A.46a}$$

where $\boldsymbol{J}$ is the *Jordan canonical form* of the matrix $\boldsymbol{A}$. In equation (A.46a), $\boldsymbol{J}$ is a diagonal matrix, the main elements of which are the eigenvalues:

$$\boldsymbol{J} := \begin{bmatrix} s_1 & & & 0 \\ & s_2 & & \\ & & \ddots & \\ 0 & & & s_n \end{bmatrix} \tag{A.46b}$$

As $\boldsymbol{V}$ is nonsingular, equation (A.46a) can be premultiplied by $\boldsymbol{V}^{-1}$. The desired transformation is therefore

$$\boldsymbol{V}^{-1}\boldsymbol{A}\boldsymbol{V} = \boldsymbol{J} \tag{A.46c}$$

Premultiplication by $\boldsymbol{V}$ and postmultiplication by $\boldsymbol{V}^{-1}$ gives the inverse transformation.

$$\boldsymbol{A} = \boldsymbol{V}\boldsymbol{J}\boldsymbol{V}^{-1} \tag{A.46d}$$

For real symmetric but otherwise arbitrary matrices, the modal matrix is *orthonormal*; i.e., its columns are normalized ($\boldsymbol{v}_i'\boldsymbol{v}_i = 1$) and mutually orthogonal ($\boldsymbol{v}_i'\boldsymbol{v}_j = 0$ for $i \neq j$) (Hildebrand, 1952). In compact form this is expressed as

$$\boldsymbol{V}'\boldsymbol{V} = \boldsymbol{I} \tag{A.46e}$$

It follows that

$$V^{-1} = V' \tag{A.46f}$$

Therefore in this special case the inverse modal matrix can be obtained by a simple transposition.

The eigenvalues of the canonical form $\boldsymbol{J}$ in (A.46b) and (A.46c) are equal to $s_1, \ldots, s_n$, i.e., equal to those of $\boldsymbol{A}$. The eigenvalues are not changed by transformation with the modal matrix. This also holds for the corresponding transformation with any nonsingular $n \times n$ matrix $\boldsymbol{T}$ (the similarity transformation). Let

$$\boldsymbol{B} = \boldsymbol{T}^{-1}\boldsymbol{A}\boldsymbol{T} \tag{A.47}$$

The two $n \times n$ matrices $\boldsymbol{A}$ and $\boldsymbol{B}$ are called *similar* to each other. We have

$$\begin{aligned} \det(s\boldsymbol{I} - \boldsymbol{B}) &= \det(s\boldsymbol{I} - \boldsymbol{T}^{-1}\boldsymbol{A}\boldsymbol{T}) \\ &= \det\{\boldsymbol{T}^{-1}(s\boldsymbol{I} - \boldsymbol{A})\boldsymbol{T}\} \\ &= \det(\boldsymbol{T}^{-1}) \det(s\boldsymbol{I} - \boldsymbol{A}) \det \boldsymbol{T} \\ &= \det(s\boldsymbol{I} - \boldsymbol{A}) \end{aligned}$$

because $\det(\boldsymbol{T}^{-1}) \det \boldsymbol{T} = 1$.

Similar matrices have the same characteristic equation and thus the same eigenvalues.

## A.6 QUADRATIC FORMS

The *bilinear form* of an $n$-vector $\boldsymbol{x}$ and an $m$-vector $\boldsymbol{y}$ with respect to an $n \times m$ matrix $\boldsymbol{Q}$ is a number $b$ such that

$$b := \boldsymbol{x}'\boldsymbol{Q}\boldsymbol{y}$$

This concept can either be interpreted as the scalar product of the two $n$-vectors $\boldsymbol{x}$ and $\boldsymbol{Q}\boldsymbol{y}$ or as the $m$-term scalar product of $\boldsymbol{Q}'\boldsymbol{x}$ and $\boldsymbol{y}$. If $\boldsymbol{x} = \boldsymbol{y}$ and $\boldsymbol{Q}$ is of the type $n \times n$, we have the *quadratic form* $q$:

$$\begin{aligned} q &:= \boldsymbol{x}'\boldsymbol{Q}\boldsymbol{x} \\ &= \quad q_{11}x_1^2 + q_{12}x_1x_2 + \ldots + q_{1n}x_1x_n \\ &\quad + q_{21}x_2x_1 + q_{22}x_2^2 + \ldots + q_{2n}x_2x_n \\ &\quad \vdots \\ &\quad + q_{n1}x_nx_1 + q_{n2}x_nx_2 + \ldots + q_{nn}x_n^2 \end{aligned} \tag{A.48}$$

The mixed products $x_i x_j$ all occur in pairs. The natural interpretation of a quadratic form is the scalar product of a transformed vector with itself:

$$(\boldsymbol{Ax})'(\boldsymbol{Ax}) = \boldsymbol{x}'\boldsymbol{A}'\boldsymbol{Ax} = \boldsymbol{x}'\boldsymbol{Qx} \tag{A.49}$$

where $\boldsymbol{Q} = \boldsymbol{A}'\boldsymbol{A}$. $\boldsymbol{Q}$ turns out to be symmetric here. However, in other cases there is no loss of generality in making $\boldsymbol{Q}$ symmetric. The value of the quadratic form (A.48) does not change if every element $q_{ij}$ is replaced by the arithmetic mean of $q_{ij}$ and $q_{ji}$. In this way, any $\boldsymbol{Q}$ in (A.48) can be symmetrized.

A *real* quadratic form or the corresponding matrix is called *positive definite* if

$$\boldsymbol{x}'\boldsymbol{Qx} \geqslant 0 \quad \text{for all } \boldsymbol{x} \neq \boldsymbol{0} \tag{A.50a}$$

(A.50b)

and

$$\boldsymbol{x}'\boldsymbol{Qx} = 0 \quad \text{only for } \boldsymbol{x} = \boldsymbol{0}$$

If $\boldsymbol{x}'\boldsymbol{Qx} \geq 0$ for all $\boldsymbol{x} \neq \boldsymbol{0}$, $\boldsymbol{Q}$ is called *positive semidefinite* (nonnegative definite). If the inequality in (A.50a) is $\boldsymbol{x}'\boldsymbol{Qx} < 0$, then $\boldsymbol{Q}$ is called *negative definite*.

If, for example, $\boldsymbol{Q}$ is a diagonal matrix, (A.48) becomes

$$q = \boldsymbol{x}'\boldsymbol{Qx} = q_{11}x_1^2 + q_{22}x_2^2 + \ldots + q_{nn}x_n^2 \tag{A.51}$$

This $q$ is positive definite if and only if all main elements $q_{ii}$ (here, equal to the eigenvalues of $\boldsymbol{Q}$) are positive. If $\boldsymbol{Q} = \boldsymbol{A}'\boldsymbol{A}$ with real $\boldsymbol{A}$, then $q = \boldsymbol{x}'\boldsymbol{A}'\boldsymbol{Ax} = (\boldsymbol{Ax})'(\boldsymbol{Ax})$. This is a sum of squares and as such is always positive semidefinite. Now $q$ is positive definite if and only if $\boldsymbol{A}$ is nonsingular, because only then $\boldsymbol{Ax} \neq \boldsymbol{0}$ for all $\boldsymbol{x} \neq \boldsymbol{0}$.

Every real symmetric matrix $\boldsymbol{Q}$ has a diagonal Jordan canonical form $\boldsymbol{J}$ and an orthonormal eigenvector matrix $\boldsymbol{V}$ (see equations (A.46b) or (A.46e)). Substitution of (A.46f) in (A.46d) gives $\boldsymbol{Q} = \boldsymbol{VJV}'$. Thus

$$q = \boldsymbol{x}'\boldsymbol{Qx} = \boldsymbol{x}'\boldsymbol{VJV}'\boldsymbol{x} = \boldsymbol{y}'\boldsymbol{Jy} \tag{A.52}$$

where $\boldsymbol{y} := \boldsymbol{V}'\boldsymbol{x}$. Because $\boldsymbol{V}$ is regular, $\boldsymbol{y}$ is different from $\boldsymbol{0}$ if and only if $\boldsymbol{x} \neq \boldsymbol{0}$. The quadratic form (A.52) is therefore positive definite if and only if all main elements of $\boldsymbol{J}$ (i.e., all eigenvalues of $\boldsymbol{Q}$, are positive).

A second, eigenvalue-independent criterion is as follows: every real symmetric $n \times n$ matrix $\boldsymbol{Q}$ or the corresponding quadratic form is positive definite if and only if all $n$ main segment determinants[‡] of $Q$ are positive (Dietrich and Stahl, 1978).

The *autocorrelation matrix* of an $n$-component real random vector $z$ is defined as the expectation of the dyad $zz'$. Thus the autocorrelation matrix $E\{zz'\}$ is real and symmetric. For all deterministic real $n$-vectors $\boldsymbol{x}$ we have

$$\boldsymbol{x}' E\{zz'\}\boldsymbol{x} = E\{(\boldsymbol{x}'z)(z'\boldsymbol{x})\} = E\{(\boldsymbol{x}'z)^2\} \geq 0$$

An autocorrelation matrix is therefore always positive semidefinite.

## A.7 VECTOR NORMS

The norm of a vector is the generalization of the absolute value of a number. Thus the vector norm is applied to areas such as the comparison of the magnitude of several vectors, the distance between two vectors, inequalities, estimates, convergence considerations, quality indices *et cetera*.

*Definition A.1*. The *norm* of a real or complex vector is a real number denoted by $\|\boldsymbol{x}\|$ which satisfies the following properties:

$$\left.\begin{array}{ll} \|\boldsymbol{x}\| > 0 & \text{for all } \boldsymbol{x} \neq \boldsymbol{0} \\ \|\boldsymbol{x}\| = 0 & \text{only for } \boldsymbol{x} = \boldsymbol{0} \end{array}\right\} \quad \text{(positive definite)}$$

$$\|c\boldsymbol{x}\| = |c| \times \|\boldsymbol{x}\| \quad \text{for all numbers } c \text{ and vectors } \boldsymbol{x}$$

$$\|\boldsymbol{x} + \boldsymbol{y}\| < \|\boldsymbol{x}\| + \|\boldsymbol{y}\| \quad \text{(triangle inequality)}$$

As usual $|c|$ is the absolute value of the number $c$. Norms can be specified in different ways; they are distinguished by a subscript.

*Examples A.8*. The simplest norms for $n$-dimensional vectors are

$$\|\boldsymbol{x}\|_1 := |x_1| + |x_2| + \ldots + |x_n| \tag{A.53}$$

$$\|\boldsymbol{x}\|_2 := (|x_1|^2 + |x_2|^2 + \ldots + |x_n|^2)^{1/2} \tag{A.54}$$

---

[‡]The $i$th main segment determinant of $Q$ is the minor of $i$th order which remains after deletion of the $(i + 1)$th and all following rows and columns (i.e., a principal minor in the top left-hand corner of $Q$).

For complex vectors,

$$\|\boldsymbol{x}\|_2 = (\boldsymbol{x}^*\boldsymbol{x})^{1/2}$$

For real vectors, (A.54) takes the form of the square root of the ordinary scalar product:

$$\|\boldsymbol{x}\|_2 = (\boldsymbol{x}'\boldsymbol{x})^{1/2} = (x_1^2 + x_2^2 + \ldots + x_n^2)^{1/2}$$

If $x_1$, $x_2$, $x_3$ are the components of the vector $\boldsymbol{x}$ with respect to a Cartesian coordinate system in three-dimensional space, then this norm represents the Euclidean length of $\boldsymbol{x}$ (Pythagoras' theorem). By analogy, the norm $\|\boldsymbol{x}\|_2$ is also called the Euclidean length in higher-dimensional spaces. Other norms are

$$\|\boldsymbol{x}\|_\infty := \max_i |x_i| \tag{A.55}$$

$$\|\boldsymbol{x}\|_Q := (\boldsymbol{x}'\boldsymbol{Q}\boldsymbol{x})^{1/2} \tag{A.56}$$

In (A.56), $\boldsymbol{x}$ and $\boldsymbol{Q}$ are real, whereas $\boldsymbol{Q}$ is also symmetric and positive definite (see (A.50)). The norm (A.56) is a generalization of the Euclidean length. In particular, if $\boldsymbol{Q} = \boldsymbol{A}'\boldsymbol{A}$ ($\boldsymbol{A}$ nonsingular), we have

$$\|\boldsymbol{x}\|_Q = \|\boldsymbol{A}\boldsymbol{x}\|_2.$$

## A.8 INTEGRATION AND DIFFERENTIATION WITH RESPECT TO SCALARS

In what follows, the integral and derivative of a matrix with respect to a scalar are defined. Vectors as single-row or single-column matrices are naturally included here. The independent scalar variable is denoted by $t$, which does not necessarily represent time.

The *integral* $\int \boldsymbol{A}(t)dt$ of an $m \times n$ matrix $\boldsymbol{A}(t)$ with respect to $t$ is an $m \times n$ matrix consisting of the integrals of the original elements; in other words,

$$\boldsymbol{B} = \int \boldsymbol{A}(t)\,\mathrm{d}t \quad \text{means} \quad b_{ik} = \int a_{ik}(t)\,\mathrm{d}t \quad \text{for all } i,k \tag{A.57}$$

This definition applies equally to definite and indefinite integrals and to single and multiple integration.

*Example A.9.* Laplace transformation of an $n$-vector $\boldsymbol{x}(t)$:

$$\mathcal{L}\{\boldsymbol{x}(t)\} := \int_0^\infty \boldsymbol{x}(t) \exp(-st)\mathrm{d}t := \begin{bmatrix} \int_0^\infty x_1(t) \exp(-st)\mathrm{d}t \\ \vdots \\ \int_0^\infty x_n(t) \exp(-st)\mathrm{d}t \end{bmatrix} \tag{A.58a}$$

*Example A.10.* Expectation of an $n$-component random vector $\boldsymbol{x}$:

$$E\{\boldsymbol{x}\} := \int_{-\infty}^{+\infty} \cdots \int_{-\infty}^{+\infty} \boldsymbol{\xi} f(\boldsymbol{\xi})\mathrm{d}\xi_1 \ldots \mathrm{d}\xi_n$$

$$:= \begin{bmatrix} \int_{-\infty}^{+\infty} \cdots \int_{-\infty}^{+\infty} \xi_1 f(\boldsymbol{\xi})\mathrm{d}\xi_1 \ldots \mathrm{d}\xi_n \\ \vdots \\ \int_{-\infty}^{+\infty} \cdots \int_{-\infty}^{+\infty} \xi_n f(\boldsymbol{\xi})\mathrm{d}\xi_1 \ldots \mathrm{d}\xi_n \end{bmatrix} = \begin{bmatrix} \int_{-\infty}^{+\infty} \xi_1 f_1(\xi_1)\mathrm{d}\xi_1 \\ \vdots \\ \int_{-\infty}^{+\infty} \xi_n f_n(\xi_n)\mathrm{d}\xi_n \end{bmatrix} \tag{A.58b}$$

where $f(\boldsymbol{\xi})$ is the joint density and $f_i(\xi_i)$ is the marginal density. The expectation of a vector $\boldsymbol{x}$ is therefore equal to the vector consisting of the expectations of the components $x_i$. The same applies to the expectation of a matrix.

The *derivative* $\mathrm{d}\boldsymbol{A}(t)/\mathrm{d}t$ of a $m \times n$ matrix $\boldsymbol{A}(t)$ with respect to $t$ is an $m \times n$ matrix consisting of the derivatives of the original elements, i.e.,

$$\boldsymbol{B} = \frac{\mathrm{d}}{\mathrm{d}t}\boldsymbol{A}(t) \quad \text{means} \quad b_{ik} = \frac{\mathrm{d}}{\mathrm{d}t}a_{ik}(t) \quad \text{for all } i,k \tag{A.59}$$

Differentiation with respect to time is usually denoted by a dot and this is also true for matrices.

The *product rule* for matrices is as follows:

$$\frac{\mathrm{d}}{\mathrm{d}t}\{\boldsymbol{A}(t)\boldsymbol{B}(t)\} = \boldsymbol{A}(t)\boldsymbol{B}(t) + \boldsymbol{A}(t)\boldsymbol{B}(t) \tag{A.60}$$

This can be proved by applying the familiar scalar form of the product rule to a general element of the product $\boldsymbol{A}(t)\boldsymbol{B}(t)$ (see equation (A.12)).

## A.9 DIFFERENTIATION WITH RESPECT TO VECTORS

In this section we will first consider scalar-valued and then vector-valued variables, which in turn are functions of $n$ independent variables $x_1, \ldots, x_n$. The $n$-tuple of these variables $x_i$ is interpreted as the vector $\boldsymbol{x}$.

We will now consider some particular partial derivatives with respect to these $x_i$.

### A.9.1 The Gradient

The *gradient* $\partial f/\partial \boldsymbol{x}$ of a scalar function $f(\boldsymbol{x})$ with respect to the $n$-vector $\boldsymbol{x}$ is an $n$-component row vector consisting of the first partial derivatives of $f$ with respect to $x_k$:

$$\frac{\partial f}{\partial \boldsymbol{x}} := \left[\frac{\partial f}{\partial x_1}, \frac{\partial f}{\partial x_2}, \ldots, \frac{\partial f}{\partial x_n}\right] \tag{A.61}$$

Sometimes the gradient is also defined as a column vector.

*Example A.11*. The gradient of a linear form ($\boldsymbol{a}$ constant):

$$\frac{\partial}{\partial \boldsymbol{x}}(\boldsymbol{a}'\boldsymbol{x}) = \boldsymbol{a}' \tag{A.62}$$

*Example A.12*. The gradient of a quadratic form ($\boldsymbol{Q}$ symmetric and constant; compare (A.48)).

$$\frac{\partial}{\partial \boldsymbol{x}}(\boldsymbol{x}'\boldsymbol{Q}\boldsymbol{x}) = 2\boldsymbol{x}'\boldsymbol{Q} \tag{A.63}$$

### A.9.2 The Hessian Matrix

The *Hessian matrix* $\partial^2 f/\partial \boldsymbol{x}^2$ of a scalar function $f(\boldsymbol{x})$ with respect to the $n$-vector $\boldsymbol{x}$ is an $n \times n$ matrix consisting of the second partial derivatives of $f$ with respect to the $x_i$:

$$\frac{\partial^2 f}{\partial \boldsymbol{x}^2} := \begin{bmatrix} \frac{\partial^2 f}{\partial x_1^2} & \frac{\partial^2 f}{\partial x_1 \partial x_2} & \cdots & \frac{\partial^2 f}{\partial x_1 \partial x_n} \\ \frac{\partial^2 f}{\partial x_2 \partial x_1} & \frac{\partial^2 f}{\partial x_2^2} & \cdots & \frac{\partial^2 f}{\partial x_2 \partial x_n} \\ \vdots & & & \\ \frac{\partial^2 f}{\partial x_n \partial x_1} & \frac{\partial^2 f}{\partial x_n \partial x_2} & \cdots & \frac{\partial^2 f}{\partial x_n^2} \end{bmatrix} \tag{A.64}$$

The Hessian matrix is symmetric. The $i$th row of the Hessian matrix is the derivative of the gradient with respect to the component $x_i$ (compare (A.61)). The Hessian matrix of a linear form is $\boldsymbol{0}$. The Hessian matrix of a quadratic form is produced by differentiating (A.63) successively with respect to all the $x_i$, where $\boldsymbol{x}'Q = x_1\boldsymbol{q}^1 + x_2\boldsymbol{q}^2 + \ldots + x_n\boldsymbol{q}^n$:

$$\frac{\partial^2}{\partial \boldsymbol{x}^2}(\boldsymbol{x}'\boldsymbol{Q}\boldsymbol{x}) = 2\boldsymbol{Q} \tag{A.65}$$

The gradient and the Hessian matrix are particularly useful in the Taylor expansion of a scalar function with vector-valued argument:

$$f(\boldsymbol{x}) = f(\boldsymbol{a}) + \left.\frac{\partial f}{\partial \boldsymbol{x}}\right|_{\boldsymbol{x}=\boldsymbol{a}} \times (\boldsymbol{x} - \boldsymbol{a}) + \frac{1}{2}(\boldsymbol{x} - \boldsymbol{a})' \times \left.\frac{\partial^2 f}{\partial \boldsymbol{x}^2}\right|_{\boldsymbol{x}=\boldsymbol{a}} \times (\boldsymbol{x} - \boldsymbol{a}) + \ldots \tag{A.66}$$

The second term on the right-hand side is the scalar product of the gradient at $\boldsymbol{a}$ with the difference $\boldsymbol{x} - \boldsymbol{a}$; the third term is a quadratic form of $\boldsymbol{x} - \boldsymbol{a}$ with respect to the Hessian matrix. If $f$ is to have an extreme value at the point $\boldsymbol{a}$, then the gradient must vanish there. A minimum is present if, in addition, the quadratic form around $\boldsymbol{a}$ is positive definite. For a maximum, this quadratic form must be negative definite.

### A.9.3 The Jacobian Matrix

The *Jacobian matrix* $\partial \boldsymbol{f}/\partial \boldsymbol{x}$ of an $m$-term function $\boldsymbol{f}(\boldsymbol{x})$ with respect to the $n$-vector $\boldsymbol{x}$ is an $m \times n$ matrix consisting of the first partial derivatives of the components $f_i$ with respect to the elements $x_k$:

$$\frac{\partial \boldsymbol{f}}{\partial \boldsymbol{x}} := \begin{bmatrix} \frac{\partial f_1}{\partial x_1} & \frac{\partial f_1}{\partial x_2} & \cdots & \frac{\partial f_1}{\partial x_n} \\ \frac{\partial f_2}{\partial x_1} & \frac{\partial f_2}{\partial x_2} & \cdots & \frac{\partial f_2}{\partial x_n} \\ \vdots & & & \\ \frac{\partial f_m}{\partial x_1} & \frac{\partial f_m}{\partial x_2} & \cdots & \frac{\partial f_m}{\partial x_n} \end{bmatrix} \tag{A.67}$$

The $i$th row of the Jacobian matrix is the gradient of $f_i$ with respect to $\boldsymbol{x}$. The determinant of a square Jacobian matrix is called the *functional determinant*. The Jacobian matrix is used in the Taylor expansion of a vector function with vector argument:

$$\boldsymbol{f}(\boldsymbol{x}) = \boldsymbol{f}(\boldsymbol{a}) + \left.\frac{\partial \boldsymbol{f}}{\partial \boldsymbol{x}}\right|_{\boldsymbol{x}=\boldsymbol{a}} \times (\boldsymbol{x} - \boldsymbol{a}) + \ldots \tag{A.68}$$

The Jacobian matrix also occurs in the formation of a gradient by using the *chain rule*

$$\frac{\partial}{\partial \boldsymbol{x}} g\{\boldsymbol{f}(\boldsymbol{x})\} = \frac{\partial g}{\partial \boldsymbol{f}} \frac{\partial \boldsymbol{f}}{\partial \boldsymbol{x}} \tag{A.69}$$

This can be proved by applying the chain rule to the general element $\partial g/\partial x_k$.

It has been shown once again in the preceding sections that the analogies between scalar and vector relationships can be clearly expressed by a suitable choice of vector matrix symbols.

## REFERENCES


Bellman, R. (1960). *Introduction to Matrix Analysis,* McGraw-Hill, New York.

Bodewig, E. (1959). *Matrix Calculus,* Interscience, New York.

Dietrich, G. and Stahl, H. (1978). *Matrizen und Determinanten und ihre Anwendung in Technik und Oekonomie* (5th Ed.), Deutsch, Thun.

Fadeeva, V.N. (1959). *Computational Methods of Linear Algebra,* Dover Publications, New York.

Gantmacher, F.R. (1958, 1959). *Matrizenrechnung,* VEB Deutscher Verlag der Wissenschaften, Berlin, Vol. 1, 1958, Vol. 2, 1959.

Hildebrand, F. (1952). *Methods of Applied Mathematics,* Prentice-Hall, Englewood Cliffs, NJ.
Souriau, J.M. (1948). *Une Méthode pour la Décomposition Spectrale et l'inversion des Matrices. C.R. Acad. Sci. Paris,* **227**, 1010–1011.
Zadeh, L.A. and Desoer, C.A. (1963). *Linear System Theory,* McGraw-Hill, New York.
Zurmuehl, R. (1965). *Praktische Mathematik* (5th edn), Springer, Berlin.
Zurmuehl, R. and Falk, S. (1984, 1986). *Matrizen und ihre Anwendungen* (5th Ed.), Part 1, Grundlagen, 1984, Part 2, Numerische Methoden, 1986, Springer, Berlin.

## BIBLIOGRAPHY

Gantmacher, F.R. (1959). *Theory of Matrices,* Vols. I and II, Chelsea, New York.
Gantmacher, F.R. (1986). *Matrizenrechnung,* Springer, Berlin.

# INDEX